Computer Science and Statistics: Proceedings of the 13th Symposium on the Interface

Edited by William F. Eddy

Springer-Verlag
New York Heidelberg Berlin

Computer Science and Statistics: Proceedings of the 13th Symposium on the Interface

Computer Science and Statistics: Proceedings of the 13th Symposium on the Interface

Edited by William F. Eddy

With 122 Figures

Springer-Verlag
New York Heidelberg Berlin

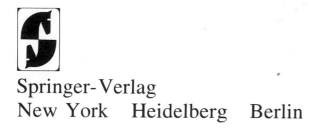

William F. Eddy
Carnegie-Mellon University
Department of Statistics
Pittsburgh, PA 15213

The Computer Science and Statistics 13th Symposium on the Interface was held March 12 and 13, 1981, at Carnegie-Mellon University, Pittsburgh, Pennsylvania.

Library of Congress Cataloging in Publication Data

Computer Science and Statistics : Symposium on the Interface (13th : 1981 : Pittsburgh, Pa.)
Computer Science and Statistics : Proceedings of the 13th Symposium on the Interface.

Bibliography: p.
Includes index.
1. Mathematical statistics—Data processing—Congresses. I. Eddy, William F. II. Carnegie-Mellon, University.
QA276.4.C58 1981 519.5'028'54 81-13608
AACR2

Printed in the United States of America

9 8 7 6 5 4 3 2 1

ISBN 0-387-90633-9 Springer-Verlag New York Heidelberg Berlin
ISBN 3-540-90633-9 Springer-Verlag Berlin Heidelberg New York

Availability of Proceedings

12th (1979)	Jane F. Gentlemen Dept. of Statistics University of Waterloo Waterloo, Ontario Canada N2L 3G1
11th (1978)	Institute of Statistics North Carolina State Univ. P.O. Box 5457 Raleigh, North Carolina 27650
10th (1977)	David Hogben Statistical Engineering Laboratory Applied Mathematics Division National Bureau of Standards U.S. Dept. of Commerce Washington, D.C. 20234
9th (1976)	Prindle, Weber, and Schmidt, Inc. 20 Newbury St. Boston, Massachusetts 02116
8th (1975)	Health Sciences Computing Facility, AV-111 Center for Health Sciences Univ. of California Los Angeles, California 90024
7th (1974)	Statistical Numerical Analysis and Data Processing Section 117 Snedecor Hall Iowa State Univ. Ames, Iowa 50010
4,5,6th (1971, 1972, 1973)	Western Periodicals Company 13000 Raymer Street North Hollywood, California 91605

Future Interface Symposia

14th Interface Symposium (1982)
Rensselaer Polytechnic Institute, Troy, New York
Chairmen: Karl Heiner & John Wilkinson

15th Interface Symposium (1983)
Houston, Texas
Chairmen: James Gentle & Homer Walker

Financial Supporters of the 13th Interface Symposium

U.S. Office of Naval Research, U.S. Dept. of the Navy

U.S. National Institutes of Health (NIH):

Division of Computer Research and Technology, NIH

Division of Research Resources, NIH

Fogarty International Center, NIH

Contents

Preface

The 13th Symposium on the Interface continued this series after a one year pause. The objective of these symposia is to provide a forum for the interchange of ideas of common concern to computer scientists and statisticians. The sessions of the 13th Symposium were held in the Pittsburgh Hilton Hotel, Gateway Center, Pittsburgh.

Following established custom the 13th Symposium had organized workshops on various topics of interest to participants. The workshop format allowed the invited speakers to present their material variously as formal talks, tutorial sessions and open discussion.

The Symposium schedule was also the customary one. Registration opened in late afternoon of March 11, 1981 and continued during the opening mixer held that evening. The formal opening of the Symposium was on the morning of March 12. The opening remarks were followed by Bradley Efron's address "Statistical Theory and the Computer." The rest of the daily schedule was three concurrent workshops in the morning and three in the afternoon with contributed poster sessions during the noon break. Additionally there were several commercial displays and guided tours of Carnegie-Mellon University's Computer Center, Computer Science research facilities, and Robotics Institute.

I would like to thank all my friends and colleagues at Carnegie-Mellon University, the Pittsburgh Hilton Hotel and elsewhere for their assistance with planning and execution of the Interface Symposium. In particular I would like to thank Bob Anderson, Chuck Augustine, Jon Bentley, Tom Clearwater, Lionel Galway, Margie Krest, John Lehoczky, Mike Mack, Mark Schervish, Jim Smith, Luke Tierney and Joe Verducci for their help with local arrangements. I would like to thank David Allen, Morrie DeGroot, Herman Friedman, Jim Gentle, Ron Helms, Beat Kleiner, Nancy Mann, Marcello Pagano, Manny Parzen, Gordon Sande, Mike Tarter, Bob Teitel, Ron Thisted and Bruce Weide for their help organizing the program. I would like to thank Tom Bajzek, Margie Cain, Wen Chen, Steve Fienberg, Jay Kadane, Diane Lambert, and Patty Porter for their help with various other matters.

The Interface Symposium depends on the cooperation of various professional organizations. I would like to thank AFIPS, AMS, ASA, ACM, IMS, SIAM and the Institute for Statistics and Applications at the University of Pittsburgh for their assistance.

Finally I would like to thank the financial supporters of the Interface: the Office of Naval Research and the Division of Computer Research and Technology, the Division of Research Resources and the Fogarty International Center, all of the National Institutes of Health.

William F. Eddy
Pittsburgh, Pennsylvania

Keynote Address

Bradley Efron, Stanford University, Stanford, California

STATISTICAL THEORY AND THE COMPUTER

Bradley Efron and Gail Gong, Stanford University

1. Introduction. Everyone here knows that the modern computer has profoundly changed statistical practice. The effect upon statistical theory is less obvious. Typical data analyses still rely, in the main, upon ideas developed fifty years ago. Inevitably though, new technical capabilities inspire new ideas. Efron, 1979B, describes a variety of current theoretical topics which depend upon the existence of cheap and fast computation: the jackknife, the bootstrap, cross-validation, robust estimation, the EM algorithm, and Cox's likelihood function for censored data.

The discussion here mainly concerns the bootstrap, Efron 1979A, though we also mention the jackknife and cross-validation. The plan is this: to show by example how a simple idea combined with massive computation can solve a problem that is hopelessly beyond traditional theoretical solutions.

We begin with a brief description of the bootstrap, and then go on to the main example, a medical-related problem in discriminant analysis. Two simulation studies are presented. We conclude with a few predictions about the future of statistical theory.

2. The Bootstrap. The basic idea of the bootstrap, Efron 1979A, is illustrated by the following problem: given an independently, identically distributed sample from an unknown distribution F

$$X_1, X_2, \ldots, X_n \overset{iid}{\sim} F,$$

we wish to estimate the standard error of some statistic $\hat{\rho} = \hat{\rho}(X_1, X_2, \ldots, X_n)$. Let $\sigma(F)$ be this standard error. Of course $\sigma(F)$ also depends on the sample size n and the form of the statistic $\hat{\rho}(\cdot, \cdot, \ldots, \cdot)$, but both of these are assumed known, so only the unknown F is listed as an argument.

The bootstrap estimate of $\sigma(F)$ is simply

$$\hat{\sigma}_{BOOT} = \sigma(\hat{F}) \tag{2.1}$$

where $\hat{F}$ is the empirical probability distribution putting mass $1/n$ on each observed value $X_i = x_i$,

$$\hat{F}: \text{mass } \frac{1}{n} \text{ on } x_i, \quad i = 1, 2, \ldots, n. \tag{2.2}$$

In other words, $\hat{\sigma}_{BOOT}$ is the nonparametric maximum likelihood estimate of $\sigma(F)$.

The function $\sigma(\hat{F})$ usually must be evaluated by a Monte Carlo algorithm: (1) Construct $\hat{F}$; (2) Sample $X_1^*, X_2^*, \ldots, X_n^* \overset{iid}{\sim} \hat{F}$ and compute $\hat{\rho}^* = \hat{\rho}(X_1^*, X_2^*, \ldots, X_n^*)$; (3) Repeat step 2 some large number B times, independently, obtaining "bootstrap replications" $\hat{\rho}^*(1), \hat{\rho}^*(2), \ldots, \hat{\rho}^*(B)$. Then

$$\hat{\sigma}_{BOOT} = \left[\sum_{b=1}^{B} \{\hat{\rho}^*(b) - \hat{\rho}^*(\cdot)\}^2/(B-1)\right]^{1/2}, \tag{2.3}$$

where $\hat{\rho}^*(\cdot) = \Sigma\hat{\rho}^*(b)/B$. As $B\to\infty$, (2.3) goes to $\hat{\sigma}_{BOOT}$ as defined in (2.1). In fact there usually is no practical reason to go beyond B in the range 50 - 1000, see Efron 1980C.

Table 1 refers to the following example: F is bivariate normal with true correlation $\rho = .5$, $n = 14$, and the statistic is the sample correlation coefficient $\hat{\rho}(X_1, X_2, \ldots, X_n)$. In this case it can be calculated that $\sigma(F) = .218$. Four different estimates of $\sigma(F)$ are compared. A full description of the other methods appears in Efron 1980B. Here we will only say that all the methods depend on the same simple idea underlying the bootstrap, and that among the nonparametric methods the bootstrap performs best.

	Exp. Value	Std. Dev.	$\sqrt{MSE}$
Bootstrap (B = 512)	.206	.063	.064
Jackknife	.223	.085	.085
Delta Method	.175	.058	.072
Normal Theory	.217	.056	.056
True Value $\sigma(F)$	.218		

Table 1. Four estimates of $\sigma(F)$, the standard error of the correlation coefficient $\hat{\rho}(X_1, X_2, \ldots, X_n)$; $n = 14$ and F is bivariate normal, true $\rho = .5$. Based on a simulation study, 200 trials, see Efron 1980B.

3. A Problem in Discriminant Analysis.[†] Among 155 acute chronic hepatitis patients, 33 were observed to die from the disease, while 122 survived. Each patient has associated a vector of 20 covariates, and on the basis of this data it is desired to produce a rule for predicting whether or not a given patient will die. If an effective prediction rule were available, it would be useful in choosing among alternative treatments.

Let $x_i = (y_i, t_i)$ represent the data for patient i, i=1, 2, ..., 155. Here t_i is the 1×20 vector of predictors, and $y_i = 1$ or 0 as the patient died or lived. Table 2 shows the data for the last 11 patients. Negative numbers represent missing values. Variable 1 is the constant 1, included for convenience. The meaning of the 19 other predictors, and their coding in Table 2, will not be explained here.

A prediction rule was constructed in three steps: (1) Letting

	#	145	146	147	148	149	150	151	152	153	154	155
prognosis	20	-3	2	-3	2	2	2	-3	2	2	-3	2
protein	19	-1	54	-1	35	-1	-1	50	-1	-1	48	42
albumin	18	2.4	4.2	3.4	2.8	4.0	4.0	3.3	4.3	4.1	4.1	3.1
SGOT	17	114	193	120	528	152	30	242	142	20	19	19
alk phos	16	-1	75	65	109	89	120	-1	126	95	84	100
bilirubin	15	1.90	1.20	4.20	1.70	0.90	0.60	7.60	0.90	0.80	1.50	1.20
varices	14	2	2	1	-3	2	2	1	2	2	1	2
ascites	13	1	2	2	-3	2	2	1	2	2	2	1
spiders	12	1	2	1	-3	2	2	1	2	1	1	1
spleen palp	11	2	2	1	-3	2	2	2	2	2	1	1
liver firm	10	2	2	1	-3	-3	2	2	1	1	2	2
liver big	9	2	2	2	-3	2	2	2	2	1	2	2
anorexia	8	1	2	2	1	2	2	1	2	2	2	2
malaise	7	1	2	2	1	2	2	1	2	1	2	2
fatigue	6	1	1	1	1	2	2	1	1	1	1	1
antiviral	5	2	2	2	2	2	2	2	2	2	2	2
steroid	4	2	1	2	1	1	2	2	2	1	1	2
sex	3	1	1	1	1	1	1	1	1	1	2	1
age	2	45	31	41	70	20	36	46	44	61	53	43
constant	1	1	1	1	1	1	1	1	1	1	1	1
	y	1	0	1	1	0	0	1	0	0	0	1

Table 2. The last 11 liver patients. Negative numbers indicate missing values.

$$\pi(t_i) = \text{Prob}\{\text{patient } i \text{ dies}\},$$

an $\alpha = .05$ test of H_0: $\beta_j = 0$ versus H_1: $\beta_j \neq 0$ in the logistic model $\log \pi(t_i)/(1-\pi(t_i)) = \alpha + \beta_j t_{ij}$, was run, separately, for $j=2, 3, \ldots, 20$. Fourteen predictors indicated prediction power by rejecting H_0: $j=1, 18, 13, 15, 12, 14, 7, 6, 19, 20, 11, 2, 5, 3$. Except for $j=1$, the constant, these are listed in order of achieved significance level, so $j=18$ attained the smallest alpha. (2) These fourteen predictors were tested in a forward stepwise multiple logistic regression model, which added predictors one at a time until no further addition achieved significance level $\alpha = .10$. Six predictors survived this step, $j=1, 13, 20, 15, 7, 2$. (3) A final forward stepwise multiple logistic regression model on these six variables, stopping at $\alpha = .05$, gave the 5 final predictors, $j=1, 13, 15, 7, 20$.

At each step of this procedure, only those patients having no relevant data missing were included in the hypothesis tests. At step 2 for example, a patient was included only if all fourteen variables were available.

The prediction rule was based on the estimated logistic regression

$$\log \pi(t_i)/(1-\pi(t_i)) = \sum_{j=1}^{\infty} \hat{\beta}_j t_{ij}, \tag{3.1}$$

where $\hat{\beta}_j$ was the maximum likelihood estimate at step 3 above, (so $\hat{\beta}_j=0$ if $j\neq 1, 13, 15, 7$, or 20). The prediction rule was

$$\eta_{\tilde{x}}(t_i) = \begin{matrix} 1 \\ 0 \end{matrix} \text{ if } \sum_j \hat{\beta}_j t_{ij} \begin{matrix} \geq c \\ \leq c \end{matrix} \quad (c = \log \frac{32}{123}). \tag{3.2}$$

Here we have written $\eta_{\tilde{x}}(\cdot)$ to indicate that the rule was based on the entire "training set" $\tilde{x} = (x_1, x_2, \ldots, x_n)$, $n = 155$. Of the 155 patients, 133 have all four variables 13, 15, 7, 20 present. When rule (3.2) was applied to them, it misclassified 21 patients, for an *apparent error rate* of $21/133 = .158$.

4. Overoptimism Analysis.

Apparent error rates are usually overoptimistic since they are computed from the same data used to form the prediction rule. We would like to get some idea of how overoptimistic .158 is, but theoretical calculations look hopeless given the complexities of the construction. As we shall see, complexity doesn't bother the bootstrap approach, which is based on brute computation.

For y and η each 1 or 0, define

$$Q[y,\eta] = \begin{matrix} 1 \\ 0 \end{matrix} \text{ if } \begin{matrix} y\neq\eta \\ y=\eta \end{matrix},$$

and let

$$\begin{aligned} O(\tilde{x},F) &= \text{true minus apparent error rate} \qquad (4.1) \\ &= EQ[Y_0, \eta_{\tilde{x}}(T_0)] - \hat{E}Q[Y_0, \eta_{\tilde{x}}(T_0)]. \end{aligned}$$

In writing (4.1) we assume that there is a true distribution F on $\mathbb{R}^{21}$ governing the distribution of any one patient's data $X = (Y,T)$, and that the training set $\tilde{x}$ is the observation of an i.i.d. sample $X_1, X_2, \ldots, X_n \overset{iid}{\sim} F$, $n = 155$. The term $EQ[Y_0, \eta_{\tilde{x}}(T_0)]$ is the true probability that a new point $X_0 = (Y_0,T_0)$ drawn independently from F will result in a misclassification using $\eta_{\tilde{x}}(\cdot)$. The term $EQ[Y_0, \eta_{\tilde{x}}(T_0)]$ is the same probability, except with $X_0 \sim \hat{F}$, and so equals the apparent error rate. (We are temporarily ignoring the fact that predictions are only made for patients having no missing predictors. This will help in the theoretical discussion. In fact the definition of $O(\underset{\sim}{x},F)$ actually used in the bootstrap analysis was exactly that which gave the apparent error rate .158 in Section 3.)

The expected overoptimism is

$$\omega(F) = EO(\underset{\sim}{X},F) .$$

The bootstrap estimate, as at (2.1), is simply $\omega = \hat{\omega}(\hat{F})$. This has to be calculated by Monte Carlo. Construct $\hat{F}$ putting mass $1/n$ on each x_i. (2) Sample $X_1^*, X_2^*, \ldots, X_n^* \overset{iid}{\sim} \hat{F}$ and compute $O(\underset{\sim}{X}^*,\hat{F})$. (Notice that this is an observable quantity since we know $\hat{F}$, even though (4.1) is unobservable.) (3) Repeat step 2 independently B times, obtaining $O^*(1), \ldots, O^*(B)$, and take $\hat{\omega} = \frac{1}{B}\sum_{b=1}^{B} O^*(b)$.

This program was carried out with $B = 500$, yielding $\hat{\omega} = .045$. Adding this to the apparent error rate .158 gives an estimated actual error rate of .203 for the rule $\eta_{\underset{\sim}{x}}(\cdot)$ fitted in Section 3. The histogram of all 500 bootstrap values $O^*(b)$ is shown in Figure 1.

```
 0   -9:
 0   -8:
 2   -7:0
 1   -6:0
 2   -5:0
 5   -4:00
10   -3:00000
12   -2:000000
17   -1:00000000
30    0:000000000000000
41    1:00000000000000000000
53    2:00000000000000000000000000
58    3:00000000000000000000000000000
44    4:0000000000000000000000
48    5:000000000000000000000000
48    6:000000000000000000000000
48    7:000000000000000000000000
37    8:000000000000000000
18    9:000000000
13   10:000000
 6   11:000
 3   12:0
 2   13:0
 2   14:0
 0   15:
 0   16:
 0   17:
```

Figure 1. Histogram of the bootstrap overoptimism values $O^*(b)$, b=1, 2, ..., 500, for the prediction rule described in Section 3. Each symbol represents two bootstrap values.

95% of these fall in the range $0 \leq O^*(b) \leq .12$. This indicates that the unobservable true overoptimism $O(\underset{\sim}{x},F)$ is very likely to be positive, though there is a lot of variability about the central value .045.

Each bootstrap replication required going through the entire fitting procedure of Section 3, substituting the bootstrap training set $\underset{\sim}{X}^*$ in place of the actual $\underset{\sim}{x}$. Even stopping at $B = 100$, which is quite adequate for approximating $\hat{\omega}$ in this case, still requires a formidable amount of computation - about 12 hours of time on a PDP 11/45. On the other hand, this is not prohibitively expensive, particularly in an age of proliferating small-but-powerful computers.

We could have used cross-validation instead of the bootstrap to estimate $\omega(F)$. To do so we delete one x_i at a time from the training set, compute the prediction rule $\eta_{\underset{\sim}{x}(i)}(\cdot)$ based on the remaining 154 points, and estimate $\omega(F)$ by

$$\omega^{\dagger} = \frac{1}{n}\sum_{i=1}^{n} Q[y_i, \eta_{\underset{\sim}{x}(i)}(t_i)] - \frac{1}{n}\sum_{i=1}^{n} Q[y_i, \eta_{\underset{\sim}{x}}(t)].$$

Theoretical calculations presented in Section 7 of Efron 1980C show a close connection between the bootstrap and cross-validation. The results of two simulation studies appear in Table 2. In these studies an estimated Fisher's linear discriminant function is being used for prediction, and the true distributions of the predictors is multivariate normal. The bootstrap estimate $\hat{\omega}$ is far less variable than $\omega^{\dagger}$, but tends to be biased downward. Both methods have faults if they are used as bias corrections to the apparent error rate, though the bootstrap results are moderately better. It seems possible that an estimate superior to both $\hat{\omega}$ and $\omega^{\dagger}$ can be found, but this is still speculation.

The bootstrap provides more than just an estimate of (F). For example, the empirical standard deviation of the bootstrap values estimates $Sd\{O(\underset{\sim}{X},F)\}$, the standard deviation of the overoptimism. These estimates are seen to be quite reliable in Table 2, e.g. $.104 \pm .014$ estimating $Sd\{O\} = .114$ in Study A.

Figure 2 shows another use of the bootstrap replications. The variables selected by the three-step procedure of Section 3, applied to the bootstrap training set $\underset{\sim}{X}^*$, are shown for the last 25 of the 500 replications. No theory exists for interpreting Figure 2, but the results certainly discourage confidence in the causal nature of the variables actually selected.

5. Some Predictions. This example raises many interesting theoretical questions, some of which have been not been answered yet. (1) Is the bootstrap a good way to estimate $\omega(F)$? Perhaps

		Apparent Error	Actual Over Optimism $O(\underset{\sim}{x},F)$	Bootstrap Est. $\hat{\omega}$ (B=200)	Cross-Val. Est. $\omega^{\ddagger}$	Bootstrap Est. of Sd{O}
Study A	Exp. Val.	.264	.096	.080	.091	.104
	(Std. Dev.)	(.123)	(.114)	(.028)	(.073)	(.014)
Study B	Exp. Val.	.069	.184	.103	.170	.087
	(Std. Dev.)	(.076)	(.100)	(.031)	(.094)	(.012)

Table 2. Two simulation studies comparing over-optimism estimates, for prediction rules based on Fisher's estimated linear discriminant function. Each study consisted of 100 trials of $X_1, X_2, \ldots, X_{14} \overset{iid}{\sim} F$, where $X = (Y,T)$, $P\{Y=0\} = P\{Y=1\} = .5$, and $T|y \sim \mathcal{N}(\mu_y, I)$. Study A: dimension = 2, $\mu_0 = (-.5, 0) = -\mu_1$. Study B: dimension = 5, $\mu_0 = (-1, 0, 0, 0, 0) = -\mu_1$. See Efron 1980C, Section 7, for further details.

1	13	7	20	15			
1	13	19	6				
1	20	16	19				
1	20	19					
1	14	18	7	16	2		
1	18	20	7	11			
1	20	19	15				
1	20						
1	13	12	15	8	18	7	19
1	15	13	19				
1	13	4					
1	12	15	3				
1	15	16	3				
1	15	20	4				
1	16	13	2	19			
1	18	20	3				
1	13	15	20				
1	15	13					
1	15	20	7				
1	13						
1	15						
1	13	14					
1	12	20	18				
1	2	20	15	7	19	12	
1	13	20	15	19			

Figure 2. Variables selected in the last 25 bootstrap replications of the prediction rule for the medical data. The variables selected by the actual data were 1, 13, 15, 7, 20. Variable 13 was selected in 37% of the 500 bootstrap replications. Likewise, 15 was selected 48%, 7 selected 35%, and 20 selected 59% of the time.

some form of smoothing, on either the t_i or the y_i, would reduce the biases evident in Table 2. (2) If we are interested in estimating the true error rate, should we make a bias correction of any sort? Bias correction is only marginally helpful in Study A, where SD{O} is large compared to $\omega = E\{O\}$, but it is quite effective in Study B. Statistics like those in the last column of Table 2 may be helpful in deciding. (3) How bad should we feel about the spread of selected variables shown in Figure 2? Is there a theory of "standard errors" for variable selection which would allow us to interpret Figure 2?

It is safe to predict that theoreticians will strive to answer the questions raised by the use of computer-intensive statistical methods. The answers themselves seem likely to be less mathematically complete than we are accustomed to. A typical *Annals* paper of the 1990's may be a combination of theoretical and numerical evidence, the latter gathered from the computer of course, more in the tradition of applied science than pure mathematics. In particular, much of what is now done by asymptotic theory may be replaced by Monte Carlo studies.

Two specific predictions: variable selection is likely to receive renewed theoretical interest. (Table 2 is just too interesting to ignore!) This is an area which has defied satisfactory theoretical solution, but is of obvious importance to the practitioner. New tools and new methods should inspire new, hopefully more successful, theory. There will be a hard reassessment of the role of parametric models in statistics. In those situations where their only rationale is computational ("These calculations produce closed-form solutions in the normal case"), there may be a strong impulse to substitute newer computational methods, relying more on the computer and less on parametrics, as in the analysis of Section 4. If we don't *have* to use parametric models then why *should* we use them? The right answer of course is when they help us extract a lot more information from our data without making unrealistic assumptions. Cox's regression method for censored data, Cox 1972, is an example of this approach at its most useful. We can expect, or at least hope, to see further examples of such harmless-but-useful parametric theories.

References

Cox, D. R. (1972). "Regression Models and Life Tables (with discussion)," *Journal of the Royal Statistical Society, Series B*, **34**, 187-220.

Efron, B. (1979A). "Bootstrap Methods: Another Look at the Jackknife," *Annals of Statistics*, **7**, 1-26.

Efron, B. (1979B). "Computers and the Theory of Statistics: Thinking the Unthinkable," *SIAM Review*, **21**, 460-480.

Efron, B. (1980A). "Censored Data and the Bootstrap," Technical Report No. 53, Department of Statistics, Stanford University. (To appear in *Journal of the American Statistical Association*.

Efron, B. (1980B). "Nonparametric Estimates of Standard Error: The Jackknife, the Bootstrap, and Other Methods," Technical Report No. 56, Department of Statistics, Stanford University. (To appear in *Biometrika*.)

Efron, B. (1980C). "The Jackknife, the Bootstrap, and Other Resampling Plans," Technical Report No. 63, Department of Statistics, Stanford University.

---

†This data set was assembled by Dr. Peter Gregory of the Stanford Medical School, who also carried out the regression analysis described in this section. We are grateful to him for allowing the use of the data, and his analysis, in this article.

Workshop 1

AUTOMATED EDIT AND IMPUTATION

Organizer: Gordon Sande, Statistics Canada

Chair: Graham Kalton, Institute for Survey Research

Invited Presentations:

Developing an Edit System for Industry Statistics, Brian Greenberg, Bureau of Census

Design of Experiments to Investigate Joint Distributions in Microanalytic Simulations, Richard Barr, Southern Methodist University

DEVELOPING AN EDIT SYSTEM FOR INDUSTRY STATISTICS

Brian Greenberg, U.S. Bureau of the Census

ABSTRACT

All survey data, in particular economic data, must be validated for consistency and reasonableness. The various fields, such as value of shipments, salary and wages, total employment, etc., are compared against one another to determine if one or more of them have aberrant values. These comparisons are typically expressed as so-called ratio edits and balance tests. For example, historical evidence indicates that the ratio between salary and wages divided by total number of employees in a particular industry usually lies between two prescribed bounds. Balance tests verify that a total equals the sum of its parts. When a data record fails one or more edits, at least one response item is subject to adjustment, and the revised record should pass all edits. An edit system being developed at the Census Bureau has as its core a mathematical based procedure to locate the minimal number of fields to impute and then to make imputations in the selected fields.

Subject-matter expertise will be called upon to enhance the performance of the core edit, especially in the area of imputation. Among the factors that will be brought to bear are patterns of response error, relative reliability of the fields, and varying reliability of the edits.

Keywords and phrases: Edit, Imputation, U.S. Census Bureau

Section 1. INTRODUCTION

This paper describes a mathematically based edit procedure, referred to here as the **core edit**, which is to be incorporated into a broader system for the editing of continuous data. When provided with an establishment survey response and the prior year response from the same establishment, the core edit locates the minimal number of fields the changing of which will make the adjusted record consistent—with respect to a given edit set. The core edit then makes imputations in those fields so chosen to yield a consistent record. This core edit makes provision for a substantial amount of subject-matter input which can easily be altered from one user to another. This system is being implemented and tested within the framework of a project group at the U.S. Census Bureau, jointly sponsored by the Statistical Research Division and the Industry Division, to develop a pilot edit for the Annual Survey of Manufactures (ASM). In this setting we have 48 field to field edits, 32 year to year edits, and 20 fields.

The project group consists of one methodologist to develop the mathematical model; a programmer to interface the core edit with the rest of the system and to access ASM data for testing; and two subject-matter specialists to evaluate the given edits and their implications, to incorporate their expertise concerning the ASM where called for in the program, and to evaluate the output of the edit. Subject-matter expertise is brought to bear to enhance the performance of the core edit, especially in the area of imputation. Among the factors that are considered are patterns of response error, relative reliability of the fields, and effective imputation strategies on a field to field bases.

Section 2. THE EDITS

We will assume throughout that the data are continuous, and for each record a prior year record exists for the establishment under consideration. Let there be N fields, let F_i, where i=1,...,N, denote these fields, and let $A(F_i)$ and $B(F_i)$ be the current and prior year values respectively in field F_i for a given record. We will also assume, for the present, that all edits arise either from a field to field ratio test or from a year to year ratio test.

A field to field ratio test involving field F_k and field F_h is the requirement that

$$(1) \qquad L_{kh} \leq A(F_k)/A(F_h) \leq U_{kh}$$

and a year to year ratio test involving fields F_k and F_h is the requirement that

$$(2) \qquad P_{kh} \leqslant (A(F_k)/A(F_h))/(B(F_k)/B(F_h)) \leqslant Q_{kh}.$$

Each ratio gives rise to two edits, namely, from (1):

$$L_{kh}A(F_h) - A(F_k) \leqslant 0$$

$$A(F_k) - U_{kh}A(F_h) \leqslant 0$$

and from (2):

$$P_{kh}(B(F_k)/B(F_h))A(F_h) - A(F_k) \leqslant 0$$

$$A(F_k) - Q_{kh}(B(F_k)/B(F_h))A(F_h) \leqslant 0.$$

The $B(F_i)$ are constant for any given record, and although the field to field edits are the same for each record, the year to year edits vary from record to record. We assume further that the prior year record passes all field to field edits.

If there is no prior year record available for some establishment, the procedure to be described remains the same except for the fact that only the field to field edits are considered.

Section 3. **GENERATION OF IMPLIED EDITS**

In their classic paper [1], Fellegi and Holt describe a procedure where given a set of edits one generates a family of implied edits (i.e., logical consequences of the given edits), which, when used with a set covering procedure on the subsets of fields that mutually fail an edit, enables one to locate the minimal number of fields to impute in a given record. This procedure, as a rule, is not practical because the number of implied edits grows quite large. In the ASM setting, however, it is easy to get a complete set of implied edits by a different means. Given the ratio tests

$$L_{kh} \leqslant A(F_k)/A(F_h) \leqslant U_{kh}$$

$$L_{hj} \leqslant A(F_h)/A(F_j) \leqslant U_{hj}$$

the implied ratio is

$$L_{kh}L_{hj} \leqslant A(F_k)/A(F_j) \leqslant U_{kh}U_{hj} \;.$$

The implied ratio gives rise to two implied edits, and one can see that a complete set of implied edits arises in this fashion. Accordingly, there are at most N(N-1) edits in the complete edit set. It is at this stage that inconsistencies in the edit set will surface (i.e., a lower bound for some ratio will exceed its upper bound), and therefore at this stage the implied edits are subject to subject-matter specialists review.

The generation of the implied year to year edits is similar to that for the field to field edits, but the generation is formal. That is, given the year to year ratio tests

$$P_{kh}B(F_k)/B(F_h) \leqslant A(F_k)/A(F_h) \leqslant Q_{kh}B(F_k)/B(F_h)$$

$$P_{hj}B(F_h)/B(F_j) \leqslant A(F_h)/A(F_j) \leqslant Q_{hj}B(F_h)/B(F_j)$$

the implied ratio is

$$P_{kh}P_{hj}B(F_k)/B(F_j) \leqslant A(F_k)/A(F_j) \leqslant Q_{kh}Q_{hj}B(F_k)/B(F_j) \;.$$

Formal inconsistencies are checked for; that is, $P_{kh}P_{hj} > Q_{kh}Q_{hj}$ would indicate an inconsistency.

For <u>each record</u> we obtain a distinct set of edits, namely, the union of the complete set of field to field edits and the complete set of year to year edits—the latter being determined by the prior year record. No effort at this juncture is made to check for inconsistencies. That is not to say that they are not important, but rather, their presence will become manifest in the running of the record under consideration. For, if there are any inconsistencies, neither the record itself nor any attempted imputation would be able to pass the complete set of edits, and the record would be expelled from the core edit with the message that an imputation is impossible. A prior year record failing some field to field edits is one potential source of inconsistency, and for that reason, we assumed above that the prior year record passes all field to field edits.

One can view the original edit set as a graph in which the fields correspond to nodes, and an edit between two fields to an arc between the corresponding nodes. If this graph is connected, when we generate the implied edits we, in essence, complete the graph. If the original graph is not connected, we work with the connected components and, in effect, partition the set of fields into subsets each of which is edited independently of the other.

Section 4. LOCALIZING THE FIELDS TO IMPUTE

After all implied edits have been generated, they are entered into the core edit, and we are ready to begin processing records. We enter a record and a prior year record for the same establishment, generate the record dependent year to year edits, and subject the record to all the edits. We create the **failed edit graph** in which the nodes correspond to fields failing at least one edit, and an arc between two nodes indicates that the corresponding fields fail an edit between them. For example, if the only edit failures in a record involve fields F_1 and F_3, F_1 and F_7, and F_1 and F_{12} the failed edit graph would be:

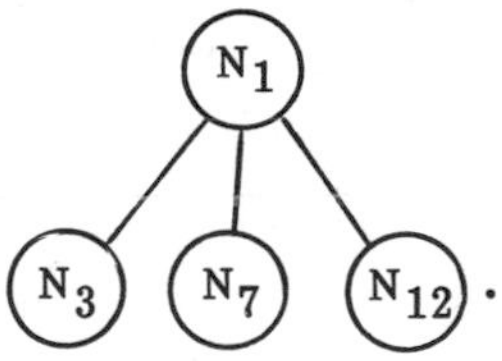

The failed edit graph is disconnected by removing one node at a time and by removing all arcs incident with a deleted node. When no arcs are left, the graph is said to be totally disconnected, and those nodes removed correspond to a set of fields to be imputed. The union of the set of fields corresponding to nodes remaining and those nodes not in the graph in the first place are mutually consistent with respect to our edit set since no two of them are involved in a failed edit. Since the edits are assumed consistent, we can make imputations in those fields deleted from the failed edit graph in such a manner that the adjusted record passes all edits, see [1]. If we choose a minimal set of nodes to delete in disconnecting the failed edit graph, we have chosen a minimal set of fields to impute.

Section 5. ASSIGNING COSTS TO THE FIELDS

Instead of seeking to remove a minimal set of nodes from the failed edit graph, one often prefers to determine costs, c(i), where i=1,...,N, and minimize a suitably chosen cost function. In this setting, if S is a subset of nodes whose removal will disconnect the failed edit graph, we form

$$c(S) = \sum_{i=1}^{N} c(i)\,d(N_i)$$

where

$$d(N_i) = \begin{cases} 1 & \text{if } N_i \text{ is in S} \\ 0 & \text{if } N_i \text{ is not in S.} \end{cases}$$

One then minimizes c(S), where S ranges over all subsets of the node set of the failed edit graph which disconnect that graph.

The costs, the c(i), are usually determined by subject-matter specialists reflecting their understanding of the relative reliability of the various fields.

Section 6. IMPUTATION STRATEGY

Suppose the set of fields has been chosen for imputation, say K fields: $F_{a(N-K+1)},\ldots,F_{a(N)}$, and the remaining fields are to remain as reported: $F_{a(1)},\ldots,F_{a(N-K)}$. The imputations will be made sequentially commencing with field $F_{a(N-K+1)}$ in the following manner. Consider all edits involving field $F_{a(N-K+1)}$ and those fields that are considered reliable, namely $F_{a(1)},\ldots,F_{a(N-K)}$, to obtain an interval in which $F_{a(N-K+1)}$ must lie. That is, we have:

$$L_{a(N-K+1)a(j)}A(F_{a(j)})$$

$$\leq A(F_{a(N-K+1)}) \leq$$

$$U_{a(N-K+1)a(j)}A(F_{a(j)})$$

and

$$P_{a(N-K+1)a(j)}A(F_{a(j)})(B(F_{a(N-K+1)})/B(F_{a(j)}))$$

$$\leq A(F_{a(N-K+1)}) \leq$$

$$Q_{a(N-K+1)a(j)}A(F_{a(j)})(B(F_{a(N-K+1)})/B(F_{a(j)}))$$

for all $j=1,\ldots,N-K$. This is a set of $2(N-K)$ closed intervals with a non-emply intersection; call the intersection $I_{a(N-K+1)}$. That is, the darker area below

is common to all intervals determined above.

Any number in $I_{a(N-K+1)}$ will be an acceptable imputation for $A(F_{a(N-K+1)})$, and the $A(F_i)$, where $i=1,\ldots,N-K+1$, will be mutually consistent. We know that $I_{a(N-K+1)}$ is not empty whenever the edit set is consistent. Accordingly, if $I_{a(J)}$ is empty for some J, then the edit set for the establishment under consideration is not consistent, and the record is sent out of the core edit for alternate review.

We next obtain an imputation for $A(F_{a(N-K+2)})$ using the values $A(F_{a(j)})$, where $j=1,\ldots,N-K+1$ and the intervals they determine as above, and we continue this process until all fields are imputed.

Section 7. IMPUTATION MODULES

If field F_k is targeted for imputation, when the time to choose $A(F_k)$ arrives we are presented with an interval $[Alow(F_k), Aup(F_k)]$ in which $A(F_k)$ must lie, and any number in that interval is an acceptable value for $A(F_k)$. Often special information can be provided by subject-matter personnel, based on their understanding of systematic response error, that can be used in selecting a value in that interval. For this purpose we have created "imputation modules" which are simply brief subroutines. When the program is ready to impute $A(F_k)$, it calls subroutine IMPFK and passes to it $Alow(F_k)$ and $Aup(F_k)$. If there are no special instructions, a default value, for example, $(Alow(F_k)+Aup(F_k))/2$ is imputed for $A(F_k)$.

It may be known that in field F_m respondents often report the item in pounds rather than per instructions in tons. The response $A'(F_m)$ (in pounds) would likely be rejected (i.e., that field would be targeted for imputation), and in subroutine IMPFM, if $A'(F_m)/2000$ is in the acceptable region, it would be imputed for $A(F_m)$. In another selected field, F_p, it may be known that a non-response almost always signifies the respondent's desire to indicate a zero. Thus, if $Alow(F_p)$ equals zero, then that will be imputed for $A(F_p)$. As a final example, if $A(F_n)$ has to be imputed and $A(F_m)$ does not, and if it is known that the ratio $A(F_n)/A(F_m)$ is fairly constant from year to year, then a good candidate for imputation for $A(F_n)$ would be $(B(F_n)/B(F_m))A(F_m)$ if it lies in the acceptable region. If not, a secondary choice might be $r_{n,m}\, A(F_m)$ where $r_{n,m}$ is the average industry wide ratio between field F_n and field F_m. Of course, if neither of these is suitable the default option is always available.

Even with several options, these subroutines are quite short and are easy to interpret, create, and modify to suit analytical needs. They proved to be an enhancement to the imputation process and offer considerable flexibility.

Section 8. BALANCE TESTS

Balance tests are used to check that a total equals the sum of its parts; for example, the sum of the reported quarterly shipments must equal the reportly yearly shipment. Balance tests are utilized in the imputation modules, for if $A(F_k)+A(F_h)=A(F_m)$ is a balance test, and if, for example, $A(F_h)$ is to be imputed with $A(F_k)$ and $A(F_m)$ fixed, then if $A(F_m)-A(F_k)$ is in $[Alow(F_h), Aup(F_h)]$ we make that imputation. If this is not the case, then either $A(F_m)$ or $A(F_k)$ has to be imputed also, and at this point in the project development this record would be sent from the core edit to a subject-matter analyst for further review.

If $A(F_k)$ and $A(F_h)$ both have to be imputed, then one of them, say $A(F_k)$, would be first imputed in IMPFK, perhaps using the choice $A(F_k)=(B(F_k)/B(F_m))A(F_m)$ if that is acceptable, and then imputing $A(F_h)$ as above. If $A(F_k)$, $A(F_h)$, and $A(F_m)$ are to be imputed, then $A(F_m)$ would be imputed in IMPFM, and we get to the others as indicated above.

At this stage of program development, balance checks are not tested for, and if an input record passed all ratio edits, it would pass even though it might fail a balance test. The editing for balance checks will be the next feature to be incorporated into the edit system.

Section 9. CONCLUSIONS

A version of the core edit is up and running at the Census Bureau and is being tested using the various specifications, subject-matter expertise, and data bank of the Annual Survey of Manufactures. When a record is run, the following information is printed out:

(1) the current year record and the prior year record,

(2) a list of all failed edits,

(3) a list of the minimal set of fields to impute,

(4) for each field to imputed, the range of acceptable values as well as the chosen imputation, and

(5) if a successful imputation is impossible, a message to that effect and some indiction of the reasons.

The program is performing satisfactorily and alterations are continually being implemented. New features are being added within the framework of the system, and refinements are being made to enhance the performance.

APPENDIX - A HEURISTIC PROCEDURE TO LOCATE THE WEIGHTED MINIMAL SET OF FIELDS TO IMPUTE

The following weighting scheme for the various fields and the following heuristic to determine the nodes to delete in the failed edit graph have proved quite effective in determining the set of fields to target for imputation. For each field F_i we determine a weight w(i), where i=1,...,N, to reflect the tightness of those edits involving field F_i. In particular, if an aberrant response in field F_i is likely to be detected by a relatively large number of edits then we choose w(i) to be low, and conversely, if only a few edits will detect an aberrant response in field F_j we choose w(j) to be high. That is, the weight of any field is inversely proportional to the tightness of the edits involving that field, and we choose the minimal weight to be greater than or equal to .51.

Thus, if the failed edit graph for a certain record is simply

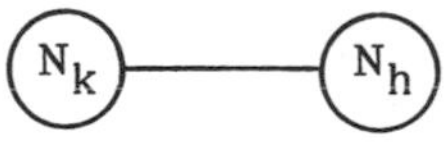

,

and if w(h)=1.85 and w(k)=.65, we would delete field F_h and target it for imputation. The rationale for this is as follows: if F_k is the "bad field," then it would likely have more edit failures (since it has a low weight), whereas if F_h is the "bad field," it is expected that not many edit failures would be detected (since it has a high weight). All else being equal, the higher the weight the more likely a field will be targeted for deletion. The weight w(i) is roughly inversely proportional to the likelihood of field F_i occuring in a failed edit if field F_i has an aberrant value, or said differently; the higher w(i), the greater the significance of each edit failure involving field F_i.

As noted before, if the edit graph were

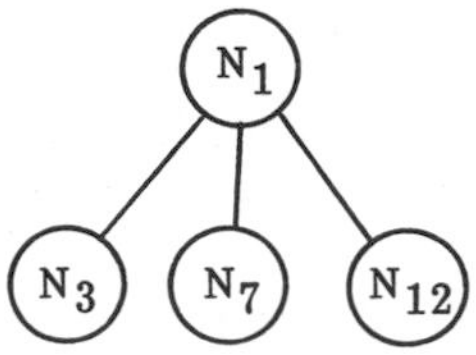

,

we would want to impute field F_1 because it occurs in more failed edits. Accordingly, we define the **weighted degree** for node N_i to be WD(i)=w(i)deg(i), where i=1,...,N, and deg(i) is the degree of node N_i.

In the heuristic we used to delete the failed edit graph, nodes are removed one at a time, and when a node is removed so are all arcs leading into it. Nodes of degree zero are discounted, and let M be the number of nodes of positive degree in the failed edit graph at any time—updated after each deletion. If N_p is the node of least weighted degree and has weighted degree less than or equal to one, remove the node adjacent to N_p. We note that there is exactly one node adjacent to a node of weighted degree less than or equal to one since all weights are greater than or equal to .51. Continue doing this until all nodes have weighted degree greater than one, and if at any time there are no nodes of positive weighted degree left, stop. Next remove the node of greatest weighted degree equaling or exceeding M-1. Again examine if there are any nodes of weighted degree less than or equal to one, and proceed as above until all remaining nodes have weighted degree between one and M-1. At this point, let N_k be the node of least weighted degree and delete the node adjacent to N_k having greatest weighted degree (incorporating suitable tie-breakers if needed). Search again for nodes of degree less than or equal to one, and so on, and continue this process until there are no nodes left. The set of nodes deleted corresponds to the weighted minimal set of fields to impute.

In the applications of this algorithm the weights had minor effect on the number of nodes in the set of nodes chosen to disconnect the failed edit graph—as opposed to letting each node have weight equal to one. Rather, the weighting broke ties in determining which of several sets of equal cardinality would be chosen, and it directed the choice of nodes to be deleted at each decision step of the algorithm. Although the heuristic described here is not optimal for arbitrary graphs where all weights equal one, the performance up to now on failed edit graphs has proved very satisfactory. This is due primarily to the structure of the failed edit graph and the relation between the weighting scheme and the heuristic.

If one should wish to use any other (weighted) node covering algorithm in the core edit to determine the set of fields to impute or some other procedure using mathematical programming and a suitable cost function, it can easily be done since this step in the core edit is carried out in an external subroutine. There is some relationship between the weights as discussed above and the costs assigned to the fields as discussed in Section 5 of the text. Since the fields that are considered more reliable usually are involved in tighten edits, one possible cost function is c(i)=1/w(i), i=1,...N, where the w(i) are the weights determined as above. The objective would be to minimize c(S), as defined in Section 5.

Reference

[1] Fellegi, I.P. and Holt, D., "A Systematic Approach to Automated Edit and Imputation," Journal of the American Statistical Association, (March 1976), 17-35.

DESIGN OF EXPERIMENTS TO INVESTIGATE JOINT DISTRIBUTIONS IN MICROANALYTIC SIMULATIONS

Richard S. Barr, Southern Methodist University

Abstract

The heart of a microsimulation model is the microdata on which it operates. Often, a single survey is insufficient to meet a model's informational needs and two separate survey files are merged. This paper focuses on the statistical problems accompanying the file merge process. Empirical evidence of data distortions induced by some procedures is presented as is an experimental design to investigate the impact of various merging methods on microdata statistics.

Key Words

microsimulation, microdata, file merging, imputation, survey analysis.

Introduction

The concept of microanalytic simulation models was developed by Guy Orcutt in the mid-1950's [5]. Today, these models abound in governmental agencies and research organizations and are used widely for policy analysis and projection of program needs. Examples include the various versions of the Transfer Income Model (TRIM), the behavioral model DYNASIM, and the tax policy simulations at the U.S. Department of the Treasury and at Brookings Institute.

The heart of these models are sample survey files, or microdata. These files consist of data records for a representative set of decision units (individuals, households, taxpayers, firms, etc.) which are processed by the simulator individually with data collected to identify aggregates, distributions, and interactions. By working at the record level, this modelling technique is very flexible and can accommodate as much detail as desired.

Microdata Files

While the recording unit may vary, microdata files usually represent the national population or a major subset such as taxpayers or Social Security system participants. Various sampling schemes are used in collecting the data, hence each record includes a weight indicating the number of population units it represents and these weights often differ among records.

Microdata files are created as byproducts of ongoing governmental programs, from legislative mandate, and as special commissioned studies. For example, both the I.R.S.'s Statistics of Income (SOI) and Social Security Earnings (SSA) files are drawn from data collected in the process of program implementation and control. The U.S. Constitution mandates the taking of a decennial census, subsets of which are used as microdata, and the Current Population Survey (CPS) is performed monthly to determine the unemployment rate, as required by law. The Survey of Income and Education (SIE) was a special study, as are numerous university-based surveys.

For the model designer and user, there are several pertinent characteristics of microdata files. First, they are expensive to create, on the order of $10 millions each. Hence, their construction is not a trivial undertaking. Second, several versions are often created to "correct" the data, for underreporting for example, through editing procedures. Third, a variety of sampling designs may be used, including stratified, clustered, and simple randomized, in order to combine information richness with brevity.

Fourth, the end product of these sometimes elaborate machinations is a multi-attribute representation of the underlying population, including all interactions and distributions of the reported data items. The distributional and interaction details are especially important for microanalytic models since they operate at the record level and base their computations on combinations of item values. Finally, by virtue of taking a middle ground between a census and population aggregates, these files are efficient from both a computational and information-content standpoint.

Limitations of Individual Samples

As illustrated by files such as the SOI and CPS, microdata are often collected primarily for the construction of aggregates or for program implementation, analysis, and control. Their use as general research data bases or in microanalytic simulation models is of secondary concern in the sample survey designs, an aspect which creates problems for these applications.

As models are built and policy proposals are analyzed, data are often required which (a) are not part of the current program, study, or system, as when new tax deductions are considered, or (b) are of superior quality since sample items are deemed to be unreliable, as with business income on the CPS.

The model user has four choices available: (1) commission a new study, at great expense and investment of time, (2) ignore the variables in question, and jeopardize the validity of the model's results, (3) impute the missing or unreliable items into an existing file, using methods which often ignore the distributional and interaction characteristics of the variables in question, or (4) merge a pair of microdata files to combine the information from two surveys. This last tack, file merging, is currently in widespread use and is investigated in this paper.

Microdata File Merging

The basic idea behind file merging, or matching, is to combine one file A with another file B to form a composite file C with all data items from the two original files. This is accomplished by selecting pairs of records to match based on data items which are common to both files. The schemes for performing the matching process fall into two general categories: exact and statistical matching.

"Exact" matching uses unique-valued common items to mate records for the same individual in both files. By using a unique identifier, such a social security number, the matching process is theoretically a simple sort and merge operation. Problems with this approach include: insignificant overlapping of samples causing few records to be matched, absence of or error in the "unique" identifiers, confidentiality restrictions which preclude legal linking of records, and the expense of handling a large number of exceptions.

Statistical merging (also called synthetic, stochastic, or attribute matching or merging) mates similar records using several common items with non-unique values. By matching like records, file C contains records which may be composites of two different persons, but whose attributes are similar enough for research purposes. There are a variety of statistical merging schemes in use today, as discussed below.

In choosing a methodology, exact matching is obviously preferable. But where such a match is not possible, statistical merging may be employed.

Statistical Merging

A pictorial description of statistical merging is presented in Figure 1. In this drawing, a_i represents the weight of the i-th record in file A and b_j the weights of the j-th record in file B. File C, the merged file, contains composite records formed by matching a record in file A with a record in file B, and assigning a merge record weight of x_{ij}. An interrecord dissimilarity measure c_{ij}, or distance function, is used to choose pairs of records for matching. The "distance" between a pair of records is usually determined from a user-defined function which compares corresponding common items and assigns a penalty value for each item pair which differs significantly. These penalties are summed to create a measure of dissimilarity, with a zero distance meaning all common items are identical or "close enough."

There are two general categories of statistical merges: unconstrained and constrained. In an unconstrained merge, file A is designated the base file and file B the augmentation file. Each base file record is matched with the most similar record in the augmentation file; the selected file B record is appended to the base file record and the base record's weight is used for x_{ij}. This is, in essence, sampling with replacement since some augmentation file records may not be matched while others may be used repeatedly. This is a very popular technique as evidenced by its use by Ruggles and Ruggles of Yale and NBER [9], Radner of the Social Security Administration [8], Minarik at Brookings Institute [7], Statistics Canada [4], and the Bureau of the Census.

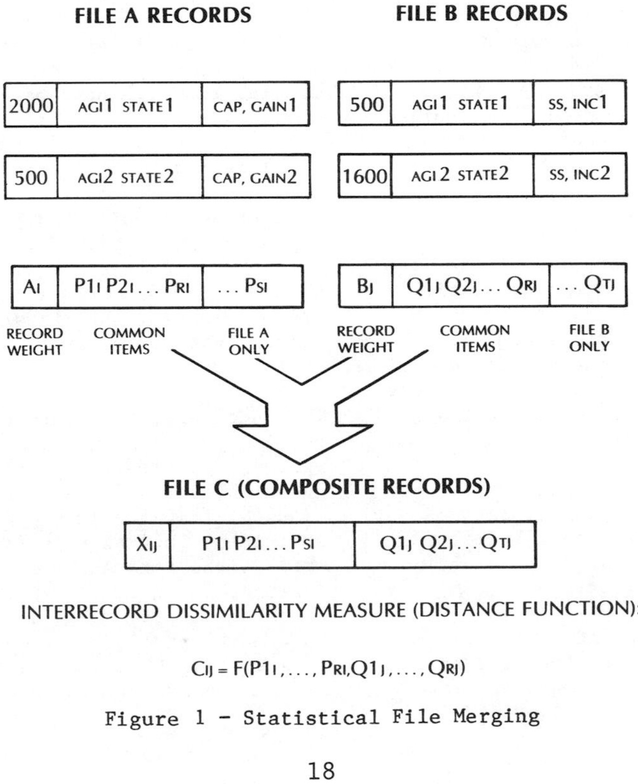

Figure 1 - Statistical File Merging

In contrast, a constrained merge uses matching without replacement. The merging algorithm enforces constraints on the record weights in both files to ensure that each record is neither under- nor over-matched relative to the number of population units represented. Mathematically, the constrained merge model is as follows.

$$\sum_{i=1}^{m} a_i = \sum_{j=1}^{n} b_j \qquad (1)$$

$$\sum_{i=1}^{n} x_{ij} = a_i, \; i=1,\dots,m, \qquad (2)$$

$$\sum_{i=1}^{m} a_{ij} = b_j, \; j=1,\dots,n, \qquad (3)$$

$$x_{ij} >= 0, \text{ for all } i \text{ and } j. \qquad (4)$$

Constraint (1) reflects the assumed equivalent underlying population sizes for the two files, although files A and B have m and n records, respectively. Some minor adjustments may be needed to accomplish this in practice. Again, x_{ij} is the merged record weight for matching record i in file A with record j in file B, and the records are not matched if $x_{ij} = 0$. Constraints (2) and (3) allow any record to be matched one or more times but such that the merge file weights must sum to the original record weights. Negative weights are precluded by (4). This merging algorithm is currently used Mathematical Policy Research.

Pictorially, the constrained merge proces is depicted in Figure 2 where the leftmost set of circles, or nodes, represent file A records with their respective weights, the rightmost nodes the file B records and weights, and the connecting arcs the possible record matches. A set of x_{ij} merge record weights are shown which meet constraints (1)-(4).

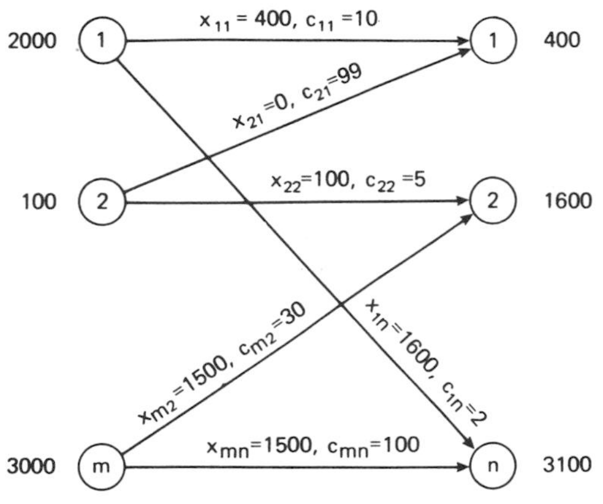

Figure 2 - Constrained Merge Model

This merge technique can be further refined by requiring the procedure to

$$\text{minimize} \sum_{i=1}^{m} \sum_{j=1}^{n} c_{ij} x_{ij} \qquad (5)$$

subject to (1)-(4). This model, originally proposed by Turner and Gilliam [12] and later derived by Kadane [6], seeks to find the best constrained match, the one with the minimum aggregate distance between matched records. This optimal constrained merge procedure requires the solution of a linear programming problem of extremely large dimensions, and is currently used by the U.S. Department of the Treasury [1,2,3].

Statistical Aspects of Merging Techniques

Unconstrained procedures utilize (5), subject to (1), (2), and (4). thus, by dropping constraint set (3), the composite records match up well at the record level. However, file B item statistics in the composite file are distorted by their implicit reweighting of the augmentation records through over- and under-matching. This reweighting has a strong impact on important extreme values, variances, covariances, and other distributional aspects of the file B items, as shown below.

While the constrained procedures may not match up as well at the record level, their merged files contain all of the information from the original files and preserve all statistical properties of the A and B data items. Further, if optimization is applied, the best overall constrained match is insured.

All of these aspects influence the results of the microanalytic models which use the merge file.

Underlying Merge Rationale

When two files are merged, we assume that two files (X1,Y) and (X2,Z) are drawn from the same population, where X1 and X2 are the sets of common items in file A and B, respectively, and Y and Z the sets of items unique to file A and B, respectively (the alignment assumption). The objective of merging is to form a file (X1,X2,Y,Z) which corresponds statistically to a sample of (X,Y,Z) taken from the same population. We do this in order to make inferences about (Y,Z) and (Y,Z|X) relationships, since we can already make (X,X), (X,Y), and (X,Z) inferences from the two original files.

Quality Considerations

What is missing is a strong theoretical justification for merging or an explanation of what is being accomplished by a merge. What is needed is both a measure for the "accuracy" of (X1,X2,Y,Z) in replicating (X,Y,Z), and a means for making such decisions as: are two files mergable? Is the composite file acceptable?

Typical reported measures of match accuracy are: counts of X1-X2 item agreements, item means, and percentage agreement by common item. The notion is that a file which matches well on the X-items matches well on the Y-Z relationships. Rarely reported are (1) comparisons of covariances, such as cov(Y) in unconstrained matches and cov(Y,Z) versus expected cov(Y,Z), (2) conditional and joint frequences for augmentation variables, and (3) other Y-Z studies. Ruggles, Ruggles, and Wolff [10] are the only contributors in this area. Moreover, the following empirical data points out potential data problems measured by these types of statistical measures.

Preliminary Empirical Data

In a recent set of experiments to investigate the effect of merging techniques on resultant file quality, subsets of the 1975 SOI and 1975 SIE were chosen, based on a nine-state geographic region. There were 7144 SOI records and 6283 SIE records, each represetning 12.7 million tax filling units, approximately 15 percent of the population. The X-variables used in the distance function were: age, race, sex, marital status, family size, wage income, business income, property income, spouse's income, and adjusted gross income. The two files were merged three ways: unconstrained with SOI as base file, unconstrained with SIE as the base file, and optimally-constrained, all using the same distance function.

In Table I, the distribution of SOI wages and business income is shown for both the original file and the unconstrained merge file using the SIE as the base. Not only are the means not in agreement but the distributions are altered, and dramatically in the case of business income. Of course, in the constrained merge, the distributions were identical to the originals.

To evaluate the unconstrained procedure's effect on covariance structure, the variance-covariance matrices of several common items were compared with the original matrices. The median percentage differences, by item, are shown in Table II.

In some cases, the median error is as small as 7 percent, but in others these second-order statistics differ greatly. Analysis of the constrained merge verified the expectation of zero error.

Table II. Variance-Covariance Differences

Common Variable	Median Variance-Covariance Error Relative to Original: Unconstrained SIE	Unconstrained SOI
Age	31.2%	26.4%
Family size	35.5	17.4
Wages	7.3	23.9
Business income	72.1	38.4
Farm income	97.7	88.8
Property income*	78.5	850.4
Spouse income	73.5	31.0
Adjusted gross income	9.9	24.4

*Interest + dividend + rental income.

Open Research Questions

Despite the widespread use of merging as a data enrichment technique, there is a paucity of much-needed research in this area. Consider the following questions

File Alignment

Under what considerations can we accept that two files are valid samples from the same population? Any two files can and are merged, often from different populations. How can this underlying assumption be tested?

Constrained Versus Unconstrained Techniques

When does either procedure create a match file which is statistically equivalent to a valid (X,Y,Z) sample drawn from the population? A goodness-of-match criterion is needed not only to answer this question but to compare alternative matching algorithms. Regression coefficient analysis suggests itself; however, the normality assumption often will not hold.

Table I. SOI Item Distributions

Income Class ($000's)	Total SOI Wages ($ Millions) Original	Unconstrained	Total SOI Business Income ($ Millions) Original	Unconstrained
< 0	0	0	- 712	- 86
1-5	12,218	11,699	857	607
5-10	22,535	21,124	836	1,194
10-15	24,745	26,639	617	1,497
15-20	10,326	23,882	677	909
20-30	21,133	24,930	808	1,402
30-50	9,597	10,141	785	964
50-100	3,371	2,741	777	1,108
100-200	1,010	136	298	601
> 200	244	0	111	0
Total	115,784	121,291	5,055	8,187
Mean:	$9,108	$9,542	$398	$644

Covariance of (Y,Z|X)

What is the effect of omitting or including cov(Y,Z|X) in the matching methodology? Do correlated X-variables carry along their correlated Y and Z variables properly?

Distance Functions

How do the various dissimilarity measures affect the resultant merge file? What is a "correct" distance function ? (See [6].) In practice, distance functions usually reflect the data aspects of greatest importance in the target microanalytic model or database.

For other research questions and issues surrounding merging activities, see [11] by the Federal Committee on Statistical Methodology.

Upcoming Experimentation

In order to benchmark the various merging schemes and study the statistical aspects of the merge process, a series of experiments are currently in progress. The 1975 SIE file has been chosen to be a test population from which a selected series of samples are drawn. Each record item will be declared to be in set X, Y, or Z based on data type and correlations with other data items.

The resultant set of files will be merged pairwise in various combinations using a variety of distance functions, merge schemes, and levels of data bias and error. The experimental design is structured to study the effect, if any, of these parameters on Y-Z relationships, standard statistical tests, and measures of "goodness" of the match. The study will also investigate the sensitivity of the various merge algorithms to the distance function used and the introduction of bias and error.

By designating the original file to be the population, the actual (X,Y,Z) is known, unlike the usual case in practice. This availability of the complete population provides the researcher with an accurate standard for comparison with any merge file.

It is anticipated that these experiments will provide new information on the statistical properties of the merge procedures in use today. New data will be generally available for the testing of hypotheses. It is also hoped that this empirical study will bring new insight into the area of microdata merging for the benefit of the statistical, econometric, and computer science communities.

References

[1] Barr, Richard S. and J. Scott Turner, "A New Linear Programming Approach to Microdata File Merging," in 1978 Compendium of Tax Research, Office of Tax Analysis, U.S. Department of the Treasury, (1978) 131-149.

[2] Barr, Richard S. and J. Scott Turner, "Merging the 1977 Statistics of Income and the March 1978 Current Population Survey," Analysis, Research, and Computation, Inc. (1980).

[3] Barr, Richard S. and J. Scott Turner, "Optimal File Merging through Large-Scale Technology," Mathematical Programming Studies, to appear.

[4] Colledge, M. J., J. H. Johnson, R. Pare, and I. G. Sande, "Large Scale Imputation of Survey Data," 1978 Proceedings of the American Statistical Association, Survey Research Methods Section, (1979) 431-436.

[5] Greenberger, Martin, Matthew Crenson, Brian Crissey, Models in the Policy Process, Russell Sage Foundation, New York (1976).

[6] Kadane, Joseph, "Some Statistical Problems in Merging Data Files," in 1978 Compendium of Tax Research, Office of Tax Analysis, U.S. Department of the Treasury, (1978) 159-171.

[7] Minarik, Joseph J., "The MERGE 1973 Data File," in R. H. Haveman and K. Hollenbeck, Microeconomic Simulation Models for Public Policy Analysis, Academic Press (1980).

[8] Radner, Daniel B, "An Example of the Use of Statistical Matching in the Estimation and Analysis of the Size Distribution of Income," ORS Working Paper No. 18, Social Security Administration, (1980).

[9] Ruggles, Nancy and Richard Ruggles, "A Strategy for Merging and Matching Microdata Sets," Annals of Economic and Social Measurement 3, 2 (1974) 353-371.

[10] Ruggles, Nancy, Richard Ruggles, and Edward Wolff, "Merging Microdata: Rationale, Practice, and Testing," Annals of Economic and Social Measurement 6, 4 (1970) 407-428.

[11] Subcommittee on Matching Techniques, Federal Committee on Statistical Methodology, "Report on Exact and Statistical Matching Techniques," Statistical Policy Working Paper 5, U.S. Department of Commerce (1980).

[12] Turner, J. Scott and Gary E. Gilliam, "Reducing and Merging Microdata Files," OTA Paper 7, Office of Tax Analysis, U.S. Treasury Department (1975).

Workshop 2

Do Statistical Packages Have a Future?

Organizer: Ronald A. Thisted, University of Chicago

Chair: Barbara Ryan, Pennsylvania State University

Invited Presentations:

The Effect of Personal Computers on Statistical Practice, Ronald A. Thisted, University of Chicago

Statistical Software Design and Scientific Breakthrough: Some Reflections on the Future, Norman H. Nie, SPSS, Inc. and University of Chicago

Some Thoughts on Expert Software, John Chambers, Bell Laboratories

THE EFFECT OF PERSONAL COMPUTERS ON STATISTICAL PRACTICE

Ronald A. Thisted, The University of Chicago

Abstract

Statistical computer packages for large computers, both in batch and interactive environments, have focused on traditional statistical analyses and the computations associated with them. With few exceptions, statistical software development has emphasized numerical computations and data management, and few statistical procedures have been designed specifically for use on computing machines. Personal computers offer new and fundamentally different computing tools around which new methods for exploratory data analysis can be developed. These methods make essential use of the personal machine's capabilities for data presentation and display.

Key Words: Personal computers, statistical computing, statistical packages, data analysis, graphics, animation, music.

1. Introduction

The title of this Workshop raises the question, "Do statistical packages have a future?" Let me begin by giving my own answer. I have no doubt that statistical packages as we know them today will continue to exist and to be used extensively. But they are not the wave of the future. (Personal computers may not be either, but I will come to that.) Innovations in statistical computing and data analysis will not likely come from the developers of packages as part of their current products.

What have been the major breakthroughs in data-analytic computing since the advent of the digital computer? Let us trace in broad outline the major steps in the evolution of statistical computer packages (SCPs). The precursors of today's SCPs were awash in the primordial soup of special-purpose statistical programs written at computer installations around the country. The first SCPs were merely collections of programs and/or subroutines for statistical calculations such as IBM's Scientific Subroutine Package and the early BIMED programs. These collections were distributed on a national level, and their development was supervised with varying degrees of care and interest by their creators. A major advance was the consolidation of isolated programs into integrated program packages with common formats for command languages and data entry, and with provisions for the same data set to be examined using multiple statistical procedures. These integrated packages were far more flexible than the early collections of programs, and they contributed greatly both to the productivity and to the disposition of data analysts. Another important step on the evolutionary path was the introduction of interactive computing and with it the shift from batch-oriented packages to interactive ones. Once again, SCPs became more versatile, easier, and more fun to use.

That brings us to today. Note that unlike the dinosaurs, none of the species whose origins we have outlined above have become extinct, nor do any show signs of becoming so. In each case, the new generations -- often spurred by advances in computer hardware development -- have made it possible for us to explore data in new ways rather than making our old ways obsolete.

What will be the next major innovation in computing for data analysis? Whatever it may be, it will use computer hardware in new ways to make data analysis more fruitful and (we hope) more enjoyable as well. (A reader has accused me of advocating "hedonistic statistics," to which charge I plead guilty.) There have been many changes in recent days on the hardware front, on any of which innovative approaches to data analysis might be based. Some examples which I shall not discuss further are the "megacomputers-on-a-chip", advances in voice synthesis and recognition, progress in robotics, and network technology. Instead I shall consider one aspect of today's computing scene that has not yet been tapped: the increasingly powerful and decreasingly expensive microcomputer, or "personal computer".

First, we shall consider the capabilities and limitations of such machines as the Apple II, the TRS-80, and the PET. Then, as preparation for looking into the future, we shall briefly examine the nature of the step from batch to interactive computing insofar as it has affected the practice of statistics. We shall then suggest some new directions based upon the strengths of personal computers.

2. Personal Computers

In this section we shall examine some of the strengths and weaknesses of "personal" computers: small computers that cost less than a new automobile and that might find a place in the home or small business. A typical computer system would consist of a central processing unit (CPU) with perhaps 64K bytes of random-access memory, a keyboard for input, a display such as a video monitor, a printer or terminal for hard-copy output including graphics, one or two floppy disks for storing programs and data, and one or more high-level programming languages such as Pascal or FORTRAN. Optionally, one might have in addition a communications facility with a modem for linking to other computers. A system with all of the features listed above can be purchased for less than $5000, and a usable system for less than $3000.

Personal computers (PCs) are not powerful number crunchers. Indeed, the most popular PCs are based on CPUs such as the Z80 and 6502 chips which have, in effect, eight-bit words and no floating point instructions. Consequently, most numerical computations are relatively slow. (A floating-point multiplication takes about 3-4 msec on the Apple II, and a logarithm can take from 20-35 msec.) It would also be difficult to argue that PCs will increase user access to computing, since a home terminal and a modem are less expensive than a PC. So what is special about the personal computer?

Personal computers differ from most computing equipment with which we are familiar in two respects. First, since the computer is located close to the peripheral devices, the system can operate at very high data rates. Instead of the typical 30 characters per second at which terminals communicate over telephone lines with computers, a PC can communicate with nearby devices at rates up to 64 times as fast (19200 baud). An example of a program designed expressly to take advantage of high data rates is EMACS, the screen-oriented text editor developed by Project MAC. The program is intended to operate at least at 9600 baud. One day's use of EMACS will provide convincing evidence that designing for high-speed data display can produce markedly different systems. Second, PCs come in attractive packages which may include built-in color graphics instructions, moderately high resolution graphics on the standard screen display, and perhaps even a loudspeaker. In addition, optional equipment such as tone generators, real-time clocks, marvelous devices called graphics tablets, floating-point arithmetic hardware, light pens, joysticks, and game paddles (for a quick game of Pong or Space Invaders) can simply be plugged into most personal computers.

The PC package often makes unconventional (often unfamiliar) input and output media available that have only rarely been exploited in the analysis of data, and the possibility of high data rates makes conceivable displays which before were too time-consuming to produce, say, at a standard terminal. These are the aspects of personal computers that we may expect statistical methods using them to exploit.

3. From Batch to Interactive Data Analysis

Using a personal computer is much like using an interactive computer through a terminal; instead of sharing a large computer with many users, one has a small computer dedicated to oneself Moreover, the style in each case is essentially interactive (although what amount to batch compilers do exist on PCs). In order to consider how personal computing might improve upon interactive computing, it is helpful to look first at the question, "To what extent is interactive statistical computing an improvement on batch?"

A set of instructions to a batch statistical package (let us call this a "program") consists of a sequence of commands, which often must be given in a prescribed order, and which taken together describe a set of data and a collection of statistical computations to be performed on those data. The entire analysis must be specified fairly rigidly in advance, and the user must translate a script for this analysis into a program for the SCP. The user transmits the program to the computer and then waits for the computer to play out the script. The program itself has a rather linear structure, and any variations on a theme must be anticipated and planned for when the program is constructed. Once the program has been submitted, neither the user nor the computer system has any flexibility about the analysis to be done.

By contrast, the program for an interactive SCP need not be determined entirely (or even largely) before the program is begun. The structure is nonlinear in the sense that nearly any command is permissible at any time. Moreover, the user can decide that a particular theme in the analysis is pleasant and can begin improvisations on that theme. There is more room for creativity and for virtuosity in doing interactive data analysis than in using batch statistical packages. Batch analysis relies more heavily on preconceptions about the data than does interactive computing. Interactive computing enables us to select our next command after having seen the results of its predecessor. Whereas in writing programs for batch SCPs the choice of the next command can be based only upon the expected outcomes of previous commands, in writing programs for interactive SCPs, that choice can depend on the actual outcome of earlier commands.

Batch and interactive SCPs share some common structural features. In each case, programs consist of the sequence: command, computation, presentation (i.e., output), inference (based on actual or expected output), and then command once again. Interactive SCPs are essentially batch SCPs reduced to smaller component chunks, which can be invoked in any order and as frequently as desired. Now this reduction was an important advance for data analysis, but it did not take us very far structurally from our batch roots. Indeed, one can take almost any interactive package and construct a sequence of commands to it that could be submitted en masse for execution. And one can mimic interactive data analysis using a batch system if one has sufficiently rapid turnaround. (However, for some batch packages the chunks are so large that they cannot be digested quickly, making it impossible to mimic truly interactive computing even with instantaneous response. Chunk size is an important characteristic distinguishing batch from interactive SCPs.)

There are important differences in the approach one takes to computing in these two modes, however. The most costly resource in a data analysis is the analyst's time and level of frustration. In batch computing, there is a large fixed overhead of human time and energy required which is not proportional to the size of the problem. This overhead includes preparing the data in an appropriate format, preparing the program, checking the input, constructing appropriate JCL (or equivalent), submitting the job, waiting for it to be run, picking up the output from the job -- often at a site remote from the analyst's work area, correcting minor errors, and the like. This implies that it pays in batch to do as many separate tasks as possible on each run, and to prefer more output to less at each step. One way to view interactive computing is that it serves to reduce this human overhead tremendously. Certain costs such as logging on to the computer, invoking a package, saving and filing final results, correcting errors at early stages of a computation need be incurred but once. This makes it practical to do smaller chunks of the computation at one time, because in an interactive system the price per chunk is proportional to the size of the chunk.

The expected cost of an error (measured in investigator-hours), either in preparing the program or in conceptualizing the problem, is enormously larger for batch than for interactive computing.

Since batch output is usually produced on high-speed printers, it can be voluminous. Here, the time required to print large amounts of paper is miniscule compared to the time required by other elements of the batch process. One can afford to print large amounts of information, and it is part of the batch philosophy to do so. Output from interactive packages is necessarily less voluminous, because either a) it is produced in real-time and is being transmitted at a low data rate (300 or 1200 baud) to a user at a hard-copy terminal, or b) it is being transmitted to a display terminal with only limited display capacity (24 lines is typical). For interactive computing, condensed output is essential.

For many data analysts, myself included, the frustrations of batch computing have largely been eliminated by excellent interactive and improved batch packages and operating systems. Frustrations remain, but they are different in nature. Few of us feel limited by hardware or software, and there are few computations that we would like to do but cannot. And most objections directed at current statistical packages could be remedied without doing violence to their philosophies of design.

So it is time to ask the questions: "What is there beyond interactive computing? What might be the next breakthrough? What would it take to make our data analysis work easier, more fruitful, or in any sense better?"

4. Beyond Interactive Computing

The traditional view (for statisticians at any rate) is that computers exist to eliminate numerical drudgery. Both batch and interactive SCPs reflect this viewpoint. For instance, both kinds of packages produce results that are easily presented on the printed page (or at least that portion of it that the screen of a display terminal represents). However we need not -- and ought not -- think of computers as merely today's successor to the Millionaire calculator. Rather, we should view computers as general symbol manipulators. We statisticians should then re-think the symbols we use to represent statistical problems, and we should re-think the ways in which we manipulate these symbols. Statisticians need new symbols, or new ways to manipulate symbols, or both, that are better suited to computers with fast display and uncoventional facilities for input and output than to 24 by 80 grids of characters. (See Muller's 1980 discussion of "Break-integrated processing," which is related to this point.)

In some sense, every mode of data analysis involves the display or exhibition of data. (Now I am an exhibitionist as well as a hedonist!) The symbols that we are used to using for representing data sets in batch analyses include tables, summary statistics, and rough plots; interactive analyses emphasize plotting even more, since graphs are quite helpful in deciding what to do (or what to look at) next. There has been some work in displaying data using high-resolution graphics, possibly with the use of color. The primary emphasis within statistical packages, however, has been to make display graphs of high quality generally available (examples: SAS Graphics, Tell-A-Graf (Integrated Software Systems 1978), SPSS Graphics Option). The high-resolution capability is used to produce publication quality versions of the same old pictures. Similarly, color is used to take the place of less exotic (and possibly less expensive) shadings on bar graphs. What used to be dotted and dashed lines in a plot now are solid red and solid green. This technology is useful and welcome, but it is hardly a breakthrough. Indeed, I have not yet seen an innovative data analytic technique based on high-resolution graphics or color graphics. The "color correlation matrix" developed at Los Alamos is a contender, but it is not yet clear that this method has led to any insights into the structure of a data set that conventional methods would not have yielded with comparable effort. The "chloroplethic" two-variable maps developed by the Bureau of the Census (1974) are spectacular and controversial (Wainer and Francolini 1980, Meyer 1981, Abt 1981, Wainer 1981). Their primary utility appears to be in communicating results rather than in assisting in data analysis.

What sorts of things are batch and interactive SCPs incapable of producing? New procedures must produce nonlinear output, that is, not the sort of thing that can be shipped out, one line at a time, to a standard hard-copy terminal. An example of a fresh recent idea that is not amenable to the standard SCP, is the Chernoff Face. Designed for clustering problems, it is an example of a procedure for representing data that is nonlinear in this sense: it is not suitable for line-at-a-time output to a standard terminal. What other sorts of methods might there be that are not designed for the printed page? Some groupings follow.

1. Animation. In one sense, animation provides a logical extension to Chernoff's faces, which are cartoon depictions of multivariate data points. What could be more natural then, if the data have a natural sequence (such as time), to watch how the corresponding faces evolve over time? We can imagine the ups and downs of, say, Chrysler Corporation reflected in the grins and grimaces of its financial reports.

More generally, we can imagine some representation for our data (such as a face or a scatterplot) evolving in a manner determined by one aspect (such as a time variable) of the data. All we need is a display device coupled with a computer whose display rate is rapid enough not to bore us.

Perhaps the best way to plot a time series is vertically or horizontally on the display surface (=page?) in such a fashion that the real time required to plot each point is proportional to the time interval from the preceding point. (An approximation to this method can be achieved using IDA on a fast display terminal. Since IDA plots series vertically, one point per line, one can just watch the series "roll by." Some people claim that it is a great help in detecting trends or systematic changes in the variance of the series.)

A more intriguing use of animation is to take a data set of moderate size, and to consider the scatter of points in three dimensions (more generally, the projection of the data into any three-dimensional flat). Why not use animation to put the data analyst "into the driver's seat" (or rather, into a cockpit), and let the analyst "fly" through the data using stick and rudder, much as Luke Skywalker and Han Solo fly through asteroid belts in Star Wars? Of course, the data analyst would be allowed to fly at a more leisurely pace. (Yet the image of the daredevil statistician zooming through the stack-loss data at breakneck speed is an attractive one. . . .)

What is particularly exciting about animation in data analysis is not that the idea is new. People with personal IBM 360/91's can soar through data-galaxies using PRIM-9; those with a VAX and only moderately expensive hardware can use PRIMH (discussed by Peter Huber in another session of this Conference). But now there are animated flight simulators for personal computers that give perspective views of a "world", that run in real time, and that should be adaptable to uses such as this. Moreover, hardware and software technologies that will improve the quality and lower the cost of animation are advancing rapidly. (See Myers 1981.)

2. Sound. What does white noise sound like? Is it possible to use sound in ways that will enhance our ability to detect patterns in data? Some attempts have been made to use chords or sequences of notes (arpeggios) to represent multivariate data points, with an ear toward using them in clustering problems in a manner analogous to Chernoff's Faces (Wilson 1980).

An alternative is to use pitch as a "third dimension" in a plot. Of course, for this to be effective, one must be able to identify the pitch corresponding to individual points, and that implies that the points must appear (or shine, or twinkle, or vanish) one at a time in the display as their respective pitches are sounded. What we have, then, is a "tune" associated with each "picture". It is tempting to speculate that with training one can learn to detect relevant melodies, just as one now learns how to look at a normal probability plot.

Suppose, for instance, that we plot the residuals from a regression as usual, but as we plot each point we hear a tone indicating the size of the residual (possibly standardized), or we hear a tone indicating its leverage. This may be a useful supplement to our standard procedures, and may help us to find outliers (at least those that matter) or to detect serial correlation more effectively. Indeed, certain patterns of autocorrelation have characteristic melodies, and with practice the trained ear can detect departures from white noise of a certain sort by listening to these tunes.

This discussion raises important psychological questions that have to do with the perception of patterns, the audition of tones, and the psychology of music. What are the best ways, for instance, to translate data into pitches and durations for tones? In a sense, the translation of data levels into frequencies by linear scaling is something like examining data on a logarithmic scale, since octaves are perceived as linear. Do the fundamental harmonic structures in music correspond to anything interesting in data structures? If so, how can we best extract such structure from a data set and present it effectively to the ear? If not -- and this may be more intriguing -- how can we use harmony to represent interesting features in the data? Of course we need not restrict ourselves to the standard twelve-tone scale of Western music, particularly if the computer is generating the tones for us directly. These issues are psychological ones that cannot be settled by the pure speculation of statisticians and philosophers. We must experiment with auditory presentation of data so that we can learn just what sorts of procedures are effective.

I would argue, however, that we should not wait for a complete cognitive psychology of music before experimenting with ways of listening to data. Indeed, the manner in which the ear analyzes sound is not yet well understood (Zwislocki 1981). There is still no theory of graphics, and there may never be. Yet progress has been and continues to be made in creating more useful and more effective visual displays.

In this area, which I call "sound data analysis," personal computers sit ready as instruments for experimentation. Programs written for the Apple II, for instance, can control a built-in speaker and generate single tones. Thus, one can use them to listen to a time series as one plots the series in high-resolution graphics in real time. For an extra few hundred dollars one can slip a circuit board into the Apple that combines a synthesizer with software capable of producing multi-voice compositions. It is undoubtedly a flight of fancy to imagine listening to an economic history of the United States with inflation in the horns, investment in the strings, wars in the percussion section, and perhaps unemployment taken by the oboe. But the hardware exists, is inexpensive and available, and the economic symphony is fanciful only because the techniques of data analysis required to make it real have not yet been developed.

3. Color. To date, I have not seen any truly innovative use of color for data analysis. Most effective uses of color have simply substituted colors for shadings and textures or have used color as a substitute for alphabetic labels. This is what color is (currently) best at -- highlighting, or nominating. The chloroplethic maps from the Census Bureau use color in just this way, to label regions of a map without using a plethora of tags and reference lines. There are difficulties with using color in other ways in data analysis. Unlike pitches, there seems to be no universally perceived natural order to the color spectrum. Moreover, just as we know little about the psychology of music and its appreciation, a satisfactory theory of color perception is equally remote.

Those who develop techniques for data analysis based upon color distinctions should keep in mind, of course, that many people (including statisticians) are color-blind, either partially or completely. (Of course, many people are tone-deaf as well!)

5. All Things Great and Small

Up to this point we have discussed the personal computer (PC) in isolation, standing alone as a data-analytic tool. It need not be so lonely. Small computers have limited capacity, both in terms of computing power and memory. There is no reason, though, why the PC should not communicate with the BC (Big Computer) via a standard modem and telephone connection. The manuscript of this paper, for instance, was prepared entirely on a microcomputer, formatted, and then shipped across telephone lines to a large mainframe for further processing and typing. One can envision the PC as simply an ordinary terminal to the big machine, or in a variety of more demanding roles. Data sets can be moved back and forth from one machine to the other. If one thinks of the BC as a peripheral device of the personal computer (rather than the reverse) one is led to a number of interesting possibilities. The personal computer (which isn't terribly good at heavy computation) could use the BC as a subprocessor for heavy computations. It is possible now under program control for a personal computer to dial another computer's telephone number, to establish communication, and to send instructions or data to the answering computer. Consider the following scenarios:

A. The personal computer dials a large computer and automatically loads a data set, and then hangs up. The analyst examines several plots, explores the consequences of several transformations, removes a few points, and then decides on some full-blown regression analyses. At this point the program on the personal computer prepares input for a SCP on the large computer, dials the big machine, and then submits a job to perform the desired analyses.

B. The data analyst uses his personal computer as a terminal to login to an interactive system, on which he runs several factor analyses. On the basis of these preliminary runs, he decides on a factor model. The analyst then constructs a batch program to produce a large amount of reference output corresponding to the selected factor model. Later, the analyst examines alternative rotations of the axes by shipping summary results to the personal computer and then using the graphics capabilities of the PC to display alternative rotations.

C. A small computer is used to examine alternative plots to be used as illustrations in a manuscript. After examining several hundred plots on the screen, and after deciding on suitable ones to illustrate the points of the paper, the analyst gives a command to the package he is using on the PC which causes the data to be formatted and shipped across a telephone line in a format suitable for direct input into a high-resolution publication-quality graphics package on a large machine.

Some of these scenarios are fanciful, others (such as scenario C) exist and are in use today. It is possible that the software products available for use on personal computers and statistical computer packages on large computers will have built-in "hooks" for communicating with each other along the lines outlined here.

6. Conclusion

Statistical packages -- both interactive and batch varieties -- will continue to exist as we know them. Personal computers will have an impact, however, in the exploratory analysis of small to moderate data sets. When rapid turnaround is not required (as it is not when large volumes of output are being requested), batch SCPs will continue to be most useful. For compute-intensive procedures (such as factor analysis) which produce moderate amounts of printed output, interactive packages on timesharing computers will be the choice. For data exploration that largely involves data exhibition, personal computers will be valuable, particularly when the computational requirements are small relative to the data transfer requirements. When the necessary computations are merely those needed to display the data (rather than to compute numerical summaries or to reduce the data), personal computers can be expected to shine.

The key to effective use of personal computers in data analysis is to of the unique strengths of such devices (high data rates and unconventional input and output media) and which avoid conspicuous weaknesses (such as the slowness of floating-point computation). The computations performed by most statistical packages currently in existence could be done on a Marchand calculator or Babbage's Analytical Engine (given enough time). The personal computer affords us an opportunity to develop new techniques for data analysis that could not be done -- even in principle -- without the computing technology that we now have available.

Acknowledgements

I am grateful to my colleagues Maureen Lahiff, Harry Roberts, Stephen Stigler, and David Wallace for having read an early version of the manuscript and for making a number of helpful comments and suggestions. This work was supported in part by grant number MCS 80-02217 from the National Science Foundation.

References

Abt, Clark C. (1981). Letter, American Statistician, 35, 57.

Integrated Software Systems Corporation (1978). Tell-A-Graf Reference Manual.

Meyer, Morton A. (1981). Letter, American Statistician, 35, 56-7.

Muller, Mervin E. (1980). "Aspects of Statistical Computing: What Packages for the 1980's Ought to Do," American Statistician, 34, 159-168.

Myers, Ware (1981), "Computer Graphics: Reaching the User," Computer (March), 7-17.

U. S. Bureau of the Census and Manpower Administration (1974). Urban Atlas. Tract Data for Standard Metropolitan Areas, Series GE 80.

Wainer, Howard (1981). Reply to Letters. American Statistician, 35, 57-8.

Wainer, Howard, and Francolini, Carl M. (1980), "An Empirical Inquiry Concerning Human Understanding of Two-Variable Color Maps," American Statistician, 34, 81-93.

Wilson, Susan R. (1980), "Musetrics," Department of Statistics, Australian National University, unpublished.

Zwislocki, Jozef, J. (1981). "Sound Analysis in the Ear: A History of Discoveries," American Scientist, 69, 184-92.

STATISTICAL SOFTWARE DESIGN AND SCIENTIFIC BREAKTHROUGH: SOME REFLECTIONS ON THE FUTURE

Norman H. Nie, University of Chicago

Data analysis software has played a major role in the recent evolution of certain types of scientific disciplines which are characterized by weak theory and intercorrelated independent variables. The evolution of these fields of inquiry has depended as much upon data analysis packages for their progress as astronomy has upon the telescope, cellular biology the microscope, and particle physics the accelerator. Three new developments in the capabilities and organization of these software packages are pending or will emerge in the foreseeable future, and are discussed in terms of their potential impact on accelerated scientific discovery in the fields of inquiry that such software packages serve. They are: research-oriented graphics, true conversational analysis, and voice controlled software. These developments may help produce a revolution in scientific insight in a number of disparate fields.

Data analysis, statistics, graphics, voice recognition, interactive conversational

Laymen and scientists take for granted the intimate relationship between the development of tools of investigation and subsequent scientific breakthroughs. Examples of scientific revolutions that follow directly from the development of new tools of observation and manipulation are numerous and persuasive. Rapid extensions in astronomical knowledge as well as new models of planetary movement follow directly from the development of the telescope. The first revolution in cellular biology is predicated on the invention of the microscope. The current knowledge explosion in microbiology is dependent in part upon the development of the electron microscope and X-ray crystallography. Recent advances in the understanding of the fundamental nature of matter are intimately related to the evolution of various types of particle accelerators. The list can be endlessly expanded.

In each case the tools yield a powerful increase in the scientist's ability to observe or manipulate variables, and thus they permit researchers either to generate new data or to see heretofore unrecognizable patterns in observations. In this way the evolution of technological tools and the progress of science go hand in hand. The development of modern statistics, computer hardware, and data analysis software also have played, and continue to play, a similar role in the development of various scientific disciplines.

Specific types of statistical software developments have and are likely to effect "limited revolutions" in a specific subset of scientific disciplines--those that might be termed underdeveloped, non-experimental, and new empirical disciplines. Typically they share the following characteristics:

1. They are primarily inductive in mode of investigation, though they often aspire to the scientific maturity of deductive hypothesis testing.

2. The observations (the data) are often, although not universally, obtained by some survey or monitoring process in the real world rather than generated in the laboratory--in most cases with a minimum of experimental control. Moreover, and most importantly, the independent variables which are candidates for key causal factors are usually numerous and intercorrelated in complex ways. Thus, the importance of each independent variable as well as the full chain of causation is difficult to establish.

3. Typically, the number of observations is large. It is impossible to use the characteristics of individual observations to aid in working through causal sequences, since these disciplines tend to deal with phenomena embedded in open, complex systems of variables.

4. Finally, there is a paucity of well-developed bodies of theory and law-like generalizations in many of these fields. In other words, crisp, clean, critical hypotheses based on well-developed and accepted theories are difficult to come by.

Taken together, these characteristics imply a lengthy and iterative data analysis process where the major findings must be pushed, pulled, and teased out of the data. Findings are rarely obvious, and they almost never merge full-blown during the first few statistical "views" of the data. While these disciplines stand the furthest from classic notions of deductive model-building or those where "critical" experiments apply, they nevertheless attempt, and certainly aspire, to conform to the scientific canons of evidence and inference.

While those disciplines share a common research methodology, they cover a wide variety of substantive fields and phenomena. The characteristics discussed above typify research in the new empirical social sciences, where data emanate from sample surveys, or the analysis of records kept by the normal process of social bookkeeping, such as election returns, budgets, census information, congressional votes, and so on. The aforementioned characteristics are typical of clinical investigations in the health field as well as other types of investigation in the biological sciences, such as wildlife and forestry management. Such research characteristics are also commonly found at the interstices where social, biological, and physical phenomena meet, such as traffic management and engineering, urban planning, analysis of safety systems, pollution research, and a variety of communications-type system behaviors.

Over the last quarter century such fields of inquiry have burgeoned, based on the assumption that scientific methodology can lead to newer and sounder knowledge, and on the appearance of a number of "tool" developments. First, many of the research problems in these fields cannot be framed without the cumulative development of statistics that has occurred over the last century. Second, because of the volume of data often involved in these types of investigations, the cumulation of knowledge is dependent upon the evolution of computer hardware. The number of calculations required for even the simplest kind of statistical tabulations of very large files makes the computer a necessary precondition for this type of work. Finally, this type of research has been highly dependent upon the evolution of "packaged" data analysis software that contains not only the generalized form of a wide variety of statistical algorithms but also the tools for organization, management, and manipulation of large volumes of data. These software packages are often built around natural languages for the expression of relationships as well as quality output procedures, both of which act to give the researcher the best chance to visualize and understand patterns in the data. If individual research endeavors were forced to develop such software de novo, neither the hardware nor the statistical algorithms would be of much use.

Statistical data analysis packages such as the SPSS Batch System, SAS, and BMDP are some of the critical tools behind an explosion of knowledge. Although these packages have their origins in very different disciplines, their statistical and methodological structure, as well as the variety of their user base, testifies to the commonality of the process of investigation in all those fields that have weak theory, minimum experimental control, volumes of data, and numerous related independent variables. The constellation of tools entailing statistical methods and high speed computation, married to research-oriented software for data manipulation and statistical analysis, literally makes these fields possible.

Packages such as these (and the list is not meant to be exhaustive) permit investigators and their colleagues to control the data analysis process through regular and easy access to the data. Hypotheses about the pattern of relationships can be tested quickly, new measures can be constructed easily, and files can be merged and manipulated with a minimum of outside personnel or large, tangential excursions into development of special-purpose software. It is precisely this rapid and facile interchange between research concepts and the data that yields the iterative process from which solid findings can emerge. Those who engage in this process know that it is neither mechanical nor straightforward; ideas feed computer runs, the results of which produce new ideas. There are moments of rapid progress when critical pieces of the puzzle fall into place, interspersed with long periods of false starts and little apparent forward progress. The investigative methodology is basically sound, but all who play in this ball park know that the characteristics of the software (the smoothness of the motions and the repertoire of plays available) are central to how the game proceeds.

Each year these packages become more powerful as new statistical methods are added, better methods for handling data are introduced, and more convenient operations are implemented, so that the researcher can spend less time wrestling with the tools and more time and energy on the substance of the intellectual problem. And, as the hardware becomes more powerful, the size and complexity of possible research designs increase, allowing the analyses to be more reflective of the intricate multivariate world in which we live.

Data analysis software of the future is likely to continue to affect the process of investigation in the new empirical sciences and to accelerate the explosion of systematic knowledge in these areas. Three major foreseeable changes in statistically-oriented data analysis software are direct graphical representations of complex relational patterns, true conversational or interactive data analysis, and voice-controlled systems.

Graphics for the Researcher

Two graphics packages are presently available from statistical software producers (SAS Institute and SPSS Inc.) that, with parallel but slightly different designs, permit users to produce color and/or black and white graphics and charts directly from unaggregated raw data. By displaying research findings in graphical form, it is easier to communicate complex research results to the consumers of the research, be they readers of scientific journals or administrators attempting to understand critical data. The intention of the graphics package producers is to make the process of converting data into visual representation easier and cheaper for the researcher. The packages were not initially designed or intended to increase the researcher's investigative powers.

There is, however, a growing body of literature in statistics that stresses the importance of graphical representation in the research process itself. Tukey and others have pointed to the important potential role of graphical representation as an intellectual aid in the understanding and interpretation of data. Patterns of relationships, particularly when they are intricate and complicated, are much more likely to be quickly recognized and correctly understood when viewed as a geometrical representation of reality rather than a numerical or mathematical representation. Numerical quantities and mathematical notations represent a very powerful and precise symbol system for the codification of information. However, (Einstein excepted) they are far from the way in which most human beings conceptualize the world. We normally and naturally think in words and pictures--and only with considerable difficulty and a good deal of training do a few of us learn to think in mathematical symbol systems. However, word-oriented languages, be they spoken or written, are notoriously imprecise and have considerable difficulty codifying and defining complex relational realities. Graphic representations, many now think, combine the best of both worlds. If well thought out and carefully prepared, they can be as precise as an abstract symbol system and yet take advantage of human perceptual superiority in deciphering pictorial or geometrical representations. Moreover, many statisticial, i.e., relational, patterns seem to translate more naturally into visual rather than verbal form. This is the essence of the old dictum that a picture is worth a thousand words.

To demonstrate this argument, I place before you a mathematical symbol sentence.

$$\frac{(x-h)^2}{a^2} + \frac{(y-k)^2}{b^2} - \frac{(z-l)^2}{c^2} = 0$$

Figure 1

It precisely describes a physical object in the real world. Many of you have considerable training and qualify as professional mathematical symbol interpreters. Yet I would wager that more than one of you is still attempting to determine the precise nature of the object. However, as I now put up the slide of the graphic equivalent of the symbol sentence, instant recognition is apparent without any symbol training and even among those who could not even provide a name for the object.

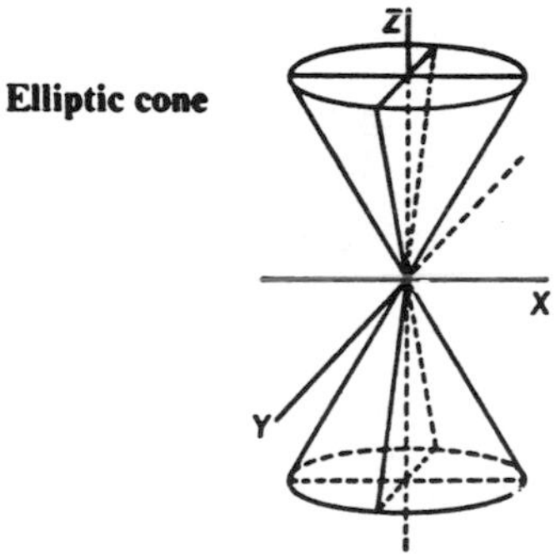

Figure 2

As software producers become more sophisticated in translating data into powerful graphics, and as the hardware and operating systems on which they reside become more powerful, I strongly suspect direct graphical representation will replace, in many instances, numerical display of statistical quantities. I believe that patterns in data that currently take hours to digest or are not apparent at all can be easily understood in geometric form. This is not a revolutionary idea; it is already being done. Histograms and scatterplot diagrams, as well as case lists of regression outliers, are but a few of the current examples.

In further support of my belief that graphical, not numerical, representation facilitates the quick apprehension of patterns of relationships within the data, I present two examples of output produced from the same data.

Partisan affiliation, 1952-1974

	1952	1956	1958	1960	1962	1964	1966	1968	1970	1972	1974
Weak Partisans	40.0%	39.0%	41.0%	40.0%	41.0%	39.0%	43.0%	41.0%	38.0%	39.0%	36.0%
Strong Partisans	37.0%	37.0%	39.0%	37.0%	37.0%	38.0%	28.0%	30.0%	30.0%	26.0%	26.0%
Independents	23.0%	24-0%	20.0%	23.0%	22.0%	23.0%	29.0%	30.0%	33.0%	35.0%	38.0%

Figure 3

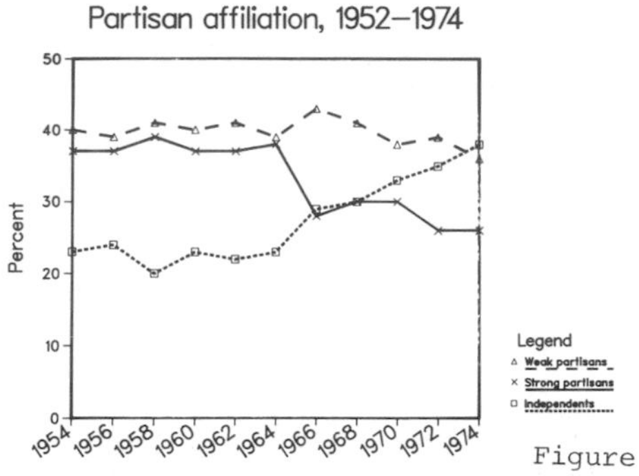

Figure 4

While it is absolutely necessary that one read each number to accurately analyze what is going on in the data in Figure 3, the same relationships depicted in Figure 4 are immediately recognizable and describable by both researchers and consumers.

The prospects for the direct generation of graphical representation for researchers rather than consumers is on the horizon. Its true power and potential awaits the second major innovation of the future: conversational computing.

True Conversational Data Analysis

Perhaps the most significant evolution on the horizon is true interactive or conversational data analysis; that is, fully prompting, immediate response software that functions efficiently with the large number of cases and variables that typify the research problems in the fields of inquiry under discussion. It is not simply a batch system in a conversational environment.

Conversational computing has a large number of advantages over batch computing. Much more work can be done because turnaround time is a matter of seconds rather than hours or days. Syntax errors are non-fatal and instantly correctable, saving hours of non-productive waiting. The convenience of submitting a job in the same environment where data analysis and research take place permits freer and spontaneous interaction with the data. The mechanics of running a conversational system interfere less with substantive issues and findings.

I am convinced that the overall effect of these characteristics of conversational computing is the acceleration of more accurate and perhaps otherwise unobtainable research insights. Like graphical representation, conversational data analysis is a powerful intellectual research aid, not just a more convenient way to do computing.

The mechanical and convenience improvements are, however, an important part of the foundation of this phenomenon. The lowering of the mechanical barriers, the change of physical location from remote to 'home' environment, the use of a machine that looks and behaves much like a typewriter, and most importantly, the design of languages and speed of movement between idea and test, will encourage the use of conversational systems by senior researchers. This has an important consequence; it puts the substantive researcher in direct contact with method and data. Knowledge about the data and the nuances of its behavior and pertubations can be directly understood by the researcher rather than filtered through several layers of research assistants. Watching what happens when particular variables are brought in and out of regressions or what the results are of changing factor rotation methods is very different than looking at the best solutions transmitted up through the research staff. Direct immersion by the senior investigator cannot help but produce better and more informed findings.

Something even more important than this happens when a researcher at any level is involved in good conversational data analysis. The system has all the power of a teaching machine--quick reinforcement for positive findings, quick closure for blind alleys, etc. There is a condensation and concentration of interaction with technique and data that is far more likely to yield a coherent sense of the whole picture. My colleague at the National Opinion Research Center, Andrew Greeley, one of the first and most dedicated users of our SCSS Conversational System, has referred to conversational computing as "an altered state of research consciousness."

Data analysis is much like a complex conversation between two Talmudic scholars. It is a tortured and complicated conversation in which the first party (the researcher) poses a number of complex and related interrogatory questions and the second party (data and technique) answers with a puzzling set of partial and equivocal replies. The first party then elaborates and refines another set of questions, and the conversation goes on. Through all the eras of batch computing, the biggest imped iment to the "conversation" has been time delay and human filters. Even in the best of all batch computing environments, questions must be asked piecemeal and responses come back hours or days later in equally piecemeal form. All of this works strongly against getting a coherent grasp of what the data are saying and hence what is going on out there in the substantive world.

Data analysis with batch computing, to use another analogy, is like attempting to follow the plot of a TV dramatization of a Russian novel, with commercial breaks interrupting the continuity of the story.

The following is a piece of a closure drawing that represents a common, everyday object, labeled "Monday Morning Output" for the sake of analogy. Can you figure out what it is?

MONDAY
MORNING
OUTPUT

Figure 5

Maybe some more pieces of the drawing would help--see "Monday Afternoon's Output."

MONDAY
AFTERNOON'S
OUTPUT

Figure 6

Here is "Tuesday Morning's Output".

TUESDAY
MORNING
OUTPUT

Figure 7

When presented piecemeal over time, it is quite difficult to merge images together to form a coherent whole, or even to conjecture what might lie within. However, almost anyone can recognize the image after a moment or two when all of its parts are presented together.

CONVERSATIONAL
SESSION

Figure 8

Conversational data analysis compresses the process of pattern recognition.

The software tools alone cannot solve all the scientific puzzles of the new empirical disciplines. Grounded theory, cumulative evidence, and creative insights are the backbone of any system of knowledge. Tools such as graphical representation and conversational data analysis are only the perfection of the lenses through which we view the puzzle.

Voice Control Systems

In the last several years, major breakthroughs have been made in the development of hardware and software for voice recognition. There are now available on the market for under $200 integrated circuitry chips that can be "trained" to recognize up to 64 words from a single speaker. By training I mean that the words are entered alphabetically and then read into the system vocally so that the digital conversion of the sound waves can be matched up with the words. A number of these chips together could currently be used to drive a conversational system such as SCSS without the researcher's having to spell out or type out a single command. The cost of these chips, like all others that have come before them, will certainly decline dramatically in the next several years as their power and capacity increase. The system would come preprogrammed with its full vocabulary ready for voice activation. The user's first task would be to enter and pronounce each variable name; from that point on the entire analysis would be voice-driven. Incorporating such a system into the statistical software may be a much easier task than direct graphics or the production and elaboration of true conversational systems--the real work is being done by others. Like true conversational computing, voice-activated systems further speed and simplify the man-machine-data interface, smoothing out and speeding up the Talmudic conversation. The combined impact of powerful graphical display with the immediacy of voice-activated conversational analysis will not only multiply the quantity of work that can be done in a given unit of time but will create, as Greeley thinks it has already, a truly heightened, altered state of research consciousness, where data, computer, statistical technique, and software become true extensions of the research intellect.

Some Thoughts on Expert Software

John M. Chambers

Bell Laboratories
Murray Hill, New Jersey 07974

ABSTRACT

Current successes in making computers *available* for data analysis will intensify the challenge to make them more *useful.* Cheaper, smaller and more reliable hardware will make computers available to many new users with data to analyse: few of them will have much statistical training or routine access to professional statisticians. We have a moral obligation to guide such users to good, helpful and defensible analysis. The best of current statistical systems allow users to produce many statistical summaries, graphical displays and other aids to data analysis. We have done relatively well in providing the *mechanics* of the analysis. Can we now go on to the *strategy*?

This paper examines *expert software* for data analysis; that is, software which tries to perform some of the functions of a statistician consulting with a client in the analysis of data. Is such software needed? Is it possible? What should it do? What should it not do? How should it be organized? Some tentative answers are proposed, with the hope of stimulating discussion and research. A comparison with expert software for other applications is made. A recent experiment at Bell Laboratories is used as an example.

1. Should we?

The desirability, indeed the obligation, to explore expert software for data analysis rests largely on developments in computing: computers are becoming rapidly cheaper and more widely available, while improved statistical software packages make doing data analysis on these machines feasible. A prediction in Chambers (1980), made initially in 1978, that serious, large-scale computers would be available to serve the needs of around 20 or so users for a purchase price of around $100,000, has essentially been overtaken by fact. A number of manufacturers are delivering or preparing to deliver machines in this price range with a serious scientific capability (32 bit addressing, plenty of memory and disk space). These are serious machines for data analysis. In addition, their usefulness, even to statisticians, will be at least as much in good tools for word processing, electronic communication and other applications as for statistics. The UNIX† operating system running on the DEC VAX-11/750 is a specific example of such a configuration.

A variety of statistical packages are available for such systems, some of them very powerful. For example, we have developed the S system for interactive data analysis (Becker and Chambers, 1981), within the UNIX environment, featuring extensive graphics, simple general expressions, on-line documentation, quick feedback to the user and easy extensibility of the system. A number of other statistical systems provide many similar features.

† UNIX is a Trademark of Bell Laboratories.

We may conclude from these developments that computerized data analysis will be increasingly widely available. Who will be doing this analysis? Many kinds of people, but usually *not* statisticians: economists, market researchers, accountants, sociologists, drug testers, pollsters, scientific experimenters of all disciplines, etc. In the past, the computing facilities might have been too expensive, not locally available or too complicated to use. Now, these barriers are disappearing.

How well are we providing for these potential users? The *mechanics* of data analysis are being made easier but we have not yet done much to help with the *strategy*. The user's guide to the SPSS system, for example, contains a cautionary note, including the following:

> "The statistical procedures in SPSS have little ability to distinguish between proper and improper applications of the statistical techniques. They are basically blind computational algorithms that apply their formulas to whatever data the user enters."

This warning is responsible and important (and, needless to say, not restricted to SPSS). However, in the new environment we must do more than caution; we must provide better guidance than "blind computational algorithms" wherever possible.

Good strategy in data analysis usually comes now, if at all, from an interaction between a client and an expert statistical consultant. The client brings the data, some questions to be posed and, frequently, considerable knowledge about the application. The consultant brings *expertise* (knowledge and experience) in data analysis. Through an iterative process of analysis, inference and dialogue the client and the expert seek to understand the data. Sometimes client and expert are one person, if the client has sufficient statistical background or the expert sufficient relevant experience. Infrequently so, however, and my previous argument suggests that the population of clients will grow rapidly. There is no practical scenario in which the population of experts will grow correspondingly.

Here is the moral imperative: Statistical software in its present form, made widely available by cheap computing, will precipitate much uninformed, unguided and simply incorrect data analysis. We are obliged to do something to help. The approach that I believe to be most important tries to bring some of the functions served by expert consultation into the statistical package itself. Consultation with a statistician would then be supplemented by analysis and dialogue guided by the software. (We should also face the likelihood that, for some users, *only* the software consultation will be available.) Section 4 discusses some alternative approaches briefly.

Systems with specialized knowledge have been considered for several applications (section 3). The term *expert software* is frequently used to describe them. I have argued here that expert software for data analysis would respond to an important need. The next step is to ask whether such software is a practical possibility and, if so, what form it is likely to take.

2. Can we?

An appreciation of the general requirements and challenges involved in expert software will benefit from studying a scenario for such a system in use. The following outline is suggested by an experimental system studied during the summer of 1980 at Bell Laboratories, described in more detail in Chambers, Pregibon and Zayas (1981). The functions performed by the system, however, have much wider applicability and are likely to be present, at least implicitly, in most future systems.

A client brings to the system a problem, with requests for data analysis stated in fairly general terms; for example, "Fit this data to a linear model involving these other variables". There now begins an open-ended interaction between client and system, which can be broken down into repetitions of three steps:

(1) computation of *statistics* consisting of summaries or transformations of the data;

(2) *inference* (or application of *action rules* in the terminology of expert software), based on the values of the statistics;

(3) *dialogue*, including graphical display or other explanation, between the software and the client to determine the next step in the analysis.

This process feeds back into itself. The dialogue will lead to alterations in the analysis, precipitating more statistics, more inference and more dialogue.

It would be nice to see at this point an explicit scenario, preferrably based on an example from a real problem. Unfortunately, such an example risks being both misleading and distracting: misleading in that it implies we know how to generate fully operational expert software systems, and distracting in that it diverts our discussion to the details of what statistical advice such a system is giving. We are just beginning to understand how to organize expert software. The details of the statistical strategy will come later and, importantly, must be easily adaptable in any case to respond to the needs and opinions of the user community. With these caveats, consider the following fragment of an imaginary session.

Our client presents some data on, say, bond yields for a particular series of bonds. These are to be fit by a model involving a number of other variables (say, economic variables like interest rates, stock prices, etc.). A preliminary chat with the software defines the problem to this extent. The software now computes some diagnostics and returns with a message something like:

> The analysis of the predictor (x) variables shows that some of the variables are seriously linearly dependent. The coefficients in the model associated with these variables will be poorly defined.

The software goes on to suggest some actions, involving both statistical steps and decisions by the client, that may alleviate the current problem. In this example, the software may present a list of variables that could be dropped to reduce the collinearity. The client may choose to drop one of those variables, may ask for explanatory text or plots or may take some other action entirely.

Statistics and inference have been separated above for both practical and philosophical reasons. Practically, the statistical calculations are what current packages know how to do. There will be much need for new, relevant statistical summaries but the overall flavor is familiar. In contrast, the kind of rule system involved in the inference is much less familiar. The ideas, if not the details, owe more to other applications of expert software than to statistical computing (see Chambers, Pregibon and Zayas, 1981).

Philosophically, too, the statistics represent safer and more familiar territory. We can and do disagree on the best measures for statistical properties like collinearity, but there is general agreement that looking at such statistics is a good idea. Far more controversial is the suggestion that software should propose answers and actions based on the statistics. There is a real danger that such software could emphasize narrow, mechanistic approaches to data analysis, just the approaches that those interested in exploratory analysis are anxious to avoid.

Nevertheless, we cannot simply ignore the need for inference. The users described in section 1 will require better advice than current statistical packages provide. I do not believe most passive solutions (reporting diagnostics or multiple answers, for example) will in practice go far enough. The challenge is to incorporate carefully designed inferential rules along with good graphics, flexible options and the kind of dialogue that will encourage the client to look deeply at the data.

Inference that proposes a range of actions must be distinguished from "automatic" data analysis. Nothing in this paper should be interpreted as taking from the client the ultimate responsibility for interpretation and decision. Keeping the balance of control in the right place depends largely on the dialogue software. We must: (1) present the statistical and inferential results clearly to users with relatively little *statistical* expertise; (2) allow the user easy and unrestricted facilities for questioning the results and making decisions; and (3) do all this in a friendly and natural environment.

The use of inferential software is probably the hardest step to accept philosophically. The design of the dialogue software (including the presentation of the inference to the client) is probably the hardest step to implement computationally. One approach is outlined in Chambers, Pregibon and Zayas (1981), using special files to describe the individual inferential rules and software similar to that for artificial intelligence applications (a production system) to handle the dialogue. However, we are very inexperienced at this area of software design. Some useful examples may be found in applications such as commercial query systems and natural language processors, but much effort and, probably, some failures can be expected.

3. What Other People Do.

The development of expert software and related techniques in the general area sometimes called artificial intelligence have received increasing attention lately (even to the extent of two articles in the New York Times Magazine (Stockton, 1980)!). Among the more serious applications are integrated circuit design, geological exploration, pilot assistance and medical diagnosis. A brief look at one system for medical diagnosis will show some interesting parallels and distinctions, when compared to data analysis.

MYCIN (Shortliffe, 1976) uses knowledge about certain infectious diseases to suggest a diagnosis, based on a dialogue with a treating physician. Over its range of applications, MYCIN is claimed to show diagnostic capabilities on the level of highly qualified human experts. It has not been used in clinical practice but as of 1980 work was reported in the Stanford University Heuristic Programming Project (Stanford, 1980) towards actual clinical use of several successor programs (e.g., in cancer therapy). A dialogue with MYCIN consists primarily of a sequence of questions posed by the program, to which the physician supplies answers; for example,

> Has John Doe recently had symptoms of persistent headache or other abnormal neurologic symptoms (dizziness, lethargy, etc.)?

The physician will respond (typically, YES or NO; occasionally with numerical results of clinical tests). During the dialogue, the program will print explanations for some of the questions (why they are being asked) and will report probable infections. The session ends with a detailed recommendation for therapy (possibly more than one recommendation).

The physician using the system can obtain an explanation of the reason for a question by responding WHY. In fact, such requests are handled by printing out some of the inferential rules (in our terminology) which led to the current place in the dialogue and which will be used at this stage. An associated software system, TEIRESIAS, develops the inferential rules based on input from human experts.

The applications to medicine and data analysis have several important parallels and a few crucial differences. In both cases, the client is using the specialized knowledge of the expert software to provide analysis and recommendations to supplement the client's own knowledge. The software's role is advisory; it does make recommendations, but the client must make actual decisions. During the dialogue, the client can ask for explanations of the current situation. The inferential rules are designed to draw plausible conclusions based on the results of selected diagnostic results (the medical systems do have a limited capability for quantitative analysis; e.g., testing whether a result lies in a specified range of values).

The differences might be summarized by saying that data analysis is a more heterogeneous, quantitative and iterative process than the form of medical diagnosis modelled by MYCIN. The diagnostic results in MYCIN are provided by the client; in our case, they will include both information from the client and, more frequently, the results of extensive statistical calculations. The form of explanation in MYCIN is only one of two different types of explanation needed in data analysis. In addition to showing the inferential rules, as in MYCIN, the statistical expert software must be able to display the quantitative results in a meaningful way (usually graphically). For example, an inferential rule involving a number of statistics that check for unusual observations (outliers) would need both an explanation of the rule itself and plots to show the presence, severity and identification of the detected outliers. There will be need for greater flow of information from the system to the client than in MYCIN.

The explanation will be more difficult in another sense as well. The client and expert in MYCIN share a common terminology and medical knowledge. MYCIN's advantage over its client is largely the result of incorporating highly specialized knowledge *within* the medical field. In contrast, the need in data analysis is to communicate *across* very different areas of specialization (the client's field of application and statistics). The ability of a human statistical consultant to bridge this gap, both in explanation and in detecting significant features in the data, will be a difficult function to imitate in the software. One idea to explore would be a client-specific glossary of terms, developed by separate dialogues with the client, that could be used to rephrase explanations and questions for different applications.

MYCIN is quite positive and assertive in its therapeutic recommendation. The inferential rules for most data analytic situations are, I feel, not sufficiently clear-cut to support so assertive a role. (One might ask, of course, how accurate rules of medical inference are in prediction.) Furthermore, few data analyses should be viewed as leading to one simple conclusion. The approach outlined in this paper is to provide step-by-step advice and recommendations as specific significant diagnostic results are obtained. The client plays a more frequent role as decision maker and is more clearly the final arbiter. The expert software proposes; the client disposes.

4. Other Approaches.

Expert software for data analysis involves a wide range of new problems in statistics and computing. We need new statistical summaries better tuned to guiding the client, and better understanding of how to make inferences from these statistics. We need good (conventional) statistical software to compute the statistics and organize the results for use by the expert software. We need software to organize the inferential rules of the system and, most important, to help the statistician study and modify these rules. We need software to manage the dialogue with the client in the most productive manner possible, and to communicate the results of the dialogue to the statistical and inferential software.

The newer and more difficult software for inference and dialogue is, as suggested in section 3, related to other applications of expert software and artificial intelligence. The statistical challenges, also, need to be related to other approaches to good data analysis. Fortunately, there is more reinforcement than conflict in the different approaches. At the present stage, no one should be discouraged from a serious attempt to develop better practical data analysis for the class of general user discussed here. We will need all the help we can get. The following are a few current topics in data analysis, with thoughts about their relation to expert software.

An obviously close relative, which might be described as *diagnostic presentation*, is exemplified by the book of Belsley, Kuh and Welsch (1980) on regression diagnostics. Some major statistical techniques like linear regression have associated diagnostics; i.e., statistical summaries that detect various potential difficulties with the technique (e.g., collinearity or outliers). Important diagnostics should be included in statistical packages. The user of the regression function in the package would be notified of the results of the diagnostics (perhaps only if the values are considered significant), *without a specific request*. Some scattered diagnostic warnings have appeared in many statistical packages, usually triggered by numerical problems rather than statistical considerations. Much wider and more consistent implementation of diagnostics is needed.

Obviously, diagnostics will be essential for the scenario of this paper. We need many good diagnostics to provide the statistics that support any attempt at inferential rules. I do think, however, that diagnostic presentation is best seen as a way station on the route to expert software. Trying to make the diagnostics really useful gets us back to exactly the needs that expert software addresses. Given a large collection of possible diagnostics, some inferential rules must be applied to do better than just displaying all the results. If the user is not assumed to be statistically sophisticated, the diagnostics must be translated into terms the user understands (the beginnings of dialogue); in addition, this usually involves suggesting what might have caused the diagnostic results (more inference). Presentation of some diagnostics needs information about the data that can only come from dialogue with the user (serial effects, for example, would be interpreted differently if the user claims that the order of observations could not be relevant).

Diagnostic presentation is important. Even in expert software, we should be prepared to show diagnostic results without specific inferential rules, if we feel unable to formulate the rules. Happily, good work on diagnostic presentation not only supports expert software, but actively pushes us towards it.

A slightly more distant relationship is with *robust estimation*. Robust methods (more generally, methods that claim to produce reasonable answers when classical statistics fail) are an important support to diagnostic results. Serious discrepancies between robust and "ordinary" estimates generate diagnostics, both overall (the difference in the regression coefficients) and specific (which observations were seriously down-weighted by a robust analysis). Such diagnostics are convenient and, often, fairly easy to explain. We will use them extensively in expert software. Robust methods are *not* an alternative to diagnostic results. It is a caricature of the purpose of robust estimation to imagine a form of automatic analysis which used robust estimation without attempting any inference when the robust and ordinary estimation disagree.

An idea related to robust analysis, but applicable elsewhere, is to return *multiple answers* to the user (Tukey, 1980). For example, given robust and classical estimates of some quantity (say, a regression fit), the software would produce both answers. An example unrelated to robustness would be the comparison of a regression model with various subset regressions, particularly if collinearity is present. The preparation and comparison of multiple related answers is another useful statistical step towards expert software. In fact, I would interpret most of the discussion in Tukey (1980) as either suggestions for or input to inferential rules in the scenario of section 2. Multiple answers to similar questions showing significant disagreement form an important diagnostic. If no good inferences can be made from the disagreement, the multiple answers may themselves be used as a diagnostic. As with other diagnostics, however, we should try to express the results in more understandable, relevant and helpful inference and dialogue.

References.

Becker, Richard A., and Chambers, John M. (1981). *S: A Language and System for Data Analysis.* Bell Laboratories Computer Information Service, Murray Hill, NJ.

Belsley, David A., Kuh, Edwin and Welsch, Roy E. (1980) *Regression Diagnostics.* John Wiley and Sons, New York.

Chambers, John M. (1980). Statistical computing: History and trends. *American Statistician,* **34**, 238-243.

Chambers, John M., Pregibon, Daryl and Zayas, Edward. (1981). Expert software for data analysis: An initial experiment. (to be presented at the 41st session of the International Statistical Institute, December 1981).

Shortliffe, E. H. (1976). *Computer-based Medical Consultations: MYCIN.* American Elsevier, New York.

Stockton, William (1980). Creating computers that think. *New York Times Magazine,* December 7 and 14, 1980.

Stanford University (1980). *Heuristic Programming Project 1980.* Computer Science Department, Stanford.

Tukey, J. W. (1980). Styles of data analysis, and their implications for statistical computing. *COMPSTAT 1980,* Physica-Verlag, Vienna.

Workshop 3

Fourier Transforms in Applied Statistics

Organizer: Marcello Pagano, Harvard University

Chair: H. Joseph Newton, Texas A&M University

Invited Presentations:

How Fast is the Fast Fourier Transform?, Persi Diaconis, Stanford University

Polynomial Time Algorithms for Obtaining Permutation Distributions, Marcello Pagano, Harvard University and Sidney Farber Cancer Institute, and David L. Tritchler, Cornell University and Memorial Sloan Kettering Cancer Institute

Efficient Estimation for the Stable Laws, Andrey Feuerverger, University of Toronto, and Philip McDunnough, University of Toronto

HOW FAST IS THE FOURIER TRANSFORM?

Persi Diaconis, Stanford University

Abstract. The average running time for several FFT algorithms is analyzed. The Cooley-Tukey algorithm is shown to require about $n^{1.61}$ operations. The chirp algorithm always works in $O(n \log n)$ operations. Examples are given to show that padding by zeros to the nearest power of 2 can lead to real distortions.

Key Words. Fast Fourier transform, analysis of algorithms, probabilistic number theory.

I. The Discrete Fourier Transform.

Given n numbers $x_0, x_1, \ldots, x_{n-1}$, let $q_n = e^{2\pi i/n}$. The discrete Fourier transform (DFT) of $x_0, \ldots, x_{n-1}$ is defined as the following collection of n numbers:

$$(1) \quad \phi(k) = \sum_{j=0}^{n-1} x_j q_n^{jk}, \quad k = 0, 1, 2, \ldots, n-1 .$$

The DFT is used in a variety of applied problems; nice surveys are in [1] and [2].

II. Speed.

We will use a very primative measure of how fast an algorithm is: the number of multiplications and additions required. In this problem it has turned out that this is a reasonable way to compare competing algorithms. If an algorithm is good using this measure, it proves good using more realistic measures such as actual running time for an implementation on a real machine or the number of bit operations. See Chapter 1 and Section 3 of Chapter 7 in [1]. To count the number of operations required to calculate $\phi(k)$, it is usual to suppose that the roots of unity q_n^{jk} are "free", e.g., stored on a list. Then, for fixed k; the sum in (1) takes n multiplications and n additions. To compute $\phi(k)$ for all n values of k requires n^2 multiplications and n^2 additions. Let us say that $2n^2$ "operations" are required for direct evaluation.

III. The Fast Fourier Transform (FFT).

The FFT is a collection of algorithms for computing the DFT. A clear, friendly description of the best known algorithm - the Cooley-Tukey algorithm - is in [2]. I will not try to review the ideas with any care. The basic idea is if the series length n is composite: $n = n_1 \times n_2$; then the computation can be broken into computing the DFT of n_1 series of length n_2 and n_2 series of length n_1. If the shorter series are computed directly, using an n_i^2 algorithm, the number of operations is approximately

$$2n_1 n_2^2 + 2n_2 n_1^2 = 2n(n_1 + n_2).$$

We will see in Section VI, that the shorter series can be computed more efficiently. Suppose that we compute a series of length n_i in $T(n_i)$ operations. Suppose that $n = n_1 \times n_2 \ldots n_r$. Then, inductive use of the basic idea shows that the number of operations required is about

$$(2) \quad T(n) \doteq n \sum_{i=1}^{r} T(n_i)/n_i .$$

In the original Cooley-Tukey algorithm, n was split into primes $n = \prod_{p|n} p$, where the product is taken with multiplicity. For series of prime length, direct computation was suggested, so $T(p) = 2p^2$. Then, (2) yields the following approximation to the number of operations required for the Cooley-Tukey algorithm

$$2n \sum_{p|n} p = 2n\, A(n)$$

where $A(n) = \Sigma_{p|n}\, p$, so $A(12) = 2+2+3 = 7$. When $n = 2^k$, this is $4n \cdot k = 4n \log_2 n$ operations required. This gives rise to the oft-quoted statement: "the FFT works in about $n \log n$ operations". As we have just seen, tbe Cooley-Tukey algorithm works in $2n\,A(n)$ operations, and if n is prime, this is $2n^2$. A more careful derivation of the operation counts used here is in [3].

IV. Average Case Analysis.

If an FFT program is to be implemented as a standard part of a computer package, then the running time over a wide range of values of n is of interest. One measure used in the analysis of algorithms is the average running time. For example, consider the average number of operations for the direct n^2 algorithm, averaged over all numbers below a fixed point x. This is

$$\frac{1}{x} \sum_{n \le x} 2n^2 \sim \frac{2x^2}{3} .$$

The Cooley-Tukey algorithm uses $2n\,A(n)$ operations as defined in Section III above. In [4] I determined the average running time as

$$(3) \quad \frac{1}{x} \sum_{n \le x} 2n\, A(n) \sim \frac{\pi^2}{9} \frac{x^2}{\log x} .$$

Following Section 18.2 of [5], if $f(n)$ is any function and $g(n)$ is a "simple" function of n such that

$$f(1) + f(2) + \ldots + f(n) \sim g(1) + \ldots + g(n) ,$$

we say that $f(n)$ is of average order $g(n)$. Thus (3) implies that $2n\,A(n)$ is of average order $\pi^2/3\; n^2/\log n$. This does not seem like much improvement over $2n^2$ operation for the direct computation.

In this problem, the average is a bad measure of the size of the running time; the distribution is somewhat long-tailed with the relatively sparse primes dominating. In [4] I determined the approximate behavior of the proportion of numbers less than x with $2n\,A(n) \leq n^{1+y}$. The result is

$$(4) \quad \frac{1}{x}\,|n \leq x : 2n\,A(n) \leq n^{1+y}| \sim L(y)$$

where the limiting distribution function $L(y)$ is supported on $[0,1]$ and, for example, $L(.33) = .05$, $L(.61) = .5$, $L(.95) = .95$. Thus, about half the numbers have $2n\,A(n) \leq n^{1.61}$. The median $n^{1.61}$ is a reasonable summary of how fast the Cooley-Tukey algorithm is. Approximations like (3) and (4) for other versions of the FFT such as the Tukey-Sande algorithm and Good's algorithm are derived in [4].

V. Padding by Zeros.

The most common reaction to the discussion above is "can't I just pad out my series by zeros to the next highest power of 2?" Often this trick works splendidly; an example being the computation of convolutions as in Chapter 7, Section 1 of [1]. Sometimes padding can lead to serious distortions. Some examples of this are in Section 2 of [4]. For a further example, I am indebted to Charles Stein: consider the problem of estimating the trend in a time series. We observe $X_i = \mu_i + Z_i$ where Z_i are assumed to be (approximately) a mean 0, circularly stationary, Gaussian series. An example yielding such a structure is weather data for $i = 1, 2, \ldots, 365$ where X_i is an average over several years of daily data. Under the above assumptions, the n transformed variables $\phi(k)$ are independent Gaussian. We can simultaneously estimate the means of the $\phi(k)$ using some form of Stein-like the shrinking technique and then transform back using an inverse Fourier transform. In this problem it seems crazy to use a transform of length 512 since this would result in a dependent, singular distribution.

VI. The Chirp Transform.

There are several methods of computing the exact DFT of n numbers in $O(n \log n)$ operations. The easiest variant is the so called chirp transform. The idea is to transform the sum defining $\phi(k)$ into a convolution. Thus

$$\phi(k) = \sum_{j=0}^{n-1} x_j\, q_n^{jk} .$$

Let $jk = \{j^2 + k^2 - (j-k)^2\}/2$. Then

$$\phi(k) = q_n^{k^2/2} \sum_{j=0}^{n-1} x_j\, q_n^{j^2/2}\, q_n^{-(j-k)^2/2} .$$

A convolution of the sequence $x_j\, q_n^{j^2/2}$ with the sequence $q_n^{(j-k)^2/2}$. Since convolutions can be computed exactly in $O(n \log n)$ operations, the chirp transform does the job. Some other algorithms, history, and references are in [4].

The change of variables underlying the chirp transform works for any n, prime or not. Thus, for given n, there are two options open: Direct use of the chirp transform versus splitting n into prime factors, using the chirp transform on a series of prime length p, and then pasting the series together using the Cooley-Tukey algorithm. We study this in the final section.

VII. To Split or Not to Split?

We can derive good approximations to the number of operations required for direct use of the chirp transform or the full factorization approach coupled with a $O(p \log p)$ transform for series of prime length p. Call these $C(n)$ and $F(n)$ respectively. Detailed formulas for $C(n)$ and $F(n)$ are given in [4]. Here we only summarize the main results. Both functions are $O(n \log n)$. However, the constant terms hidden in the O-notation have an effect. For example, when $n = 1000$, $C(n) = 67{,}584$ and $F(n) = 46{,}200$. In [4] it is shown that neither approach dominates. For example,

$$\frac{1}{x} \sum_{n \leq x} F(n) \sim \frac{3}{\log 2}\, x \log_2 x$$

while

$$\frac{1}{x} \sum_{n \leq x} C(n) \sim a(x)\, x \log_2 x$$

where $a(x)$ is a bounded function which oscillates between 4 and 4.5 $(3/\log 2 = 4.33)$.

The algorithm of choice thus depends on n. For any specific n, it is easy to determine using the formulas given in [4]. For routine use, the direct chirp transform is easier to program and never seems worse by a factor of two. If a series of specific length is to be used many times or a very long series is to be transformed, it seems wise to compare.

References

[1] Aho, A., Hopcraft, J., and Ulman, J. (1974). The Design and Analysis of Computer Algorithms, Addison-Wesley, Reading, Massachusetts.

[2] Bloomfield, P. (1976). Fourier Analysis of Time Series: An Introduction. Wiley, New York.

[3] Rose, D. (1980). Identities of the fast Fourier transform. Lin. Alg. Appl. 29, 423-443.

[4] Diaconis, P. (1980). Average running time of the fast Fourier transform. Jour. of Algorithms 1, 187-208.

[5] Hardy, G. H., and Wright, E. M. (1960). The Theory of Numbers, 4th ed. Oxford, London.

POLYNOMIAL TIME ALGORITHMS FOR OBTAINING PERMUTATION DISTRIBUTIONS*

Marcello Pagano, Harvard University and Sidney Farber Cancer Institute
David L. Tritchler, Cornell University and Memorial Sloan Kettering Cancer Institute

Polynomial time algorithms are presented for finding the permutation distribution of any statistic which is a linear combination of some function of either the original observations or the ranks. The algorithms require polynomial time as opposed to complete enumeration algorithms which require exponential time. This savings is effected by first calculating and then inverting the characteristic function of the statistic.

Key words: Permutation distributions; Fast Fourier Transform; Non-parametrics.

Introduction

The early work on permutation distributions was done by Pitman (1937, 1938) and he attributes the rationale for permutation distributions to Fisher (1935). In order to explain the tests obtained from these distributions we can present the example used by Pitman. Namely, given the two samples, 1.2, 2.3, 2.4, 3.2 and 2.8, 3.1, 3.4, 3.6, 4.1, decide whether they are samples from two different populations. We concentrate on differences in location, so to facilitate the explanation of the method, rescale the two samples. They are then 0, 11, 12, 20 and 16, 19, 22, 24, 29. We can ask the question, if from these nine numbers we consider all possible subsamples of size four, what proportion will have as large, or larger, a difference between the two sample means than the one observed? Consider all ${}_9C_4$ = 126 possible samples. To answer the above question, we can order these 126 samples by the sum of the elements in the first sample. Table 1 displays eight of these samples, the four most extreme at either end of the spectrum.

Sample I				Σx_i	$\lvert\Sigma(x_i-17)\rvert$
0	11	12	16	39	29
0	11	12	19	42	26
0	11	12	20*	43	25
0	11	12	22	45	23
		⋮			
16	22	24	29	91	23
19	20	24	29	92	24
19	22	24	29	94	26
20	22	24	29	95	27

*observed sample

Table 1 The eight most extreme of the 126 samples of size 4 from the original 9 observations.

From Table 1 we see that three of the samples are as extreme, or more so, on the one end of the spectrum, as the observed sample. If we are interested in deviations in both directions, then subtract the mean, 17, from each observation and look at the absolute sum of the deviations. In this case there are five cases as extreme or more extreme. The p-value is thus 5/126 = .0397.

In general, one has two samples, $x_1,\ldots,x_m$ and $y_1,\ldots,y_n$ which can be combined to form, $\underline{z}^T = (x_1,\ldots,x_m, y_1, \ldots,y_n)$. The statistic of interest can be written

$$(1) \qquad S = \sum_{j=1}^{N} a(z_j)\, I_j \qquad , N = n + m$$

where, in the above case, $a(x) = x$. We wish to find the distribution of S as the indicator, binary variates I index all ${}_NC_m$ possible subsamples of size m. Each of these subsamples is considered equally likely.

Other functions $a(\cdot)$ can be used to effect most powerful unbiased tests (see Lehmann (1959)), but these must be chosen prior to analyzing the data. Parzen's (1979) approach is to derive the $a(\cdot)$ from the data. Another data dependent score function is given in Lambert (1977). Her motivation is to make the permutation test robust.

The obvious approach to calculate the distribution of S is the one which requires the total enumeration of all ${}_NC_m$ subsamples. In the worst case (m=n) this increases as $4^n/n^{\frac{1}{2}}$. The effect of this increase is evident in Table 2 which gives the time, in seconds, required to obtain the distribution of S, as a function of m when n=m, on a DEC 2040 computer.

*This research was supported in part by grants from the National Institutes of Health, DHS, CA-28066 and CA-09337.

m	Time in seconds
5	.2
6	.5
7	1.7
8	6.5
9	26.2
10	105.7
11	425.8
12	1,728.8
15	21,600.0*

*Computation terminated before the answer was obtained.

Table 2 Time required to obtain the permutation distribution of S for two samples of size m.

The obvious computational problems with permutation distributions led researchers to find other methods. Pitman found the asymptotic distributions for the test statistics, but for small samples the problem remained unsolved. Wilcoxan (1945) replaced the argument z_i in (1) by the rank of z_j amongst $z_1,\ldots,z_N$. This breakthrough led to other nonparametric tests that had a number of advantages. One of these is that it is practical to tabulate their distributions in a number of important situations.

But even with nonparametric tests based on ranks there are still some situations that defy tabulation, such as, for example, stratified samples, censored observations, and ties. In the next section we present an algorithm to calculate the distribution of S in both cases (when $a(\cdot)$ is a function of the original observations or when $a(\cdot)$ is a function of the ranks) in polynomial time. As a result it is feasible to obtain distributions that would previously have been infeasible. Furthermore, these distributions are useful not only for testing purposes but also for point estimation, confidence intervals, etc. (see, for example, Kempthorne and Doerfler (1969), Hollander and Wolfe (1973), and Tritchler (1980)).

A Polynomial Time Algorithm

The proposed algorithm has two main steps:

Step 1 Calculate the characteristic function of S at equispaced points on $[0,2\Pi]$.

Step 2 Invert the characteristic function to yield the permutation probability distribution of S.

To calculate the characteristic function of S denote $a(z_i)$ or $a(R_j)$ generically by a_j. R_j is the rank of z_j in $z_1,\ldots,z_N$. This enables us to handle simultaneously both situations, when $a(\cdot)$ is a function of the observations and when $a(\cdot)$ is a function of the ranks. Furthermore, define

$$P_m(N) = \{\underline{j} = (j_1,\ldots,j_m): 1 \le j_1 < \ldots < j_m \le N\}.$$

Then, if $\phi(m,N,\theta)$ is the characteristic function of S evaluated at θ considering all samples of size m from N observations,

$$\phi(m,N,\theta) = \sum_{\underline{j}\varepsilon P_m(N)} \left(\prod_{k=1}^{m} \exp(i\theta a_{j_k})\right) / {}_N C_m \ .$$

In order to calculate $\phi(\cdot)$, define $\psi(\cdot)$ by

$$\psi(m,N,\theta) = {}_N C_m \ \phi(m,N,\theta).$$

Then $\psi(\cdot)$ obeys the difference equation

$$\psi(j,k,\theta) = \psi(j,k-1,\theta) + \exp(i\theta a_k)\ \psi(j-1,k-1,\ )$$

for $j=1,\ldots,k$ and $k=1,\ldots,N$. This is a generalization to characteristic functions of the formula in Mann and Whitney (1947) for finding the distribution of the Wilcoxon statistic. From this difference equation we see that the characteristic function can be found in approximately 2mN (complex) multiplications and additions.

In order to use the characteristic function to find the distribution of S we rely on the following development.

Let X be a discrete random variable with distribution $Pr(X=j)=P_j$, $j=0,1,\ldots,U$ and characteristic function

$$\phi(\theta) = \sum_{j=0}^{U} P_j \exp(ij\theta) \ , \quad \theta \varepsilon [0,2\pi] \ .$$

Since X is defined on a finite integer lattice, in order to find the P_j we may use the basic theorem in Fourier series.

Theorem

For any integer $Q > U$ and $j=0,\ldots,U$,

$$P_j = \frac{1}{Q} \sum_{k=0}^{Q-1} \phi\left(\frac{2\pi k}{Q}\right) \exp\left(\frac{2\pi ijk}{Q}\right) \ .$$

That is, knowing the characteristic function at these Q equispaced points on $[0,2\pi]$ is equivalent to knowing it everywhere. Furthermore, if it is known at these points, one may use a Fast Fourier Transform (FFT) (See Singleton (1969) for example) to invert it. Thus one may choose Q to take full advantage of the FFT.

This theorem cannot be used directly for all S defined in (1) because these variables generally are not integer valued. In order to make them integer valued one can modify their definition by replacing the function $a(\cdot)$ by

$$a'(\cdot) = [L(a(\cdot) - \alpha)/(A - \alpha)] \ , \quad (2)$$

the square brackets indicating the integer value of what they enclose, where α is the minimum value of $a(\cdot)$ in the sample at hand, A a strict upper bound for $a(\cdot)$ and L a multiplier. The first two constants are easily found. The last determines both the value of Q and the precision with which the statistic is determined and should thus be chosen judiciously; a simple choice for Q is $Q = mL$. To be more specific, the larger we make L the closer $a'(\cdot)$ will be to $a(\cdot)$ but, at the same time, the larger Q will be. At a minimum, especially for small samples, L must be large enough so that two distinct values of $a(\cdot)$ yield two distant values of $a'(\cdot)$. To this end suppose the two closest unequal $a(\cdot)$ in a particular sample are γ units apart. Then choose L so that

$$L \geq (A - \alpha)/\gamma \, . \qquad (3)$$

For most practical purposes there is no difference between an L of 10^3 and an L of 10^4. Further considerations, such as the use of a binary computer and the FFT, might suggest that L also be chosen as a power of two, for example.

However L is chosen, the result is that the algorithm will yield the exact distribution of the statistic S defined with $a'(\cdot)$ replacing $a(\cdot)$ in (1). The approximation is one whose effect we can monitor. For example, if $\alpha = 0$, $A = 1$, and we choose L to be 10^3, then we have $a(\cdot)$ correct to three digits. Running the program again but with a larger L will reveal the effect of this approximation. Of course, if in this example the scores are only given to three digits then we have the exact distribution.

Table 3 gives the time required to find the distribution function, at its jump points, of an S statistic by using the algorithm suggested in this section. The times for the proposed algorithm are given for both the Wilcoxon test and the exponential scores test. The FFT used is that of Singleton (1969). The exponential scores are non-integer valued and the transformation (2) (with L chosen to satisfy (3)) increase Q (which was set equal to mL) and hence the execution time. In this instance the values of

$$\alpha = N^{-1}, \; A = \sum_{k=N-m+1}^{N} k^{-1}, \text{ and } \gamma = (N-1)^{-1}$$

are available analytically. For the Wilcoxon test Q was chosen to be $mN = 2m^2$. The calculations were performed on a DEC-2040.

	Proposed Algorithm	
m	Wilcoxon	Exponential Scores
5	.5	.5
6	.6	.8
7	.8	1.2
8	1.0	1.9
9	1.3	2.7
10	1.8	3.7
11	2.3	5.3
12	3.2	7.1
15	10.4	15.4
20	19.8	46.7
30	85.7	213.8

Table 3 Timings, in Seconds, for Two Samples of Size m

If one compares Tables 2 and 3 the savings afforded by the proposed algorithm are quite evident.

Generalizations

The above algorithm can be extended to handle single samples, censored observations, stratified samples (see Pagano and Tritchler (1980)) and to calculate confidence intervals (see Tritchler (1980)). It can also be generalized to multiple samples. For example in the three sample case, define S_1 to be the sum of the scores in the first sample, S_2 the sum of the scores in the second sample; that is

$$S_1 = \sum_{j=1}^{m_1} a_j \, , \; S_2 = \sum_{j=m_1+1}^{m_1+m_2} j \, .$$

Then the bivariate ψ function obeys

$$\psi(j,k,\ell,\theta_\ell,\theta_2) = \exp(i\theta_1 a_\ell)\, \psi(j-1,k,\ell-1,\theta_1,\theta_2) + \exp(i\theta_2 a_\ell)\, \psi(j,k-1,\ell-1,\theta_1,\theta_2) + \psi(j,k,\ell-1,\theta_1,\theta_2)$$

for $j,k=1,2,\ldots$ and $j+k\leq\ell = 1,2,\ldots$. From this function, as before, we can obtain the distribution of (S_1,S_2).

An example of the above is given by the Wilcoxon statistic. Table 4 exhibits part of the frequency distribution of the Wilcoxon statistic for three samples each of size three. The rest of the distribution can be obtained by symmetry.

		6	7	8	9	10	11	12	13	14	15	...
		⋮	⋮	⋮	⋮	⋮	⋮	⋮	⋮	⋮	⋮	⋮
	15	1	1	3	6	6	9	16	14	18	12	...
	16	1	2	4	6	8	12	15	14	18	18	...
	17	2	2	4	6	10	10	15	14	14	14	...
	18	3	3	5	6	10	12	16	15	15	16	...
S_1	19	3	2	5	6	8	9	12	10	12	9	...
	20	3	3	4	6	6	8	10	10	8	6	...
	21	3	2	4	6	6	6	6	6	6	6	...
	22	2	2	3	4	4	5	5	4	4	3	...
	23	1	1	2	2	3	2	3	2	2	1	...
	24	1	1	2	3	3	3	3	2	1	1	...
Marginal		20	20	40	60	80	100	140	140	160	160	...

Table 4 Frequency distribution of Wilcoxon statistic for three samples of size (3,3,3). Each entry in the frequency table should be divided by 1,680.

Given the distribution of the Wilcoxon statistic it is then easy to obtain the distribution of the Kruskal-Wallis statistic. It is interesting to note that since the surface evident in Table 4 is not unimodal the Wilcoxon and Kruskal-Wallis statistics induce different partitions of the sample space. For example the point (15,12) has a probability of 16/1680 and is more likely than the point (15,13) which has a probability of 14/1680. But the value of the Kruskal-Wallis statistic is 6 for the point (15,12) and 8/3 for the point (15,13). This implies that the two test statistics force different sample outcomes into the "significance" region.

References

Fisher, R.A. (1935). The Design of Experiments. Oliver E. Boyd, Edinburgh, London (1st edition 1935, 7th edition 1960).

Hollander, M. and Wolfe, D.A. (1973). Nonparametric Statistical Methods. Wiley, New York.

Kempthorne, O. and Doerfler, T.E. (1969). The behavior of some significance tests under experimental randomization. Biometrika, 56, 231-248.

Lambert, D.M. (1977). P-values: asymptotics and robustness. University of Rochester Doctoral Dissertation.

Lehmann, E.L. (1959). Testing Statistical Hypotheses. Wiley, New York.

Mann, H.B. and Whitney, D.R. (1947). On a test of whether one of two random variables is stochastically larger than the other. Annals of Mathematical Statistics, 18, 50-60.

Pagano, M. and Tritchler, D. (1980). On obtaining permutation distributions in polynomial time. Sidney Farber Cancer Institute Technical Report No. 118Z, Boston, Massachusetts.

Parzen, E. (1979). Nonparametric statistical data modeling. Journal of the American Statistical Association, 74, 105-121.

Pitman, E.J.G. (1937/38). Significance tests which may be applied to samples from any population. Journal of the Royal Statistical Society, B4, 119-130, 225-237. Biometrika, 29, 322-335.

Singleton, R.C. (1969). Multivariate complex Fourier transform, computed in place using mixed-radix Fast Fourier Transform algorithm. Stanford Research Institute.

Tritchler, D. (1980). Algorithms for permutation confidence intervals. Sidney Farber Cancer Institute Technical Report No. 149Z, Boston, Massachusetts.

Wilcoxon, F. (1945). Individual comparisons by ranking methods. Biometrics, 1, 80-83.

EFFICIENT ESTIMATION FOR THE STABLE LAWS

Andrey Feuerverger and Philip McDunnough

University of Toronto

ABSTRACT This paper is concerned with Fourier procedures in inference which admit arbitrarily high asymptotic efficiency. The problem of estimation for the stable laws is treated by two different approaches. The first involves FFT inversion of the characteristic function. A detailed discussion is given of truncation and discretization effects with reference to the special structure of the stable densities. Some further results are given also concerning a second approach based on the empirical characteristic function (ecf). Finally we sketch an application of this method to testing for independence, and also present a stationary version of the ecf.

KEY WORDS Fourier transform; empirical characteristic function; stable laws; efficient estimation; stationarity.

1. INTRODUCTION

This paper is concerned with Fourier procedures for inference, in particular, procedures which can be shown to admit arbitrarily high efficiency. A general treatment of such questions was given by the authors in [1] and [2]. While these methods offer promising approaches to a wide variety of problems, the most striking of these is perhaps the problem of inference for the stable laws. This application was investigated in some detail in [3]. In this paper we provide some additional results concerning this particular application, and address a number of related matters. Some necessary background is reviewed below, but only briefly, and the reader is referred to [1-3] for a more complete discussion and background.

With respect to the stable distributions there are in fact two distinct contexts for efficient Fourier procedures. (Superficially these contexts appear to be unrelated; a deeper examination however reveals certain unexpected and striking connections.) The first context involves the use of the fast Fourier transform (FFT) to numerically invert the stable characteristic functions. DuMouchel [4] showed that the usual asymptotic results of maximum likelihood theory apply to this case, and was the first to implement the procedure [5,6]. In the second context, inference is based directly on the empirical characteristic function and no inversion takes place. The asymptotic efficiency result for such procedures is due to the authors [1,2].

These two general approaches are the only ones known to us, to be fully efficient (or, at least, arbitrarily highly efficient) for estimating the parameters of the stable laws. We comment however that an order statistics approach is known for the general scale - location problem. See, for example chapter 9 of [7]. Possibly the stable shape and skewness parameters could also be efficiently estimated via order statistics; however such results are not available as yet. A new approach to inference via density-quantile functions was introduced recently in [8] which appears promising for the case of the stable laws, however these implications have yet to be investigated. Finally, we mention that work of related interest appears in [9] and [10].

2. THE STABLE DISTRIBUTIONS

The family of stable distributions consists precisely of the limits of suitably scaled and centered sums of independent and identically distributed random variables. In the case that the random variables have second moment we obtain the normal distribution which is, of course, the best known stable distribution. However the wider class of distributions which share in the central limiting property forms a family having in fact four parameters: location, scale, shape and skewness. These distributions possess densities which are unimodal and analytic, however these functions are available in closed form only in very special cases. Expansions due to Feller and Bergstrom are available for the densities, but these present considerable computational difficulties. Attempts at inference consequently have often been based on the characteristic function for the stable laws which have in fact a known and simple form.

By virtue of their definition the stable laws clearly possess a natural interest for certain statistical applications and for questions of robustness in particular. In this connection we refer to the recent work [11], especially chapter 2. There, working with the student densities which are proportional (except for scale and location) to $(1+x^2)^{-(\nu+1)/2}$, $\nu > 0$, it is shown that normality-based analyses produce misleading results when applied to student data with $\nu = 3$ degrees of freedom, but that student $\nu = 3$ analyses give essentially correct results in the normal case. Regrettably, a corresponding analysis based on $\nu = 2$ is not available, however one might anticipate roughly comparable results. Now, since in the range $0 < \nu = \alpha < 2$ the stable and student densities have common asymptotic character $c\cdot|x|^{-\alpha-1}$ for the tails, the broad underlying suggestion here is that the stable laws with values of the shape parameter near, or in fact *very* near $\alpha = 2$ may be useful in certain robustness contexts. (We remark that as $\alpha \to 2$ the stable densities become Gaussian in a uniformly continuous manner, although the asymptotic character of the tails undergoes a discontinuous change at $\alpha = 2$. Further, we note that the central portion of the $\alpha \simeq 2$ cases are much more Gaussian in shape than the student cases with small ν.) Other applications of the stable laws as well as references to these may be found in [3].

The characteristic function (cf) for the stable distributions (with scale and location omitted) may be written as

$$\phi(t) = \exp\{-|t|^{\alpha} - i\beta^* h(t,\alpha)\} \tag{2.1a}$$

where

$$h(t,\alpha) = \frac{t(|t|^{\alpha-1} - 1)}{\alpha - 1} \tag{2.1b}$$

Note $h(t,\alpha)$ is continuous on R^2 and defined by continuity as $t\ln|t|$ when $\alpha = 1$. The parameters α and β^* are, respectively, the shape and skewness. The parameterization (2.1) was derived in [3] and has the advantage of removing the singularity which is present at $\alpha = 1$ in the classical representations, and also of providing densities which vary continuously in the supremum norm over the domain of the $(\alpha, \beta^*, \mu, \sigma)$ parameterization. Here μ and σ are location and scale as defined by the representation

$$c(t) = e^{i\mu t}\phi(\sigma t) \tag{2.2}$$

for the cf. Further, the domain for the parameterization (2.1), shown in figure 1 has a shape which is more in keeping with the known behaviour characteristics of the densities, especially near $\alpha = 0,2$. The asterisk in β^* serves to distinguish (2.1) from the classical parameterizations; the four parameters are not in agreement among the various representations, except for α which is common to them all. Further properties of the stable laws may be found in [3] and the references cited there.

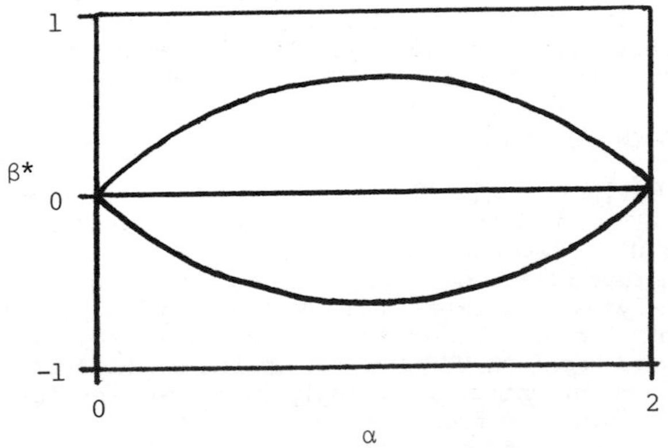

Figure 1: Domain of the (α,β^*) parameterization for the stable laws.

3. THE FFT APPROACH

Implementation of the FFT in the context of maximum likelihood for the stable laws was carried out in [5,6] and later in [3] in independent work. The problem of evaluating the integral

$$f(x) = \frac{1}{2\pi}\int_{-\infty}^{\infty} \phi(t)e^{-itx}dt \tag{3.1}$$

by means of the FFT is treated in some detail in these references. See also [12]. Here we shall comment further on only two aspects of this problem. These concern the fact that the FFT involves first a truncation and secondly a discretization of the integral (3.1).

Consider first the effects of truncation. Under typical circumstances $\phi(t)$ will be effectively zero outside of $(-M, M)$, the truncation region. When this is not so, we may consider replacing $\phi(t)$ by $\phi(t)g(t)$ in the integrand (3.1) where $g(t)$ is a tapering function:

$$\begin{cases} g(0) = 1 \\ g(t) = 0, \quad |t| > M \end{cases} \tag{3.2}$$

The effect of this, of course, is that we then obtain not $f(x)$ but rather the convoluted form $\int_{-\infty}^{\infty} f(y)\hat{g}(x-y)dy$ where $\hat{g}(\cdot)$ is the Fourier transform of $g(\cdot)$. For this convolution to also be a density it is sufficient, though not necessary, that $(\hat{g}, g)$ be a density and characteristic function pair; though we do have the necessity if the convolution is required to be a density for arbitrary f. One possibility would be to set

$$g(t) = \begin{cases} (1 - \frac{|t|}{M})\cos\frac{\pi t}{M} + \frac{1}{\pi}\sin\frac{\pi|t|}{M}, & |t| \le M \\ 0, & |t| > M \end{cases}$$

corresponding to

$$\hat{g}(x) = \frac{2\pi M}{(M^2x^2 - \pi^2)^2}(1 + \cos Mx)$$

$$= \pi M \left[\frac{\sin\frac{1}{2}(Mx - \pi)}{\frac{1}{2}(M^2x^2 - \pi^2)}\right]^2, \quad -\infty < x < \infty .$$

The function $\hat{g}(x)$ is the density function corresponding to the distribution having smallest second moment amongst distributions whose characteristic functions $g(\cdot)$ vanishes outside $(-M, M)$. (See [13].) This density has almost all of its mass between $\pm 3\pi/M$ and very weak lobes at $\pm 5\pi/M$, $\pm 7\pi/M, \ldots$.

For the stable laws in particular, the following alternative resolution of the truncation problem is worth noting. Firstly, we remark that inversion by means of the discrete Fourier transform is problematical only for smaller values of α. Essentially this is because $e^{-|t|^\alpha}$ has a steep gradient near $t=0$ and yet falls slowly as $|t| \to 0$. Therefore a numerical grid is called for which is both dense and extensive and hence unmanageable for small α. Now, according to a result of Zolotarev (see theorem 5.8.4 of [14]) the values of any stable density in the range $\frac{1}{2} < \alpha < 1$ may be obtained directly from a corresponding stable density with shape parameter α^{-1} and vice versa. Now, arguing as in §5.8 of [14] we may in fact establish that the Bergstrom - Feller expansion for $\alpha > 1$ continues to be Fourier related to the expression for the characteristic function, even when $\alpha > 2$ (though of course the expression will no longer correspond to a distribution in this case). This observation leads to an extended version of Zolotarev's result (see (2.11) of [3]) which is valid in the range $0 < \alpha < \frac{1}{2}$. An exception at $\alpha = \frac{1}{2}$ does not seem to be removable, however in any case, inversion problems become severe only inside the range $\alpha < \frac{1}{2}$. But these may be replaced with inversions involving α^{-1} hence eliminating this problem.

The resolution in [5] p. 35 should be noted here. It involves replacing the truncated $\phi(t)$ by a "wrapped summation"

$$\tilde{\phi}(t) = \sum_{k=-\infty}^{\infty} \phi(t+2\pi kM), \quad t \in [-M, M]. \tag{3.3}$$

While small α may involve much "wrapping" it should be noted that when this is done, the effects of truncation will in fact be eliminated! (Here we are assuming that x is restricted to the grid points of the discrete transform.)

Finally, concerning the problem of truncation, it should be noted that this is a problem only for small α and that these cases are of lesser statistical interest in general. In our work we used 10% cosine tapers and found this gave satisfactory results provided α was not less than about 0.5.

We turn now to the problems associated with the discretization of (3.1). Due to discretization we will obtain not $f(x)$ but rather an aliased version

$$\sum_{j=-\infty}^{\infty} f(x+jN\Delta x) \tag{3.4}$$

where Δx is the x-spacing, and N the number of points associated with the discrete transform. There are several ways to deal with this problem. First, accurate de-aliasing may be achieved by means of the asymptotic ($|x| \to \infty$) expansions for the density since $N\Delta x$ typically is not small. (In fact these same expansions are required for the tails where inversion is inaccurate numerically.) Alternately, an approach due to Filon is adopted in [5]. See also [13].

Our approach to de-aliasing was based on the empirically discovered observation that the aliasing error may, for all practical purposes, be regarded as constant! To see how this may be, consider typical values such as $N = 1024$ and $\Delta x = 0.1$ so that the term $N\Delta x$ appearing in (3.4) takes the value 102.4. Now this is much larger than the useful range of the inversion; in fact, after some numerical study, we used $|x| = 7.5$ as the cut-off value beyond which the FFT inversion was ignored and the asymptotic expansions were used. Consequently alias correction is required only in a zone such as $|x| < 7.5$ in which the correction can be regarded as constant. This is checked by computing (3.4) with the term $j = 0$ omitted. For example, substituting the Cauchy form $(1+x^2)^{-1}$ for f in (3.4) the $j = \pm 1$ terms add up to .000191 at $x = 0$ and to .000194 at $x = 7.5$ and the remaining terms $j = \pm 2, \pm 3, \ldots$ are even less discordant. As the alias is much less than the actual density within $|x| < 7.5$, this degree of accuracy is highly adequate. A rough check for the cases $\alpha > 1$ can be made using a form such as $(1+|x|^3)^{-1}$ for f. The corresponding sums of the $j = \pm 1$ terms at $x = 0$ and $x = 7.5$ are .00000186 and .00000192. The cases $\alpha < 1$ can be expected to behave better still. The only question arises at $\alpha = 2$ and values of α *exceedingly* close to 2 where the cubic character of the tail may not yet assert itself for moderate x. But here the alias is of such low order that it may be taken as zero. Thus over the full α range, and the x range of interest the alias may be regarded as constant. Since an exact expression is available for the stable density at $x = 0$, the constant alias may be determined at once as the difference at $x = 0$ between the exact and the FFT determined density! (This difference contains numerical error, but this, in turn, has much lesser order than the alias.)

Some Monte Carlo results for maximum likelihood estimation for the stable laws are provided in [15] and [3].

4. INFERENCE IN THE FOURIER DOMAIN

Suppose $X_1, X_2, \ldots, X_n$ are iid variates with density in $\{f_\theta(x)\}$ where θ for simplicity here is real univariate parameter. (The multiparameter extension is straight-forward.) The likelihood equation can be written in the form

$$\int_{-\infty}^{\infty} \frac{\partial \log f_\theta(x)}{\partial \theta} \, dF_n(x) = 0 \tag{4.1}$$

where $F_n(x)$ is the empirical cdf, or in a suggestive alternate form

$$\int_{-\infty}^{\infty} \frac{\partial \log f_\theta(x)}{\partial \theta} \, d(F_n(x) - F_\theta(x)) = 0 \tag{4.2}$$

where $F_\theta(x)$ is the cdf and we are using the fact that the expectation of a score function is null.

Now define the following transformed quantities: the characteristic function:

$$c_\theta(t) = \int e^{itx} \, dF_\theta(x); \tag{4.3}$$

the empirical characteristic function (ecf)

$$c_n(t) = \int e^{itx}\, dF_n(x) = \frac{1}{n}\sum_{j=1}^{n} e^{itX_j}; \qquad (4.4)$$

and the inverse transform of the score

$$w_\theta(t) = \frac{1}{2\pi}\int \frac{\partial \log f_\theta(x)}{\partial\theta} e^{-itx}\, dx . \qquad (4.5)$$

Under fairly general conditions (see [2]) we may apply a form of the Parseval theorem to (4.2) to obtain

$$\int_{-\infty}^{\infty} w_\theta(t)\,(c_n(t) - c_\theta(t))\,dt = 0 \qquad (4.6)$$

which is equivalent to (4.2) and is in fact a Fourier domain version of the likelihood equation. As the score function is not generally integrable, the "weights" $w_\theta(t)$ must be regarded as generalized functions, however (see [2]) this presents no real difficulties. The result (4.6) immediately suggests that the ecf will have valuable applications, not only as an ad hoq tool, but as a tool which supports asymptotically efficient inference procedures. This viewpoint is explored in some detail in [1] and [2] where detailed reviews of the properties of the ecf may also be found. In particular the covariance properties of the ecf process $Y_n(t) = \sqrt{n}\,(c_n(t) - c(t))$ depends directly on $c(t)$ and not the density, and as well the process has, asymptotically, certain properties of a Gaussian process.

One consequence of these results, in the context of estimation for the stable laws, is that the stable parameters may be estimated with arbitrarily high asymptotic efficiency using the so-called *harmonic regression procedure*. This involves the choice of a grid $0 < t_1 < t_2 <\ldots< t_k$ of points and fitting the real and imaginary parts of $c_\theta(t) = e^{i\mu t}\cdot e^{-|\sigma t|^\alpha}$ where $\theta = (\mu, \sigma, \alpha)$ (in the symmetric case) to $c_n(t)$ at the selected points using weighted nonlinear least squares with the asymptotically optimal covariance matrix. The asymptotic covariance structure of these estimates approaches the Cramer-Rao bound as the selected grid $\{t_j\}$ becomes fine and extended. This procedure (see [2], [3]) is *very considerably* simpler to implement than FFT based maximum likelihood.

The question of optimal selection of the points $\{t_j\}$ can be studied (for k not too large) by means of straightforward regression-type calculations, and table (4.1) provides the asymptotic values of n·var for the parameters μ, α, σ evaluated at $\mu = 0$, $\sigma = 1$ for the symmetric laws for $\alpha = 1.2(.2)1.8$. These values are provided for the procedures based on k = 2, 3, 4, 6, 10, 20, 40 *equally spaced points* $t_j = j\tau$, $j = 1,\ldots k$, a real and an imaginary value being taken at each point. Below each value of the asymptotic variance we give the value of τ, the optimal pair being reported in each particular case. Table (4.1) extends and is complementary to table 1 of [2].

Table 4.1: Asymptotic variances N·VAR and optimal uniform spacing intervals for estimating symmetric stable parameters using K ecf points.

	k	2	3	4	6	10	20	40
$\alpha = 1.2$	μ	2.52	2.41	2.36	2.31	2.28	2.27	2.26
		.49	.39	.33	.26	.18	.11	.06
	α	5.78	3.76	3.07	2.53	2.17	1.94	1.84
		.42	.34	.29	.23	.16	.10	.06
	σ	1.81	1.74	1.71	1.68	1.66	1.64	1.63
		.75	.61	.50	.36	.23	.13	.07
$\alpha = 1.4$	μ	2.46	2.40	2.38	2.36	2.35	2.34	2.34
		.41	.33	.27	.21	.14	.09	.05
	α	4.98	3.54	3.05	2.68	2.44	2.30	2.25
		.43	.34	.29	.22	.15	.09	.05
	σ	1.39	1.37	1.35	1.33	1.32	1.31	1.31
		.77	.51	.39	.28	.18	.10	.06
$\alpha = 1.6$	μ	2.35	2.33	2.32	2.31	2.31	2.31	2.30
		.36	.27	.22	.16	.10	.05	.03
	α	3.94	3.07	2.79	2.58	2.47	2.41	2.39
		.43	.33	.27	.20	.13	.07	.04
	σ	1.12	1.08	1.07	1.05	1.05	1.05	1.04
		.55	.39	.31	.22	.14	.08	.04
$\alpha = 1.8$	μ	2.21	2.20	2.20	2.20	2.20	2.19	2.19
		.29	.21	.17	.19	.15	.06	.06
	α	2.50	2.14	2.03	1.97	1.94	1.92	1.91
		.42	.31	.25	.17	.11	.09	.06
	σ	.85	.83	.83	.82	.82	.82	.81
		.43	.31	.24	.17	.16	.09	.06

If we do not restrict to *uniform* spacing, substantial improvement is possible but optimal grids can be obtained feasably only if k is small. For k = 2, 3, 4 and 5 the α-optimal spacings were determined in [3]. Table (4.2) here is intended to indicate the intrinsic complexity of the k-dimensional surfaces using the case k = 4. In table (4.2) an alternative ridge is shown where the Fisher information is locally maximized. The suboptimal cases, only examined from $\alpha = 1.2$ to 1.9, are shown in square brackets. It may be seen that at $\alpha = 1.5$ the two ridges are approximately equal in height, while maxima for $\alpha < 1.5$ occur along one of the ridges and for $\alpha > 1.5$ along the other ridge shown. The asymptotic variances attained (for α) are shown in the last column.

Table 4.2: α-optimal spacings for k = 4 showing an alternative ridge

α	t_1	t_2	t_3	t_4	$N \cdot VAR(\hat{\alpha})$
1.0	.03	.16	1.48	2.43	1.4637
1.1	.04	.18	1.39	2.22	1.7207
1.2	.05 [.022]	.21 [.09]	1.30 [.23]	2.03 [1.50]	1.9710 [2.0106]
1.3	.07 [.030]	.24 [.11]	1.23 [.25]	1.87 [1.42]	2.1974 [2.2245]
1.4	.08 [.038]	.26 [.13]	1.17 [.28]	1.72 [1.34]	2.3774 [2.3908]
1.5	.048 [.09]	.15 [.28]	.31 [1.11]	1.28 [1.58]	2.4831 [2.4832]
1.6	.058 [.10]	.17 [.30]	.33 [1.06]	1.20 [1.44]	2.4691 [2.4800]
1.7	.070 [.12]	.19 [.33]	.36 [1.00]	1.12 [1.29]	2.3079 [2.3246]
1.8	.082 [.13]	.21 [.34]	.38 [.94]	1.04 [1.13]	1.9464 [1.9630]
1.9	.100 [.15]	.23 [.35]	.40 [.86]	.90 [.96]	1.3038 [1.3144]

5. A TEST FOR INDEPENDENCE

In this section we consider briefly use of the ecf as a natural tool for testing independence. Suppose (X_j,Y_j), $j = 1,2,\ldots,n$ are iid observation pairs; define

$$\hat{c}_{XY}(s,t) = \frac{1}{n}\sum_{j=1}^{n} e^{i(sX_j+tY_j)} ,$$

$\hat{c}_X(s) = \hat{c}_{XY}(s,o)$, $\hat{c}_Y(t) = \hat{c}_{XY}(o,t)$; let c_{XY}, c_X, c_Y be the corresponding non-empirical c.f.'s. We consider testing $H_o{:}c_{XY}(s.t) = c_X(s)\, c_Y(t)$ all s,t, first in the case that the marginals c_X, c_Y are known. For a "discrete" procedure, we need k pairs of points (s_j,t_j), $j = 1,2,\ldots,k$. Let $\hat{\underset{\sim}{z}}$ be the $2K \times 1$ vector of real and imaginary components of $\hat{c}_{XY}(s,t) - c_X(s)\, c_Y(t)$. A natural test is then based on the quantity $D^2 = \hat{\underset{\sim}{z}}'\, \underset{\sim}{\Sigma}^{-1} \hat{\underset{\sim}{z}}$ where $\underset{\sim}{\Sigma}$ is an estimate of $N \cdot VAR(\hat{\underset{\sim}{z}})$ evaluated under H_o and may be determined from

$$E_{H_o}\{\hat{c}_{XY}(s_1,t_1)\, \hat{c}_{XY}(s_2,t_2)\}$$

$$= c_X(s_1+s_2)c_Y(t_1+t_2) - \frac{n-1}{n} c_X(s_1)c_X(s_2)c_Y(t_1)c_Y(t_2).$$

Under H_o, $n\,D^2$ will be asymptotically χ^2_{2k}. The usual advantages of Hotelling's test for the mean of a multivariate normal carry over asymptotically; in particular, the test based on D^2 is asymptotically UMP invariant among tests based on the $\hat{c}_{XY}(s_j,t_j)$.

When the marginals are unknown, we may consider a procedure based on $\phi_n(s,t) = \hat{c}_{XY}(s,t) - \hat{c}_X(s)\hat{c}_Y(t)$. Since

$$E\hat{c}_X(s)\,\hat{c}_Y(t) = \frac{n-1}{n}\, c_X(s)\; c_Y(t) + \frac{1}{n} c_{XY}(s,t)$$

we will have $E\phi_n(s,t) = 0$ for all s,t if and only if X,Y are independent. The covariance structure of $\phi_n(s,t)$ may be obtained by a tedious but straightforward calculation and this leads again to a D^2 type testing procedure. An alternative procedure for this case may be developed using linear model notions based on the idea that independence corresponds to $c_{XY}(s,t)$ having structure which is of unit rank, i.e. a product of a function of s and one of t; these functions may further be constrained to lie in the subspace of characteristic functions.

6. A STATIONARY VERSION OF THE ECF

For certain applications (e.g. [16]) it is convenient to have a stationary version of the ecf. To this end, it is helpful to think of $c_n(t)$ as being a sum (or average) of complex sinusoids having angular frequencies X_j. A striking feature now of the ecf is that these sinusoids all have phase zero and hence are aligned at $t = 0$ to give $c_n(0) = 1$.

This alignment however does not bear on the informational content of $c_n(t)$. Thus, the original X_j may be recovered from $c_n(t)$ upon examining the function $\langle c_n(t)\, e^{-itx} \rangle$ where the operator $\langle\ \rangle$ is defined as $\lim_{T\to\infty}\frac{1}{2T}\int_{-T}^{T} dt$ and is taken here and below to act on t. However if a non-aligned version were used, such as

$$s_n(t) = \frac{1}{n}\sum_{j=1}^{n} e^{i(tX_j+\Phi_j)} \qquad (6.1)$$

with arbitrary phase angles Φ_j, then the X_j and the $\Phi_j \pmod{2\pi}$ as well are likewise recoverable from the function $\langle s_n(t)\, e^{-i(tx+\phi)} \rangle$.

Suppose now that in (6.1) the Φ_j are independent random variables with uniform distribution on $[-\pi,\pi]$, and independent of the X's. Then $s_n(t)$ will be a sum (average) of i.i.d. stationary complex processes and will itself be a stationary complex process with

$$E\, s_n(t) = 0$$

$$n \cdot E\, s_n(t_1)\, \overline{s_n(t_2)} = c(t_1 - t_2)$$

and

$$E\, s_n(t_1)\, s_n(t_2) = 0\ .$$

We remark that the characteristic function $c(t)$ itself may be cast in a stationary form through Wiener's generalized harmonic analysis approach. See [17]. Thus $c(t)$ is determined by almost every realization of a stationary complex Gaussian process $\{s(t),\ -\infty < t < \infty\}$ defined through

$$E\, s(t) = 0$$

$$E\, s(t_1)\, \overline{s(t_2)} = c(t_1 - t_2)$$

and in fact the determination can be made through $\langle s(t+u)\, s(t) \rangle = c(u)$. Recall that the complex Gaussian distribution requires $E\, s(t_1) s(t_2) = 0$. The process $\{s(t)\}$ may be referred to as the stationary characteristic function process (scf) and $\{s_n(t)\}$ where the Φ_j are iid uniform $(-\pi,\pi]$ may be referred to as the empirical scf process (escf).

Now the finite dimensional distributions of $\sqrt{n}\ s_n(t)$ are seen easily to converge to those of $\{s(t)\}$ while the matter of weak convergence may be treated as in [18] when $E|X_j|^{1+\delta} < \infty$ for some $\delta > 0$, and more generally as in [19]. We mention that weak convergence is not in general required for purposes of application; in particular, we may adopt lemma (5.2) of [3] to the escf process with a parallel proof.

Kendall and Kent [16, 20] have defined the quantogram

$$\xi_n(t) = n^{-\frac{1}{2}} \sum_{j=1}^{n} e^{i(t+T_n)X_j}$$

where $T_n \uparrow \infty$, and this may be regarded as an alternative empirical version of $s(t)$. In fact, for large T_n and for X_j from an absolutely continuous distribution we may make the approximate identification $T_n X_j \equiv \phi_j \pmod{2\pi}$. Actually, $\xi_n(t)$ is just the ecf viewed in a domain increasingly remote from the origin. It may be understood in the context of almost periodic functions theory (e.g. [21]) that under broad conditions $\xi_n(t)$ and $s_n(t)$ will be essentially equivalent processes as $T = T_n \to \infty$.

Finally we note a negative result for the processes defined in this section. Consider the statistical use of the escf, and suppose this is based on the asymptotic properties of $s_n(t_1),\ldots,s_n(t_k)$. Then it may be readily established that the Fisher information for these variates increases with k, but essentially does not depend on n. Therefore, except for special applications, the question of how best to use the escf is unresolved.

ACKNOWLEDGEMENTS The computations described herein were carried out (in FORTRAN) on the University of Toronto IBM 360/165. The manuscript was typed by Celia Galati. The research of both authors is supported by NSERC, Canada.

REFERENCES

[1] FEUERVERGER, A. and MCDUNNOUGH, P. (1981). On the efficiency of empirical characteristic function procedures. J. Royal Statist. Soc. B, 43, to appear.

[2] FEUERVERGER, A. and MCDUNNOUGH, P. (1981). On some Fourier methods for inference. J. Amer. Statist. Assoc., Vol. 76, June, to appear.

[3] FEUERVERGER, A. and MCDUNNOUGH, P. (1981). On efficient inference in symmetric stable laws and processes. *Proceedings of the International Symposium on Statistics and Related Topics*, Ottawa, May 1980. Eds. Saleh, A.K. Md E., et. al. North Holland Publishing Co.

[4] DUMOUCHEL, W. H. (1973). On the asymptotic normality of maximum likelihood estimates when sampling from a stable distribution. Annals of Statist. 1, 948-957.

[5] DUMOUCHEL, W. H. (1971) Stable distributions in statistical inference, Ph.D. dissertation, Yale University.

[6] DUMOUCHEL, W. H. (1975). Stable distributions in statistical inference: 2 - Information from stably distributed samples, J. Amer. Statist. Assoc. 70, 386-393.

[7] DAVID, H. A. (1970). *Order Statistics.* Wiley, New York.

[8] PARZEN, E. (1979). Nonparametric statistical data modelling. J. Amer. Statist.

Assoc., Vol. 74, 105-121.

[9] FENECH, A. P. (1976). Asymptotically efficient estimation of location for a symmetric stable law. Annals Statist. 4, 1088-1100.

[10] DE HAAN, L. and RESNICK, S. I. (1980). A simple asymptotic estimate for the index of a stable distribution. J. Royal Statist. Soc. B, 42, 83-87.

[11] FRASER, D.A.S. (1979). *Inference and linear models*, McGraw Hill, New York.

[12] DAVIES, R. B. (1973). Numerical inversion of a characteristic function. Biometrika, 60, 415-417.

[13] BOHMAN, H. (1960). Approximate Fourier analysis of distribution functions. Ark. Mat. 4, 99-157.

[14] LUKACS, E. (1970). *Characteristic functions*, 2nd ed., Hafner, New York.

[15] DUMOUCHEL, W. H. (1974). Stable distributions in statistical inference: 3 - Estimation of the parameter of a stable distribution by the method of maximum likelihood. Unpublished report. Presented to NBER-NSF Conference on Bayesian Econometrics, Ann Arbor, April 1974.

[16] KENDALL, D. G. (1977). Hunting Quanta. 2nd ed. *Proc. Symp. to Honour Jerzy Neyman*, Warszawa, 1974, Polish Acad. Sci., 111-159.

[17] WOLD, H.O.A. (1948). On prediction in stationary time series. Annals Math. Statist. 19, 558-567.

[18] FEUERVERGER, A. and MUREIKA, R.A. (1977). The empirical characteristic function and its applications. Annals Statist. 5, 88-97.

[19] CSORGO, S. (1981). Limit behaviour of the empirical characteristic function. Annals Probab., to appear.

[20] KENT, T. J. (1975). A weak convergence theorem for the empirical characteristic function. J. Appl. Probab. 12, 515-523.

[21] BOHR, H. A. (1947). *Almost periodic functions*, Chelsea, New York.

Workshop 4

Algorithms and Statistics

Organizer: Bruce W. Weide, Ohio State University

Chair: Bruce W. Weide, Ohio State University

Invited Presentations:

Applications of Statistics to Applied Algorithm Design, Jon Louis Bentley, Carnegie-Mellon University

Algorithms with Random Input, George S. Lueker, University of California at Irvine

Recent Results on the Average Time Behavior of Some Algorithms in Computational Geometry, Luc Devroye, McGill University

APPLICATIONS OF STATISTICS TO APPLIED ALGORITHM DESIGN[1]

Jon Louis Bentley
Department of Computer Science
Carnegie-Mellon University
Pittsburgh, Pennsylvania 15213

Abstract -- The field of Applied Algorithm Design is concerned with applying the results and techniques of Analysis of Algorithms to the real problems faced by practitioners of computing. In this paper we will study the applications of probability and statistics to that endeavor from two viewpoints. First, we will study a general methodology for building efficient programs that employs the tools of data analysis and statistical inference, probabilistic analysis of algorithms, and simulation. Second, we will see how these techniques are used in a detailed study of an application involving the Traveling Salesman Problem, and in a brief overview of several other applications.

Index Terms -- Analysis of algorithms, probabilistic algorithms, software engineering, traveling salesman problem.

1. Introduction

Although it is not the only important issue in developing a software system, efficiency can play a major role in the success of a system. One of the primary tools a programmer can use to achieve efficiency is the selection of algorithms to accomplish the programming task at hand. Lewis and Papadimitriou [1978] and Bentley [1979] both survey the field of algorithms, and Knuth [1973] and Aho, Hopcroft and Ullman [1974] are standard textbooks in the area. Those references describe how the field of algorithms can be (and has been) applied with considerable success to the problems of developing efficient software systems. Unfortunately, those references also describe a number of cases in which algorithms that are always efficient (even in the worst case) are not known to exist for many important problems.

For that reason among others, a subarea of algorithms has recently received a great deal of interest: *probabilistic algorithms*. This term describes algorithms with performance that may not be good in the worst case, but is good on the average. Bentley [1981a], Janko [1981] and Lueker [1981] all survey the field of probabilistic algorithms. Although a number of interesting theoretical results have already been achieved in this young field, very few of those have been tested in practice.

The purpose of this paper is to describe a methodology for applying algorithm design to real problems that is especially appropriate for applying probabilistic algorithms. Section 2 describes a paradigm of Applied Algorithm Design in general terms. That general paradigm is then illustrated by one example in detail in Section 3, and by several small examples in Section 4. Finally, conclusions are offered in Section 5.

2. A Paradigm of Applied Algorithm Design

In this section we will study the general paradigm of applied algorithm design. We will first describe the overall process by refining the diagrams of Figure 1, and then study the components of the diagram individually.

The most simple view of applied algorithm design is the user's view shown in Figure 1a. The user is a programmer who has a real problem and must produce a real program to solve it. A host of programming methodologies have recently been developed to help the user with this process; those methodologies emphasize the rigorous development of correct and easily maintainable programs. For most applications, the user should proceed using those tools and ignoring this paradigm, because for most applications efficiency is not a major concern. Only if efficiency is known to be a concern should the user bring the problem of Figure 1a to an algorithm designer.

The algorithm designer's first view of the problem is that

[1] This research was supported in part by the Office of Naval Research under Contract N0014-76-C-0370.

shown in Figure 1b. He cuts away the mass of irrelevant detail in the real problem and abstracts from it an easily understood abstract problem. He gives the best algorithm he can to solve it, which can then be translated into a real program for solving the real problem. At this point, though, all he can say about the algorithm and the program is how well they perform in the worst case, and this is often not good enough for the user's requirements.

If a better program is desired than can be attained in the worst-case sense, then we must know "what the data is usually like". There are many answers to that question, and two kinds are shown in Figure 1c. The user brings the algorithm designer a set of real data that capture what the user thinks will be typical of inputs to the program later. The algorithm designer studies that data and induces from it a probabilistic model that captures its relevant points. Given the abstract problem and the probabilistic model of the data, the algorithm designer can design an algorithm that performs very well on the average.

Once the algorithm designer has the fast program he must test it empirically to make sure that the predicted results are in fact obtained; this process is shown in Figure 1d. He runs the real program on real data to see how closely the predictions conform to both the analytic results and the performance of the algorithm on data randomly generated from the model. If the three results do not all match then we must decide whether the error was introduced by the simulation process, by an inappropriate probabilistic model, or by an erroneous analysis.

The complete paradigm of applied algorithm design is shown in Figure 1e, and all of the edges between objects are labelled with the names of the corresponding activities. The upper part of the diagram could be called "reality" and has traditionally been the realm of program developers; likewise, the lower part could be called "theory" and has traditionally been the realm of algorithm analysts. Thus the job of the applied algorithm designer is to bridge the gap between theory and practice by mapping a user's real-world problem to a theoretical domain, solving it there, and then mapping that theoretical solution back to a real-world solution.

2.1. The Entities

The above introduction provides a brief overview of the entire paradigm; we will now focus more closely on its parts by investigating the entities involved in the process (which are the nodes in the graph of Figure 1e). This discussion will be rather abstract; we will see a concrete example of these objects in Section 3. The following list describes the entities.

- **The Real Problem.** This is the problem that the user brings to the algorithm designer; it is the input to the process of applied algorithm design. It is important to note, though, that the algorithm designer should pay great attention to the statement of the problem to ensure that what the user says he wants is what he in fact does want. This step has much more to do with psychology than engineering, but is crucial enough to make or break the rest of the process.

- **The Real Program.** This is the output of the process of algorithm design; it is a functional program that runs in the context of a larger system. We will see that it is usually appropriate to build at least two "real" programs in the process of applied algorithm design: the second program that is integrated into the system is preceded by a prototype program that is less concerned with the human interfaces and robustness of a final product and more concerned with ease of testing, modification, and instrumentation.

- **The Abstract Problem.** The real problem that the user has is often hidden beneath a level of detail that, while important for the real solution, tends to obscure the underlying computational structure of the problem. If the algorithm designer is to see the problem clearly, then he must develop a crisp abstract problem that is simple enough to present the problem clearly yet not so simple as to obscure the true complexity of the problem. The abstract problem should be described in the form of "given this, compute that, subject to these constraints"; Aho, Hopcroft and Ullman [1974] provide a number of examples of succinct and informative problem definitions.

- **The Algorithm.** The algorithm that solves the problem is a general description of a solution process that is not imbedded in any particular programming language; it is a clean solution to the abstract problem. Although it does not solve the details of the real problem (which were eliminated to achieve the abstract problem), it must be easy to incorporate those details into the algorithm as it is made more precise in a program. The algorithm should be described in a very high-level language both to facilitate development of the algorithm and to leave open as many opportunities as possible for the translation to the particular system on which the algorithm will be coded. Many examples of such algorithms can be found in Aho, Hopcroft and

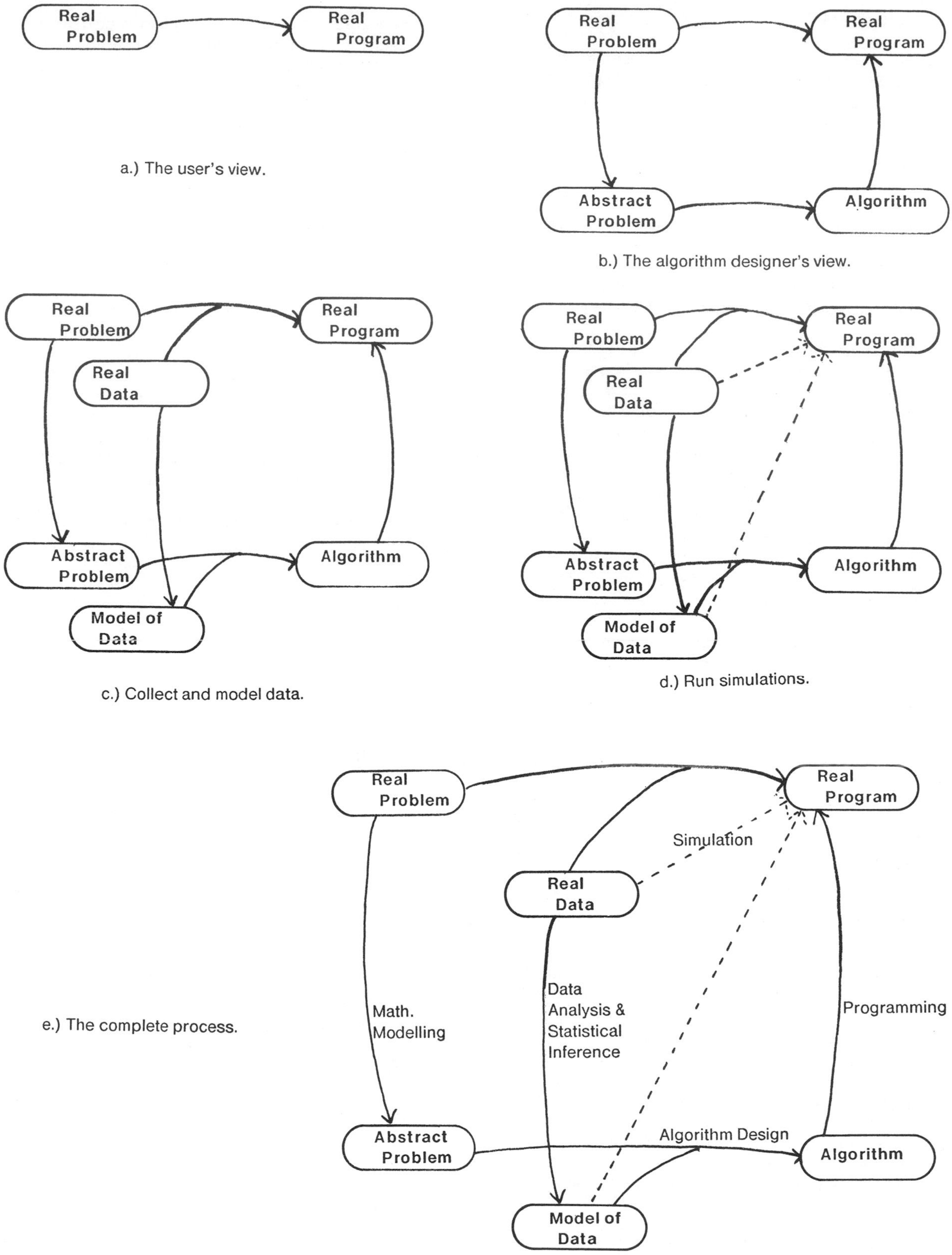

Figure 1. The paradigm of Applied Algorithm Design.

Ullman [1974].

- **The Real Data.** If the best worst-case algorithm the algorithm designer can find does not satisfy the user's needs, then they will have to study the real input data to find patterns there that they can exploit to yield faster algorithms. To this end, the user must supply the algorithm designer with a set of input data that he suspects is typical of the input data that will be seen during the operation of the program in production. It is essential that the data presented to the algorithm designer at this step be representative of the production-time data. This data should be presented in a simple format so that the prototype programs can process it quite easily.

- **The Model of the Data.** The purpose of the data model is to provide a mathematical description of what the data is usually like; we use that description to analyze the performance of our algorithm on the average. The typical technique we use for the description is to give a probabilistic model of the inputs, although we sometimes use other methods (we will mention one in Subsection 4.2). There are two knowledge sources that we bring to bear in creating such a model. The first is an *a priori* knowledge of the real-world phenomenon that generates the data, and the second is the *a posteriori* observation of the real data supplied by the user. We usually use the *a priori* knowledge to conjecture a form of the probabilistic model and then use the *a posteriori* data to test whether the model fits the data and, if so, estimate the model's parameters.

2.2. The Activities

In the above subsection we studied the nodes of the graph of Figure 1e, which are the objects that the algorithm designer studies. In this subsection we will take a different view of the process and study the edges of the graph, which are the activities of the algorithm designer.

- **Mathematical Modelling.** There are two conflicting goals in creating a mathematical model of a real-world phenomenon: the model must be simple enough to deal with mathematically, but complex enough not to lose any of the critical features of the phenomenon. While many applied mathematics texts offer a few paragraphs on modelling, it seems that this skill is still very much an art. The only way to learn this is to construct many models, try to apply them, and then see whether the models are either too complex to deal with mathematically or too simple to describe reality.

- **Data Analysis and Statistical Inference.** Given our *a priori* knowledge of the structure of the process that generates the data and the *a posteriori* data, it is our goal to give a probabilistic model that describes the data at hand. The first step in this task is to conjecture a preliminary model for the data; to accomplish this we use both our understanding of the structure of the problem and the techniques of exploratory data analysis described by Tukey [1977]. We then use the classical techniques of statistics to estimate the parameters of the model and to test the hypothesis that the data did in fact come from the model.

- **Algorithm Design and Analysis.** Once the algorithm designer has the abstract problem and the probabilistic model of the data, he can use the tools of algorithm design and analysis to develop a fast algorithm for the problem. Knuth [1973] and Aho, Hopcroft and Ullman [1974] are classical texts in this field, and Bentley [1979] provides a survey of algorithm design. In many applications we are much more concerned with the average performance of our algorithms under probabilistic assumptions about the input than with their worst-case performance. Techniques for designing and analyzing probabilistic algorithms are discussed by Bentley [1981a] and Lueker [1981].

- **Programming.** A number of methodologies address the issues involved in producing executable code in a particular computer language from the high-level specification of an algorithm, and exactly these techniques should be brought to bear as we code the real program from the algorithm. If the underlying algorithm is particularly efficient, then that alone is often sufficient to achieve the final program. Sometimes, however, we must use additional methods to make the program even more efficient; Bentley [1981b] has described a set of techniques that he calls "Writing Efficient Code" that should often be brought to bear at this stage of the development. Those techniques will not change the asymptotic performance of an algorithm (that is, they won't reduce a quadratic algorithm to a linear-time algorithm), but they can often squeeze the coefficient in front of the dominant term to be much lower (it is not uncommon to see speedups of an order of magnitude).

- **Simulation.** Once the real program is functional, we should use it to perform two kinds of simulation. Namely, we should execute it on both the real data and on data randomly generated from our probabilistic model. Our simulations must be careful to give the right kind of output: enough to provide sufficient detail, yet not too much detail (which makes it difficult to extract the relevant facts from the mass).

3. A Case Study in Detail

In this section we will study in detail one problem that was a bottleneck in a large geopolitical database system. In the first design of the system a simple approach had been used on the problem; that approach was easy to develop and maintain, but led to performance that was unacceptably slow. The following subsections will show how the paradigm of the previous section was applied to this real problem; the result of this application was a savings of over a factor of six in CPU time and of over thirty percent in plotter time.

3.1. The Problems

The real problem faced by the programmer working on the database system was to schedule a mechanical plotter to produce maps of election districts. There are two parts of such maps. The first part is the set of lines that form the boundaries of districts; these required relatively little time to draw. The expensive task was to draw a sequence of marks (usually x's) that represent each of the precincts inside the district. Each mark was colored to reflect a certain attribute; for instance, if the map was to represent the voting characteristics of the area, then a red mark might represent a district with >50% Republican registration, yellow might stand for 45-50%, green for 40-45%, and blue for <40%. A great deal of plotting time was spent drawing these marks (some maps had more than 1000 precincts). The time spent actually drawing the marks themselves could not be reduced, but there was a large factor of waste in moving the pen from one mark to the next. Thus the real problem was to plot the marks in an order that had minimum total move time.

The abstract problem corresponding to this has been studied for almost a century: it is called "the traveling salesman problem". We are given a set of N cities in the plane, and we must schedule the tour of a salesman so that he visits all cities with minimum distance traveled. (A more general case occurs when the distances between cities is given by a graph; we will not be concerned with that case.) An excellent introduction to this fascinating problem can be found in Lewis and Papadimitriou [1978]. The abstract problem can be cleanly stated as follows.

The Traveling Salesman Problem (TSP).

Input: A set of N points in the plane.

Output: A permutation of the integers {1..N} that describes a tour of the N cities. For instance, the output (4,2,3,1) describes the tour that starts at city 4, then goes to cities 2, 3, and 1, and then closes the tour by returning to city 4 (note that the last step is implicit in the output).

Minimize: The total distance required by the tour; that is, the distance from the first city to the second plus from the second city to the third, and so on.

Note that while the real problem involves a large number of details (such as the exact set of commands used by the particular plotter and the CPU/plotter interface), the abstract problem is mathematically very simple. If we solve the abstract problem, then we will not have solved the real problem in its entirety (we still have to issue commands to the plotter), but we are very close to a solution.

At this point the user might expect the algorithm designer merely to go to his algorithms cookbook and pull out a solution to the traveling salesman problem. Unfortunately, the problem does play a prominent role in algorithms texts, but a negative one. Specifically, the TSP is a member of a class called the "NP-complete problems" that people suspect can never be solved efficiently; Lewis and Papadimitriou discuss the TSP in this context. This tells us that we should give up the quest of finding an efficient algorithm that produces the optimal tour in all circumstances and concentrate instead on finding a tour that is usually close to optimal (that is, an *approximate* tour).

3.2. The Input

If we are to give an algorithm that performs well on most input, then we must have some idea of the appearance of "most input". There is numerous real data on the distribution of precincts within cities. In this particular application, the user provided the algorithm designer with tapes representing 318 precincts in San Diego County (California) and 1122 precincts in the State of New Mexico. Examination of this data showed that two simple distributions were able to describe much of dispersion of points. Specifically, a uniform distribution accurately described completely developed areas (such as the residential portions of large cities), and a bivariate normal distribution described developing areas around a central city. Other patterns in the data were not described in these distributions and arose frequently enough to be included in an accurate model of the input. Population concentrations along highways, rail ways, mountain ranges and borders dictate that linear components should be included. The demographic "donut phenomenon" of similar populations in circular discs centered around cities is also common (for instance, precincts with similar percentages of Republican registration tend to cluster in bands at some distance from the city center); we therefore must include point sets distributed uniformly within an anulus. Although we were not able to give a complete probabilistic model of the inputs, we can give the following set of models that together describe most of the distributions encountered in practice.

- Uniform on the unit square.
- Bivariate normal.
- Uniform on a line.
- Uniform on an anulus.

(A more interesting derivation of a probabilistic model can be found in Subsection 4.1.)

3.3. The Algorithms

We will now discuss algorithms for the TSP. Because the problem is known to be NP-complete, we will not seek an algorithm that always gives optimal tours, but rather try to find algorithms that usually produce tours with lengths not far from the length of the optimal tour. In this subsection we will consider two such approximation algorithms: the minimum spanning tree (MST) and nearest neighbor (NN) algorithms. In the actual investigation of the problem a strip-based algorithm due to Beardwood, Halton and Hammersley was also considered, but was found to be inferior to the other two; it will therefore not be discussed in this subsection. As a further step towards decreasing the length of this subsection, we will not give citations to the literature for some results. Such citations can be found in Bentley and Saxe [1980] and Lewis and Papadimitriou [1978].

Our first task is to describe both of the algorithms. The NN algorithm starts at an arbitrary starting city and then successively visits the closest unvisited city until all cities have been visited. The MST algorithm is more complicated: we first build the minimum spanning tree of the point set, and then perform a simple traversal of the tree to yield a traveling salesman tour. Details of that algorithm are given by Lewis and Papadimitriou.

It is easy to see that both the MST and NN algorithms can produce tours that are not optimal; we must now describe exactly how far from optimal they can be. Rosenkrantz, Stearns and Lewis showed that the length of the MST tour is never more than twice the length of the optimal tour, and that the NN tour of an N-point set is never more than a factor of (lg N)/2 longer than optimal (and can come close to that bound). The last bound is unsettling: if we have 1000 points, then the tour can be a factor of five longer than optimal. By this measure, the MST tour appears to be far superior to the NN tour.

Because it appeared that there was more to say about the worst-case performance than the above ratios, the algorithm designers took a different view of the problem and investigated *absolute* bounds, in which we examine the worst-case performance of the heuristic in an absolute sense, rather than as a ratio to the optimal. Specifically, the question studied was "if N points are placed on the unit square, then how long a tour can the two approximations produce?" Bentley and Saxe showed that the NN algorithm never produces a tour of length more then $\sim 2.15N^{1/2}$, and that the MST's bound is at least $1.15N^{1/2}$. In this absolute worst-case sense, the NN algorithm is quite competitive with the MST algorithm.

Because neither of the two above tests clearly indicates which algorithm is appropriate in practice, we conducted a number of simulation experiments to compare the two. On data uniform on the unit square, the MST tours were observed to have length $\sim .95N^{1/2}$ while the NN tours had length $\sim .92N^{1/2}$. When various techniques were used to improve the tours (involving choosing several starting points for the traversals), the NN tours reached better values more quickly. On uniform data sets, therefore, the NN tours started better and improved faster than MST tours. Similar tests conducted on bivariate normal data sets gave similar results.

The performance of the two algorithms on linear and anular data sets was determined analytically. Both the MST and NN algorithms yield optimal tours on linear sets. For point sets along the perimeter of a circle the NN tours are somewhat better than the MST tours; as the width of the anulus grows the two tours become closer in length.

The final test on the performance of the heuristics was on the real data from San Diego and New Mexico. The mean tour length for San Diego was 14.74 for 30 MST tours and 15.19 for 10 NN tours; the minimum lengths were 14.44 and 14.89, respectively. The mean lengths for New Mexico were 16.57 for 30 MST tours and 14.88 for 5 NN tours; the minimum lengths were 16.29 and 14.22. Thus the NN algorithm was slightly worse than the MST on the rather uniform data of San Diego, but much better on the highly clustered data of New Mexico.

So far we have considered only the *efficacy* of the two

heuristics MST and NN (that is, how good a tour they produce); we will now briefly study their *efficiency* (how quickly the tours can be found). Both the tours can be found by fairly straightforward algorithms in $O(N^2)$ time; the NN algorithm requires only about a dozen lines of code, and the MST algorithm requires about a hundred. Shamos showed that the MST tour can be found in O(N lg N) worst-case time, and Bentley, Weide and Yao [1980] showed that under certain assumptions, it can be found in linear expected time. Bentley and Saxe showed that the NN tour can be found in $O(N^{3/2} \lg N)$ worst-case time or $O(N \lg^2 N)$ expected time.

3.4. The Programs

A large prototype program was developed that embodied both the MST and NN heuristics; that program was used to gather the statistics mentioned above. Both the heuristics were implemented by the relatively simple quadratic-time algorithm. The straightforward quadratic algorithms required approximately 47 seconds of PDP-KL10 time to solve a 1000-city problem; the techniques of writing efficient code were used to reduce that time to approximately 8 seconds (a detailed discussion of the NN program can be found in Bentley [1981b]).

With the weight of the above evidence, the NN program was chosen to be implemented in the production system. It produced tours that were less than half the length of the tours previously used (which were those given by the order in which a person entered them and therefore rather short). While the previous program would spend 15 minutes drawing and 30 minutes moving on typical maps; the new method spent only 15 minutes moving (for a total of 30 minutes per map rather than 45). Furthermore, it appears that the optimal move time is about 12 minutes, so it really wouldn't pay to attempt to reduce the move time further. The program initially took two minutes to compute the tour of a 1000-point city (on an HP-1000), but incorporating the changes of Bentley [1981b] reduced that to about twenty seconds, which was quite acceptable.

3.5. Summary of the Problem

The above text is a brief description of a rather long task. The algorithm designer (the author) spent about fifty hours on the project over two calendar weeks, not including the time spent later to prove the theorems. At the end of those two weeks, the user had a high-level language description of a simple program that performed quite well (in both efficacy and efficiency), and the algorithm designer had a host of open research problems!

4. Other Case Studies

In the previous section we saw how the paradigm of Applied Algorithm Design was used to solve one particular real-world problem. We will now briefly survey several other problems, at decreasing levels of detail.

4.1. VLSI Design Rule Checking

In this subsection we will study a problem that arose in a design system for Very Large Scale Integrated circuitry (VLSI); the details of this case study can be found in Bentley, Haken and Hon [1980]. The "bottom line" of this example is that problems can be solved in ten minutes that previously required several hours, and problems can be solved in several hours that previously would have required weeks of computing time.

- **The Real Problem.** The physical components on a VLSI chip have certain structural relations that dictate that the corresponding mathematical objects in a geometric description of the chip must have certain relationships. For instance, if two polygons represent pieces of wire on the chip, then the polygons must either be far enough apart to ensure that they will not overlap when fabricated, or they must overlap enough to ensure that the fabricated chip will indeed have an electrical connection at that point; any other relation between two polygons is illegal. This is an example of a VLSI *design rule*; most fabrication processes have a set of several dozen similar design rules that must be satisfied to ensure that designs are "syntactically correct". The real problem of VLSI design rule checking is to provide a program that is given a description of a VLSI design and checks to ensure that none of the set of rules is violated.

- **The Abstract Problem.** We will make two fundamental simplifications in abstracting a tractable mathematical problem. The first is that all polygons are rectangles with their sides parallel to the coordinate axes; this was already required in the design system in which the program was to be embedded. The second simplification is much more important: we will declare that the actual form of the several dozen design rules isn't all that important, and assert that the crucial phenomenon is the pairwise intersection of rectangles. That leaves us with the following computational problem.

  > Given a set of N rectangles in the plane with their sides parallel to the coordinate axes, report (by calling a certain procedure) all pairs of rectangles that intersect.

- **The Real Data.** The real data consisted of

complete descriptions of sixteen designs of VLSI circuits. The designs were chosen as the largest of a set of thirty available designs (the largest were chosen because future designs, for which the system was being constructed, were going to be much larger than the prototypes). They represented a number of different design systems and design styles, and varied in number of rectangles from approximately 10,000 to 110,000. A number of statistics were gathered for each design, and a number of plots were prepared that summarized different important features.

- **The Model of the Data.** The statistics gathered from the real data surprised most of the experts who had guessed about the structure of the designs. Specifically, the designs were much more regular than the VLSI experts had guessed. This regularity facilitated a very simple and very accurate family of probabilistic descriptions of the designs.

- **The Algorithms.** The most straightforward algorithm for rectangle intersection compares all N(N-1)/2 pairs of rectangles; this requires $O(N^2)$ time. Such algorithms were actually used in some systems, and processing chips with 100,000 elements commonly required a weekend of CPU time. Bentley and Wood [1980] gave an algorithm that requires only O(N lg N) time, but it was very difficult to code and prove correct, and had a large constant factor. Bentley, Haken and Hon [1980] were able to exploit the regular structure of VLSI designs as reflected in the probabilistic model to achieve an algorithm that requires only O(N) average time for the rectangle intersection problem, under the assumption that the inputs come from the probabilistic model mentioned above.

- **The Real Program.** The algorithm of Bentley, Haken and Hon was realized in two real programs. The first was a prototype Pascal program that solved only the rectangle intersection problem, and is described in Bentley, Haken, and Hon. This program demonstrated that the abstract algorithm had the predicted performance when applied to real data, and also allowed several important parameters to be fine-tuned. The final production program was coded by Dorothea Haken in the C programming language. The workhorse of the program was a subroutine that performed rectangle intersection, which was used to implement logical operations on layers of the chip. The final program was able to check a one-thousand device chip in ten minutes. Its linear-time performance shows that for every one-hundred devices on the chip (which usually correspond to about one-thousand rectangles), the program will require one minute of CPU time.

4.2. Polygon Inclusion

Bentley and Carruthers [1980] describe algorithms for tasks that arose in two distinct application domains. The real problems arose in a geopolitical database system and an Artificial Intelligence computerized vision system; in both cases the abstract problems involved testing the inclusion of points in polygons. Bentley and Carruthers describe a set of statistics on polygons from various applications, and a model of polygons that covers all of the polygons they observed in practice. They then analyzed previous polygon algorithms under the model, and predicted that they would work quite well in practice. (The previous algorithms had actually been discarded from some theoretical investigations because of their extremely poor worst-case behavior, which in fact never arose in practice.) A prototype program was able to solve tasks in thirty seconds that required over nine hours when solved by existing programs.

4.3. Other Applications

Work on Applied Algorithm Design has been underway in the Computer Science Department at Carnegie-Mellon University since late 1979. The previous sections described some of the major projects that have been undertaken; we will now describe several others.

- A prototype algorithm has been developed for assigning VLSI designs to multiproject chips. It solves the abstract problem of two-dimensional bin packing, which is often called the "cutting stock" problem. The performance of the program is comparable to that of humans on the task.

- Good expected-time algorithms for the geometric problem of "range searching" have been developed and implemented.

- The techniques of Applied Algorithm Design were applied to a sorting problem in a particular application. A canned system sort would have required over five minutes to sort the data, while exploitation of certain properties of the data allowed the sort to be accomplished in several seconds.

- Bentley, Weide and Yao [1980] describe both a number of theoretical algorithms as well as a concrete algorithm for solving several geometric problems. The theorems state that only that the algorithms will perform well on uniform data, but they provide empirical evidence that shows that the algorithms perform very well on data that actually arises in geopolitical database applications.

5. Conclusions

Probabilistic analysis of algorithms has had a rapid growth in recent years in the theoretical computer science community; one of its primary flaws, however, has been that it has had little impact on software applications. In Section 2 of this paper we saw a paradigm for applying probabilistic algorithms, and in Sections 3 and 4 we saw that paradigm profitably applied to several real problems.

This paper describes the current status (as of April 1981) of an ongoing research project in the Computer Science Department of Carnegie-Mellon University. This project has enunciated a paradigm and has applied it profitably to many real problems. Two directions open for further work are the application of this paradigm to other problems, and the elaboration of the paradigm as a natural outgrowth of its use. Those activities will provide powerful tools for practitioners, and interesting problems for researchers.

References

Aho, A. V., J. E. Hopcroft and J. D. Ullman [1974]. *The Design and Analysis of Computer Algorithms*, Addison-Wesley, Reading, MA.

Bentley, J. L. [1979]. "An introduction to algorithm design," *IEEE Computer Magazine 12*, 2, February 1979, pp. 66-78.

Bentley, J. L. [1981a]. "Probabilistic analysis of algorithms," to appear in *Proceedings of the ORSA-TIMS Special Interest Meeting on Applied Probability -- Computer Science, The Interface*.

Bentley, J. L. [1981b]. "Writing efficient code," Carnegie-Mellon University Computer Science Technical Report, April 1981.

Bentley, J. L. and W. A. Carruthers [1980]. "Algorithms for testing the inclusion of points in polygons," *Eighteenth Annual Allerton Conference on Communication, Control and Computing*, pp. 11-19, October 1980.

Bentley, J. L., D. Haken, and R. Hon [1980]. "Statistics on VLSI designs," Carnegie-Mellon Computer Science Technical Report CMU-CS-80-111. (A preliminary version is "Fast geometric algorithms for VLSI tasks," *COMPCON Spring '80*, pp. 88-92, IEEE.)

Bentley, J. L. and J. B. Saxe [1980]. "An analysis of two heuristics for the Euclidean traveling salesman problem," *Eighteenth Annual Allerton Conference on Communication, Control and Computing*, pp. 41-49, October 1980.

Bentley, J. L. and D. Wood [1980]. "An optimal worst-case algorithm for reporting intersections of rectangles," *IEEE Transactions on Computers C-29*, 7, pp. 571-577, July 1980.

Bentley, J. L., B. W. Weide and A. C. Yao [1980]. "Optimal expected-time algorithms for closest-point problems," *ACM Transactions on Mathematical Software 6*, 4, December 1980, pp. 563-580.

Foster, M. J. [1980]. Personal communication of M. J. Foster of Carnegie-Mellon University.

Janko, W. [1981]. "Bibliography of probabilistic algorithms," in preparation.

Karp, R. M. [1977]. "Probabilistic analysis of partitioning algorithms for the traveling-salesman problem in the plane," *Mathematics of Operations Research 2*, 3, August 1977, pp. 209-224.

Knuth, D. E. [1973]. *The Art of Computer Programming, volume 3: Sorting and Searching*, Addison-Wesley, Reading, MA.

Lewis, H. and C. H. Papadimitriou [1978]. "The efficiency of algorithms," *Scientific American*, January 1978, pp. 97-109.

Lueker, G. [1981]. "Algorithms with random inputs," to appear in *Proceedings of Computer Science and Statistics: Thirteenth Annual Symposium on the Interface*, Pittsburgh, PA, March 1981.

Rabin, M. O. [1976]. "Probabilistic algorithms," in *Algorithms and Complexity: New Directions and Recent Results*, J. F. Traub, Ed., Academic Press, New York, NY, pp. 21-39.

Tukey, J. W. [1977]. *Exploratory Data Analysis*, Addison-Wesley, Reading, MA.

Weide, B. W. [1978]. Statistical Methods in Algorithm Design and Analysis, Ph.D. Thesis, Carnegie-Mellon University, August 1978.

ALGORITHMS WITH RANDOM INPUT

George S. Lueker, University of California at Irvine

Abstract

Randomness arises in connection with algorithms and their analysis in at least two different ways. Some algorithms (sometimes called coin-flipping algorithms) provide their own randomness, perhaps through the use of random number generators. Sometimes, though, it is useful to analyze the performance of a deterministic algorithm under some assumption about the distribution of inputs. We briefly survey some work which gives a perspective on such problems. Next we discuss some of the techniques which are useful when carrying out this type of analysis. Finally, we briefly discuss the problem of choosing an appropriate distribution.

Keywords: Probabilistic analysis of algorithms, Boole's inequality, Chebyshev's inequality, independence, optimization problems, random graphs, bin packing.

I. Background

In the early 1970's, the notion of NP-completeness was developed and explored. Two major papers dramatically influenced research in algorithms over the next decade. Cook's work [Co71] showed the problem of testing whether an unquantified boolean expression was satisfiable to be NP-complete. This implied that one could test an arbitrary boolean expression for satisfiability in polynomial time if and only if the class $\mathcal{P}$ of problems solvable in deterministic polynomial time was the same as the class $\mathcal{NP}$ of problems solvable in nondeterministic polynomial time. Thus the problem of finding efficient algorithms for boolean satisfiability was shown to be much more profound than it might at first appear to be. Karp [Ka72] showed that a large number of problems, including many optimization problems of practical significance, were also NP-complete. Thus all of these problems could be solved efficiently (i.e., in polynomial time) or none of them could.

These discoveries were sufficient to discourage all but the boldest of researchers from continuing to try to find polynomial time solutions for such problems. Nonetheless, the great importance and interest of the problems continued to draw attention. One major line of research that was greatly encouraged by the notion of NP-completeness was the development of algorithms which gave approximate answers, and were provably within some error bound. It is perhaps surprising to see the wide range of behavior among NP-complete problems with regard to approximation. For the 0-1 knapsack problem, Ibarra and Kim [IK75] have shown that for any ε greater than 0, there exists an $O(n \log n)$ algorithm whose result is within a factor of $(1+\varepsilon)$ of the true optimum; their algorithm is a clever combination of dynamic programming and the greedy method. For the traveling salesman problem, with the triangle inequality, Christofides [Ch76] has produced an algorithm with an error bound of 50%. For the problem of coloring graphs with a minimum number of colors, Garey and Johnson [GJ76b] have shown that the existence of a coloring algorithm which guarantees the use of at most r times the true minimum number of colors, for any fixed $r<2$, would imply that $\mathcal{P}=\mathcal{NP}$. See [GJ76a, GJ79] for more about approximation of NP-complete problems.

Perhaps partly because some NP-complete problems do not seem to yield even to approximation algorithms, researchers have turned to yet another approach--probabilistic analysis. At this point there are two possible approaches. We can assume some probability distribution of inputs, or we can assume that our algorithm itself operates in a partly random way, and works well for any input, on the average. Following [Ya77], we call the former approach distributional, and the latter approach randomized. There is a famous number-theoretic example of a randomized algorithm. A fundamental open question is whether one can test whether a number is prime in an amount of time bounded by some polynomial in the length of its binary representation. (In [Mi75] it is shown that this can be done if the extended Riemann hypothesis (ERH) is true, but the status of the ERH is still unknown. Progress on efficient algorithms which give a definite yes-no answer is continuing to be made [Ad80], but still no algorithm with a polynomial time bound is known.) [SS77, RA76] show how to construct an algorithm which runs quite rapidly and which can be made to have an arbitrarily small probability of giving a wrong answer. This algorithm provides a practical way of testing large numbers for primality, and has had a substantial impact. (For example, see [RSA78] for a discussion of its relation to encryption.)

Miller [Mi75] discusses the class of randomly decidable problems, denoted $\mathcal{R}$. Roughly, a problem is said to be in $\mathcal{R}$ if there is an algorithm, with built-in randomness, such that

a) if the correct answer is no, the algorithm always says no, and
b) if the correct answer is yes, the algorithm says yes on at least half of its computations.

Thus the problem of recognizing composites is in this class.

Now the distributional approach has some clear advantages when compared to the randomized approach. For example, if a user has just one problem instance to solve, and the algorithm performs very badly on it, it is of little consolation to know that the algorithm does well on most other problem instances. Further, even if the user does intend to solve many problems, it may be difficult to predict in advance what are reasonable assumptions about the nature of the input distribution. (We will return to this issue in section III.)

Sometimes it is possible to convert an algorithm which relies on an input distribution into a randomized algorithm. Quicksort [Ho61, Kn73, Se75] provides a simple and well-known example. Suppose that, when performing a partition, we always choose the first element of the array segment to be partitioned as the partition element. Then the algorithm will run on $O(n \log n)$ time if all permutations of the input are equally likely. However, some inputs will cause the algorithm to use $\Theta(n^2)$ time; in particular, some worst cases occur if the input elements are already in ascending or descending order. This algorithm can easily be converted into a randomized algorithm with an average complexity of $\Theta(n \log n)$ on any input by simply choosing the partition element at random (with equal probability) from the array segment to be partitioned.

A more interesting example of this phenomenon is provided by the work of Carter and Wegman [CW77] on hashing. Hashing is often thought of as a scheme in which we choose a hash function and hope that the distribution of inputs is one on which our choice works well. In [CW77] an elegant randomized scheme is provided whereby one chooses a hash function at random, thereby letting this choice provide the necessary randomness.

It might seem to be worthwhile to seek randomized algorithms for NP-complete problems. Unfortunately, however, it is not hard to see that if any NP-complete problem is in $\mathcal{R}$, then all NP-complete problems are [AM77]. (If we are seeking approximate solutions, the situation may change somewhat. I am not aware of any result indicating that randomized algorithms are not a useful tool for finding good approximate solutions to NP-complete problems. On the other hand, I do not know of good examples of such algorithms, either.) In this paper we primarily consider the distributional approach.

(We should mention that while NP-completeness has greatly stimulated interest in probabilistic analysis, there are probabilistic questions of great interest in computer science which are not directly related to NP-complete problems; for example, there are interesting open problems related to random search trees.)

II. Some methods for analyzing average case performance

II.1 Order statistics

Often an optimization problem or a related algorithm involves looking at the maximum or minimum of several random variables. Thus the properties of order statistics are of interest. [Da70] is standard reference on these; Weide [We78] provides examples of their application in analysis of algorithms. Here we will give a very simple example, similar to part of [Bo62], which illustrates some of the ideas and possible problems. We will use the following well-known fact.

Lemma. The expected minimum of r independent uniform draws from [0,1] is 1/(r+1).

We now consider a simple greedy algorithm for the traveling salesman problem. We will assume the input is a complete graph on n vertices with edge weights chosen uniformly and independently from [0,1]. For convenience, assume the vertices are numbered 1 to n. We will use the following algorithm TSP_GREEDY below.

```
procedure TSP_GREEDY;
begin
   start at vertex 1;
   repeat
      follow the cheapest edge which leads
            from the current vertex to an
            unvisited vertex
   until all vertices have been visited;
   return to vertex 1;
end;
```

Note that on the first iteration we choose the minimum of n-1 uniform variables, so the expected weight of the traversed edge is 1/n. On the second iteration we choose the minimum of n-2 uniform variables, so the expectation is 1/(n-1). (Note that none of these edges have yet been considered, so their weights are not conditioned in any way.) By similar reasoning we conclude that the average cost of all edges traversed during the repeat loop is

$$\sum_{i=1}^{n-1} (n-i+1)^{-1} = H_n - 1 \sim \ln n,$$

where H_n denotes the harmonic numbers. Now consider the edge traversed when we return to vertex 1. It is not valid to consider this edge as having a weight distributed uniformly on [0,1] since we have already considered this edge and thus conditioned its weight; in particular, we know that it was not chosen when we initially sought the cheapest edge leading from vertex 1, so we would guess that it tends to be a bit larger than 1/2. This fact that the operation of an algorithm conditions the quantities considered can be a major difficulty in the analysis of an algorithm. For the current problem, it is possible to determine the exact expected weight of the edge back to vertex 1, but here we will simply note that it must lie in [0,1] so the expected total weight of the tour produced by this greedy algorithm must be asymptotic to ln n. See [Ka79] for an analysis of much more interesting algorithms for a closely related traveling salesman problem.

II.2 Dealing with dependence

A common problem that arises during a probabilistic analysis is the lack of independence. For example, as we saw in the preceding example, the behavior of an algorithm at some point may be conditioned by previous parts of the execution. In this section of the paper we will consider a few of the most basic ways for dealing with this problem.

II.2.A. Boole's inequality

If $E_1, E_2, \ldots, E_k$ are events, then regardless of any dependencies among them it is true that

$$P\{E_1 \cup E_2 \cup \ldots \cup E_k\} \le P\{E_1\} + P\{E_2\} + \ldots + P\{E_k\}.$$

This is sometimes called Boole's inequality [Fe68].

One application of this method arises in finding lower bounds on the solutions to optimization problems. Many such problems can be cast in the following form. Let I represent a problem instance. Let Π represent the set from which we make a choice as we try to find an optimum. For example, for the TSP, Π is the set of all hamiltonian tours in the given graph. If $\pi \in \Pi$ is a particular choice, let $F(\pi, I)$ denote the cost of the solution to instance I if we make choice π. Thus the minimum is

$$F_{min}(I) = \min_{\pi \in \Pi} F(\pi, I).$$

Now let $\mathcal{I}$ be a random structure which represents the input distribution, and suppose the random variable $F(\pi, \mathcal{I})$ turns out to be the same for all $\pi \in \Pi$. (This can often happen, by symmetry.)

Then

$$\begin{aligned} P\{F_{min}(\mathcal{I}) \le x\} &= P\{\textstyle\bigcup_{\pi \in \Pi} F(\pi, \mathcal{I}) \le x\} \\ &\le \sum_{\pi \in \Pi} P\{F(\pi, \mathcal{I}) \le x\} \\ &\qquad \text{(by Boole's inequality)} \\ &= |\Pi| \; P\{F(\pi, \mathcal{I}) \le x\} \end{aligned}$$

It is not surprising that this bound is valid; it is perhaps surprising that it can sometimes yield interesting bounds on the optimum. See, for example, [Do69, FH78, Lu81b]. (In [Do69], the argument was phrased in a more combinatorial way, but the idea was quite similar.)

Another interesting way to use Boole's inequality can be phrased roughly as follows. (This can be useful even in some cases where the original problem statement does not involve probabilistic notions.) Suppose we wish to show that a structure S meeting a set C of constraints exists. Let $\mathcal{S}$ denote a random structure. For any constraint $c \in C$, let $P(c)$ be the probability that $\mathcal{S}$ does not meet constraint c. Now if

$$\sum_{c \in C} P(c) < 1,$$

it must be, by Boole's inequality, that $\mathcal{S}$ has a positive probability of meeting all of the constraints. Thus a structure of the desired type must exist. This type of argument is often associated with the name of Erdös; see [ES74] for examples. In [AKLL79] a similar argument is used to show the existence of reasonably short universal traversal sequences. The argument which was used in [Ad78] to show that any problem in $\mathcal{R}$ admits polynomial circuits can be viewed in this way. Finally, in [Lu81b] an argument of a similar spirit is used in analyzing a greedy algorithm for finding light cliques in a weighted graph.

II.2.B. Chebyshev's inequality

Another useful tool for dealing with a lack of independence is Chebyshev's inequality. See for example [Bo62]. Here we will present an example that was used in a partial analysis of the 0-1 knapsack program in [Lu81a]. (The analysis is extremely similar to the analysis of cliques in random graphs [ER60, GM75].) Consider the following problem.

How small can we choose ε so that given 2k independent uniform draws $W_1, W_2, \ldots, W_{2k}$ with mean 0 and variance 1, we can be reasonably hopeful of finding some subset of k of the W_i whose sum lies in $[-\varepsilon, \varepsilon]$?

Let $G_n(x)$ be the cumulative probability distribution function for the sum of n such independent uniform variables. Let $Y_k(\varepsilon)$, or simply Y_k, be a random variable telling the number of distinct sets of k of the W_i whose sum lies in $[-\varepsilon,\varepsilon]$. Now note that

$$E[Y_k] = \binom{2k}{k} [G_k(\varepsilon) - G_k(-\varepsilon)]$$

$$\sim 2\binom{2k}{k} \frac{\varepsilon}{\sqrt{2\pi k}}$$

Unfortunately, it is not the case that a reasonably large expectation for a nonnegative random variable guarantees a reasonably large probability that it is non-zero. Fortunately, the following corollary of Chebyshev's inequality is useful at this point.

$$P\{Y_k=0\} \le \frac{E[Y_k^2]}{E[Y_k]^2} - 1.$$

The computation of $E[Y_k^2]$ is quite messy and is omitted from this paper. It may be shown that if $\varepsilon = \alpha k 4^{-k}$, then

$$\frac{E[Y_k^2]}{E[Y_k]^2} - 1 \sim \sqrt{\frac{4}{3}} + \frac{\pi}{\sqrt{2}\alpha} - 1.$$

If $\alpha=7$, this is just under 1/2; thus letting $\varepsilon=7 k 4^{-k}$ gives at least a 50-50 chance of finding the desired sum for large enough k. On the other hand, if $\varepsilon=o(k 4^{-k})$, then $E[Y_k]$ approaches zero, so the probability of finding the desired sum becomes very small.

II.2.C. Tinkering with the input distribution to obtain the desired independence

Sometimes it is possible to make a small change in the input distribution which eliminates some of the interdependence between the random variables which need to be considered as the algorithm proceeds. We will give two examples.

Here is a method which was used, for example, by Karp [Ka77] in an analysis of the traveling salesman problem. Suppose we have a set of n points distributed uniformly over a unit square. Let S_1 and S_2 be two disjoint squares contained within the unit square. Note that the distributions of points within S_1 and S_2 are not independent. In particular, if n points lie in S_1, we know that none lie in S_2. Now suppose instead that we distribute Π_n points uniformly over the unit square, where Π_n denotes the Poisson distribution with mean n. Then what happens in S_1 will be independent of what happens in S_2.

Here is a second example of a way to introduce independence. Consider the following bin-packing problem.

Given weights $x_1,x_2,\ldots,x_n$ and a capacity b, find a packing of the x_i into bins of capacity b such that

a) no bin's capacity is exceeded, and

b) as few bins as possible are used.

A number of probabilistic analyses of this problem have appeared, notably [CSHY80, Fr80].

Assume the x_i are uniform and independent on [0,1]; for convenience assume they are sorted in increasing order. Thus x_i is the ith order statistic of n uniform draws from [0,1]. Also, for notational convenience, define $x_0=0$ and $x_{n+1}=1$. Let b be 1.

In an analysis in [Lu81c], the differences x_i-x_{i-1}, for $i=1\ldots,n+1$, turn out to be quite important. One easily sees, however, that these differences are not independent. One could make use of the Poisson distribution as in the previous example, but here another trick is more convenient. Let z_i, for $i=1,\ldots,n+1$, be independent draws with density function e^{-x} for $x\ge 0$. Define

$$\hat{x}_0 = 0$$

$$\hat{x}_{i+1} = \hat{x}_i + z_{i+1}$$

$$\hat{b} = \hat{x}_{n+1}$$

Now it is clear that the differences $\hat{x}_{i+1} - \hat{x}_i$ are independent. Using this fact, it is shown in [Lu81c] that a certain algorithm uses an average of $n/2+O(\sqrt{n})$ bins for this distribution.

Of course, we have changed the problem, and may really be interested in the solution for the original distribution; fortunately, however, it is not hard to reinterpret the results in terms of the original distribution. The behavior of the algorithm being considered (and many other natural packing algorithms) is unaffected by a simultaneous scaling of the x_i and b. In particular, suppose we set

$$\tilde{x}_i = \hat{x}_i/\hat{b}$$

$$\tilde{b} = \hat{b}/\hat{b} = 1.$$

Then by [Fe66, section III.3], the distribution of the $\tilde{x}_i$ is just the original distribution (order statistics of uniform draws) so the $n/2+O(\sqrt{n})$ behavior must hold for this distribution also. Note that in this problem we obtained the desired independence almost for free!

III. Some remarks about input distributions

We have been discussing analysis of the average behavior of algorithms with specified input distributions, but have said little about the choice of the input distribution. What should we do if we do not know the distribution of the input?

At least two useful approaches exist. One which has been used, for example, by [Be81], is to actually go into the real world and try to investigate the nature of the problem data. This seems to be a very useful strategy, and potentially enables one to tune the algorithm closely to the application.

Another interesting strategy is to investigate what might be termed the distributional robustness of the algorithm. For example, it is well-known that address calculation sort runs in average $\Theta(n)$ time on input drawn from a uniform distribution. A variety of people, including Karp [Kn73, exercise 5.2.1.38], Weide [We78], and Devroye [De79, De81] have shown that while one standard version might use more than $\Theta(n)$ time on other distributions, it is possible to produce a version of the algorithm which runs in $O(n)$ time for a remarkably wide class of input distributions. (Note that a single algorithm, which does not need to have any information about the distribution, is meant here.)

Another example is Willard's work on interpolation search [Wi81]. He shows that while a slight change from a uniform input distribution can destroy the $O(\log \log n)$ behavior of a straightforward algorithm, a modified algorithm will run in $O(\log \log n)$ time under a fairly wide class of distributions, without prior knowledge of the distribution.

Acknowledgements

Dave Johnson made a useful comment to me about randomized algorithms and approximation. Scott Huddleston, when I mentioned the use of partial sums of exponential distributions for introducing independence in a scheduling problem, suggested that I also use this trick when analyzing bin packing.

References

[Ad78] L. Adleman, "Two Theorems on Random Polynomial Time," Proc. Nineteenth Annual Symposium on Foundations of Computer Science (1978), pp. 75-83.

[Ad80] L. Adleman, "On Distinguishing Prime Numbers from Composite Numbers," Proc. 21st Annual Symposium on Foundations of Computer Science (1980), pp. 387-406.

[AM77] L. Adleman and K. Manders, "Reducibility, Randomness, and Intractibility," Proc. Ninth Annual ACM Symposium on Theory of Computing (1977), pp. 151-163.

[AKLL79] R. Aleliunas, R. M. Karp, R. J. Lipton, and L. Lovász, "Random Walks, Universal Traversal Sequences, and the Complexity of Maze Problems," Proc. 20th Annual Symposium on Foundations of Computer Science (1979), pp. 218-223.

[Be81] J. L. Bentley, "Applications of Statistics in Applied Algorithm Design," this conference.

[Bo62] A. A. Borovkov, "A Probabilistic Formulation of Two Economic Problems," Soviet Mathematics, 3:5 (1962), pp. 1403-1406.

[CW77] J. L. Carter and M. N. Wegman, "Universal Classes of Hash Functions," Proc. Ninth Annual ACM Symposium on Theory of Computing (1977), pp. 106-112.

[Ch76] N. Christofides, "Worst Case Analysis of a New Heuristic for the Travelling Salesman Problem," in Algorithms and Complexity: New Directions and Recent Results, J. F. Traub, ed., Academic Press, New York, 1976.

[CSHY80] E. G. Coffman, Jr., K. So, M. Hofri, and A. C. Yao, "A Stochastic Model of Bin-Packing," Information and Control 44:2 (February 1980), pp. 105-115.

[Co71] S. A. Cook, "The Complexity of Theorem Proving Procedures," Proc. Third Annual ACM Symposium on Theory of Computing (1971), pp. 151-158.

[Da70] H. A. David, Order Statistics, John Wiley and Sons, New York, 1970.

[De79] L. Devroye, "Average Time Behavior of Distributive Sorting Algorithms," Technical Report No. SOCS 79.4, March 1979.

[De81] L. Devroye, "On the Average Complexity of Various Sorting and Convex Hull Algorithms," this conference.

[Do69] W. E. Donath, "Algorithm and Average-value Bounds for Assignment Problems," IBM J. Res. Dev. 13 (1969), pp. 380-386.

[ER60] P. Erdős and A. Rényi, "On the Evolution of Random Graphs," Publ. Math. Inst. Hung. Acad. Sci. 5A (1960), pp. 17-61.

[ES74] P. Erdős and J. Spencer, Probabilistic Methods in Combinatorics, Academic Press, New York, 1974.

[Fe68] W. Feller, An Introduction to Probability Theory and Its Applications, Vol. I, Third Edition, John Wiley and Sons, New York, 1968.

[Fe66] William Feller, An Introduction to Probability Theory and Its Applications, Volume II, John Wiley and Sons, New York, 1966.

[FH78] M. L. Fisher and D. S. Hochbaum, "Probabilistic Analysis of the Euclidean K-Median Problem," Report 78-06-03, Wharton Department of Decision Sciences, University of Pennsylvania, May, 1978.

[Fr80] G. N. Frederickson, "Probabilistic Analysis for Simple One- and Two-Dimensional Bin Packing Algorithms," Information Processing Letters 11:4,5 (12 December 1980), pp. 156-161.

[GJ76a] M. R. Garey and D. S. Johnson, "Approximation Algorithms for Combinatorial Problems: An Annotated Bibliography," in Algorithms and Complexity: New Directions and Recent Results, J. F. Traub, ed., Academic Press, New York, 1976.

[GJ76b] M. R. Garey and D. S. Johnson, "The Complexity of Near-Optimal Graph Coloring," JACM 23:1 (January 1976), pp. 43-49.

[GJ79] M. R. Garey and D. S. Johnson, Computers and Intractability: A Guide to the Theory of NP-Completeness, W. H. Freeman and Company, San Francisco, 1979.

[GM75] G. R. Grimmett and C. J. H. McDiarmid, "On Coloring Random Graphs," Math. Proc. Camb. Phil. Soc. 77 (1975), pp. 313-324.

[Ho61] C. A. R. Hoare, "Algorithm 63: Quicksort," CACM 4:7 (July, 1961), pp. 321-322.

[IK75] O. H. Ibarra and C. E. Kim, "Fast Approximation Algorithms for the Knapsack and Sum of Subset Problems," JACM 22 (1975), pp. 463-468.

[Ka72] R. M. Karp, "Reducibility among Combinatorial Problems," in Complexity of Computer Computations, R. E. Miller and J. W. Thatcher, eds., Plenum Press, N. Y., 1972, pp. 85-104.

[Ka77] R. M. Karp, "Probabilistic Analysis of Partitioning Algorithms for the Traveling-Salesman Problem in the Plane," Math. Op. Res. 2:3 (August 1977), pp. 209-224.

[Ka79] R. M. Karp, "A Patching Algorithm for the Nonsymmetric Traveling-Salesman Problem," SIAM J. Comput., 8:4 (November 1979), pp. 561-573.

[Kn73] D. Knuth, The Art of Computer Programming, Vol. 3: Sorting and Searching, Addison-Wesley, Reading, Mass., 1973.

[Lu81a] G. S. Lueker, "On the Average Difference Between the Solutions to Linear and Integer Knapsack Problems," Technical Report #152, Department of Information and Computer Science, University of California, Irvine, September 1980; presented at Applied Probability--Computer

Science, The Interface, Boca Raton, Florida, January 1981.

[Lu81b] G. S. Lueker, "Optimization Problems on Graphs with Independent Random Edge Weights," Technical Report #131, Department of Information and Computer Science, University of California, Irvine; SIAM J. Comput., to appear.

[Lu81c] G. S. Lueker, "An Average-Case Analysis of Bin-Packing," manuscript.

[Mi75] G. L. Miller, Riemann's Hypothesis and Tests for Primality, Ph.D. Thesis, University of California at Berkeley, 1975.

[Ra76] M. O. Rabin, "Probabilistic Algorithms," in Algorithms and Complexity: New Directions and Recent Results, J. F. Traub, ed., Academic Press, New York, 1976.

[RSA78] R. Rivest, A. Shamir, and L. Adleman, "A Method for Obtaining Digital Signatures and Public Key Cryptosystems," CACM 21:2 (February 1978), pp. 120-126.

[Se75] R. Sedgewick, "Quicksort," Report STAN-CS-75-492, Computer Science Department, Stanford University, May 1975.

[SS77] R. Solovay and V. Strassen, "A Fast Monte-Carlo Test for Primality," SIAM J. Comput. 6:1 (March 1977), pp. 84-85.

[We78] B. W. Weide, Statistical Methods in Algorithm Design and Analysis, Ph.D. Thesis, Carnegie-Mellon University, Pittsburgh, Pennsylvania (August 1978); appeared as CMU Computer Science Report CMU-CS-78-142.

[Wi81] D. E. Willard, "A Log Log N Search Algorithm for Nonuniform Distributions," Applied Probability--Computer Science, the Interface, Boca Raton, Florida, 1981.

[Ya77] A. Yao, "Probabilistic Computations: Toward a Unified Measure of Complexity," Proc. Eighteenth Annual Symposium on Foundations of Computer Science (1977), pp. 222-227.

RECENT RESULTS ON THE AVERAGE TIME BEHAVIOR OF SOME ALGORITHMS IN COMPUTATIONAL GEOMETRY

Luc Devroye*, McGill University

ABSTRACT

We give a brief inexhaustive survey of recent results that can be helpful in the average time analysis of algorithms in computational geometry. Most fast average time algorithms use one of three principles: bucketing, divide-and-conquer (merging), or quick elimination (throw-away). To illustrate the different points, the convex hull problem is taken as our prototype problem. We also discuss searching, sorting, finding the Voronoi diagram and the minimal spanning tree, identifying the set of maximal vextors, and determining the diameter of a set and the minimum covering sphere.

KEYWORDS Algorithms. Average time. Computational geometry. Convex hull. Sorting. Searching. Closest point problems. Divide and conquer.

1. INTRODUCTION.

There has been an increasing interest in the study and analysis of algorithms in computational geometry (a recent survey paper by Toussaint (1980) had 168 references). Most of the emphasis has been placed on the study of the worst-case complexity of various algorithms under several computational models. It is well-known that many algorithms perform considerably better on the average than predicted by the worst-case analyses. In this note, we would like to point out a few recent developments in the analysis of the average complexity of some algorithms. To keep the general discussion simple and yet insightful we make a couple of convenient assumptions.

The Assumptions.

The input data $X_1,\dots,X_n$ can be considered as a sequence of independent identically distributed R^d-valued random vectors with common density f. (Here the unrealistic assumption is that real numbers can be stored in a computer.)

An algorithm takes time $T=T(X_1,\dots,X_n)$, a Borel measurable function of the input data, and it halts with probability one, i.e. $T < \infty$ almost surely.

The common operations (+,-,/,*,mod,compare,move) take time uniformly bounded over all vaues of the operands. For example, a*b or a mod b take time bounded by a constant not depending upon a or b. (Once again, this is unrealistic, because the multiplication or comparison of two real numbers takes infinite time.)

Average Time.

We are interested in the average time E(T) taken by certain algorithms. Obviously, E(T) depends upon f and n only because the averaging is done over all random samples of size n drawn from the density f.

Fast Average Time Algorithms.

In many applications, fast average time algorithms can be obtained by the bucketing principle: find the smallest rectangle C covering $X_1,\dots,X_n$; divide C into equal-sized rectangles (buckets), and solve the problem by travelling from bucket to bucket while performing some local operations. We will also discuss the dramatic savings in average time that can be obtained by the proper application of the divide-and-conquer principle. Finally, the quick elimination (or: throw-away) principle may allow us to further reduce the average time: here one takes a superficial look at the data, and eliminates useless points. The more involved work is then performed on the reduced data sequence.

We will not discuss all the fast average time algorithms. For example, to find the convex hull of $X_1,\dots,X_n$ in R^2 under certain computational models, at least c n log n time is needed (Avis, 1979; Yao, 1979). Jarvis' algorithm (Jarvis, 1973) takes average time 0(nE(N)) where N is the number of convex hull points. In the design of this algorithm, no special care is taken to obtain fast average time behavior. Nevertheless, for certain distributions E(N)=0(1) (see Carnal (1970); this is true for multivariate t-distributions, etc.), so that Jarvis' algorithm runs in linear average time for a fairly large class of distributions. In this note, we will be satisfied with a short inexhaustive and biased survey of deliberate attempts at reducing the average time of algorithms and of the probability theoretical and mathematical tools needed in the ensuing analysis.

2. THE BUCKETING PRINCIPLE.

Let C be the smallest closed rectangle covering $X_1,\dots,X_n$ and let C be divided into m^d equal-sized rectangles (buckets) where $m=\text{int}(n^{1/d})$. Bucket memberships can thus be obtained for all data points in time 0(n). Often one keeps track of these memberships by using m^d linked lists, one per bucket, so that 0(n) space is used.

The obvious application in R^1 involves sorting $X_1,\dots,X_n$. Here one empties the buckets from left to right and performs a subsequent sort within each bucket, if necessary. If this subsequent sort is

* The author is with the School of Computer Science, McGill University, 805 Sherbrooke Street West, Montreal, Canada H3A 2K6.

a comparison-based sort (e.g. heapsort, bubble sort, shell sort, merge sort or quicksort) with average time $g(n)$ (this number is independent of f, since we have a comparison-based sort), then the overall average time for sorting is

$$E(T) = O(n) + E(\sum_{i=1}^{n} g(N_i))$$

where $N_1,\ldots,N_n$ are the cardinalities of the n buckets. Devroye and Klincsek (1981) addressed the question of when $E(T)=O(n)$. They showed that when $g(u) \uparrow \infty$ as $u \to \infty$, $g(u)/u^2$ is nonincreasing, and g is convex, then $E(T)=O(n)$ if and only if f has compact support and

$$\int g(f(x))\, dx < \infty. \qquad (1)$$

Notice that they put no continuity or boundedness assumptions on f. Akl and Meijer (1980) found that for sufficiently smooth densities, bucket sort (with slight ad hoc improvements) compares favorably with even the best version of quicksort.

Consider now the following generalization of the previous result: travel from bucket to bucket, performing within the j-th bucket operations taking average time bounded between $ag(N_j)$ and $bg(N_j)$ when N_j is given. Here $0 < a \le b < \infty$. Once again, the average time of the entire algorithm is

$$E(T) = O(n+E(\sum_{i=1}^{m^d} g(N_i))).$$

Assume that $g(u)/u \uparrow \infty$ and $g(u)/u^K \downarrow 0$ for some positive K as $u \to \infty$. Also, assume that g is convex. Then $E(T)=O(n)$ if and only if f has compact support and (1) holds (Devroye, 1981a). These basic results have several applications. We cite just two examples.

Examples.

1. Searching in constant average time.

Assume that $X_1,\ldots,X_n$ are stored in the bucket data structure given above, and that we are presented with X_Z (where Z is uniformly distributed over $1,\ldots,n$). We have to determine the index i such that $X_i=X_Z$. This is the classical problem of successful search. If the X_i's are stored in the buckets in order of arrival, then the average search time is comparable to

$$E(\frac{1}{n} \sum_{i=1}^{m^d} N_i^2+1)$$

(note: two sequences a_n, b_n are comparable when $a_n=O(b_n)$ and $b_n=O(a_n)$). By a simple extension of the previous result, we see that $E(T)=O(1)$ if and only if f has compact support and $\int f^2(x)\, dx < \infty$. If within each bucket the data are organized into a binary search tree rather than a linked list, by considering one of the coordinates of the X_i's as the key for sorting, then $E(T)=O(1)$ if and only if f has compact support and

$$\int f(x) \log_+ f(x)\, dx < \infty.$$

2. Convex hull algorithms that are based upon sorting.

The convex hull of $X_1,\ldots,X_n$ is a subsequence $X_{i_1},\ldots,X_{i_k}$ of $X_1,\ldots,X_n$ such that for all X_{i_j} there exists a hyperplane through X_{i_j}, and all X_q's, $q \ne i_j$, belong to the same closed halfspace determined by this hyperplane. In R^2, it can be obtained from $X_1,\ldots,X_n$ as follows: (i) Find a point x that belongs to the interior of the convex hull of $X_1,\ldots,X_n$. Sort all that X_i's according to the polar angles of X_i-x (by using the bucket sort described above). This yields a polygon P. (ii) Visit all vertices of P in turn by pushing them on a stack. Pop the stack when non-convex-hull points are encountered. In essence, this is Graham's algorithm (1972) with a modification in the sorting method that is used. Step (ii) takes time $O(n)$. The average time taken by (i) is $O(n)$ when the density of the polar angle of X_i-x is square integrable. Since x itself is a random vector, one must be careful before making any inference about f. Nevertheless, it is sufficient that f is bounded and has compact support. End of examples.

The previous applications have one feature in common: the times taken by the algorithms on individual buckets just depend upon the number and/or position of the data points within these buckets (and not on, say, the number of data points in neighboring buckets). In more involved problems, we cannot avoid looking at neighboring buckets. For example, consider the class of "closest point problems" in R^d (Shamos and Hoey, 1975) such as: find all nearest neighbor pairs, construct the Voronoi graph, find the minimal spanning tree, etc. (see Bentley and Friedman (1979) for other applications). Shamos (1978) and Weide (1978) discuss many applications of the bucketing principle, and Bentley, Weide and Yao (1980) give a fairly comprehensive treatment of the average time analysis of bucketing algorithms for closest point problems. We take the liberty to cite a couple of examples from their study:

Examples.

1. The all-nearest-neighbor problem.

All nearest neighbor pairs can be found in $O(n \log n)$ time (worst-case) (Lipton and Tarjan, 1977). Weide (1978) proposed a bucketing algorithm in which for a given X_i, a "spiral search" is started in the bucket of X_i, and continues in neighboring cells, in a spiraling fashion, until no data point outside the buckets already checked can be closer to X_i than the closest data point already found. Bentley et. al. (1980) showed that Weide's algorithm halts in average time $0(n)$ when there exists a bounded open convex region B such that the density f of X_1 is 0 outside B and satisfies

$$0 < \inf_B f(x) \leq \sup_B f(x) < \infty .$$

2. The Voronoi diagram.

The Voronoi diagram in R^2 can be found in time $0(n \log n)$ (worst-case) (Shamos (1978), Horspool (1979), Brown (1979)). Bentley et. al. (1980) have a bucketing algorithm that uses sprial search and has some additional features. The Voronoi diagram can be found in average time $0(n)$ when d=2 and the density f of X_1 satisfies the condition of Example 1. From the Voronoi diagram, the convex hull can be obtained in linear time (Shamos, 1978).

3. The minimal spanning tree.

For a graph (V,E), Yao (1975) and Cheriton and Tarjan (1976) give algorithms for finding the minimal spanning tree (MST) in worst-case time $0(|E| \log \log |V|)$. The Euclidean minimal spanning tree (EMST) of n points in R^d can therefore be obtained in $0(n \log \log n)$ time if we can find a supergraph of the EMST with $0(n)$ edges in $0(n \log\log n)$ time. Yao (1977) suggested to find the nearest neighbor of each point in a critical number of directions; the resulting graph has $0(n)$ edges and contains the MST. This nearest neighbor search can be done by a slight modification of the algorithm in Example 1. Hence, the EMST can be found in average time $0(n \log \log n)$ for any d and for all distributions given in Example 1. The situation is a bit better in R^2. We can find a planar supergraph of the EMST in average time $0(n)$ (such as the Delaunay triangulation (the dual of the Voronoi diagram), the Gabriel graph, etc.) and then apply Cheriton and Tarjan's (1976) $0(n)$ algorithm for finding the MST of a planar graph. Thus, in R^2 and for the class of distributions given in Example 1, we can find the EMST in linear average time. End of examples.

Finally, we should mention a third group of bucketing algorithms, where special buckets are selected based upon a global evaluation of the contents of the bucket. For example, assume that not more than a_n buckets are selected according to some criterion (from the approximately n original buckets) in time $0(n)$, and that only the data points within the selected buckets are considered for further processing. If N is the number of selected points, then we assume that "further processing" takes time $0(g(N))$ for a given function g. Because the global evaluation procedure is not specified, we should assume the worst case, and this leads to the study of the order statistics of the cardinalities of the buckets. The following results can be found in Devroye (1981b). When M is the maximum of n i.i.d. Poisson (1) random variables, then $E(M) \sim \log n/ \log \log n$. The same is true if M is the maximum of $N_1,\ldots,N_n$, where N_i is the cardinality of $[\frac{i-1}{n}, \frac{i}{n})$ and the data is $U_1,\ldots,U_n$, a sequence of i.i.d. uniform (0,1) random variables. Using tight bounds on the upper and lower tails of M, one can show that

$$E(g(N)) = 0(g(a_n \frac{\log n}{\log \log n}))$$

where $a_n \geq 1$, g is nondecreasing, $g(x)=0(1+x^\beta)$ (some $\beta > 0$), $\sup_{x>0} \frac{g(cx)}{g(x)} < \infty$ (all $c > 1$), and the X_i's have a bounded density f with compact support.

Example. The convex hull in R^2.

Shamos (1979) suggested to construct the convex hull in R^2 in the following fashion: mark all the nonempty extremal buckets in each row and column (the extremes are taken in the northern and southern directions for a column, and eastern and western directions for a row); mark all the adjacent buckets in the same rows and columns; apply Graham's $0(n \log n)$ convex hull algorithms to all the points in the marked buckets. It is clear that $a_n=0(\sqrt{n})$ and that the average time of the algorithm is $0(n)+ 0(E(g(N))$ = where $g(n)=n \log n$. This is

$$0(n) + 0(\sqrt{n} \frac{(\log n)^2}{\log \log n}) = 0(n).$$

Furthermore, the average time spent on determining the bucket menberships divided by the total average time tends to 1. End of example .

3. THE DIVIDE-AND-CONQUER PRINCIPLE.

A problem of size n can often be split into two similar subproblems of size approximately equal to n/2, and so forth, until subproblems are obtained of constant size for which the solutions are trivially known. For example, quicksort (Sedgewick (1977, 1978)) is based on this principle. The average time here is $0(n \log n)$, but, unfortunately enough, since the sizes of the subproblems in quicksort can take values 0,1,2,... with equal probabilities, the worst-case complexity is $0(n^2)$. One can start in the other direction with about n equal-sized small problems, and marry subsolutions in a pairwise manner as in mergesort. Because of the controlled subproblem size, the worst-case complexity becomes $0(n \log n)$ (Knuth, 1975). Both principles will be referred to as divide-and-conquer principles. They have numerous applications in computational geometry with often considerable savings in average time. The first general discussion of their value in the design of fast average time algorithms can be found in Bentley and Shamos (1978).

Let us analyze the divide-and-conquer algorithms more formally. Assume that $X_1,\dots,X_n$ are R^d-valued independent random vectors with common distribution, and that we are asked to find $A_n = A(X_1,\dots,X_n)$, a subset of $X_1,\dots,X_n$, where $A(.)$ satisfies:

1) $A(x_1,\dots,x_n) = A(x_{\sigma(1)},\dots,x_{\sigma(n)})$, for all $x_1,\dots,x_n \in R^d$, and all permutations $\sigma(1),\dots,\sigma(n)$ of $1,\dots,n$.

2) $x_1 \in A(x_1,\dots,x_n) \Rightarrow x_1 \in A(x_1,\dots,x_i)$ for all $x_1,\dots,x_n \in R^d$, and all $i \le n$.

The <u>convex hull</u> satisfies these requirements. If $Q_1(x),\dots,Q_{2^d}(x)$ are the open quadrants centered at $x \in R^d$, then we say that X_i is a maximal vector of $X_1,\dots,X_n$ if some quadrant centered at X_i is empty (i.e., contains no X_j , $j \ne i$, $j \le n$) . The <u>set of maximal vectors</u> also satisfies the given requirements. Let N = cardinality(A_n) . For $p \ge 1$, we know by Jensen's inequality that

$$E(N^p) \ge (E(N))^p \quad .$$

In the present context, we would like an inequality in the opposite direction. For random sets A_n satisfying 1) and 2), and under very weak conditions on the behavior of $E(N)$, we have

$$E(N^p) = O((E(N))^p)$$

(Devroye, 1981c). For example, it suffices that $E(N)$ is nondecreasing, or that $E(N)$ is regularly varying at infinity. Also, if $E(N) \le a_n \uparrow$, then $E(N^p) = O(a_n^p)$. In essence, the results of Devroye (1981c) imply that under weak conditions on the distribution of X_1 , $E(N^p)$ and $(E(N))^p$ are comparable. The same is true for other nonlinear functions of N . For example, if $E(N) \le a_n \uparrow$, then $E(N \log(N+e)) = O(a_n \log a_n)$. Thus the knowledge of $E(N)$ allows us to make statements about other moments of N . Here are some known results about $E(N)$.

<u>Examples</u>. 1. <u>A_n is the convex hull. X_1 has a density f .</u>

(i) $E(N) = o(n)$ (Devroye, 1981d).

(ii) If f is normal, then $E(N) = O((\log n)^{(d-1)/2})$ (Raynaud, 1970). For $d=2$, $E(N) \sim 2\sqrt{2\pi \log n}$ (Renyi and Sulanke, 1963, 1964).

(iii) If f is the uniform density in the unit hypersphere of R^d , then $E(N) = O(n^{(d-1)/(d+1)})$ (Raynaud, 1970).

(iv) If f is the uniform density on a polygon of R^2 with k vertices, then $E(N) \sim \frac{2k}{3} \log n$ (Renyi and Sulanke, 1963, 1968).

(v) If f is a radial density, see Carnal (1970). For example, if f is radial, and $P(||X_1||>u) = L(u)/u^r$ where $r \ge 0$ and L is slowly varying (i.e., $L(cx)/L(x) \to 1$ as $x \to \infty$, all $c > 0$), then $E(N) \to c(r) > 0$. If $P(||X_1||>u) \sim c(1-u)^r$ for some c , $r > 0$ as $u \uparrow 1$, and $P(||X_1||>1)=0$, then $E(N) \sim c(r)\, n^{1/(2r+1)}$ for some $c(r) > 0$.

2. <u>A_n is the set of maximal vectors.</u> When X_1 has a density and the components of X_1 are independent then $E(N)$ is nondecreasing (Devroye, 1980) and $E(N) \sim 2^d (\log n)^{d-1}/(d-1)!$ (Barndorff-Nielsen, 1966; Devroye, 1980). <u>End of examples</u>.

A_n can be found by the following merging method. Assume for the sake of simplicity that $n = 2^k$ for some integer $k > 1$.

1. Let $A_{1i} = A(X_i)$, $1 \le i \le n$. Set $j \leftarrow 1$.
2. Merge consecutive A_{ji}'s in a pairwise manner (A_{j1} and A_{j2} ; A_{j3} and A_{j4} ; etc.) .
3. Set $j \leftarrow j+1$. If $j > k$, terminate the algorithm ($A_n = A_{k1}$) . Otherwise, go to 2.

We assume that merging and editing of A_{ji} and $A_{j\ i+1}$ with cardinalities k_1 and k_2 can be done in time bounded from above by $g(k_1)+g(k_2)$ for some nondecreasing positive-valued function g , and that $E(g(n)) \le b_n \uparrow$ where, as before, N = cardinality(A_n) . Then the given algorithm finds A_n in average time

$$O(n \sum_{j=1}^{2n} b_j/j^2) \ .$$

If the merging and editing take time bounded from below by $a(g(k_1)+g(k_2))$ and $E(g(N)) \ge s b_n$ where g and b_n are as defined above, and $a,s > 0$ are constants, then we take at least

$$\gamma n \sum_{j=1}^{n} b_j/j^2$$

average time for some $\gamma > 0$ and all n large enough (Devroye, 1981c). Thus, the divide-and-conquer method finds A_n in linear average time if and only if

$$\sum_{j=1}^{\infty} b_j/j^2 < \infty .$$

Examples.

1. The set of maximal vectors.

Merging of two sets of maximal vectors can be achieved in quadratic time by pairwise comparisons, for any dimension d. We can thus take $g(u) = u^2$ in the previous analysis if we merge in this way. If $E(N) \sim a_n \uparrow \infty$, then we can check that the divide-and-conquer algorithm runs in linear average time if and only if

$$\sum_{j=1}^{\infty} a_j^2/j^2 < \infty .$$

2. Convex hulls in R^2.

Two convex hulls with ordered vertices can be merged in linear time into a convex hull with ordered vertices (Shamos, 1978). Thus, if $E(N) = O(a_n)$ and $a_n \uparrow \infty$, then

$$\sum_{j=1}^{\infty} a_j/j^2 < \infty \qquad (2)$$

is sufficient for the linear average time behavior of the divide-and-conquer algorithm given here. When $\liminf E(N)/a_n > 0$, then (2) is also necessary for linear average time behavior. Notice here that (2) is satisfied when, say, $a_n = n/\log^{1+\delta} n$ or $a_n = n/(\log n . \log^{1+\delta} \log n)$ for some $\delta > 0$. This improves the sufficient condition $a_n = n^{1-\delta}$, $\delta > 0$, given in Bentley and Shamos (1978).

3. Convex hulls in R^d.

Merging can trivially be achieved in polynomial (n^{d+1}) time for two convex hulls with total number of vertices equal to n. When $E(N) \le a_n \uparrow$ and

$$\sum_{j=1}^{\infty} a_j^{d+1}/j^2 < \infty ,$$

we can achieve linear average time. This condition is fulfilled for the normal density in R^d and the uniform density on any hypercube of R^d. End of examples.

4. THE QUICK ELIMINATION (THROW-AWAY) PRINCIPLE.

In extremal problems (e.g., find the convex hull, find the minimal covering ellipse, etc.) many of the data points can be eliminated from further considerations without much work. The remaining data points then enter the more involved portion of the algorithm. Often the worst-case time of these elimination algorithms is equal to the worst-case time of the second part of the algorithm used on all n data points. The average time is sometimes considerably smaller than the worst-case time. We illustrate this once again on our prototype problem of finding the convex hull.

Examples.

1. The convex hull.

Assume that we seek the extrema $e_1,\ldots,e_m$ in m carefully chosen directions of R^d, form the polyhedron P formed by these extrema, and eliminate all X_i's that belong to the interior of the extremal polyhedron P. The remaining X_i's are then processed by a simple worst-case $O(g(n))$ convex hull algorithm. What can we say about the average time of these algorithms? An average time of $o(g(n))$ would indicate that the elimination procedure is worthwhile on large data sequences. We could also say that the elimination procedure achieves 100% asymptotic efficiency. In Devroye (1981d) it is shown that this happens when (i) the open halfspaces defined by the hyperplanes through the origin perpendicular to the e_i's cover R^d except possibly the origin; and (ii) X_1 has a radial density f, where

$$\alpha(u) = \inf\{t : P(||X_1||>t)=u\}$$

is slowly varying at 0 ($\lim_{t\downarrow 0} \alpha(tu)/\alpha(t) = 1$, all $u>0$), and $\alpha(u) \to \infty$ as $u \downarrow 0$. Condition (i) holds when the e_i's are determined by the $d+1$ vertices of the regular $(d+1)$-vertex simplex in R^d centered at the origin. One could also take $2d$ directions defined by $(0,0,\ldots,0,\pm 1)$, etc. . Condition (ii) is satisfied by the normal density and a class of radial exponential densities (Johnson and Kotz, 1972, pp. 298). The previous result can be sharpened in specific instances. For example, if f is normal, $g(n) = n \log n$, and (i) holds, then the average time is $O(n)$. Furthermore, the average time spent by the algorithm excluding the elimination is $o(n)$ (Devroye, 1981d).

When $d=2$ and f is bounded away from 0 and infinity on a nondegenerate rectangle of R^2 ($f=0$ elsewhere), and $e_1,\ldots,e_8$ are equi-spaced directions, then the average time of the elimina-algorithm is $O(n)$ even when $g(n) = n^2$ (Devroye and Toussaint, 1981).

Eddy (1977) has given a slightly different elimination algorithm in which the number of directions and the directions themselves depend upon the data. Akl and Toussaint (1978) report that in R^2, for certain distributions, almost all elimination algorithms achieve extremely fast average times provided that $e_1,\ldots,e_m$ are easily computed (e.g., they are axial or diagonal directions.)

2. Finding a simple superset.

Assume that we wish to find $A_n=A(X_1,\ldots,X_n)$ in the following manner: (i) Find a set $B_n=B(X_1,\ldots,X_n)$ where B_n is guaranteed to contain A_n, in average time T_n; (ii) Given that the cardinality (B_n) is equal to N, find A_n from B_n in worst-case time bounded by $g(N)$. Note that the average time of the entire elimination algorithm is bounded by

$$T_n + E(g(N)) .$$

2.1. A_n is the convex hull, B_n is the set of maximal vectors.

We discussed some distributions for which $T_n=O(n)$. In R^2, step (ii) can be executed with $g(n)=n^2$ (Jarvis' algorithm, 1973) or $g(n)=n \log n$ (Graham's algorithm, 1972). Thus, the entire algorithm takes average time $O(n)$ when $E(N \log_+ N)=O(n)$ or $E(N^2)=O(n)$, according to the algorithm selected in step (ii). When the components of X_1 are independent and X_1 has a density, then these conditions are satisfied by the results of Devroye (1980, 1981c) given in Section 3. The linearity is not lost in this case in R^d even when $g(n) = n^{d+1}$ in step (ii).

2.2. A_n is the diameter of $X_1,\ldots,X_n$; B_n is the convex hull.

$A_n=\{X_i,X_j\}$ is called a diameter of $X_1,\ldots,X_n$ when $||X_k-X_m|| \le ||X_i-X_j||$ for all $1 \le k \le m \le n$, $(k,m)\neq(i,j)$. Given B_n, some A_n can be found by comparing all $\binom{N}{2}$ distances between points in B_n (see Bhattacharya (1980) for an in-depth treatment of the diameter problem, and a survey of earlier results). But B_n can be found in linear average time for many distributions. In such cases, our trivial diameter algorithm runs in linear average time provided that $E(N)=O(\sqrt{n})$. Assume for example that f is the uniform density in the unit hypersphere of R^d, then the trivial diameter algorithm runs in linear average time if and only if (i) the convex hull can be found in linear average time, and (ii) $d \le 3$ (section 3, example 1 (iii)).

2.3. A_n is the minimum covering circle, B_n is the convex hull.

The minimum area circle in R^2 covering $X_1,\ldots,X_n$ has either three convex hull points on its perimeter, or has a diameter determined by two convex hull points. Again, it can be found (trivially) from the convex hull in worst-case time $O(n^4)$ (see Elzinga and Hearn (1972, 1974), Francis (1974) and Shamos (1978) for $O(n^2)$ algorithms and subsequent discussions). Thus, A_n can be identified in linear average time if the convex hull B_n can be found in linear average time, and if $E(N) = O(\sqrt{n})$ (or $O(n^{1/4})$, if the trivial algorithm is used).

5. REFERENCES.

[1] S.G. AKL, H. MEIJER: "Hybrid sorting algorithms: a survey", Department of Computing and Information Science, Queen's University, Technical Report 80-97, 1980.

[2] S.G. AKL, G.T. TOUSSAINT: "A fast convex hull algorithm", Information Processing Letters, vol. 7, pp. 219-222, 1978.

[3] D. AVIS: "On the complexity of finding the convex hull of a set of points", Technical Report SOCS 79.2, School of Computer Science, McGill University, Montreal, 1979.

[4] O. BARNDORFF-NIELSEN, M. SOBEL: "On the distribution of the number of admissible points in a vector random sample", Theory of Probability and its Applications, vol. 11, pp. 249-269, 1966.

[5] J.L. Bentley, J.H. FRIEDMAN: "Data structures for range searching", Computing Surveys, vol. 11, pp. 398-409, 1979.

[6] J.L. BENTLEY, M.I. SHAMOS: "Divide and conquer for linear expected time", Information Processing Letters, vol. 7, pp. 87-91, 1978.

[7] J.L. BENTLEY, B.W. WEIDE, A.C. YAO: "Optimal expected-time algorithms for closest point problems", ACM Transactions of Mathematical Software, vol. 6, pp. 563-580, 1980.

[8] B. BHATTACHARYA: "Applications of computational geometry to pattern recognition problems", Ph.D. Dissertation, McGill University, Montreal, 1980.

[9] K.Q. BROWN: "Voronoi diagrams from convex hulls", Information Processing Letters, vol. 9, pp. 227-228, 1979.

[10] H. CARNAL: "Die konvexe Hülle von n rotationssymmetrische verteilten Punkten", Zeitschrift für Wahrscheinlichkeitstheorie und verwandte Gebiete, vol. 15, pp. 168-176, 1970.

[11] P. CHERITON, R.E. TARJAN: "Finding minimum spanning trees", SIAM Journal of Computing, vol. 5, pp. 724-742, 1976.

[12] L. DEVROYE: "A note on finding convex hulls via maximal vectors", Information Processing Letters, vol. 11, pp. 53-56, 1980.

[13] L. DEVROYE: "Some results on the average time for sorting and searching in R^d", Manuscript, School of Computer Science, McGill University, Montreal, 1981a.

[14] L. DEVROYE: "On the average complexity of some bucketing algorithms", Computers and Mathematics with Applications, to appear, 1981b.

[15] L. DEVROYE: "Moment inequalities for random variables in computational geometry", Manuscript, School of Computer Science, McGill University, Montreal, 1981c.

[16] L. DEVROYE: "How to reduce the average complexity of convex hull finding algorithms", Computing, to appear, 1981d.

[17] L. DEVROYE, T. KLINCSEK: "On the average time behavior of distributive sorting algorithms", Computing, vol. 26, pp. 1-7, 1981.

[18] L. DEVROYE, G.T. TOUSSAINT: "A note on linear expected time algorithms for finding convex hulls", Computing, to appear, 1981.

[19] W.F. EDDY: "A new convex hull algorithm for planar sets", ACM Transactions of Mathematical Software, vol. 3, pp. 398-403, 1977.

[20] J. ELZINGA, D. HEARN: "The minimum covering sphere problem", Management Science, vol. 19, pp. 96-104, 1972.

[21] J. ELZINGA, D. HEARN: "The minimum sphere covering a convex polyhedron", Naval Research Logistics Quarterly, vol. 21, pp. 715-718, 1974.

[22] R.L. FRANCIS, J.A. WHITE: Facility Layout and Location: An Analytical Approach, Prentice-Hall, 1974.

[23] R. GRAHAM: "An efficient algorithm for determining the convex hull of a finite planar set", Information Processing Letters, vol. 1, pp. 132-133, 1972.

[24] R.N. HORSPOOL: "Constructing the Voronoi diagram in the plane", Technical Report SOCS 79.12, School of Computer Science, McGill University, Montreal, 1979.

[25] R.A. JARVIS: "On the identification of the convex hull of a finite set of points in the plane", Information Processing Letters, vol. 2, pp. 18-21, 1973.

[26] N.L. JOHNSON, S. KOTZ: Distributions in Statistics: Continuous Multivariate Distributions, John Wiley, New York, 1972.

[27] D. KNUTH: The Art of Computer Programming, vol. 3: Sorting and Searching, Addison-Wesley, Reading, Mass., 2nd Ed., 1975.

[28] R.J. LIPTON, R.E. TARJAN: "Applications of a planar separator theorem", 18th Annual IEEE Symposium on the Foundations of Computer Science, pp. 162-170, 1977.

[29] H. RAYNAUD: "Sur le comportement asymptotique de l'enveloppe convexe d'un nuage de points tirés au hasard dans R^n", Comptes Rendus de l'Académie des Sciences de Paris, vol. 261, pp. 627-629, 1965.

[30] A. RENYI, R. SULANKE: "Uber die konvexe Hülle von n zufällig gewählten Punkten I", Zeitschrift für Wahrscheinlichkeitstheorie und verwandte Gebiete, vol. 2, pp. 75-84, 1963.

[31] A. RENYI, R. SULANKE: "Uber die konvexe Hülle von n zufällig gewählten Punkten II", Zeitschrift für Wahrscheinlichkeitstheorie und verwandte Gebiete, vol. 3, pp. 138-147, 1964.

[32] A. RENYI, R. SULANKE: "Zufällige konvexe Polygone in einem Ringgebeit", Zeitschrift für Wahrscheinlichkeitstheorie und verwandte Gebeite, vol. 9, pp. 146-157, 1968.

[33] R. SEDGEWICK: "The analysis of quicksort programs", Acta Informatica, vol. 7, pp. 327-355, 1977.

[34] R. SEDGEWICK: "Implementing quicksort programs", Communications of the ACM, vol. 21, pp. 847-857, 1978.

[35] M.I. SHAMOS: "Computational geometry", Ph.D. Dissertation, Yale University, New Haven, Connecticut, 1978.

[36] M.I. SHAMOS: Seminar given at McGill University, 1979.

[37] M.I. SHAMOS, D. HOEY: "Closest-point problems", Proceedings of the 16th IEEE Symposium on the Foundations of Computer Science, pp. 151-162, 1975.

[38] G.T. TOUSSAINT: "Pattern recognition and geometrical complexity", Proceedings of the 5th International Conference on Pattern Recognition and Image Processing, Miami, Florida, 1980.

[39] B.W. WEIDE: "Statistical methods in algorithm design and analysis", Ph.D. Dissertation, Carnegie-Mellon University, Pittsburgh, Pennsylvania, 1978.

[40] A.C. YAO: "An $O(|E| \log\log |V|)$ algorithm for finding minimum spanning trees", Information Processing Letters, vol. 4, pp. 21-23, 1975.

[41] A.C. YAO: "On constructing minimum spanning trees in k-dimensional space and related problems", Research Report STAN-CS-77-642, Department of Computer Science, Stanford University, Stanford, 1977.

[42] A.C. YAO: "A lower bound to finding convex hulls", Technical Report STAN-CS-79-733, Department of Computer Science, Stanford, 1979.

Workshop 5

Pattern Recognition

Organizers: Michael E. Tarter, University of California, Berkeley, and Nancy R. Mann, University of California, Los Angeles

Chair: Michael E. Tarter, University of California, Berkeley, and Nancy R. Mann, University of California, Los Angeles

Invited Presentations:

Applications of Pattern Recognition Methods to Hydrologic Time Series, Sidney J. Yakowitz, University of Arizona

Recent Advances in Bump Hunting, I.J. Good, Virginia Polytechnic Institute and State University, and M.L. Deaton, Verginia Polytechnic Institute and State University

Pattern Recognition in the Context of an Asbestos Cancer Threshold Study, Michael E. Tarter, University of California, Berkeley

APPLICATIONS OF PATTERN RECOGNITION METHODS TO HYDROLOGIC TIME SERIES

Dr. Sidney Yakowitz, University of Arizona

ABSTRACT

The paper surveys hydrologic studies by the speaker and others in which pattern recognition concepts play a prominent role. Data sets gathered from measurements of riverflow and water table pressure heads motivate relatively delicate statistical questions. Techniques such as cluster analysis, feature extraction, and non-parametric regression are ingredients of the state-of-the art solutions to these questions.

1. Introduction and Scope

Pattern recognition theory is a subject which was in vogue during the 1960's, but has since suffered a decline in popularity, partly due to failure of its proponents to provide devices whose performance excelled that of more pragmatic gadgets on easier problems, and partly due to its failure to solve the harder problems. Still, the pattern recognition "bubble" did provide some benefits. It helped to focus attention on certain interesting statistical problems. Also many of us who studied pattern recognition as doctoral students were encouraged to receive better training in statistics and applied mathematics than some of our fellow students in different engineering disciplines; I, for one, am thankful for that twist of fate. Finally, I feel compelled to offer the thought that pattern recognition may yet have its day in the sun: many pattern recognition applications in the 1960's seemed hampered by lack of computer power and appropriate statistical methodology. As the cost of computation declines, and as computer-oriented statistical methodology such as presented at these Interface Conferences continues to be refined, some of the obstacles may diminish; many of the problems that pattern recognition theoreticians attacked, such as handwriting, printing, and speech recognition, automatic medical diagnosis, and language translation are still unresolved.

The intention of this study is to describe progress in hydrological studies in which pattern recognition methodology plays a prominent role. Toward presenting these developments, we will first glance at the topics that are thought to comprise pattern recognition theory. Following that, the hydrologic problems will be outlined, and then attacks on these problems will be reviewed, with particular emphasis on the way in which pattern recognition principles impinge on the attack.

Hydrology is one of many disciplines in which data pertaining to certain problems of economic significance is plentiful and even enormous, and yet the problems are really not satisfactorily resolved. One would think that, with such quantities of data on hand, satisfying answers would be forthcoming. The particular problems to be discussed come from the areas of riverflow and groundwater modelling and prediction.

2. An Abridged Synopsis of the Principles of Pattern Recognition

A caricature of the pattern recognition paradigm is shown in Figure 1.

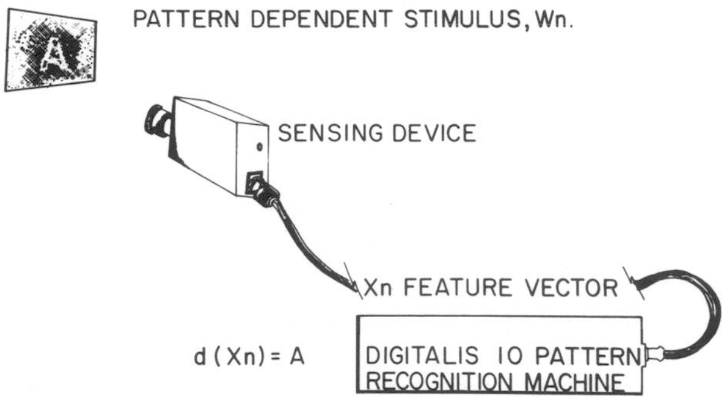

Figure 1 Pattern Recognition Paradigm

It is presumed that a sequence $\{W_n\}$ of stimuli impinge on a sensing device, which summarizes its measurements of W_n in the form of a vector $\vec{x}_n$ of relatively low dimension. Thus the sensing device converts the stimulus W_n into a vector $\vec{x}_n$, known as the "feature vector". The limitations on the dimension of $\vec{x}_n$ are imposed by computational and statistical factors. The feature vector $\vec{x}_n$ serves as input to the pattern recognition machine which is physically a digital computer, and in statistical parlance, serves simply as a decision function.

Let us now survey some of the statistical assumptions customarily made by pattern recognition specialists (e.g., books in category 1 of the reference section). They are:

1. There is a finite set $\Theta = \{A, B, \ldots, Z\}$ of "patterns."
2. At each epoch n, nature chooses a pattern θ_n. The distribution of the stimulus W_n depends on the pattern θ_n chosen.
3. The output of the pattern recognition machine is a pattern, say d_n, which is suppose to be a guess of θ_n.

There are usually a few more assumptions in force, depending on the particular application. An almost universal further assumption is that the pattern sequence $\{\theta_n\}$ is independently and identically distributed. In some cases, it is presumed that the distribution $F_\theta(x)$ of the feature vector associated with pattern θ is a member of a specified parameteric family. But an interesting aspect of pattern recognition theory is that in many cases it has avoided such assumptions and tends to push at the frontiers of nonparametric statistical methodology. To give some flavor for the practical aims of pattern recognition, I have listed in Table 1 some of the areas in which pattern recognition applications have been made.

Table 1

Some PRM Application

1. Robotics Recognize geometric shapes or printed letters.
2. Surveillance Decide whether returned sonar signals are reflected from submarine of class A, B, C, or a whale.
3. Space Communications Decode noisy digital signals from a space probe.
4. Medical Science Decide, on basis of microscope scan, whether a given cell is malignant.

We have discussed the format, assumptions, and aims of pattern recognition theory. This synopsis is concluded with an outline of some of the central methodological themes of pattern recognition. For it is these methodological aspects rather than pattern recognition problem itself which is carried over to our hydrological studies.

Many of the topics of pattern recognition theory can be classified under one of the two categories, 1.) "learning" or 2.) "feature extraction". Learning concerns the task of using past samples or results of a special testing period to infer the distribution of the feature vector (the distribution depending, of course, on the pattern which is "active"). If a correct classification θ_n of each learning sample $\vec{x}_n$ is available, then the learning is said to be "supervised", and otherwise, it is "nonsupervised". The second category, feature extraction, deals with the design of the measurement device itself; one desires a feature vector $\vec{x}_n$ which is as informative as possible with respect to the multiple-hypothesis testing problem of guessing the active pattern.

Essentially every method I know of for unsupervised learning fits into the broad framework of "cluster analysis" as explicated in Hartigan's (1975) book. Through methods of cluster analysis, given a sequence $\{\vec{x}_n\}$ of unclassified feature-vector learning samples and a number K, one partitions the samples into K sets or "clusters". Members of any given cluster should be, in some sense, "close" to one another.

Cluster analysis having been completed, one can program a pattern recognition machine as follows: Let $\vec{c}_j$ denote the center (i.e., 1st moment) of the jth cluster, and let $\vec{x}$ be a feature vector to be classified. One may choose the decision function to say that the pattern is j if $\vec{x}$ is closer in Euclidean distance to $\vec{c}_j$ than to the other cluster centers.

For purposes of both supervised and nonsupervised learning, pattern recognition theorists have comprised a market and even a source for methods of nonparametric density estimation and regression. As an example showing the use of a literature nonparametric techniques originating in the pattern recognition literature known as "potential function method", consider the following situation. There are two patterns A and B, and S_A and negative on S_B. If $\vec{x}$ is an unclassified feature vector, the associated decision rule is to guess A if and only $g(\vec{x}) > 0$.

3. Applications to River Flow Modelling

Having completed our short course on pattern recognition, let us proceed to our first applications area, namely river flow analysis. There are two central tasks with respect to river flow: modelling and forecasting. Modelling is a prerequisite for scientific design and control of reservoirs, for irrigation management, and for designation of flood plains, among other tasks. The forecast/prediction problem encompasses the tasks of anticipating floods and droughts, and therefore has obvious economic implications in terms of flood warning systems and emergency countermeasures.

My assessment is that progress with respect to either of the activities of stream flow modelling and forecasting is meagre.

Let us examine some actual historical records of flows (Figures 2 and 3).

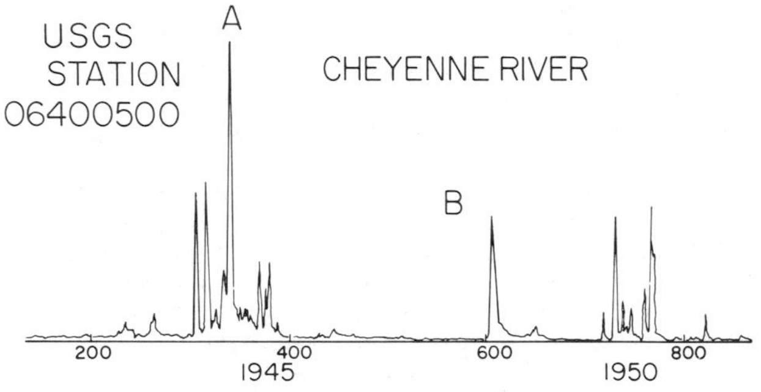

Figure 2 A Record from the Cheyenne River

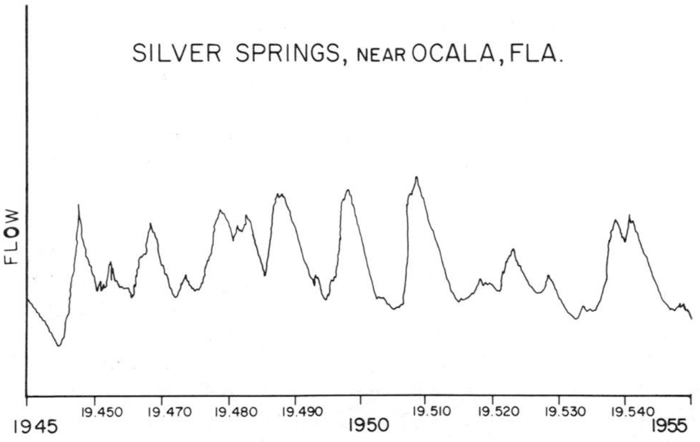

Figure 3 A Record from the Silver Springs River

Intuitive questions which come to mind in examining such graphs include: With respect to the region marked A in Figure 2, "Could one have predicted that the flow would soon take on a high value?" Or, with respect to the region marked B, "At what point could one have said, with a given level of confidence, that this was going to be a dry year?" Or, with respect to Figure 3, "Do years with small flows tend to come in runs?"

I believe I am fairly familiar with the literature on river flow, and it is my opinion that methodology for answering questions such as just posed is very weak, and essentially nonexistent. Most works on riverflow modelling adapt autoregressive (AR) or autoregressive moving-average (ARMA) models. My experience is that ARMA sample functions do not bear close resemblance to the rivers from which they are calibrated. The second most popular class of river flow models are inspired by Mandelbrot's fractile models, which seem to fall even wider of the mark of providing sample functions resembling the calibrating rivers.

One maverick approach which at least succeeds in providing simulated records bearing some visual resemblance to the historical flows, was provided by Panu and colleagues (Panu et al. (1978)), Panu and Unny (1980). Their works explicitly acknowledge pattern recognition motivations, and they employ feature selection ideas and supervised learning.

The approach of these papers begins with their designating the patterns as being different seasons of the annual cycle, specifically W = "wet" and D = "dry". The pattern "stimulus" consists of 160 daily measurements during a season, and the feature vector summarizes these measurements with a six dimensional vector. Since one can tell by the calendar which pattern is "active", one can perform supervised learning to obtain the feature vector distributions F_W and F_D associated with the two patterns, W and D.

In Figure 4, we have repeated three types of feature vector constructs which Panu and colleagues have considered.

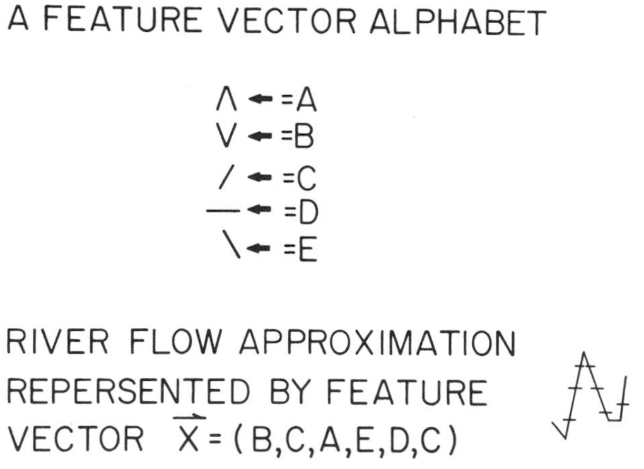

Figure 4 An Alphabet and Feature Vector For River Modelling

The desired approximation model of the river flow is then provided by F_W for wet seasons and F_D for dry. To obtain a riverflow simulation, Panu and colleagues simply obtain independent feature vector simulations:

$$\bar{X}_{1D}, \bar{X}_{2D}, \ldots \sim F_D$$

$$\bar{X}_{1W}, \bar{X}_{2W}, \ldots \sim F_W$$

and intertwine them as

$$\bar{X}_{1D}, \bar{X}_{1W}, \bar{X}_{2D}, \bar{X}_{2W}, \ldots$$

to obtain simulations as shown in Figure 5.

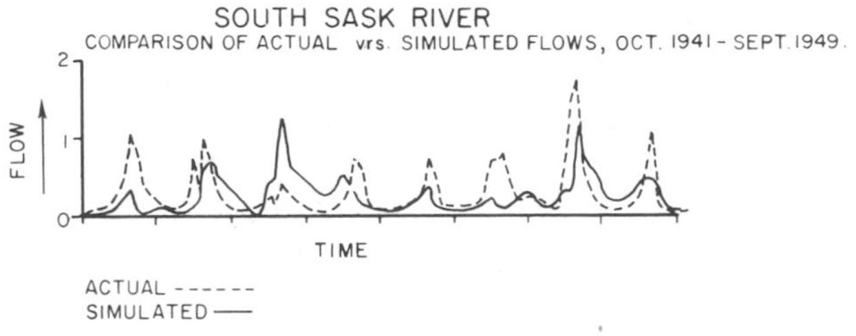

Figure 5 Record and Simulation of South Sask River

The simulations are plotted along with the profiles of the historical record data. It appears to this author that the simulated wave forms do bear some resemblance to the historical record and that it would take some numerical effort to test the hypothesis that the river approximations and the simulations come from identical processes.

I have provided a river flow model in which cluster analysis plays a crucial role. Basically, my approach assumes that a river is a discrete-time Markov process of known Markov order. One can, of course incorporate seasonality, as we did, by having a separate Markov law for each season. With this provision, the Markov assumption would seem plausible because essentially any time series can be approximated arbitrarily accurately by a Markov process of high enough order. (Denny and Yakowitz (1978) have studied the order inference problem.)

For sake of exposition, let us assume that the order is one. Then the evolution of the river is determined by the transition distribution function

$$F(x|y) = P[X_{n+1} \leq x | X_n = y],$$

which is presumed continuous in both variables. Whereas the Markov model is plausible, the serious problem toward using it is that there are no standard ways for inferring the transition distribution $F(x|y)$ from the data.

This is where cluster analysis comes in. Let $\{x_n\}$ denote the historical river record and let K be a fixed integer. The approximation $F(x|y)$ of the transition distribution function is constructed according to the three steps below:

1. Obtain K clusters $S_1, \ldots, S_K$ from the data. (I used the K-means algorithm described in Hartigan (1975)).
2. Let $F_j(x)$ be, for each cluster j, a distribution function constructed nonparametrically from

the successors to states in S_j, i.e., from $\{x_{n+1}: x_n \varepsilon S_j\}$.
3. For any state y, the approximation $\hat{F}(x|y) \triangleq F_j(x)$, where j is the index of the cluster whose center is closest to y. Thus cluster centers play the role of "patterns", and unsupervised learning is done by cluster analysis.

The rationale for this approximation is that if K is chosen so that for an increasing number of samples, the clusters become larger and more numerous, then a) $F_j(x)$ should be a good approximation for $F(x|y)$ when y is near the center of S_j, and b) For any state y, there will be a nearby cluster center. That the estimate $\hat{F}(x|y)$ does constitute a consistent estimator for $\hat{F}(x|y)$ was proven by Yakowitz (1979a). But the main heartening feature about this procedure is that, in our experience, it seems to work fairly well.

In Figure 6, we have shown a sample function generated according to the Markov law with $F(x|y)$ equal to $G(x;m(y), \sigma^2(y))$, with $G(x;m,\sigma^2)$ the distribution function for the normal law with mean m and variance σ^2, and $m(y) = \text{sign}(y)|y|^{0.8}$, and $\sigma^2(y) = (4 + 4\exp(y))^{-1}$.

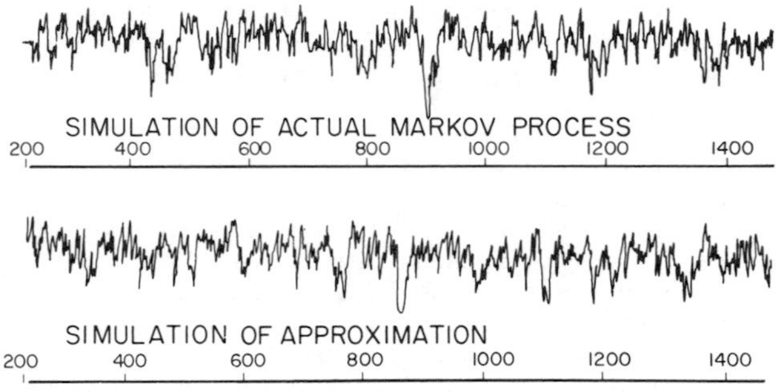

Figure 6 A Markov Process and Approximation

This is admittedly a rather arbitrary process. Some supporting statistical analysis and prediction performance comparisons with autoregressive models are presented in Yakowitz (1979a).

In another study (Yakowitz (1979b)) a 30 year daily record for the Cheyenne River (obtained from the USGS) was applied to the Markov transition distribution estimation algorithm, the Markov order having been inferred to be three. Seasonality was accounted for by dividing the year into "summer" and "winter" portions, and treating each portion as a separate Markov process. Simulations obtained using the inferred transition distribution functions are compared to the actual river flow record in Figure 7. More discussion and statistical computations are to be found in Yakowitz (1979b).

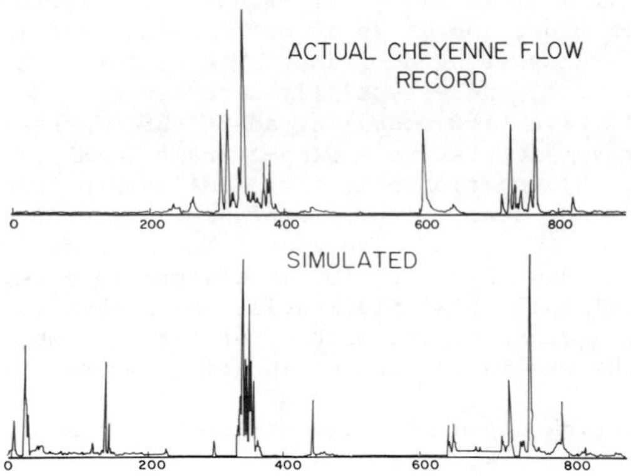

Figure 7 Record and Simulation of Cheyenne River

IV. Applications to Water Table Prediction

The problem of concern here is the following: One has annual records $w_j = (w_{j1}, \ldots, w_{jN})$ of each well j, $1 \leq j \leq J$, each record w_j giving the depth to water over the past N years. Find a prediction vector $y = (y_1, \ldots, y_J)$ such that the jth well next year (i.e., for year N+1). The computational study employed a data base of N = 28 year records for J = 808 wells in the Tucson basin. An empirical finding is that the records w_j tended to vary rather substantially from well to well, although there was some tendency for nearby wells to exhibit similar graphs. We now outline the modelling assumptions, statistical procedure, and results which appeared in Yakowitz (1976).

From examination of the records and some exploratory least squares analysis, we were led to adopt the following hypotheses:
Assumption 1: The Tucson basin can be partitioned into geographic regions such that each well in a given region can be regarded as an independent realization of a time series whose distributional law is the same for each well in the region.
Assumption 2: For purposes of linear least-squares prediction, the first differences $d_k = w_{jk} - w_{jk-1}$ of each well record in region R satisfy the third order autoregressive (AR) model:

$$d_{k+1} = C^R + a^R d_{k-1} + c^R d_{k-2} + N_{k+1} \qquad (1)$$

where the N_j's are uncorrelated unbaised noise samples with sommon variance.

The statistical procedure followed in Yakowitz (1976) was the following:
1. Let K = 7 be the number of regions, this number being chosen so as to allow some variability over the aquifer but nevertheless maintain enough wells in each region to allow AR parameter inference.
2. Approximate each 29-entry well record w_j by a five-dimensional feature vector. The feature vector for each well was obtained by averaging four and five year portions of the record.
3. Use a cluster technique to divide the feature

vectors into K clusters. We used the BMSE algorithm described in Duda and Hart (1973).

4. Regarding the records in each cluster R as independent realizations of the same process, use these records to calibrate the autoregression parameters

$$C^R, \; a^R, \; b^R, \text{ and } \; c^R \text{ in (1).}$$

Space limitations prevent us from reviewing the details of the numerical study presented in Yakowitz (1976). We will mention that the records of each cluster tended to resemble one another, and that visually, there seemed to be substantial distinction between trajectories of one cluster and those of another. A further heartening development was that the locations of wells in the same cluster (i.e., region) tended to be grouped geographically, thus lending support to Assumption 2 above. In Figure 8, we have displayed typical plots of the one-step ahead predictor against the actual observed trajectories of wells in different regions. Further computational studies and details may be found in Yakowitz (1976).

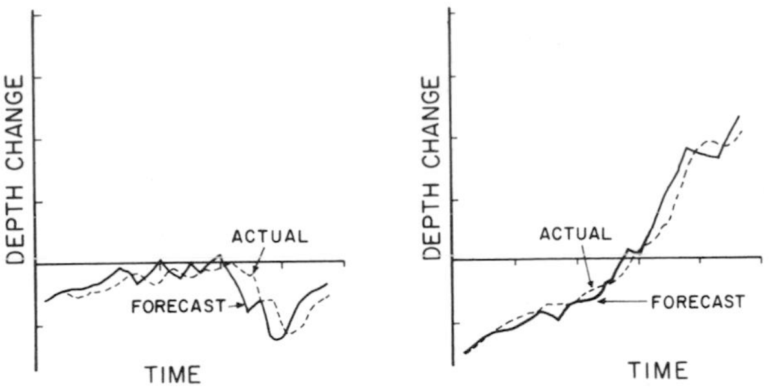

Figure 8 One Step-ahead Well Depth Prediction

The pattern recognition ideas employed in the above study included feature vector data approximation of the stimulus w (the well records), and unsupervised learning (through cluster analysis). Each geographic region of constant time series distribution was viewed as a pattern.

Acknowledgments. This work was supported by NSF grants Eng Int 78-12284, ENG 78-07358, and CME 79-05010.

REFERENCES

1. Texts on Pattern Recognition

Duda, R. and P. Hart, Pattern Classification Analysis, Wiley, New York, 1973.

Andrews, H., Introduction to Mathematical Techniques in Pattern Recognition, Wiley, New York, 1972.

Patrick, E. A., Fundamentals of Pattern Recognition, Prentice-Hall, Englewood Cliffs, N.J., 1972.

Tou, J. T., and R. C. Gonzales, Pattern Recognition Principles, Addison-Wesley, Reading, MA., 1974.

2. Books on Clustering

Hartigan, J. A., Clustering Algorithms, Wiley, New York, 1975.

Van Ryzin, J. (editor), Classification and Clustering, Academic Press, New York, 1977.

3. References on Statistical Methods for Markov Processes

Billingsley, P., Statistical Inference for Markov Processes, University of Chicago Press, Chicago, 1961.

Denny, J. and S. Yakowitz, Admissible run-contingency type tests for independence and Markov dependence, J. Amer. Statist, Assoc., 73(361) 177-181, 1978.

Yakowitz, S., Nonparametric Estimation of Markov Transition Functions, Ann. Statist., 7(3), 671-679, 1979a.

4. River Flow Models Using Pattern Recognition Ideas

Panu, U. and T. Unny, Extension and application of feature prediction model for synthesis of hydrologic records, Water Resour. Res., 16(1), 77-96, 1980.

Panu, U., T. Unny and R. Ragade, A feature prediction model in synthetic hydrology based on concepts of pattern recognition, Water Resour. Res., 14(2), 335-344, 1978.

Yakowitz, S., A nonparametric Markov model for daily river flow, Water Resour. Res., 15(5), 1035-1043, 1979b.

5. Goundwater Analysis Using Pattern Recognition Ideas

Yakowitz, S., Model-free Statistical methods for water table prediction, Water Resour. Res., 12(5), 836-844, 1976.

6. Foundations for Identification

Schuster, E. and S. Yakowitz, Contributions to the theory of nonparametric regression, with application to system identification, Ann. Statist., 7(1), 137-149, 1979.

Michael Tarter - I was going to ask you about the lag in your Markov process. Is it continuous?

Yakowitz - The data are taken daily, so it is a discrete time Markov process, but has continuous state space.

Tarter - So it's daily?

Yakowitz - Yes, and actually we tried tests of various Markov orders. There's a statistical test that Jack Denny and I described in a JASA paper [see references] for choosing order on the basis of the riverflow data.

Sam Harold - I'm from the Wilkes-Barre area where this flooding also occurs. Have you used these techniques to study peak flow in a river? I would think that is relatively important.

Yakowitz - No. Do you mean as far as predicting the peak and so forth? No, as I suggested, this is an uncharted realm altogether. I do believe that hydrologist Eric Wood, at Princeton, is trying to devise schemes for detecting when a large peak flow will occur. He is using methodology from linear statistical filtering theory. I don't believe this approach is appropriate. You're not after a linear predictor; you're after a dichotomous pattern recognition answer to the question: "Is it going to flood or is it not going to flood?" And I think it's inappropriate to use linear filtering methods for that problem without justification. But flood prediction is a wide open field and I've been encouraging people to get interested in it. Incidentally, there have been a number of famous statisticians who have worked on hydrology problems. For example, Dynkin, P. Moran, David Kendall, Prahbu, and Milton Sobel have worked on one aspect or another of riverflow and reservoir control.

I.J. Good - I have the impression that your Markov approximation is rather too flat, and I'm wondering whether this is similar to sampling the same river twice?

Yakowitz - Professor Good has caught one of the things that I was hoping people would overlook. Look toward the bottom of this river graph and you'll see that this wiggles around believably, whereas in the similation it's zero or it takes a jump. This discontinuity is due to the fact that these values occur at the centers of two different clusters. In retrospect I could have beaten that drawback, by using a nonparametric density method or just smoothing the empirical distribution functions by convolving them in some way.

Harold - We have one additional problem that's probably as important as what you're doing in terms of pattern recognition. Is anybody doing anything about the problem encountered when we know before we sample that we're going to have to use pattern recognition methods in order to try to determine what kind of patterns we have, so we need to determine required sample size?

Yakowitz - Well, as I mentioned at the beginning of my talk, pattern recognition seems to be somewhat disbanded as a field in itself. So, as far as I know, almost nothing is going on. In that connection I might mention that a feature article in a recent issue of IEEE Spectrum is on pattern recognition -- the military is still at it. While the problems that pattern recognition theorists have concerned themselves with are still being studied, there's not much general literature on that subject, per se, that I'm aware of. In the earlier literature, as far as concerns sample sizes and such, there's not a great deal available.

There's been quite a bit of study on the problem of feature extraction that I mentioned; that is, your sensing device has to encode high dimension information into low dimension vectors. There's quite a bit of literature on how to do that wisely, so that your decision function is as accurate as possible. There are ways of measuring how informative a feature vector is. These subjects, of course, have a counterpart in discriminant analysis, principle components, and so forth.

RECENT ADVANCES IN BUMP HUNTING

I.J. Good & M.L. Deaton, Va. Poly. Inst. & State U.

ABSTRACT

For speeding up the algorithm for the method of maximum penalized likelihood it is tempting to try to make use of the Fast Fourier Transform (FFT). This can be done by circularizing the data and making the x axis discrete. For circular data the circularization is of course unnecessary. Other methods are discussed and some comparisons are made.

For multivariate data one could use a multidimensional FFT by putting the data on a torus. Apart from circularization or toroidalization one can speed up the estimation of the hyperparameter by repeatedly doubling or otherwise increasing the fineness of the grid.

Roughness penalties of the form $\beta\int\{[(f)^{\xi}]''\}^2\,dx$, where f(x) is the density function, are also considered, where ξ is not necessarily ½ or 1, and corresponding algorithms are suggested. For the scattering data considered in previous work we have compared the results for $\xi = ½$ and $\xi = 1$ and find that the estimates of the bumps are almost the same but two of the former bumps have each been split into a pair.

Keywords: Probability density estimation; bump-hunting; penalized likelihood; roughness; roughness penalties; hyperparameters; tail-trouble; circularization; discretization; fast Fourier transform.

1. INTRODUCTION

Bump-hunting and probability density estimation are closely related although the latter is an estimation problem while bump-hunting can be regarded as cluster analysis, discrimination, or significance testing. That is, it is concerned with the location and evaluation of bumps. If a bump is found that was not conjectured in advance, then the activity can be regarded as a form of cluster analysis or pattern recognition, especially in bivariate or multivariate problems.

The maximum penalized likelihood method or MPL method requires the maximization of the log-likelihood minus a roughness penalty, that is, the maximization of

$$\Omega(f) = L - \Phi(f) \qquad (1.1)$$

where L is the log-likelihood, $\Phi(f)$ is a roughness penalty, f is the density function being sought, and $\Omega(f)$ is called the "overall score". (Ω was called ω by G & G′, meaning Good & Gaskins, but we shall use ω in a different sense.) The MPL method can be understood both in a non-Bayesian and in a Bayesian manner. For the latter, one can think of $e^{-\Phi}$ as being proportional to a prior density of f(.) in a function space although this density may be improper (G & G′, 1971, p. 256). This interpretation is especially pertinent for the evaluation of individual bumps by comparison of the posterior densities with the bump and with the bump removed by a process of "surgery" (G & G′, 1980, p. 49). When a bump is conjectured in advance it should be evaluated without using the factor $e^{-\Phi}$ (see G & G′, 1980, p. 49).

For so-called "raw data", consisting of i.i.d. observations $x_1, x_2, \ldots, x_N$, we have

$$L = \sum_{i=1}^{N} \log f(x_i), \qquad (1.2)$$

while for histogram data

$$L = \sum_{j=1}^{J} n_j \log \int_{B_j} f(x)dx, \qquad (1.3)$$

where n_j is the number of observations in "bin" (or interval) B_j, and J is the number of bins.

There are a variety of roughness penalties that appear reasonable at least at first sight. A wide class of roughness penalties depend on a measure of roughness, and the penalty might then be of the form

$$\text{Roughness Penalty} = \text{hyperparameter} \times \text{roughness}. \qquad (1.4)$$

The hyperparameter might also be called a *procedural parameter* when the Bayesian interpretation is not intended. A roughness penalty might be taken as a linear combination of penalties of the form (1.4). For example, G & G′ (1971) considered penalties of the form

$$4\alpha\int[\gamma'(x)]^2\,dx + \beta\int[\gamma''(x)]^2\,dx \qquad (1.5)$$

but especially the cases $4\alpha\int(\gamma')^2\,dx$ alone, for the *first method*, and $\beta\int(\gamma'')^2\,dx$ alone, for the *second method*. Here $\gamma(x) = [f(x)]^{½}$. The first and second methods are of course both examples of formula (1.4) which we regard as the standard form for a roughness penalty. It has the advantage of depending on only one hyperparameter or procedural parameter. The "hyper-razor" principle is that hyperparameters should not be "multiplied" without necessity (Good, 1980), and this principle favors the standard form (1.4) for the roughness penalty.

As a matter of notation, if the x axis is scaled so that the data have sampling variance v we write α_v^* and β_v^* for the transformed values of α and β. When v = ½ we write just α^* and β^* as in G & G′ (1980).

2. SOME PROBLEMS RELATED TO DENSITY ESTIMATION AND BUMP-HUNTING

There are a variety of questions that arise in relation to density estimation and bump-hunting, such as:

(a) Whether to treat the problem in a non-theoretical EDA manner or with the use of some explicit

theory. (Section 3.)
(b) Whether to use MPL, or a more subjectivistic Bayesian method, or a window method or mixtures of parametric distributions, or moving averages, or other methods mentioned in the discussion of G & G´ (1980). (Section 4 herein.)
(c) If mixtures are used, whether they should be mixtures of normal distributions, Cauchy distributions, or others. (Section 5.)
(d) When mixtures are used, whether a bump should be redefined as the contribution from one of the basic, e.g., normal distributions, instead of the part of the density curve between two points of inflection and concave from below, with a corresponding definition for bivariate and multivariate problems. (Section 6.)
(e) Whether the procedural parameters should be determined by the data. (Section 7.)
(f) Whether the costs of calculation or on the other hand the reliability of the results should be the primary considerations. (Section 8.)
(g) Whether tail trouble can be avoided. (Section 9.)
(h) Whether the first or second method is better. (Section 10.)
(i) Whether to treat the data as discrete. (Section 11.)
(j) If the data are discrete whether to use a Fast Fourier Transform. (Section 12.)
(k) Whether other discrete orthogonal transforms might be useful. (Section 13.)
(l) Whether to alternate between methods to achieve convergence of iterative procedures. (Section 14.)
(m) Whether it is better to use γ or f in the definition of the roughness. (Section 15.)
(n) More generally whether to replace γ by f^{ξ} for some $\xi > 0$, where ξ is not necessarily equal to ½ or 1. (Section 15.)
(o) Whether such changes have much effect on the density estimates, on the bumps found, and on their evaluation. (Section 15.)
(p) Whether any given method can be conveniently generalized to more than one dimension. (Section 16.)
(q) In more than one dimension whether the roughness penalty should be defined so as to be invariant with respect to orthogonal transformations or with respect to some other group of transformations.

These problems will be discussed in turn.

3. HOW NON-THEORETICAL SHOULD YOU BE?

The problem of how non-theoretical to be is one that arises widely in statistics, and in science generally. It is difficult to lay down any general principles other than the type II principle of rationality which is the recommendation to maximize expected utility allowing for the costs of calculation and of thinking (Good, 1971a). The application of this principle in practice depends on judgment and on the urgency of the application, the accessibility of a computer, and on the degree of interaction with the client. There are occasions when the back of an envelope is all you need and others where many thousand dollars of computer time could reasonably be well spent, partly because it is negligible compared with the cost of collecting the data.

4. WHICH METHOD TO USE WHEN THE BACK OF AN ENVELOPE IS INADEQUATE

A complete answer to question (b) may never be known, but one advantage of methods that can be interpreted from a Bayesian point of view, such as the MPL method, is that they lead to the evaluation of individual bumps. The remainder of this paper, apart from the next section, will be concerned almost exclusively with the MPL method.

5. THE USE OF MIXTURES

David Rothman (1981) has recently written a program for estimation of a probability density as a mixture of normal distributions, and has applied it to the high-energy data analyzed by G & G´ (1980). These data were from the Lawrence Radiation Laboratory and we call them the LRL data. The data were in the form of a histogram with J = 172 and N = 25752. The notion of fitting data by mixtures of normal distributions goes back to Karl Pearson (1894).

Rothman wrote a program which he used for fitting the LRL data by adding one weighted normal distribution at a time, maximizing the likelihood at each stage, and he adds that the maximization is "perhaps only local". He obtained a mixture of ten normal curves with the following weights, means, and variances:

$$\begin{aligned}
&.003819N(331.5352,\ 25.9992)\\
+\ &.038648N(476.1119,\ 72.2125)\\
+\ &.120534N(767.2134,\ 38.0514)\\
+\ &.502803N(780.1729,\ 110.0794)\\
+\ &.008206N(949.2720,\ 13.9579)\\
+\ &.255571N(1229.6401,\ 158.8688)\\
+\ &.021107N(1272.8826,\ 34.2393)\\
+\ &.040141N(1683.5839,\ 100.6415)\\
+\ &.007182N(1898.9355,\ 41.7737)\\
+\ &.001990N(1976.0613,\ 11.3769).
\end{aligned}$$

He obtained $X^2 = 188.2$ with 171 - 29 = 142 degrees of freedom, P = .005. He says "I cannot find any more modes, nor can I find any to drop. This process took many days to complete, and I wish there were an easier way". The expected bin frequencies for his run are shown in row (f) of Table 1 together with the bumps which, however, are defined in accordance with Section 6. Rothman mentioned that G & G´ (1980) has one bad-looking bin at 1985 MeV. It is possible that G & G´ filtered out a bump one bin wide here through using only 271 terms in the Fourier series, or it may be that roughness should be measured by $\int (f'')^2\,dx$ instead of $\int (\gamma'')^2\,dx$ (see Section 15). But there is some reason to think that that bump is an artifact; see G & G´ (1980, p. 48, col. ii). Also thin bumps need thinner bins.

There is some theoretical reason to believe that the LRL data could be better fitted by a mixture of Cauchy distributions instead of normal distributions.

Rothman's fit to the chondrite data, which had previously been analyzed by Leonard (1978) and by G & G´, was

$$\begin{aligned} &.1818N(22.2575, \ .8726) \\ + &.4083N(33.4249, \ .7518) \\ + &.4099N(27.9854, \ 1.1653). \end{aligned}$$

For the chondrite data N was only 22.

Because of the large number of parameters it is difficult to apply the method of mixtures of normal distributions in two or more dimensions, except perhaps for bimodal densities in two dimensions, or if it is assumed that all the covariance matrices are the same. We understood that the latter method has been tried by NASA.

6. THE DEFINITION OF A BUMP

Good & Gaskins defined a bump as the part of a curve between two points of inflection and concave from below. But when the method of mixtures is used it is natural to define a bump as a stretch between the two points of inflection of one component of the mixture. Since the meanings of the bumps in row (f) are therefore different from those in the other rows, a different notation has also been used to represent them.

In more dimensions the points of inflection must be replaced by curves or manifolds of vanishing Gaussian curvature.

7. SHOULD THE DATA DETERMINE THE PROCEDURAL PARAMETERS OR HYPERPARAMETERS?

Given enough experience of similar problems in a given field of application it might be possible to decide on the values of hyperparameters, or to give them a clear-cut prior distribution. This could be regarded as an example of von Mises's simple and unsophisticated form of the empirical Bayes method (von Mises, 1942). The method adopted by Good & Gaskins is to choose the hyperparameters in the light of the current data by using non-Bayesian criteria of goodness of fit, such as X^2 (chi-squared), a Kolmogorov-Smirnov statistic, and a runs test that depends on the signs of "observed minus expected" within bins. [Note the slip on page 49 line 1 of G & G´ (1980), where the words "observed" and "expected" were interchanged.] The idea was, roughly speaking, that tail-area probabilities should be close to ½ for fits that are neither too close to the data and too rough, or too far from the data and too smooth.

8. HOW IMPORTANT ARE COSTS OF CALCULATION?

The question (e) has already been discussed in Section 3. Here we wish to mention only that the cost is relevant not just in each application. It is also relevant in purely statistical research, because the lower the cost the more experimentation can be done for sharpening a statistical procedure. This is one reason why we have recently been concerned with reducing the running time.

9. THE NESSIE EFFECT, OR CAN TAIL TROUBLE BE AVOIDED?

All methods of density estimation have tail trouble because it is usually dangerous to extrapolate far beyond the data points. This danger is present even when parametric methods are used because it is unusual for a parametric model to be reliable far into the tails unless the model is derived from a realistic physical model of the specific application.

One reason for working with the square root $\gamma(.)$ of $f(.)$ was that any estimate for γ would lead to positive values for $f(.)$. This idea was not quite as good as it seemed at first sight because in the extreme tail $\gamma(x)$ is apt to cross the x axis a few times so that the tails of $f(x)$ can have a few small spurious humps, a dragon-tail or Nessie effect. This occurs, for example, for the LRL data tabulated by G & G´ (1980, p. 43) and also in our present Table 1. This Nessie effect has, however, been of negligible importance so far. For example, for the LRL data, although the estimated density vanished a few times in the tails beyond the data the largest value of the estimate in the tails was utterly negligible. Tail trouble is discussed further in Appendix C of G & G´ (1980). [The comment there that $f'(x)$ is discontinuous when $\gamma(x)$ vanishes is incorrect; of course $f'(x) = 0$ at such points.]

10. IS THE FIRST OR SECOND METHOD BETTER?

Runs have been done by Dr. Gaskins for the LRL data using the first method, for comparison with G & G´ (1980) where the second method was adopted. In the notation of G & G´ (1980), Table 2 shows how the best value of α^* was selected. As α^* is increased the fitted density becomes smoother and further away from the data, so X^2 increases, and meanwhile the number of bumps decreases, but there are far more bumps than when the "second method" is used for which there were only 13 bumps (in the range of 172 bins of the actual data). The number of bumps is equal to half the number of points of inflection and these are detected by changes of sign in the sequence of second differences of the estimated density function at centers of bins.

One distinction between the results for $\alpha^* = 19$ and $\beta^* = .225$ (the optimal β^*) was seen by counting the number of bins where the second difference exceeds 1/100. There are 46 such bins for $\alpha^* = 19$ and only 31 for $\beta^* = .225$. This confirms the view of G & G´ (1971, p. 261) that the first method does not capture the concept of smoothness because of kinkiness in its output.

As a measure of roughness $\int(\gamma'')^2\,dx$ seems more reasonable than $\int(\gamma')^2\,dx$ because curvature depends on second derivatives. This view is further supported by a theorem given by Scott, Tapia & Thompson (1980, p. 821). They proved that a roughness penalty proportional to $\int(\gamma^{(k)})^2\,dx$,

Table 1. The LRL Data of N = 25,752 "Events" in 172 bins, each of width 10 MeV, showing four different fits.

(a)	285	295	305	315	325	335	345	355	365	375	385	395	405	415	425	435
(b)	5	11	17	21	15	17	23	25	30	22	36	29	33	43	54	55
(c)	5	8	^{i}[12	15	18	20	23	25]	27	29	32	35	38	^{ii}[42	46	50
(d)	5	9	[13	16]	17	19	[22	24	27]	28	[31]	34	37	43	[48	52
(e)	4	[10	15	18]	19	20	[22	24]	26	28	30	33	38	[43	48	52
(f)	5	8	12	17	[21	23]	24	24	24	25	27	31	36	40	45	50

(a)	445	455	465	475	485	495	505	515	525	535	545	555	565	575	585	595
(b)	59	44	58	66	59	55	67	75	82	98	94	85	92	102	113	122
(c)	53	55]	57	60	62	65	69	74	^{iii}[79	84]	88	93	99	106	116	128
(d)	54]	54	57	[60]	62	64	69	76	[82	88	92]	95	100	108	119	132
(e)	54]	56	57	59	61	64	69	75	[81	86	90]	93	98	105	115	127
(f)	55	59	[62	65	67]	70	71	73	76	79	83	88	95	104	115	128

(a)	605	615	625	635	645	655	665	675	685	695	705	715	725	735	745	755
(b)	153	155	193	197	207	258	305	332	318	378	457	540	592	646	773	787
(c)	143	160	179	200	225	253	285	321	364	413	471	533	^{iv}[598	658	709	744
(d)	149	166	[186]	206	230	261	[295]	327	360	405	463	[525	585	644	698	730
(e)	142	160	179	201	227	257	292	332	377	428	483	[540	595	645	686	715
(f)	144	161	181	203	227	253	283	316	356	403	459	522	591	660	721	767

(a)	765	775	785	795	805	815	825	835	845	855	865	875	885	895	905	915
(b)	783	695	774	759	692	559	557	499	431	421	353	315	343	306	262	265
(c)	761	759	742	709	664	611]	557	504	455	412	376	346	321	300	282	266
(d)	737]	729	[722	699	653]	597	549	501	455	416	378	349	[329]	306	284	268
(e)	729	730	717	692	657	613	565]	517	469	425	386	353	326	302	283	266
(f)	[791	789]	763	718	661	600	541	489	444	406	373	343	317	292	268	248

(a)	925	935	945	955	965	975	985	995	1005	1015	1025	1035	1045	1055	1065	1075
(b)	254	225	246	225	196	150	118	114	99	121	106	112	122	120	126	126
(c)	^{v}[251	235	218	199]	178	158	140	126	118	113	112	113	115	^{vi}[118	121	124
(d)	253	239	[226	209	186]	160	140	127	120	118	117	118	[120]	123	[125	128
(e)	251	[236	219	200	179]	158	138	123	114	111	110	113	116	[120	124	126
(f)	238	240	[243	225]	188	155	136	126	119	115	112	110	110	111	113	116

(a)	1085	1095	1105	1115	1125	1135	1145	1155	1165	1175	1185	1195	1205	1215	1225	1235
(b)	141	122	122	115	119	166	135	154	120	162	156	175	193	162	178	201
(c)	125]	126	127	129	132	136	140	144	148	154	161	169	176	184	193	^{vii}[202
(d)	129]	128	127	128	133	[141	144	146]	148	155	163	[171	178]	181	188	[198
(e)	127]	127	127	129	132	[136	140]	144	148	154	161	169	177	185	193	[202
(f)	119	123	128	132	137	141	145	150	154	158	162	167	173	[180	189	200

(a)	1245	1255	1265	1275	1285	1295	1305	1315	1325	1335	1345	1355	1365	1375	1385	1395
(b)	214	230	216	229	214	197	170	181	183	144	114	120	132	109	108	97
(c)	210	215	217	214	208	199	188	175	163]	150	138	128	120	113	107	100
(d)	208	214	215	213	206	195]	184	[175	164]	149	135	128	[122]	115	108	102
(e)	210	215	217	215	209	200	189	176	163]	150	137	128	120	113	107	[101
(f)	210]	218	[223	[222]	215	203]	188	173	158	145	134	125	117	110	103	97

(a)	1405	1415	1425	1435	1445	1455	1465	1475	1485	1495	1505	1515	1525	1535	1545	1555
(b)	102	89	71	92	58	65	55	53	40	42	46	47	37	49	38	29
(c)^{viii}	[94	88]	81	74	68	62	56	52	48	45	43	42	41	40	39	38
(d)	[96]	89	82	[77]	69	63	58	53	48	46	[45	44]	43	[42]	40	38
(e)	95	89	82	76]	69	62	56	51	47	45	44	[43	42	40]	38	37
(f)	91	85	79	74	68	64	59	55	51	48	45	43	41	40	39	38

(a)	1565	1575	1585	1595	1605	1615	1625	1635	1645	1655	1665	1675	1685	1695	1705	1715
(b)	34	42	45	42	40	59	42	35	41	35	48	41	47	49	37	40
(c)	39	40	^{ix}[41	42	43	44	43]	43	43	43	^{x}[43	43	43	42	41	39]
(d)	38	[41	42]	43	44	[45	44]	41	41	41	[43	43	44	43]	40	[38]
(e)	37	39	[42	44	45	45	44]	42	41	41	[43	44	45	44	41]	39
(f)	38	38	39	40	40	41	42	43	43	44	44	44	44	43	42	41

(a)	1725	1735	1745	1755	1765	1775	1785	1795	1805	1815	1825	1835	1845	1855	1865	1875
(b)	33	33	37	29	26	38	22	27	27	13	18	25	24	21	16	24
(c)	37	35	33	32	^{xi}[30	28]	27	25	23	22	21	21^{xii}	[21]	20	20	20
(d)	36	35	[33]	31	30	[30]	27	[26	24]	21	21	[22	22]	21	20	[20]
(e)	36	34	33	32	[31	29	28	25]	22	20	20	[21	22	21]	20	20
(f)	39	37	35	33	30	28	25	23	21	20	19	19	19	20	21	22

(a)	1885	1895	1905	1915	1925	1935	1945	1955	1965	1975	1985	1995
(b)	14	23	21	17	17	21	10	14	18	16	21	6
(c)	20^{xiii}	[20	19	19	18	18	17	16	15	13	9]	6
(d)	19	[20	19]	18	[18	17]	15	15	[15	13	11]	6
(e)	20	[20	20	19	18]	16	15	16	[17	17	13]	7
(f)	22	[22	21]	19	17	14	12	12	17	[21]	16	6

Notes for Table 1. (a) MeV; (b) Observed bin frequencies, total 25752; (c) Estimated frequencies using $\beta^* = .225$, total 25752 (see Section 10); (d) Estimated frequencies using $\alpha^* = 19$, total 25748 (Section 10); (e) Estimated frequencies when the roughness penalty is proportional to $\int (f'')^2\,dx$, total 25752 (Section 15); (f) Rothman's fit, total 25744 (Sections 5 and 6).

The bumps have been marked by brackets for the fits (c), (d), and (e). The definition of the bumps for row (f) is given in Section 6 and they have been marked in a special notation.

where $\gamma^{(k)}$ denotes the kth derivative of $\gamma(x)$, leads to a polynomial spline of degree 2k. With k = 1 the spline would not have continuous second derivatives, but with k = 2 the second derivative would be continuous. Thus the use of the first method (k = 1) is theoretically suspect when bump-hunting, although the argument is less convincing for density estimation as such.

For comparison of the first and second methods (k = 1 and k = 2) their expected bin frequencies, and the estimated bumps, are shown in rows (c) and (d) of Table 1. Row (c) corresponds to $\beta^* = .225$ and row (d) to $\alpha^* = 19$.

When a truncated orthonormal expansion of $\gamma(x)$ is used all derivatives are continuous. But when infinite expansions are used discontinuities arise. The truncation is analogous to the rounding off of corners.

TABLE 2. SOME RUNS FOR THE LRL DATA.

α^*	X^2	$P(X^2)$	$DN^{\frac{1}{2}}$	$P(DN)^{\frac{1}{2}}$	S	P(S)	R_0	Bumps
10	125	.004	.732	.344	99	.023	.98	41
15	151	.047	1.000	.731	99	.023	.71	35
18	165	.401	1.144	.855	97	.046	.47	35
19	169	.499	1.191	.883	93	.142	.06	35
20	174	.594	1.237	.907	93	.142	.27	34
21	178	.681	1.281	.925	93	.142	.35	34
25	195	.912	1.453	.971	90	.270	.70	33
30	217	.991	1.659	.992	86	.500	.92	32
β^* =.225	191	.869	.610	.151	95	.084	.37	13

Notes for Table 2. (i) X^2 refers to the 172 real bins and has 170 degrees of freedom. (ii) D is the Kolmogorov-Smirnov one-sample statistic, while N = 25752 is the sample size. (iii) S is the number of changes of sign in the sequence of signs of "expected minus observed". Given the null hypothesis it has mean 85.5 and S.D. 6.5. (iv) We take a single tail-area $P \leqslant \frac{1}{2}$ if the corresponding statistic by itself leads one to suppose that the fit is too rough. (v) R_0, defined by G & G´, is a statistic which is derived from $P(X^2)$, $P(D\sqrt{N})$, and P(S), and is optimal when close to .5. The "best" value of α^* is 19. (vi) The numbers of bumps refer to the range of 172 real bins. (vii) Most of this table was computed by Dr. Gaskins with the help of programs used for G & G´. Almost identical results were obtained for the case $\beta^* = .225$, by means of the circularized discrete procedure (FFTMPL: see Section 12).

11. SHOULD CONTINUOUS DATA BE TREATED DISCRETELY?

Continuous data must always be treated discretely, but the question is whether to treat them even more discretely for simplicity, for example, by rounding to a thousandth instead of a billionth of the range. This rounding reduces the data to a histogram consisting of say t "bins". The derivatives of f(x) can then be replaced by finite differences. Apparently Scott, Tapia & Thompson (1980) were the first to use discretization for MPL and the method is available with the name NDMPLE in IMSL. NDMPLE runs faster than previous programs used by G & G´. It uses $\int(f'')^2\,dx$ as a measure of roughness, but this has little to do with the running time.

12. THE USE OF A FAST FOURIER TRANSFORM

For discretized data use of a Fast Fourier Transform (FFT) is natural. The FFT has a large "literature"; see, for example, Good (1958, 1960, 1971b), Cooley & Tukey (1965), Brigham (1974). G & G´ did not use an FFT because their primary aim was to achieve some results without worrying much about running time. When Fourier methods are used, continuous or discrete, it is necessary to circularize the data. Then one must extend the range far enough so that the density is judged to become negligible in the tails. This was done for the LRL and chondrite data by G & G´ (1980). After extending the range, making the data discrete, and circularizing, we have "observations" at say t points which we may regard as lying at the t vertices of a regular polygon. (For data already circular see Mardia, 1972.) We scale the data to make the period equal to 1.

One definition of a discrete Fourier transform (DFT) is given by the formula

$$a_s{}^* = t^{-\frac{1}{2}} \sum_{r=0}^{t-1} a_r \omega^{rs} \quad (0 \leqslant s \leqslant t-1) \qquad (12.1)$$

where $\omega = e^{2\pi i/t}$. The inversion formula is

$$a_r = t^{-\frac{1}{2}} \sum_{r=0}^{t-1} a_s{}^* \omega^{-rs} \quad (0 \leqslant r \leqslant t-1). \qquad (12.2)$$

There are analogues of several formulae familiar in the theory of ordinary Fourier transforms; for example, Good (1962). An FFT is a fast method for computing a DFT, the running time being proportional to t log t. There are now even faster FFT's with running time proportional to t; see, for example, McClellan & Rader (1979), Rice (1977), and Winograd (1976).

We now discuss some of the details of the discrete circularized method with period 1. Divide the range into t equal intervals or bins, each of width h = 1/t. Let the middle of the rth bin be called $x_r = (r - \frac{1}{2})/t$ and let $p_r \approx f(x_r)h$. Take h small enough to justify assuming that $\sum_{r=0}^{t-1} p_r = 1$. Let $\pi_r = p_r^{\frac{1}{2}} \approx [f(x_r)h]^{\frac{1}{2}} = \gamma(x_r)h^{\frac{1}{2}}$, though in Section 15 we write more generally $\pi_r = p_r{}^{\xi}$. Write p. for (p_r). We want to maximize Ω(p.) subject to the constraint

$$\sum_{r=0}^{t-1} \pi_r{}^2 = 1, \qquad (12.3)$$

where

$$\Omega(\text{p.}) = \sum n_r \log \pi_r{}^2 - \Phi(\text{p.}). \qquad (12.4)$$

For generality we take the roughness penalty

$$\Phi(\text{p.}) = A\sum(\pi_{r+1} - \pi_{r-1})^2 + B\sum(\pi_{r+1} - 2\pi_r + \pi_{r-1})^2$$

$$= 2A(1 - \psi_2) + 2B(3 + \psi_2 - 4\psi_1) \qquad (12.5 \;\&\; 6)$$

where

$$\psi_q = \sum_{r=0}^{t-1} \pi_r \pi_{r+q}. \qquad (12.7)$$

The approximations for the roughness are

$$\int [\gamma'(x)]^2\, dx \approx \tfrac{1}{4} h^{-1} \sum [\gamma(x_{r+1}) - \gamma(x_r)]^2$$
$$= \tfrac{1}{4} h^{-2} \sum (p_{r+1}^{\frac{1}{2}} - p_{r-1}^{\frac{1}{2}})^2, \qquad (12.8)$$

$$\int [\gamma''(x)]^2 \approx h^{-3} \sum [\gamma(x_{r+1}) - 2\gamma(x_r) + \gamma(x_{r-1})]^2$$
$$= h^{-4} \sum (p_{r+1}^{\frac{1}{2}} - 2p_r^{\frac{1}{2}} + p_{r-1}^{\frac{1}{2}})^2 \qquad (12.9)$$

so that

$$A \approx \alpha h^{-2},\ B \approx \beta h^{-4}. \qquad (12.10)$$

The Lagrange equations reduce to

$$\frac{n_r}{\pi_r} - A(2\pi_r - \pi_{r+2} - \pi_{r-2})$$
$$- B(\pi_{r+2} - 4\pi_{r+1} + 6\pi_r - 4\pi_{r-1} + \pi_{r-2}) = \lambda \pi_r \quad (12.11)$$

where λ is an "undetermined multiplier". By multiplying by π_r and summing we find that

$$\lambda = N - 2A(1 - \psi_2) - 2B(3 - 4\psi_1 + \psi_2). \qquad (12.12)$$

It might be practicable to solve these equations iteratively, but we have applied the DFT first because the Fourier coefficients will be small for s far from 0. We have, easily, after writing η_s for $\pi s/t$,

$$2\pi_r - \pi_{r+2} - \pi_{r-2} = 4t^{-\frac{1}{2}} \sum_{s=0}^{t-1} \pi_s^* \omega^{-rs} \sin^2(2\eta_s),$$

$$\pi_{r+2} - 4\pi_{r+1} + 6\pi_r - 4\pi_{r-1} + \pi_{r-2}$$
$$= 16t^{-\frac{1}{2}} \sum_{s=0}^{t-1} \pi_s^* \omega^{-rs} \sin^4 \eta_s. \qquad (12.13\ \&\ 14)$$

The DFT of (12.8) gives

$$t^{-\frac{1}{2}} \sum_r (n_r/\pi_r)\omega^{rs} - 4A\pi_s^* \sin^2(2\eta_s)$$
$$- 16B\pi_s^* \sin^4 \eta_s = \lambda \pi_s^*. \qquad (12.15)$$

We can now try the "iteration":

$$\pi_s^* [\lambda + 4A\sin^2(2\eta_s) + 16B \sin^4 \eta_s]$$
$$= t^{-\frac{1}{2}} \sum_r (n_r/\pi_r)\omega^{rs} \qquad (12.16)$$

$$\lambda = N - 2A(1 - \psi_2) - 2B(3 - 4\psi_1 + \psi_2). \quad (12.17)$$

Assume (π_r), compute λ and π_s^*; then recompute (π_r) by the inverse DFT, etc.

We have obtained convergence with this method with B = 0 and with A = 0. We then adjusted the hyperparameter to obtain a suitable value for a criterion such as chi-squared, as in G & G´ (1980). By applying this method, which we call FFTMPL, to the LRL data with $\alpha^* = 0$, $\beta^* = .225$, the run took only 5.0 seconds, when there were eight iterations, whereas the run described by G & G´ (1980) takes about four seconds per iteration. The results were almost identical: expected frequencies in the bins never differed by more than 2 for the two runs. Runs were also done for the chondrite data, in which N was only 22, and the results were identical with those obtained by G & G´ when fine enough grids were used. A comparison of running time for NDMPLE and FFTMPL was made for the chondrite data (N = 22) and for a sample with N = 1000 from the population $\frac{1}{2}N(.25, 1/256) + \frac{1}{2}N(.75, 1/256)$. The results are shown in Tables 3 and 4. For the case N = 22 FFTMPL was two or three times as fast, and for N = 1000 it was four to eight times as fast. The documentation of NDMPLE states that its cost is proportional to t^2 whereas we know that the running time of a standard FFT is proportional to $t \log t$. Judging by Tables 3 and 4, the running time of NDMPLE increases with N, but not for FFTMPL. In Tables 3 and 4 the running times are not smooth functions of t because of the nature of the systems software. The documentation of the software states that a precise comparison of running times cannot be made in a multiprogramming environment, but the evidence from Table 4 seems conclusive.

TABLE 3. A COMPARISON OF RUNNING TIMES OF NDMPLE AND FFTMPL FOR THE CHONDRITE DATA (N = 22)

NDMPLE			FFTMPL		
t	#its	Time	t	#its	Time
64	8	.38	64	12	.20
128	8	.81	128	11	.38
256	12	2.98	256	9	.66
512	11	4.79	512	11	1.65
1024	30 (b)	-	1024	30 (b)	-

TABLE 4. A COMPARISON FOR A SAMPLE WITH N = 1000.

NDMPLE			FFTMPL		
t	#its	Time	t	#its	Time
64	6	1.34	64	10	.17
128	6	1.79	128	8	.29
256	7	4.50	256	7	.53
512	8	5.40	512	8	1.26
1024	30 (b)	-	1024	30 (b)	-

Notes for Tables 3 and 4. (a) The three columns represent the numbers of grid points, the numbers of iterations, and the execution times in seconds excluding input and output times. NDMPLE and FFTMPL use different roughness penalties and for these tables we did not attempt to optimize the hyperparameters. (b) For t = 1024 convergence was not attained in thirty iterations.

13. OTHER DISCRETE ORTHOGONAL TRANSFORMS

There are other systems of orthogonal functions, such as Walsh's and Haar's, for which see, for example, Mathematical Society of Japan (1954/1977), Section 312C, and Fino & Algazi (1977). These systems might be useful for our problem.

14. CONVERGENCE OF THE ITERATIVE PROCEDURES

The information about convergence of the iterative methods is so far empirical. From Tables 3 and 4 it seems that, for FFTMPL and NDMPLE, convergence can require more than 30 iterations or perhaps can fail to occur if the number t of grid points is large enough. For example, with N = 22 we attained quick convergence with t = 512 but not with t = 1024.

In G & G´ (1980), convergence was usually attained when the Fourier method was used, and more reliably but at greater cost when Hermite expansions were used. Alternation between these two methods was sometimes beneficial. For FFTMPL it will probably be advantageous for reaching convergence if t is gradually increased, the output of one run being used for the input of the next one. This device was used by G & G´ (1980) and in NDMPLE.

15. IS IT BETTER TO USE f^{ξ} IN PLACE OF $f^{\frac{1}{2}}$?

There were historical reasons for working with $\gamma(x)$ rather than with $f(x)$. One of them is that $\int\gamma^2\,dx = 1$, so that $\gamma(.)$ is a point in Hilbert space. This theoretical elegance would be lost if we worked directly with $f(.)$, but in practice every sample has finite support. So if $f(.)$ is continuous it too can be regarded as of integrable square. Hence it would be dogmatic to insist on the superiority of $\int(\gamma'')^2\,dx$ as a measure of roughness rather than $\int(f'')^2\,dx$. Or we could try using the more general definition:

$$\text{Roughness} = \int[(f^{\xi})'']^2\,dx,$$

$$\text{Roughness Penalty} = \beta \times \text{Roughness},$$

where ξ is some positive constant. To make this generalization in the formulae of Section 12, interpret γ as f^{ξ}, replace (12.3) by $\sum \pi_r^{1/\xi} = 1$, and make the other changes corresponding to the bivariate formulae in Section 16.

Does a switch from γ to f much affect the density estimates or the location and evaluation of bumps? With the help of NDMPLE, which uses the roughness $\int(f'')^2\,dx$, we obtained estimates for the LRL data. By circumspectly selecting the hyperparameter we forced X^2 to have the value 191 that had been obtained by G & G´ using the roughness $\int(\gamma'')^2\,dx$ with $\beta^* = .225$ (see the last row of Table 2). The expected frequencies of these two runs are shown, with their bumps, in rows (c) and (e) of Table 1. The bumps are much the same except that the first and last bumps have been split into pairs, and there is one new very weak bump at 1515 to 1535 MeV.

We are now trying to develop programs using the roughness $\int[(f^{\xi})'']^2$ and the univariate forms of (16.8) and (16.11).

16. MORE DIMENSIONS AND TOROIDALIZATION

Both NDMPLE and FFTMPL should be practicable for bivariate data, and we hope to compare them.

Let the observations be (x_i, y_i) $(i = 1, 2, \ldots, N)$. Instead of circularizing the data we now "toroidalize" them. Suppose the numbers of grid points for the x and y coordinates are t and t´ and the widths of the bins are $h = 1/t$ and $h' = 1/t'$. Let the number of observations in bin (r, r') be $n_{r,r'}$ and let $x_r = (r - \frac{1}{2})/t$, $x_{r'} = (r' - \frac{1}{2})/t'$. The bivariate MPL method consists of a search for a bivariate density function $f(x, y)$ that maximizes

$$\Omega(f) = \sum_{i=1}^{N} \log f(x_i, y_i) - \Phi(f),$$

$$\approx \sum n_{r,r'} \log f(x_r, x_{r'}) - \Phi(f). \tag{16.1}$$

The integral of $f(x, y)$ within a bin (r, r') is say $p_{r,r'} \approx f(x_r, x_{r'})h\,h'$.

In G & G´ (1971, 1972) the roughness for two-dimensional data, using the "second method", is defined as

$$\iint[(\frac{\partial^2\gamma}{\partial x^2})^2 + (\frac{\partial^2\gamma}{\partial y^2})^2]dx\,dy, \tag{16.2}$$

though the two terms in the integrand could have distinct coefficients so that the roughness penalty is

$$\beta\iint(\frac{\partial^2\gamma}{\partial x^2})^2\,dx\,dy + \beta'\iint(\frac{\partial^2\gamma}{\partial y^2})^2\,dx\,dy. \tag{16.3}$$

For the sake of generality we redefine $\gamma(x, y)$ as $[f(x, y)]^{\xi}$. Then

$$\iint[\gamma(x, y)]^{1/\xi}\,dx\,dy = 1. \tag{16.4}$$

Let $\pi_{r,r'} = p_{r,r'}^{\xi} \approx [f(x_r, x_{r'})h\,h']^{\xi}$

$$= \gamma(x_r, x_{r'})(h\,h')^{\xi},$$

so

$$\sum_{r,r'}\pi_{r,r'}^{1/\xi} = 1. \tag{16.5}$$

In discrete terms we wish to maximize

$$\Omega = \sum n_{r,r'} \log \pi_{r,r'}^{1/\xi} - B\sum(\pi_{r,r'+1} - 2\pi_{r,r'} + \pi_{r,r'-1})^2 - B'\sum(\pi_{r+1,r'} - 2\pi_{r,r'} + \pi_{r-1,r'})^2 \tag{16.6}$$

subject to the constraint (16.5). Here

$$B \approx \beta(h\,h')^{1-2\xi}h^{-4},\quad B' \approx \beta'(h\,h')^{1-2\xi}h'^{-4} \tag{16.7}$$

and β and β' are mathematically independent of h and h´. Moreover $\beta = \beta'$ if the roughness is defined by (16.2). The Lagrange equations, if the $\pi_{r,r'}$ are positive, are

$$\frac{n_{r,r'}}{\xi\pi_{r,r'}} - 2B(\pi_{r+2,r'} - 4\pi_{r+1,r'} + 6\pi_{r,r'} - 4\pi_{r-1,r'} + \pi_{r-2,r'}) - 2B'(\ldots) = \frac{\lambda}{\xi}\pi_{r,r'}^{\frac{1}{\xi}-1} \tag{16.8}$$

where

$$\lambda = N - 4B\xi(3\psi_{0,0} - 4\psi_{1,0} + \psi_{2,0})$$
$$- 4B'\xi(3\psi_{0,0} - 4\psi_{0,1} + \psi_{0,2}), \qquad (16.9)$$

and $\psi_{\mu,\mu'} = \sum_{r,r'}\pi_{r,r'}\pi_{r+\mu,r'+\mu'}$. Let

$$a^*_{s,s'} = (t\,t')^{-\frac{1}{2}}\sum_{r,r'}a_{r,r'}\omega^{rs}\omega'^{r's'} \qquad (16.10)$$

where $\omega' = e^{2\pi i/t'}$. The DFT of (16.8) gives, with self-explanatory notations,

$$(t\,t')^{-\frac{1}{2}}\sum_{r,r'}(n_{r,r'}/\pi_{r,r'})\omega^{rs}\omega'^{r's'}$$
$$- 32B\xi\pi^*_{s,s'}\sin^4\eta_s - 32B'\xi\pi^*_{s,s'}\sin^4\eta'_{s'}$$
$$= \lambda(\pi^{-1+1/\xi})^*_{s,s'}. \qquad (16.11)$$

When $\xi = \frac{1}{2}$ we can solve for $\pi^*_{s,s'}$ for all (s, s'), and proceed with the iterative process of Section 12. When $\xi = 1$, where $\pi_{r,r'}$ means $p_{r,r'}$, the right side of (16.11) reduces to 0 when $(s, s') \neq (0, 0)$ and to $\lambda(t\,t')^{\frac{1}{2}}$ when $(s, s') = (0, 0)$. Hence

$$\lambda = \frac{1}{t\,t'}\sum_{r,r'}\frac{n_{r,r'}}{p_{r,r'}} \qquad (16.12)$$

and we have $t\,t' - 1$ other equations together with $p^*_{00} = (t\,t')^{-\frac{1}{2}}$ for solving for $p^*_{s,s'}$. Curiously, λ drops out when calculating the p's (when $\xi = 1$) though the equation obtained by eliminating λ between (16.9) and (16.12) might be useful for normalizing the p's at intermediate stages.

When ξ is neither $\frac{1}{2}$ nor 1, equation (16.11) contains DFT's of two distinct vectors $(\pi_{r,r'})$ and $(\pi^{-1+1/\xi}_{r,r'})$ so the iterative procedure has to be changed. For example, one might solve for one of the two DFT's, then deduce new values for the $\pi_{r,r}$'s and for the other DFT, etc. In all procedures our aim is to find values of B and B′ that would make some statistic such as X^2 have its typical value such as its expected value or its mode. If the roughness were measured in terms of the kth derivative the fourth power of the sine function in (16.11) would be replaced by the $(2k)^{th}$ power.

17. INVARIANT MULTIVARIATE ROUGHNESS PENALTIES

A tensorial definition of multivariate roughness penalties was given in G & G′ (1971) to achieve invariance under a transformation of coordinates. The invariance was there based on the concept of utility. Some further roughness formulae are given by Good (1981) that are invariant under orthogonal transformations, for example, one with integrand the square of the Laplacean. Such formulae are appropriate when x and y are spacial directions rather than say blood temperature and pressure. Only one of the proposed orthogonally invariant definitions of the penalty was additive with respect to the coordinate axes, in being the sum total roughness in each direction when these are independent. This one had the integrand $[(\partial\gamma/\partial x) + (\partial\gamma/\partial y)]^2$, which, however, corresponds to the "first method", that is, it depends only on first derivatives. Perhaps the optimal roughness penalty should depend on the application, but the one considered in our Section 16 might be reasonably general-purpose.

18. A COMPARISON OF THE ACCURACY OF METHODS

For the LRL data we hope to compare the accuracy of methods by a split-sample technique.

19. This work was supported by H.E.W., N.I.H. Grant #R01 GM 18770.

REFERENCES

Brigham, E.O. (1974). *The Fast Fourier Transform.*

Cooley, J.W. & Tukey, J.W. (1965). An algorithm for the machine calculation of complex Fourier series. *Mathematics of Computation 19,* 297-301.

Fino, B.J. & Algazi, V.R. (1977). A unified treatment of discrete fast unitary transforms. *SIAM Journal Comput. 6,* 700-717.

Good, I.J. *JRSSB 20* (1958), 361-372; *22* (1960), 372-375; *Amer. Math. Month. 69* (1962), 259-266; Twenty-seven principles of rationality. In *Foundations of Statistical Inference* (ed. Sprott & Godambe, 1971a), 124-127; *IEEE Trans. Comp. C20* (1971b), 310-317; chap. in *Bayesian Statistics* (ed. Bernardo, Lindley & Smith, 1980), and in *Trabajos de Estadística Inv. Oper.; J. Statist. Comp. Simul. 12* (1981), 142-144;

Good, I.J. & Gaskins, R.A. Nonparametric roughness penalties for probability densities. *Biometrika 58* (1971), 255-277. Global nonparametric estimation of probability densities. *Va. J. Sc. 23* (1972), 171-193. Density estimation and bump-hunting by the penalized likelihood method exemplified by scattering and meteorite data. *JASA 75* (1980), 42-73 (with discussion).

Leonard, Tom (1978). Density estimation, stochastic processes, and prior information. *JRSS B, 40,* 113-146.

Mardia, K.V.(1972). *Statistics of Directional Data.*

Mathematical Society of Japan (1954/1977). *Encyclopedic Dictionary of Mathematics* (American Mathematical Society; tr. from the Japanese).

McClellan, J.H. & Rader, C.M. (1979). *Number Theory in Digital Processing.*

Mises, R. von (1942). On the correct use of Bayes' formula. *Ann. Math. Statist. 13,* 156-165.

Pearson, Karl (1894). Contributions to the mathematical theory of evolution. *Philos. Trans. Roy. Soc. 185,* 71-110.

Rice, B.F. (1977). Faster integral convolutions. (Mimeographed, July 19; pp. 32).

Rothman, David (1981 Jan.). Private communication.

Scott, David W., Tapia, R.A. & Thompson, J.R. (1980). Nonparametric probability density estimation by discrete maximum penalized-likelihood criteria. *The Annals of Statistics 8,* 820-832.

Winograd, S. (1976). On computing the discrete Fourier transform. *Proc. National Acad. Sc., USA, 73,* 1005-1006.

Michael Tarter - Brunk and Fellner have some techniques for pre-transforming data. That's a very Bayesian kind of approach in which one could take a cumulative, especially if the components are known, transform the data on the basis of that, and essentially measure what is left over. This would have to be on the interval (0,1), after using the cumulative as a transformation. You might try that sort of thing, I don't know whether you've done it.

Good - If you give me the reference I'd like to look that up. I'm not quite sure that I fully understand it.

Tarter - Use the empirical distribution of x as a transformation ... then you don't run into your tail problems.

Good - Yes, but somehow it seems a bit glib to me because I think in applied problems there has to be tail trouble. Because if you assume there's no tail trouble it means you're relying on extrapolation.

Tarter - If your bump is in the middle and not near the tails you might as well do something so that at least you don't have to see the tail trouble. It scrunches the tails together so you don't see them.

Good - So it sweeps the tails up against the walls, so to speak, doesn't it?

Nancy Mann - Can I ask what Dave Rothman wrote to you about?

Good - He fitted a mixture of normal distributions one normal at a time until he was satisfied. About ten normals was the best number for this particular LRL data. His chi-squared had a tail-area probability of .005 which he didn't like much, but adding an eleventh normal made things worse because of the three additional degrees of freedom. I suspect that a mixture of Cauchy distributions might be more appropriate for the high energy physics data, for theoretical physics reasons. I intend to write to him and to suggest he try it if he wants to. It seems to be quite a lot of work to use mixtures.

David Scott - I enjoyed the numerical computations very much. I might mention there's possibly a reason in two dimensions for a penalty function that uses the fourth derivatives. If you look at the Hilbert space you need to have higher order derivatives for every additional dimension. So the second one may do funny things, I'm not quite sure what. For two dimensions, I think, you need at least the 4th, maybe even the 6th, it becomes quite odd-shaped otherwise.

Good - I'd like to know more about that; of course, there's no difficulty in principle in using 4th derivatives.

Scott - No, not at all. You're using approximations. The other comment - in my poster session it turns out that, for parametric models, the use of discretized data might actually be advantageous. It improves the mean-squared error, so it might not be suboptimal to discretize data when you're using nonparametric methods. Using discrete data up to a point actually decreases the mean-squared error term. At least that's true for kernel density estimation. It's probably true for the MPL method also.

Good - That is rather surprising because you seem to be throwing away information when you discretize. Discretizing is, of course, making a histogram.

Scott - The discretization grid is very fine.

Good - All observations discretize to seven decimal places, so to speak. It's simply a question of whether you use three decimal places or seven. When you use two or three you call it discretizing it, when you use seven it's more continuous. How many are used in your application?

Scott - It depends on the sample size. But for kernel methods, if you have a sample of size 1000 you could discretize to a fifth or a fourth [?] and still get estimates that are better.

Good - What do you mean by a fourth?

Scott - Make a histogram and that's the bin width. I'll put all the data at the center. That's the bin width.

Good - So how many bins do you then have? 250?

Scott - Say, 50 to 100.

Good - For some reason or another when you throw information away you improve the mean-squared error of the estimates?

Scott - It's black magic.

Tarter - I think you have to be particularly careful because data can be naturally discretized by data collection procedures. So if you yourself discretize you run into more patterns and peaks. The two forms of discretization interact with each other and provide artificial bumps.

Good - What if you use the discrete Fourier transform? Would you get the same trouble, do you think?

Tarter - You need the equivalent of Sheppard's corrections.

Mann - What are Sheppard's corrections?

Tarter - The Sheppard's corrections were once very well known by statisticians before they stopped using tabular things. They were corrections that you used on a frequency table when you computed the moments from a frequency table to get out the discretization produced by the binning. In an elementary statistics textbook the first exercise requires you to compute the mean and the variance from a frequency table.

Sidney Yakowitz - I was struck by a resemblance of your equations to ones I had seen in a different context. A roughness penalty reminds me of something called Tychinoff regularization, which is used for solving ill-posed linear problems. In fact, Grace Wahba makes use of this approach in some of her work. Do you know about this connection, and have some comments?

Good - No, I probably ought to know about it but I don't. I'm not well up on that kind of analysis.

Yakowitz - Well, conceivably there might be some use to looking into that. Without knowing the details of this field myself, I know that it does have some bearing on splines, and so forth.

Good - Yes, it sounds very much like it. And probably Sobolev spaces. When I read a paper by Scott, Tapia, and Thompson where they talk about Sobolev spaces I felt rather like the person who'd been talking prose all his life, because I'd never heard of a Sobolev space and I'd been using one.

Scott - It was deMontricher.

Good - It leads you into fairly difficult analysis.

Yakowitz - The need for this stuff in the ill-posed problem context seems to me very close to what you're trying to do, that is, not to have a terribly rough density function.

Good - It is close and, as I mentioned, Scott, Tapia, and Thompson point out that it is equivalent to the use of splines. The whole interpretation, of course, is different. The theory of splines appears to have originated in numerical analysis rather than probability theory, as far as I know. Maybe if one traced the history right back to its beginnings that might not be so. Do you know, David?

Scott - I'm sorry, I'm daydreaming.

Tarter - It originated with approximation theory.

Good - Yes, approximation theory, but that was not particularly related to probability presumably.

Parzen - Let me make a remark on that point. What has been shown starting with the work of Wahba and Kimeldorf is that there is an isomorphism between solutions in approximation theory, solutions in filtering theory of engineers, and time series parameter estimation. So that although you have these people in these different fields motivated by different criteria formulating the optimization problem they're solving, the solution to the three problems are equivalent and the consequence of that is the following: when people take data and pass a smooth curve through it, essentially they are implicitly assuming a probability model for the noise and for the, so to speak, Bayesian prior on the family of signals which they're making observations on. They ought to be aware of the fact that a truly numerical approximation procedure is essentially a solution to a statistical parameter estimation problem.

Good - I agree they should all be aware of it. I think statisticians all ought to be aware of the fact that they're all Bayesians.

Parzen - I don't agree with that at all.

Tarter - Don't reduce a problem to a previously unsolved problem.

Good - My remark is subject to a definition of a Bayesian.

John Hartigan - I have several questions, Jack. I'll say them all and you can respond to them all. They're very short. The first one: Did you mention that one way of viewing your criterion function was that it was the posterior density of the unknown density? I've seen that before in your papers and I've always wondered what measure the density was taking, because I know it's very difficult to define densities on function spaces.

Good - No.

Hartigan - Next question: I would think that asymptotically, at least, your method must be equivalent to some kind of kernel averaging of the data so that you could find your solutions, but once you have a lot of data you would think that the density that you obtain would be equivalent to some kind of kernel estimate in the neighborhood of each point. Although the kernel's shape might change as you move from point to point. But it would be very interesting to discover what that kernel was because then you could get some understanding of how your method should perform compared to the kernel method and what the kernel widths are compared to your B and so on. I think it would be nice to establish that equivalence, if you could.

Thirdly, I was curious that you discussed several powers of f as alternatives but you didn't discuss log(f), and I would think that since log(f) is already sitting there in the likelihood function, it would be natural to get log(f) into the penalty function as well. My last remark is, it's true that one always gets problems in the tails, but if the only thing that's worrying you in the tails is those messy little bumps, you'd think that a suitable choice of a penalty function would simply make sure the tails look nice. There must be some way to choose a penalty function so that the tails look nice. That completes the four questions.

Good - Well, I think if one really maximized $\omega(f)$, there wouldn't be any tail trouble in the Nessle sense. I think it's that in the process of doing the maximization we have to assume that gamma was positive in order to carry out the analysis. That wasn't quite true, you see. Gamma could go negative in our iterative procedure, though only just. I think one would clearly increase the score if one simply smoothed off the tail, and this would probably agree with what Scott, Tapia, and Thompson said. Even if you smoothed by eye! If you had gotten a wiggly tail and you smoothed it by eye you'd probably increase $\omega(f)$.

Hartigan - Are those bumps connected with individual observations?

Good - No, they're outside the range of the observations. They're outside, they're irrelevant to everything. In the LRL case, they're outside the observations and for that size of sample it's only a theoretical difficulty. For smaller samples, for a sample of size one for example, it's worse. But then you'd never be interested in a sample of size one. I'm not sure whether the tail trouble is a practical difficulty. Now the logarithm. If you take the derivative of the logarithm of f, you get f'/f. Square that. Take the expectation and get $\int (f')^2/f$. And that is equivalent to the square of gamma prime. And it's also Fisher's measure of information regarding the locality of the curve. This is all related to why we originally got interested in

$$\int [\gamma']^2 dx = \tfrac{1}{4}\int \frac{f'^2}{f}\, dx \quad .$$

That's where that 4 came from. It's not necessarily doing something entirely different if you take derivatives of the logarithm. [For contingency tables a roughness penalty proportional to entropy, which uses logarithms of course, was suggested in Ann. Math. Stat. 34 (1963), p. 931, and used in a thesis at VPI and SU by W. Palz (1977).]

Density in function space. There are respectable definitions. Our definition is not respectable because our prior is not a proper density in function space. It doesn't integrate to one; it doesn't integrate to infinity; it integrates to zero. Almost all density functions are infinitely rough.

Hartigan - It depends what your measure is.

Good - It is a measure in a reasonable sense. For example, if you expand a Hermite series to a finite number n of terms then you're on the unit sphere. Now put a uniform distribution over the sphere. Then your measure is Lebesgue on the sphere. If you proceed to the limit as $n \to \infty$ you get a successively smaller and smaller set of functions for any given roughness. For any given roughness you get less and less measure. It's a measure in that sense without going into respectable things like Wiener measure which Manny Parzen can tell us all about. That's a rough answer; you can give a more respectable prior in function space but it wouldn't use the roughness penalty.

Hartigan - Your rough answer is "It's not really a density."

Good - It is not a proper density function. We pay the price for that because Bayesian-wise we can't compare hyperparameters in a purely Bayesian manner. It is a theoretical disadvantage of that particular roughness penalty.

Hartigan - Presumably you have to find a measure that is carried by "smooth" functions. You'd only consider your density with respect to that measure carried by smooth functions. And it would be a nice smooth thing on those functions. Although indeed most densities would have probability zero, with respect to the underlying measure.

Good - The trouble is, how smooth? Every density function, if it has a continuous derivative or something, can be called smooth.

Hartigan - You'd say they're smooth if

$$\int [f']^2 dx < \infty$$

You'd define a measure carried by the function with that $< \infty$. You could do that, and then your density, you'd hope, might be reasonable.

Good - I'm not sure that that would escape the difficulty, but it would be worth trying. I'm sure that some more respectable measures can be used; in fact, I think Tom Leonard has used a proper measure.

Hartigan - The other question was about the kernel.

Good - The two methods are equivalent in the sense that they are both consistent for estimating the total probability over intervals. Let's not bother too much about ordinates of f because you have extra problems there. If you settle for a certain little interval in which the integrated probability is to tend to the right value when the sample size increases, in that sense both methods are consistent. I think at this conference there's a paper related to the tails when the kernel method is used. Ignoring tail problems, if the two methods are consistent, then they're equivalent for sufficiently large samples.

Hartigan - I would like to go an order of magnitude down and say that your method is mathematically equivalent to a kernel method with a suitably chosen kernel.

Good - Equivalent in what sense?

Hartigan - In the sense that the estimate from your method and the estimate from the kernel method differ by a smaller order of magnitude than the standard error.

Good - If they're both consistent that's going to be true, isn't it?

Hartigan - They differ by a smaller order of magnitude.

Good - Smaller than just consistency would imply?

Hartigan - Yes.

Tarter - You can tell what the kernel is exactly. If you have your estimator as a Fourier expansion all you need to do is take each coefficient of your estimator, divide it by the true coefficient of the underlying density and that becomes the Fourier coefficient of the kernel.

Good - If you know what the underlying density is.

Tarter - Of course.

Good - I think the questioner doesn't want it to depend on the underlying density.

Tarter - You can't do it.

Good - And therefore I take it that the answer to the question is "No." [Perhaps the equivalent kernels could be computed iteratively but it might not be useful to do so.]

Hartigan - Jack knows that the way to play chess against two masters is that you just take their moves and use them back and forth against the opponents.

Scott - Unfortunately, having such a nice optimization problem means you can't characterize the solution very easily. There are some interesting simulation results, at least for the discrete version. It appears from simulation results on Gaussian data that the integrated mean squared error is going down like $n^{-4/5}$. If you compare it to the optimum Epanechnikov kernel it's about 2½ times smaller. I would think there's reason for optimism, that things are a little better with the penalized likelihood than the kernel, although the rates are the same.

Good - As long as one is using the right measure of roughness, it seems to me the MPL method must be pretty well optimal because it has a Bayesian interpretation.

Scott - The difficulty with the square root of f is not that terrible, by the way. Unfortunately when you look at the derivative you get the density function in the denominator which sort of makes you tend to peak at every data point. This penalty function is some smoothness expression divided by the value of f at that point. It tends to make a peak at the data points. For example, the integrand of penalty function is going to be like $(f''/f)^2$. And to make that small one way is to make the density value large at the data point.

Good - But that doesn't happen. [The condition Sf = 1 might prevent it.]

Scott - It doesn't happen because we don't actually solve the problem exactly. We use approximations. If you look at the first penalty function the exact solution is an exponential spline. By using an approximation we throw away all the high order terms. It's not much of a difficulty.

PATTERN RECOGNITION IN THE CONTEXT OF AN ASBESTOS CANCER THRESHOLD STUDY

Michael E. Tarter
University of California at Berkeley

Abstract

Five distinct stages of a pattern recognition problem are discussed in the context of a study of drinking water asbestos health effects. A series of nonparametric regression procedures is used to examine the possiblilty of a threshold in the relationship between asbestos in drinking water and cancer. Evidence is presented which suggests that rather than there being a threshold, the dose response curve $E(Y|X)$ is linear. It is also shown that by choosing a large value (in comparison to earlier studies) for the pre-transformation location parameter of the log observed-to-expected response variable, the overall significance level of the asbestos in drinking water-cancer association is greatly increased and the resolution of the threshold pattern recognition procedure is greatly enhanced.

Keywords: Asbestos, benchmark simulation, drinking water, log transformation, maximum likelihood, nonparametric regression, threshold.

Introduction

The purpose of this paper is to use a sequence of examples to introduce and discuss the topic of pattern recognition. Successful pattern recognition is considered to involve five distinct stages of scientific inquiry:

1. Prefocus
2. Data preprocessing
3. Methodological adaptation
4. Benchmark data confirmation
5. Procedural confirmation

Context of examples: drinking water asbestos - cancer incidence study

Following studies conducted in Minnesota (please see Nicholson 1974 and Cook et al 1974) and California (please see Cooper et al 1979 and Kanarek et al 1979), the health effects of asbestos fibers in drinking water supplies have been of considerable interest to both the scientific community and the general public. Elmes (1980) and Wagner (1980) have recently presented two excellent survey articles as part of the "thematic set of papers on geology and health" which deal with the health effects of asbestos dust inhalation. The effect of other types of asbestos exposure is an open question, as is the question of the existence of threshold-type cancer dose-response relationships (please see Schneiderman et al 1979).

The mean concentration of asbestos in San Francisco Bay area census tract drinking water is estimated to be 6.8 million asbestos fibers per liter. Certain census tracts in Duluth Minnesota, the San Francisco Bay area and in Seattle Washington have estimated mean concentrations from four to eleven times greater than the mean San Francisco Bay area concentration. Since occupational asbestos exposure and respiratory as well as digestive cancers have been linked in several studies (please see Selikoff et al 1975 and Enterline et al 1975), the need for careful study of drinking water asbestos exposure seems fairly evident.

1. Prefocus: hint of threshold-type relationships

Initial studies of the San Francisco Bay area data were conducted by use of GRAFSTAT interactive graphical exploratory data analysis procedures (please see Tarter and Kronmal 1976, Tarter et al 1976 and Tarter 1978a). Two surprising data patterns were uncovered. The first, described in Tarter 1980, involves a putative population subcomponent comprised of high-exposure, low-cancer census tracts, all of which were associated with one small geographically contiguous section of San Francisco. The second pattern was brought into focus by a nonparametric regression technique. As described by Fisz 1963, p. 91 and implemented by Tarter 1979a p. 134-5, one can now use modern density estimation procedures to directly estimate the expectation $E(Y|X)$ of one variable Y given the value of another variable X (e.g., excess cancer mortality and asbestos level respectively) *without the need for the representation of $E(Y|X)$ by any prespecified type of model.*

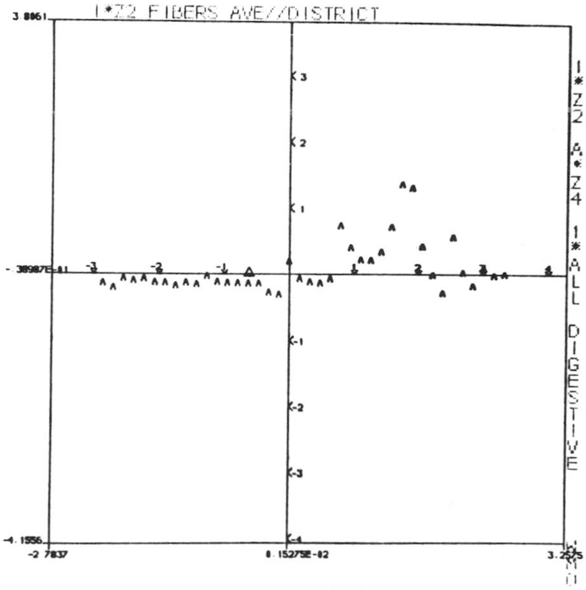

Figure 1. Original nonparametric regression curve - hint of threshold-type relationship

Figure 1 is an example of one such estimate of $E(Y|X)$. Both the fibers per district underlying X variable and the per-district six-year excess digestive cancer mortality Y variable have been adjusted (i.e., partialled) for log median income by a procedure described in Tarter 1978b. The coordinate axes of Figure 1 are marked in standard deviation units and pass through the mean *adjusted* fiber level exposure $\bar{X}$ and mean *adjusted* excess mortality $\bar{Y}$. The data median point is located at the small triangle slightly to the left of the $(\bar{X}, \bar{Y})$ point. Each "A" is a maximum likelihood estimator of the conditional density $f(y|x)$ location parameter obtained by procedures described by Tarter 1979a, section 3.

Notice that the locus of "A" points coincides almost exactly with the x axis between negative three and positive one standard deviation units of the X variable and then increases dramatically. This indicates the possible existence of a threshold value below which asbestos fiber exposure and incidence of cancer are unrelated.

2. Data preprocessing

In 1978 the Cooper-Kanarek study was the only investigation whose data could be made available for study of the existence or non-existence of a threshold effect. However, in one important sense the San Francisco Bay area was one of the worst places to conduct a study of this kind. Very few Bay area census tracts had average exposures between the extremes of high peninsular and low East Bay asbestos levels. As any photographer knows, it is difficult to obtain a clear picture when the scene being photographed contains small brightly illuminated sections and large sections which are very faintly illuminated. Unfortunately the putative threshold point is located in what could be called the most shaded section of Figure 1.

The need for increased resolution at threshold or separation points would appear to be commmon to a great many pattern recognition applications. It turned out in our asbestos study that of the two alternative methods for resolution enhancement, data preprocessing and methodological adaptation, the former was the most effective means of enhancing the resolving power of Bay Area Threshold Study estimators. Of course in other applications the adjustment of the computer-statistical procedure itself may be more critical than data quality.

The response variable used in the study which suggested the existence of a threshold was the natural logarithm of the per-district observed to expected 1969-74 mortality ratio. Specifically, consider the variate $\log(O/E)$, where O represents the observed and E the expected number of cancer cases in a census tract over a six year period. One might first simply add a small positive constant C to O/E so that if O were to equal zero the log of $(O/E)+C$ would remain finite. However,

$$\log((O/E)+C) = \log((O+CE)/E) = \log(O+CE) - \log(E)$$

Thus, since the argument of $\log(E)$ is supposed to predict O itself and not $O+CE$, it seems better to use the variate $\log((O+C)/(E+C))$ in place of $\log((O/E)+C)$.

The behavior of the variate $\log((O+C)/(E+C)) = \log(O+C) - \log(E+C)$ is shown in Figure 2. Here C takes on the value one and $X = \log(E+C)$ is plotted on the x axis against the estimated expectation of $Y = \log(O+C)$ on the y axis. The locus of points associated with zero excess mortality would in this situation be the diagonal line $y = x$, and indeed the nonparametric regression curve marked by the letter A does approximate a diagonal line. However, using a series of nonparametric estimation procedures we can estimate $\mathrm{Var}(Y|X)$, the variance of the conditional density of Y given X. The bands marked by the letters $\bar{A}$ and $\underline{A}$ are plus or minus two estimated $(\mathrm{Var}(Y|X))^{1/2}$ units from the estimate of $E(Y|X)$. Under the assumption of homoscedasticity, these bands should be parallel. Since the distance between a point and the diagonal line equals the mortality residual, parallel bands imply that each census tract contributes a roughly equal amount of statistical information about the cancer response variate.

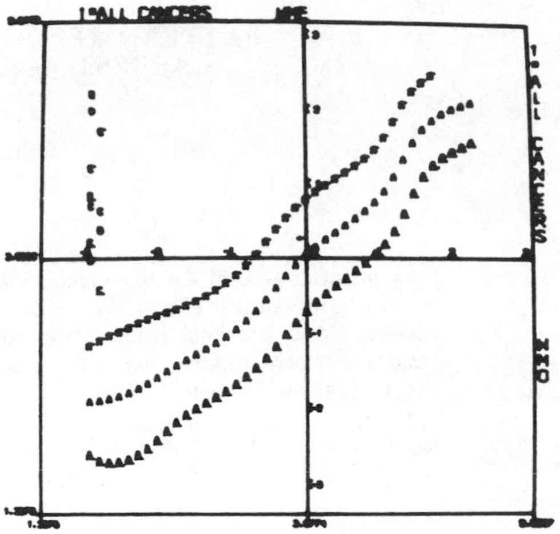

Figure 2. Nonparametric regression and conditional standard deviation bands

The bands show a marked narrowing as x and y increase, which implies that census tracts with many observed and expected cancer cases contribute less usable statistical information than do census tracts with few (or perhaps no) cancer cases.

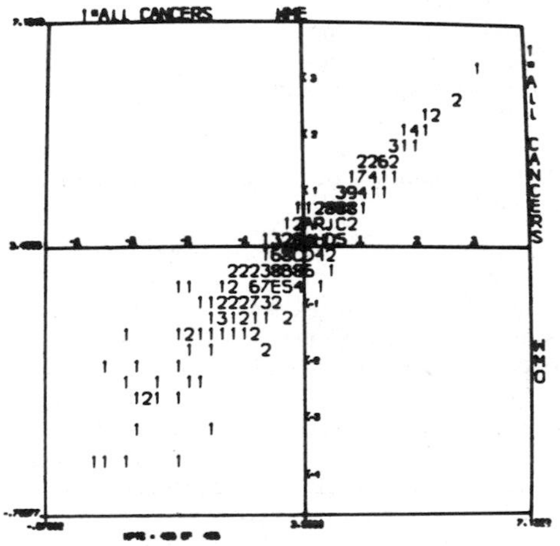

Figure 3. Census tract observed-expected frequency diagram showing substantial narrowing

When the four hundred and twenty-six census tract points $(\log(E+1), \log(O+1))$ were plotted (Figure 3), the distortion produced by setting the constant C equal to one is clearly demonstrated. It is rather amazing that even though information obtained from the large census tracts is considerably damped and information obtained from the small tracts is emphasized, the previous studies still obtained significant findings for some cancer sites.

By using procedures designed to estimate the constant C of a logarithmic transformation (see Tarter and Kowalski (1972) and Tarter (1979b)), we were able to obtain parallel estimated $E(Y|X) \pm 2(\mathrm{Var}(Y|X))^{1/2}$ bands and as a consequence greatly enhance the resolving power of the basic nonparametric regression curve which in turn demonstrated the likelihood of a linear relationship between asbestos and excess cancer mortality. The estimated optimal value of C for digestive cancer was thirty-two.

3. Methodological adaptation

The curve shown in Figure 1 could have been affected by the adjustment of the Y and X variates for log median income level. Consequently, to avoid this possible form of data contamination and to better interpret the position of a threshold value, the X axis used in our pattern recognition study was chosen to be the estimated asbestos exposure gross mean, *unadjusted* for any possible intervening variable. In Figure 4, the Y variable is the least squares residual log digestive cancer mortality for white males from 1969-74.

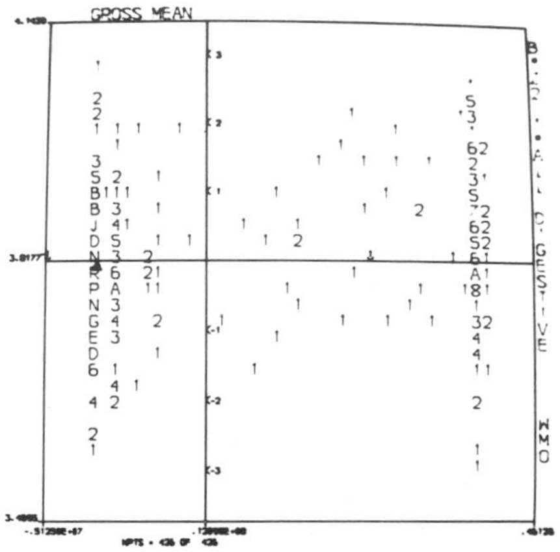

Figure 4. Basic frequency diagram after response variate optimization

In this frequency diagram each number 1 represents the coordinate of a single Bay area census tract. The x axis is scaled in terms of gross mean particles per liter asbestos exposure and the y axis represents excess digestive cancer mortality. This latter varate is identical to the dependent variable of the Kanarek et al (1980) study with the exception that 1) an optimal value was chosen for the constant of the log transformation, and 2) an optimized residual of log expected from log observed was used in place of the simple difference of log expected from log observed. The number of tracts with common exposure mortality coordinates is displayed if this number is between one and nine. Consecutive letters of the alphabet are used to tally from ten to thirty-five similar census tracts.

The coordinate axes are set to pass through the means 12,099,000 fibers per liter and $3.8177 = \log(O+32) - \hat{\beta}\log(E+32)$ mortality residual. The statistic $\hat{\beta}$ represents the least squares regression coefficient or slope of the best fit of $\log(E+32)$ to $\log(O+32)$ for the 426 Bay area census districts.

The most important methodological pattern recognition technique utilized in this study was based on the use of properties of the underlying Fourier series estimators. As previously mentioned, the data can be thought of as being comprised of two bright spots (the peaks of high and low exposure level census tracts) separated by a large shady (plains) region. To increase the resolution of the nonparametric regression estimator the left and right margins were reduced. As far as the methodology used to estimate the nonparametric regression curve is concerned, this margin reduction essentially wraps the data more tightly about a three dimensional surface, e.g., globe, and replaces the two "mountain ranges" of East Bay and San Francisco data and the plains of intermediate asbestos levels with a single high-low mountain range and a slightly enlarged "plains" region.

Problems and procedures associated with sparse-rich (plains-mountains or bright-shady) data sets are described by Brieman, Meisel and Purcell (1977) and Tarter (1979b). In the problem at hand, by in effect wrapping the scene more tightly about a globe and linking the two rich areas one could much more accurately capture detail in both sparse and rich regions. Of course, as with any complex statistical or pattern recognition procedure, unwanted side-effects could also result. Checks for such side-effects will be discussed in the last two sections of this paper.

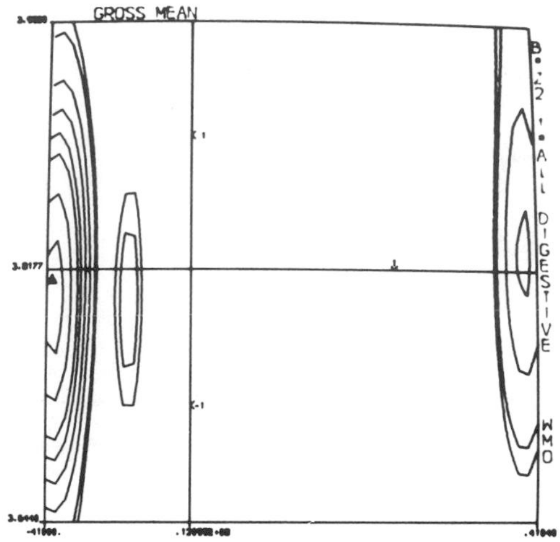

Figure 5. Density contour diagram estimated from data shown in Figure 4

Before proceeding to the basic nonparametric regression curve, it is useful to examine a topological map of the "scene" described previously. The curves shown in Figure 5 are isopleths or terraces associated with the estimated density of the exposure-mortality data. The sample median of the data is displayed as a small triangle slightly below the x-axis and to the far left of the y-axis. This implies of course that at least half of the census tracts were at or below the same low exposure value. Each curve is the set of points (x,y) such that the value of the density estimator $\hat{f}(x,y)$ equals a constant. The values of these constants are 90%, 70%, 50%, 40%, 30%, 20%, 10%, and 7.5% of the maximum value attained by $\hat{f}(x,y)$ at the highest modal value (the highest peak). Besides the leftmost mode near the median, there are two other estimated modes, the first at low-intermediate asbestos levels and the second near extremely high levels (41,040,000 fibers/liter). The fact that no isopleths occur in the vast intermediate levels indicates that the estimated density there is less than 7.5% of the highest modal value. As will be seen in the last section of this paper, the clear resolution of a linear trend in the beginning and end portion of the display and the faint but discernable trail in the middle portion of the figure is due to the fact that information is particularly scarce just where the threshold curve might hypothetically begin moving upwards. Thus, in order to capture as much valid information as possible, the display was enlarged in both the y and more importantly the x directions.

Because it requires a linear model, the usual form of parametric regression curve is useless as a way of detecting the presence or absence of a threshold. To clarify this point, consider the statistical property $E(Y|X)$ discussed by Fisz (1963, Sections 3.7-3.8) and others which underlies many regression procedures. Here the symbol E represents expected, Y excess mortality, the slash the word "given" and X asbestos level. As its name suggests, simple linear regression implies that $E(Y|X) = \alpha + \beta X$ where α and β are two constants usually estimated by least squares methods. Thus, the path that the property $E(Y|X)$ takes as X moves from west to east is assumed to be "as the crow flies", e.g., in a purely north-easterly direction. The true "trail" however may move in a purely easterly direction and might veer sharply northeast at some threshold point. Specifically, the existence of a threshold implies that for all X between 0 and a single threshold point X_0, $E(Y|X) = \alpha_1 + 0X$, and for all X between X_0 and some second value $X_1 > X_0$, $E(Y|X) = \alpha_2 + \beta_2 X$, where $\beta_2 \neq 0$. Thus, model $E(Y|X) = \alpha + \beta X$ could not describe a relationship which involved a threshold within the domain of values of X where $0 < X < X_1$, since two distinct values of β are required to describe $E(Y|X)$ over this domain.

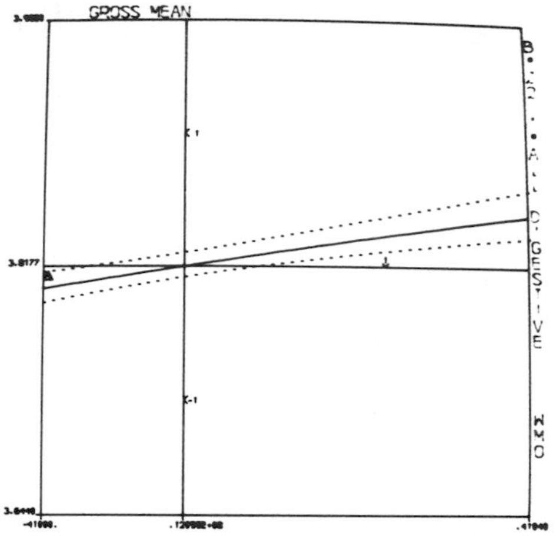

Figure 6. Conventional linear regression and confidence bands

One advantage of the simple linear model $E(Y|X) = \alpha + \beta X$ however is that the simple least squares procedure used to estimate α and β can be easily extended to provide a single set of confidence bands. The bands shown in Figure 6 specify a region within which one would be 95% confident that the true regression curve $E(Y|X)$ must lie, if that curve were indeed a line and if certain other assumptions are met.

The t statistic associated with the null hypothesis that regression slope $\beta = 0$ takes on the value 4.899 in this case. The α confidence level or p value for this test is larger than the level associated with the largest value, 4.265, given in Dixon and Massey's *Introduction to Statistical Analysis,* third edition (1969). If t had been 4.265 one would have been 99.999% confident that there was an association between asbestos and cancer. Compared to any value found in previous Kanarek et al (1980) work, the improved procedures reported here have raised what was a barely significant (at the 99% level) finding to a degree of certainty beyond any standard tabled value. Thus, even though the line shown in this figure does not graphically show a particularly large slope ($\hat{\beta} = .126\text{E}{-}8$), a careful analysis of the data shows that the hypothesis $\beta = 0$, i.e., that $E(Y|X)$ takes an entirely easterly trail, is untenable.

Figure 7 is the single most useful graph of the more than five hundred which were obtained as part of this threshold study. The locus of points marked by the letter "A" represents a nonparametrically estimated regression curve which traces the trail between low and high asbestos exposure *without any assumption concerning the general shape of this trail.* The fact that this estimator of $E(Y|X)$ is almost identical to the linear regression

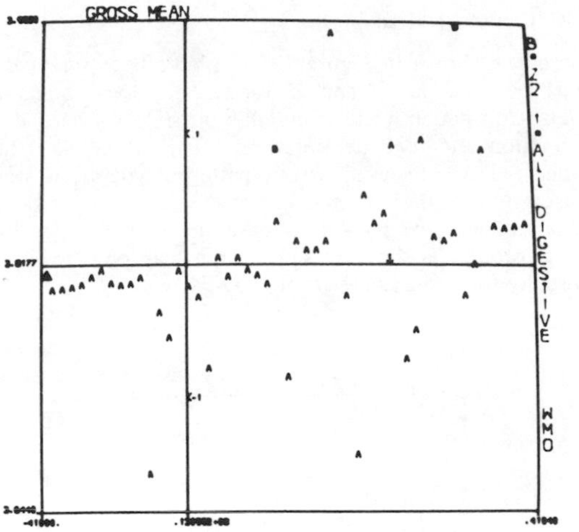

Figure 7. Nonparametric regression analog to Figure 6

$\hat{\alpha} + \hat{\beta} X$ suggests that not only is there a relationship between asbestos level and cancer mortality, but that this relationship appears to be perfectly linear, i.e., as far as the estimated $E(Y|X)$ function is concerned, *no threshold exists!*

The upward or northward movement of the trail is apparent even in areas where there are low to moderate exposure levels. Naturally, in intermediate regions where there are very little data, the curve or trail is harder to follow. At the high end, however, the trail is again clearly marked.

It of course goes without saying that functions such as $E(Y|X)$ "are insufficient to state anything concerning the cause that *produces* the state or the phenomenon in question", e.g., excess cancer mortality (please see Bunge 1959, p. 92).

4. Benchmark data confirmation

The use of a standardized test pattern can be as useful for pattern recognition purposes as it is for the alignment of electronic and photographic procedures. The crucial point to be checked in our threshold study was whether the margin reduction procedure described above could have influenced the ramp as opposed to step, i.e., diagonal as opposed to horizontal shape of the nonparametric regression curve.

Our principal check of this contingency was based on a GRAFSTAT option which allowed us to simulate an artificial sample of data which matched the characteristics of the natural data set with the exception that under successful analysis a stepped rather than ramp shaped curve would be expected to result.

One such data test pattern is shown in Figure 8. A sample of the same size as that used to produce the previous figures was generated from a three-component bivariate normal mixture with density

$$0.50N(\mu_x = -2, \mu_y = -0.5, \sigma_x = 0.1, \sigma_y = 2, \rho = 0) +$$
$$0.15N(0, 0, 1, 2, 0) + 0.35N(2, 0.5, 0.1, 2, 0)$$

where $N(\mu_x, \mu_y, \sigma_x, \sigma_y, \rho)$ represents the bivariate normal with mean vector (μ_x, μ_y) transpose and variance-covariance matrix

$$\begin{pmatrix} \sigma_x^2 & \rho\sigma_x\sigma_y \\ \rho\sigma_x\sigma_y & \sigma_y^2 \end{pmatrix}$$

The data shown in Figure 8 resembles that shown in Figure 4 with the exception that: 1) since ρ was chosen to be zero

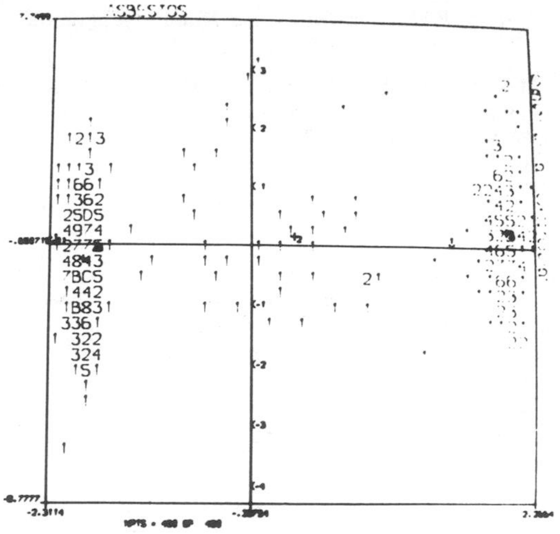

Figure 8. Benchmark data analog to Figure 4

within each of the three population subcomponents, the regression curve would be expected to be horizontal at the extreme ends of the density's support, and 2) the discreteness of the X variate in Figure 4 caused by water sampling limitations is not matched by the data shown in Figure 8.

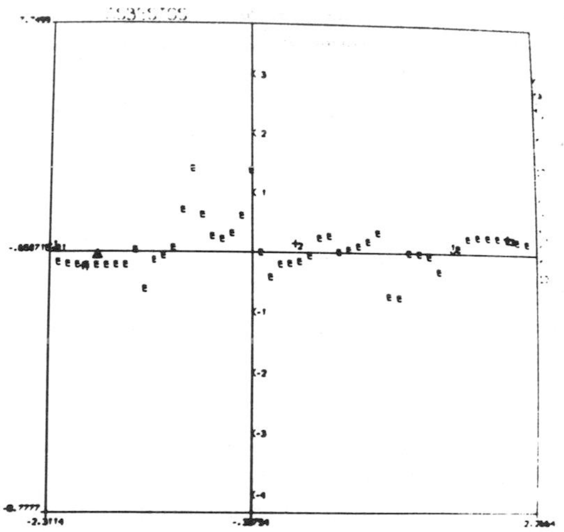

Figure 9. Benchmark data analog to Figure 7 showing stepped rather than ramped increase

As shown in Figure 9, distinct horizontal steps rather than a single ramp were obtained from this data test pattern at the beginning and end of the display. This and other experiments performed with benchmark test data shows that the sample size available would in general be sufficient to clearly distinguish between a stepped and a ramped or diagonal increase in cancer excess mortality in the Y^+ or northerly direction with increase in asbestos levels in the X^+ or easterly direction.

5. Procedural confirmation

It is very important to ascertain whether the shape of the regression curve shown in Figure 7 could have been affected by the particular type of nonparametric regression procedure used in this study. Fortunately the computational tractability of Fourier series methods made it possible to construct a large variety of alternative nonparametric regression curves using a wide variety of alternative nonparametric approaches.

Following the work of Huber (1964) there has been considerable interest in robust estimates of location based on combinations of least squares and least absolute value metrics. To extend Huber's approach to the field of nonparametric as opposed to parametric regression the following strategy was used.

The estimator of $E(Y|X)$ shown in Figure 7 is based on the application of likelihood estimation procedures developed by Tarter 1975a, 1975b and 1979a to a series nonparametric density estimator $\hat{f}(y|x)$. In this approach a sequence of weights $\{w_k\}$ $k=\pm 1, \pm 2, \cdots$ is selected which, as described in Tarter (1979a, Section 3), can provide location parameter maximum likelihood estimates intermediate to the sample mean and sample median. Specifically, the equation

$$\sum_{k\neq 0} (c_2-(-1)^k(c_1{}^2))\left\{\frac{\hat{B}_k(x)}{2\pi ik}\right\} e^{2\pi ik(y+1/2)} = 0$$

defines a locus of maximum likelihood estimators of $y = E(Y|X)$ under the assumption that for a fixed value of $X = x$, $Y - E(Y|X) = Z$ has a density proportional to $e^{-c_1{}^2z^2 - c_2|z|}$. Here

$$\hat{B}_k(x) = \frac{\sum_s \hat{B}_{s,k} e^{2\pi isx}}{f(x)}$$

where $\hat{B}_{s,k}$ is a bivariate sample trigonometric moment (please see Tarter and Kronmal 1969 and 1974) and f is the marginal density of the variate X. Thus, by varying c_1 and c_2, a researcher can obtain a sequence of nonparametric estimators of $E(Y|X)$ which are analogs to least squares with $c_2 = 0$, least absolute value with $c_1 = 0$, or any procedure based on a metric which is of the form $c_1{}^2z^2 + c_2|z|$, i.e., a linear combination of least squares and least absolute deviation.

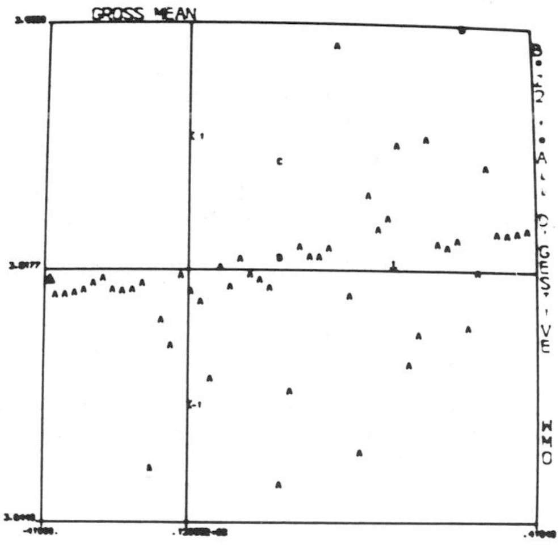

Figure 10. Nonparametric regression curve based on a metric intermediate between least squares and least absolute value

There is little need to illustrate the results of trials based on various choices of weights c_1 and c_2 for the asbestos threshold study. Although there was some shift in the position of the pathological values of $E(Y|X)$ (all located at a considerable distance from the principal diagonal locus of points), there was absolutely no discernable change in the position of the thirty five diagonal locus $\hat{E}(Y|X)$ evaluations. (Please see Figure 10, where $c_1 = c_2$, i.e., a metric was chosen intermediate between least squares and least absolute value). This finding is particularly reassuring since in another experiment, reported in Tarter

(1979b Section 5), data discreteness did tend to slightly affect the results produced by alternative nonparametric estimators.

Discreteness correction procedures described in the above paper were applied to the data used to construct Figure 7. There was some reduction in the number of pathological $\hat{E}(Y|X)$ evaluations but absolutely no change in the positions of nonpathological evaluations.

Forms of nonparametric regression other than the above were also applied to the asbestos data. One procedure based on an estimate of the mode of the conditional density of cancer mortality given asbestos level did show the possible existence of a secondary density component which might correspond to a second diagonal line whose rightmost tip intersected the rightmost tip of the line shown in Figure 7. Since the slope of this line was much greater than that of the line shown in Figure 7, if the existence of this second population component is confirmed, the trend shown by this component would provide strong evidence for the linear relationship between cancer mortality and domestic water supply asbestos.

Acknowledgements

The author would like to gratefully acknowledge the contributions of Robert Cooper, Jim Millette, Jack Murchio, Carole Leong, Bill Freeman, Joe Rodrigue, and Seemin Qayum for their comments and assistance in the preparation of this paper. This research was supported by EPA Purchase Order No. 80-175, NIH General Medical Sciences Grant GM25386-03 and NIH National Cancer Institute Grant R01 CA-21448-03.

References

Brieman, L., Meisel, W. and Purcell, E. (1977), "Variable Kernel Estimates of Multivariate Densities," *Technometrics,* 19, 135-44.

Bunge, M. (1959), *Causality,* Cambridge: Harvard University Press.

Cook, P.M., Glass, G.E. and Tucker, J.H. (1974), "Asbestiform Amphibole Mineral Detection and Measurement of High Concentrations in Municipal Water Supplies", *Science,* 185, 853-55.

Cooper, R. et al (1978), "Asbestos in Domestic Water Supplies in Five California Counties", *UCB-EHS Publications No. 78-2.*

Dixon, W.J. and Massey, F.J. (1969), *Introduction to Statistical Analysis, 3rd Edition,* New York: McGraw-Hill Book Co.

Elmes, P.C. (1980), "Fibrous Minerals and Health", *Journal of the Geological Society of London,* 137, 525-35.

Enterline, P.E. and Kendrick, M.A. (1967), "Asbestos-dust Exposure at Various Levels and Mortality", *Archives of Environmental Health,* 15, 181-86.

Fisz, M. (1963), *Probability Theory and Mathematical Statistics, 3rd Edition,* New York: John Wiley and Sons, Inc.

Kanarek, M.S. et al (1980), "Asbestos in Drinking Water and Cancer Incidence in the San Francisco Bay Area", *American Journal of Epidemiology,* 112, 54-72.

Selikoff, I.J., Churg, J. and Hammond, E.C. (1964), "Asbestos Exposure and Neoplasia", *Journal of the American Medical Association,* 188, 142-47.

Schneiderman, M.A., Decouflé, P. and Brown, C.C. (1979), "Thresholds for Environmental Cancer: Biologic and Statistical Considerations," *Annals of the New York Academy of Sciences: Public Control of Environmental Health Hazards,* ed. E.C. Hammond and I.J. Selikoff, 329, 92-130.

Tarter, M.E. and Kowalski, C. (1972), "A New Test For and Class of Transformations to Normality", *Technometrics,* 14, 735-43.

Tarter, M.E. and Kronmal, R.A. (1974), "The Use of Density Estimates Based on Orthogonal Expansions", *Exploring Data Analysis: The Computer Revolution in Statistics,* ed. W.J. Dixon and W.L. Nicholson, Berkeley: University of California Press.

____ (1976), "An Introduction to the Implementation and Theory of Nonparametric Density Estimation", *The American Statistician,* 30, 105-12.

Tarter, M.E. (1976), "A Description of the Berkeley Graphical Biometry Project", *Proceedings of the Computer Science and Statistics Ninth Annual Symposium on the Interface,* ed. D. Hoaglin and R.E. Welsch, 44-46.

____ (1978a), "Implementation of Harmonic Data Analysis Procedures", *Proceedings of the Computer Science and Statistics Eleventh Annual Symposium on the Interface,* ed. A.R. Gallant and T.M. Gerig, 234-46.

____ (1978b), "Interactive Graphical Isolation of Homogeneous Data Subgroups", *Computer Programs in Biomedicine,* 8, 81-86.

____ (1979a), "Trigonometric Maximum Likelihood Estimation and Application to the Analysis of Incomplete Survival Information", *Journal of the American Statistical Association,* 74, 132-39.

____ (1979b), "Biocomputational Methodology - An Adjunct to Theory and Applications", *Biometrics,* 35, 9-24.

____ (1980), "Comments on I.J. Good and R.A. Gaskin", *Journal of the American Statistical Association,* 75, 63-66.

Wagner, J.C. (1980), "The Pneumoconioses Due to Mineral Dusts", *Journal of the Geological Society of London,* 137, 537-45.

Workshop 6

Volume Testing of Statistical Programs

Organizer: Robert F. Teitel, Teitel Data Systems

Chair: Robert F. Teitel, Teitel Data Systems

Invited Presentations:

Volume Testing of Statistical/Database Software, Robert F. Teitel, Teitel Data Systems

Volume Testing of Statistical Software -- The Statistical Analysis System (SAS), Arnold W. Bragg, SAS Institute and North Carolina State University

Solving Complex Database Problems in P-STAT, Roald and Shirrell Buhler, P-STAT, Inc.

Volume Testing of SPSS, Jonathan B. Fry, SPSS Inc.

Volume Testing of Statistical Systems, Pauline R. Nagara, Institute for Social Research, and Michael A. Nolte, Institute for Social Research

Scientific Information Retrieval (SIR/DBMS), Barry N. Robinson, SIR, Inc.

VOLUME TESTING OF STATISTICAL/DATABASE SOFTWARE

Robert F. Teitel, TEITEL Data Systems

ABSTRACT

By volume testing is meant assessing the ability of systems to manipulate data with large values for the width, length, and depth dimensions. The depth dimension is used as a measure of the complexity of the data relationships in a non-planar data collection.

Several carefully designed problems for complex data manipulation by statistical and database systems are presented. These problems are referenced in subsequent papers in this volume on complex data manipulation capabilities of the major statistical/database systems in use today.

KEYWORDS

complex data
statistical systems
database systems
evaluation
benchmark problems
data manipulation

I. INTRODUCTION

By "volume testing" we mean the determination of the limits of a statistical or database system's ability to handle large amounts of data with complex structure. Volume, as used with rectangular solids, is calculated as width x length x depth. In the context of the evaluation of statistical or database software, the dimension width, the number of variables, and the dimension length, the number of observations, are well-understood measures. When we discuss non-planar data, or data with complex structure, we introduce another dimension to be considered in the evaluation of the processing capability of statistical or database systems. We will call this additional dimension, measuring complexity, the depth of the data.

We will not attempt a precise definition of depth for, as we will see, definitions and interpretations become somewhat murky even for such ostensibly simple concepts as width and length when dealing with complex data structures. One of the data collections in the problem set to be presented would be called "network" by most people, "hierarchical" by some, and just "complex" by others. The number of variables is usually equal to the total of the number of variables in all record types; the number of observations is either the number in the record type chosen as the unit-of-analysis or is the number of households or is the total number of records over all record types. The other data collection in the problem set appears to be a simple rectangular structure with well-defined width and length. The processing requests, however, make it clear that it is really a complex structure with an additional dimension, which we call depth.

To begin to assess the capabilities of the various statistical systems (and closely related research data management or tabulatory systems) in dealing with complex data collections, we devised some years ago a problem set consisting of two data collections and two processing requests for each data collection. This problem set has had limited circulation within the statistical computing community, and has been considered "unbiased" in the sense that it is not clearly directed at the capabilities of one or two systems. One of the data collections closely resembles the structure of the National Travel Survey of 1972, the other data collection is a pure invention. The processing requests are representative of meaningful tables from the two data collections.

II. THE PROBLEM SET

The problem set consists of the descriptions of two data collections, and two data manipulation exercises for each data collection. The data manipulation exercises are stated in terms of two-way frequency distributions or cross-tabulations. The actual tabulations are not the principal concern: the emphasis in volume testing is on the manipulation of the data structure prior to the tabulation stage of the processing.

A. The TRIPS Data Collection

TRIPS is a large collection of data consisting of four groups of variables (variously called segments, relations, levels, types). The groups of variables are related to each other as shown in the following diagram. (The table form is on the next page.)

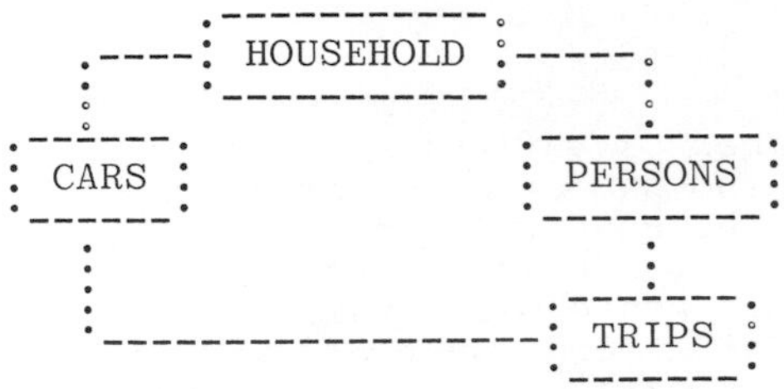

Each household contains a variable number of cars and a variable number of persons; each person contains a variable number of trips. We include zero in the definition of "variable number". In addition, each trip, if taken in a car owned by the household, will contain the identification of that car. There are no counts of persons or cars in the household record, nor is there a count of trips in the person record. The car record contains a model-year variable, the person record an age variable, and the trip record a duration variable in addition to the own-car variable.

The first of the two tabulations to be produced from the TRIPS data is a simple frequency distribution of households by the number of cars owned and by the number of persons over the age of 16 in the household.

household count	persons over 16				
	1	2	3	4	.
cars 1					
2					
3					
.					

The second tabulation to be produced from the TRIPS data is a frequency distribution of trips of at least three days' duration taken in a car owned by the household by the age of the person taking the trip and by the model year of the car.

trip count	model year of car					
	..	1970	1971	1972	1973	..
age ..						
16						
17						
18						
..						

The two tabulations, though superficially similar, require different complex data processing capabilities. Each table consists of a count of occurrences of one record type based on variable values in other record types, or on counts of other record types. The first table requires "downward" access, from household to cars and to persons; the second table "upward" access, from trips to persons and to cars.

B. The PERSONS Data Collection

PERSONS is a large data collection of various data elements on people. The data collection is relatively closed with respect to ancestry: for most people, data on their parents, grandparents, etc., as well as data on their children are contained in the data collection. The variables available for each person include the year of birth, the level of education, sex, and the identification of the mother and father, if known. The records are in identification number order, but not in any relationship order.

The first of the two tabulations to be produced from the PERSONS data collection is a simple frequency distribution of offspring by the educational level of each parent.

offspring count	father's education					
	..	elem	hs	coll	grad	..
mother's education						
..						
elem						
hs						
coll						
grad						
..						

The second table to be produced from the PERSONS data collection is a frequency distribution of the "last births" by the age of the mother at the birth of her last offspring and by the sex of that last offspring. The count of "last births" could be interpreted as either the count of mothers at last birth or as a count of the youngest child of each mother. (The table form is on the next page.)

These tabulations also require different complex data processing or access paths. The first is again a simple tabulation requiring "upward" access to the education variables of the parents. The second tabulation, if viewed as a count of mothers, requires scanning the children to determine the youngest; if viewed as a count of youngest children, it requires a scan of siblings to determine the youngest.

last birth count	sex of last offspring male	female
mother's age		
- 18		
19 - 25		
26 - 30		
..		

III. THE DISTRIBUTION TAPE

Programs have been written to generate data for the TRIPS and PERSONS data collections. The programs permit the specification of the basic data structure parameters, how many households, or how many trips per person, and the ranges of the values of each of the variables used in the tabulations. There are approximately 9000 records in each of the distributed data files.

A. The TRIPS File

The TRIPS data file consists of household observations in household identification number order. Each household observation contains, in order, a household record, all the car records, and sets of travel data in person identification number order. Each set of travel data contains a person record follwed by trip records for that person. The format of each record type is as follows.

household record:
  1- 4: household identification.
     5: record type = '1'.
  6- 9: ignore.
 10- 60: 51 one-digit variables.
 61-100: 20 two-digit variables.
101-120: 5 four-digit variables.

car record:
  1-120: as above, except for
      5: record type = '2'.
  6- 7: car within household id.
        the last four-digit variable is the model year of the car.

person record:
  1-120: as above, except for
      5: record type = '3'.
  6- 7: person within household id.
        first two-digit variable is the age of the person.

trip record:
  1-120: as above, except for
      5: record type = '4'.
  6- 7: person within household id.
  8- 9: trip within person id.
        first one-digit variable is the id of the car used for trip (0 means other conveyance used).
        last two-digit variable is the duration of the trip.

All variables of all record types should be maintained in the system file or database. The number of total variables is the width component of "volume testing".

B. The PERSONS File

The PERSONS data file consists of person observations in person identification number order. The record format for each person person record is as follows.

person record;
  1- 5: person identification number.
  6- 56: 51 one-digit variables, the first is the sex variable ('1'=female, '2'=male).
 57- 86: 15 two-digit variables, the first is the education (1..19).
 87-110: 6 four-digit variables, the first is year of birth (1800...).
111-115: id of this person's mother
116-120: id of this person's father

All variables should be maintained in the systems file or database.

IV. BIOGRAPHY

Robert F. Teitel has over 20 years of technical and managerial experience in providing computer support for social and economic research at commercial, non-profit, and educational organizations. He has performed research in the application of modern programming techniques and database design methodology to statistical computing, and has published and lectured widely on that research.

TEITEL Data Systems, located in Bethesda, MD, is a consulting and professional services organization specializing in data management, statistical computing, and related customized software design and implementation.

SCIENTIFIC INFORMATION RETRIEVAL (SIR/DBMS)

Barry N. Robinson, SIR, Inc.

SIR/DBMS is a database management system that has been geared to the unique needs of the research community. The data definition commands in SIR/DBMS are patterned after the well-known statistical package SPSS, and SIR/DBMS interfaces directly (through the creation of system files) to SPSS, BMDP and any other system that can read SPSS system files, such as SAS and P-STAT. SIR/DBMS can easily handle complex hierarchical and network data structures.

In the examples that follow, we illustrate the actual SIR/DBMS commands required to define the databases and then retrieve the desired information.

KEYWORDS: database management system, hierarchical files, network files, statistical interface, sort id's, summary records.

I. The TRIPS Database

A. Case Definition

These commands apply to the entire database. They specify:

- the name of the case identifying variable (HOUSEID)
- various limits on cases, record types, input columns
- the variables in the Common Information Record
- documentation for the database as a whole.

```
CASE ID         HOUSEID

RECTYPE COL     5
N OF CASES      1000
RECS PER CASE   150
MAX REC TYPES   7
MAX INPUT COLS  120

COMMON VARS     HOUSEID

DOCUMENT        THE TRIPS DATABASE IS ORGANIZED
                HIERARCHICALLY BY HOUSEHOLD ID.
                THE STRUCTURE IS:

                          HOUSEHOLD
                           /     \
                          /       \
                       CAR        PERSON
                                     \
                                      \
                                      TRIP
```

B. Record Definition

For each record type in the database (household, car, person, trip), you must specify a set of Record Definition commands. The SORT ID command provides the names of the additional variables required to uniquely identify a record. SORT ID´s are needed when there can be more than one record of a given type per case.

```
RECORD SCHEMA  1,HOUSE
MAX REC COUNT  1

DOCUMENT       THE HOUSE (TYPE 1) RECORD CONTAINS
               SPECIFIC INFO ABOUT THE ENTIRE
               HOUSEHOLD. THERE IS ONE TYPE 1
               RECORD PER CASE.

DATA LIST      (1)/1
               HOUSEID           1 -   4 (A)
               V1 TO V51        10 -  60 (I)
               W1 TO W20        61 - 100 (I)
               X1 TO X5        101 - 120 (I)

MISSING VALUES HOUSEID TO X5(BLANK)
VAR LABELS     HOUSEID, HOUSEHOLD ID/
END SCHEMA



RECORD SCHEMA  2,CAR
SORT ID        CARID
MAX REC COUNT  99

DOCUMENT       THE CAR (TYPE 2) RECORDS CONTAIN
               INFO ABOUT THE CARS BELONGING TO
               THE HOUSEHOLD. THERE IS ONE TYPE
               2 RECORD FOR EACH CAR IN THE
               HOUSEHOLD UP TO A MAXIMUM OF
               99 PER HOUSEHOLD (CASE).

               EACH CAR RECORD CONTAINS:
                    HOUSEHOLD ID
                    CAR ID
                    CAR MODEL YEAR
                    VARIOUS OTHER THINGS.

DATA LIST      (1)/1
               HOUSEID           1 -   4 (A)
               CARID             6 -   7 (I)
               V1 TO V51        10 -  60 (I)
               W1 TO W20        61 - 100 (I)
               X1 TO X4        101 - 116 (I)
               YEAR            117 - 120 (I)

MISSING VALUES HOUSEID TO YEAR(BLANK)
VAR LABELS     CARID,    CAR IDENTIFIER/
               YEAR,     MODEL YEAR/

END SCHEMA
```

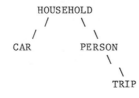

```
RECORD SCHEMA    3,PERSON
SORT ID          PERSONID
MAX REC COUNT    99
DOCUMENT         THE PERSON (TYPE 3) RECORDS
                 CONTAIN INFO ABOUT PEOPLE IN THE
                 HOUSEHOLD.  THERE IS ONE TYPE 3
                 RECORD FOR EACH PERSON IN THE
                 HOUSEHOLD UP TO A MAXIMUM OF
                 99 PER HOUSEHOLD (CASE).  EACH
                 PERSON RECORD CONTAINS:
                     HOUSEHOLD ID
                     PERSON ID
                     PERSON'S AGE
                     VARIOUS OTHER THINGS.

DATA LIST        (1)/1
                 HOUSEID            1 -   4 (A)
                 PERSONID           6 -   7 (I)
                 V1 TO V51         10 -  60 (I)
                 AGE               61 -  62 (I)
                 W2 TO W20         63 - 100 (I)
                 X1 TO X5         101 - 120 (I)

MISSING VALUES   HOUSEID TO X5(BLANK)
VAR LABELS       PERSONID,PERSON IDENTIFIER/
                 AGE,      AGE IN YEARS/
END SCHEMA
```

```
RECORD SCHEMA    4,TRIP
SORT IDS         PERSONID, TRIPID
MAX REC COUNT    99
DOCUMENT         THE TRIP (TYPE 4) RECORDS
                 CONTAIN INFO ABOUT THE TRIPS
                 TAKEN BY EACH HOUSEHOLD MEMBER.
                 THERE IS ONE TYPE 4 RECORD TYPE
                 FOR EACH TRIP TAKEN BY EACH PERSON
                 UP TO A MAXIMUM OF 99 PER HOUSE-
                 HOLD (CASE).  EACH TRIP RECORD
                 CONTAINS:
                     HOUSEHOLD ID
                     PERSON ID
                     TRIP ID
                     OWN CAR USED ON THE TRIP
                     TRIP DURATION IN DAYS
                     VARIOUS OTHER THINGS.

DATA LIST        (1)/1
                 HOUSEID            1 -   4 (A)
                 PERSONID           6 -   7 (I)
                 TRIPID             8 -   9 (I)
                 OWNCAR            10 -  10 (I)
                 V2 TO V51         11 -  60 (I)
                 W1 TO W19         61 -  98 (I)
                 DAYS              99 - 100 (I)
                 X1 TO X5         101 - 120 (I)

MISSING VALUES   HOUSEID TO X5(BLANK)
VAR LABELS       PERSONID,PERSON IDENTIFIER/
                 TRIPID,   TRIP IDENTIFIER/
                 OWNCAR,   HOUSEHOLD OWNED CAR
                           USED IN TRIP/
                 DAYS,     DURATION OF TRIP IN DAYS/
END SCHEMA
```

C. Retrieval Problem 1

In the first retrieval we generate summary records (belonging to a "flat" summary file of the database) in which every record contains the counts:

- number of cars (NCAR)
- number of people over 16 years of age (NPEOP).

At the heart of the retrieval is the PROCESS CASES ALL loop which loops through all of the cases in the database.

```
RETRIEVAL
.     PROCESS CASES ALL
.          COMPUTE NCAR = COUNT(2)
.          IF (NCAR EQ 0) NEXT CASE
.          SET NPEOP (0)
.          PROCESS REC PERSON
.               IF (AGE GT 16) NPEOP = NPEOP + 1
.          END PROCESS REC
.          IF (NPEOP GT 0) PERFORM PROCS
.     END PROCESS CASES
SPSS SAVE FILE FILENAME = PERSCAR
END RETRIEVAL
```

To determine the number of cars in a given household (case) we use the COUNT function to tell us how many type 2 records (i.e. car records) there are for the household. If this count is zero, we skip the case entirely.

To determine the number of people over 16 years in a given household, we look at all the PERSON records for that household (using the PROCESS REC loop) and count those people who are 16 years or older.

Having determined the two desired counts for the case (NCAR and NPEOP), we transmit the summary record containing these two variables to the SPSS SAVE FILE procedure. The records are only transmitted (PERFORM PROCS) if NPEOP is greater than zero.

(Note the placement of the PERFORM PROCS command. Since it has been placed within the PROCESS CASES loop, it can create, at most, one summary record per case.)

The SPSS SAVE FILE procedure creates an SPSS System File that can be read directly by SPSS.

Within SPSS, the desired result can be obtained by running CROSSTABS of NCAR by NPEOP.

D. Retrieval Problem 2

This retrieval is slightly more complicated because we want to generate a summary record for each TRIP record within a household that satisfies the selection criteria:

- the trip was at least 3 days long
- the trip was made in one of the household's cars.

Note the structure of this retrieval. Especially note that the PROCESS REC TRIP loop is nested within the PROCESS CASES ALL loop and that the PERFORM PROCS command is within the inner loop. This will enable us to generate a summary record for each qualifying TRIP record.

```
RETRIEVAL
. PROCESS CASES ALL
.    PROCESS REC TRIP
.       IF (DAYS LT 3 OR OWNCAR EQ 0) NEXT RECORD
.       MOVE VARS PERSONID, OWNCAR
.       PROCESS REC PERSON, WITH (PERSONID)
.            MOVE VARS AGE
.       END PROCESS REC
.       PROCESS REC CAR, WITH (OWNCAR)
.            MOVE VARS YEAR
.       END PROCESS REC
.       PERFORM PROCS
.    END PROCESS REC
. END PROCESS CASES
SPSS SAVE FILE FILENAME = LTRIP
END RETRIEVAL
```

The IF statement throws out trips that do not satisfy our selection criteria.

Once a TRIP record is approved by the IF statement, we gather the information we are interested in, namely:

- the age of the person who took the trip
- the year of the car that was used.

The TRIP record provides us with the ID of the person who took the trip (PERSONID) and the ID of the car that was used (OWNCAR). It is then a simple matter to retrieve the appropriate PERSON record

PROCESS REC PERSON, WITH (PERSONID)

and get the person's AGE and then retrieve the appropriate CAR record

PROCESS REC CAR, WITH (OWNCAR)

and get its YEAR of manufacture.

Once again PERFORM PROCS sends the summary record to the SPSS SAVE FILE procedure. Within SPSS, we run a crosstabulation of AGE by YEAR.

II. The PEOPLE Database

A. Case Definition

```
CASE ID         PERSONID

RECTYPE COL     1
N OF CASES      10000
RECS PER CASE   30
MAX REC TYPES   5
MAX INPUT COLS  120

COMMON VARS     PERSONID, FATHER,  MOTHER,
                EDUCAT,   BIRTHYR, SEX

DOCUMENT        THE PEOPLE DATABASE CONTAINS INFO
                ON PEOPLE AND THEIR PARENTS AND
                CHILDREN.

                THE ACTUAL INPUT CONSISTS OF
                PERSON INFO AND THE ID'S OF
                THEIR PARENTS.

                THE PERSON-TO-CHILD LINKS ARE
                CREATED BY RUNNING A SIR/DBMS
                RETRIEVAL. THE RETRIEVAL
                CREATES A SECOND RECORD TYPE
                CONTAINING THE ID'S OF
                PEOPLE'S CHILDREN.
```

B. Record Definition

```
RECORD SCHEMA  1,PERSON
MAX REC COUNT  1

DOCUMENT       THE PERSON (TYPE 1) RECORD IS
               THE BASIC RECORD IN THE DATABASE.
               IT CONTAINS:
                  PERSON IDENTIFIER
                  SEX
                  YEARS OF EDUCATION
                  YEAR OF BIRTH
                  MOTHER'S ID
                  FATHER'S ID
                  VARIOUS OTHER THINGS.

               NOTE THAT ALL THE DATA EXCEPT
               FOR THE "VARIOUS OTHER THINGS"
               ARE STORED IN THE COMMON
               INFORMATION RECORD (CIR).
```

```
DATA LIST        (1)/1
                 PERSONID             1 -   5 (I)
                 SEX                  6 -   6 (I)
                 V1 TO V50            7 -  56 (I)
                 EDUCAT              57 -  58 (I)
                 W1 TO W14           59 -  86 (I)
                 BIRTHYR             87 -  90 (I)
                 X1 TO X5            91 - 110 (I)
                 MOTHER             111 - 115 (I)
                 FATHER             116 - 120 (I)

VAR RANGES       SEX    (   1,    2)/
                 EDUCAT (   1,   19)/
                 BIRTHYR (1800, 2100)/
MISSING VALUES   PERSONID, MOTHER, FATHER
                    (0,BLANK)/
                 SEX TO X5 (BLANK)/
VAR LABELS       PERSONID, PERSON IDENTIFIER/
                 EDUCAT,   EDUCATION IN YEARS/
                 BIRTHYR,  YEAR OF BIRTH/
                 MOTHER,   ID OF PERSON´S MOTHER/
                 FATHER,   ID OF PERSON´S FATHER/
VALUE LABELS     SEX (1) FEMALE (2) MALE /
END SCHEMA



RECORD SCHEMA    2, CHILD
SORT ID          CHILDID
MAX REC COUNT    30

DOCUMENT         THE CHILD (TYPE 2) RECORDS ARE
                 CREATED BY A SIR/DBMS RETRIEVAL RUN.
                 THE PURPOSE OF THESE RECORDS
                 IS TO PROVIDE A DIRECT (FORWARD)
                 LINK BETWEEN PEOPLE AND THEIR
                 CHILDREN.

                 THERE IS ONE TYPE 2 RECORD FOR EACH
                 OF A PERSON´S CHILDREN UP TO A
                 MAXIMIUM OF 30 PER PERSON (CASE).

DATA LIST        (1)/1
                 PERSONID             1 -  5 (I)
                 CHILDID              6 - 10 (I)

MISSING VALUES   PERSONID, CHILDID (0, BLANK)
VAR LABELS       CHILDID, ID OF THE PERSON´S CHILD/
END SCHEMA
```

C. <u>Retrieval Problem 1</u>

In this retrieval we are interested in retrieving two pieces of information for our summary record:

- education level of the person´s father
- education level of the person´s mother.

Once we have this information for each person in the database, we can perform a crosstabulation of father´s education (EDUCDAD) by mother´s education (EDUCMOM) to achieve the desired result.

Note that the basic loop is the PROCESS CASES ALL loop and that the PERFORM PROCS command is within this loop. This means that we will generate no more than one summary record per person.

```
RETRIEVAL
.     PROCESS CASES ALL
.          IF (EXISTS(FATHER) EQ 0 OR
                EXISTS(MOTHER) EQ 0) NEXT CASE
.          MOVE VARS FATHER, MOTHER
.          CASE IS FATHER
.               COMPUTE EDUCDAD = EDUCAT
.          END CASE
.          CASE IS MOTHER
.               COMPUTE EDUCMOM = EDUCAT
.          END CASE
.          PERFORM PROCS
.     END PROCESS CASES
SPSS SAVE FILE FILENAME = EDUCPAR
END RETRIEVAL
```

To begin, we throw away those people whose mother or father are not contained in the database. This situation is represented by a zero (i.e. missing) value in the MOTHER or FATHER column. We use the EXISTS function to detect the missing values.

If the person´s mother and father are both in the database, we go directly to the FATHER case and retrieve his education level and then directly to the MOTHER case and retrieve her education level.

Since we have now retrieved all the information we need, we use PERFORM PROCS to write the summary record to the SPSS SAVE FILE procedure.

D. <u>Retrieval Problem 2 (Part 1)</u>

In this retrieval, we want to generate summary records containing:

- mother´s age at birth of her last child
- sex of her last child

This retrieval implies that we can move efficiently from a mother´s record to her children´s records. However, the database as it is originally constituted, only allows us to move efficiently from a child to its parents (backward-link) but not the other way (forward-link).

Therefore, we create such a forward-link by running an intermediate retrieval step and creating new records in the database linking parents to their children.

```
RETRIEVAL UPDATE
.    PROCESS CASES ALL
.         COMPUTE CHILDID = PERSONID
.         MOVE VARS FATHER, MOTHER
.         IF THEN (EXISTS(FATHER) EQ 1)
.              CASE IS FATHER
.                   NEW RECORD IS CHILD, (CHILDID)
.                   END RECORD
.              END CASE
.         END IF
.         IF THEN (EXISTS(MOTHER) EQ 1)
.              CASE IS MOTHER
.                   NEW RECORD IS CHILD, (CHILDID)
.                   END RECORD
.              END CASE
.         END IF
.    END PROCESS CASES
END RETRIEVAL
```

The procedure is called a RETRIEVAL UPDATE because we will actually be modifying the database. As we loop through the cases, we will create a new type 2 (CHILD) record for each person's parents if they exist in the database.

E. <u>Retrieval Problem 2 (Part 2)</u>

Now we are ready to perform the retrieval for mothers and their last child.

We begin with a loop over all the cases and throw out father records (SEX EQ 2) or mothers with no children (COUNT(2) EQ 0).

Next we store the mother's year of birth in YRMOM. This will be used later to determine the mother's age at the birth of her last child.

We then initialize the variables YRCHILD and SEXCHILD for the search to follow.

Then we loop through the mother-child link records (PROCESS REC CHILD) and for each one we:

- determine if the child is younger than any other encountered for this mother.
- if the child is the youngest so far, we save the child's year of birth and sex.

Finally, we compute the mother's age at the birth of her youngest child and ship the summary record off to the SPSS SAVE FILE procedure.

A crosstabulation of BIRTHAGE by SEXCHILD gives the desired table.

```
RETRIEVAL
.    PROCESS CASES ALL
.        IF (SEX EQ 2 OR COUNT(2) EQ 0) NEXT CASE
.        COMPUTE YRMOM = BIRTHYR
.        SET YRCHILD (0)/ SEXCHILD (NMISSING)
.        PROCESS REC CHILD
.             MOVE VARS CHILDID
.             CASE IS CHILDID
.                  IF THEN (BIRTHYR GT YRCHILD)
.                       COMPUTE YRCHILD = BIRTHYR
.                       COMPUTE SEXCHILD = SEX
.                  END IF
.             END CASE
.        END PROCESS REC
.        COMPUTE AGEBIRTH = YRCHILD - YRMOM
.        PERFORM PROCS
.    END PROCESS CASES
SPSS SAVE FILE FILENAME = MOTHERS
END RETRIEVAL
```

A Typical Case in the TRIPS Database

Household (Type 1)

HOUSEID	RT	
BR29	1	....

Cars (Type 2)

HOUSEID	RT	CARID		YEAR	
BR29	2	01	....	1976	....
BR29	2	02	....	1980	....

Persons (Type 3)

HOUSEID	RT	PERSONID		AGE	
BR29	3	01	....	44	......
BR29	3	02	....	42	......
BR29	3	03	....	19	......
BR29	3	04	....	16	......

Trips (Type 4)

HOUSEID	RT	PERSONID	TRIPID		OWNCAR		DAYS
BR29	4	01	01	...	02	..	14
BR29	4	03	01	...	01	..	2
BR29	4	03	02	...	01	..	3
BR29	4	03	03	...	00	..	5
BR29	4	04	01	...	02	..	10

Typical Cases from the PEOPLE Database

PERSONID	SEX	EDUCAT	BIRTHYR	MOTHER	FATHER
0401	1	16	1947	0000	0000
0428	2	19	1945	0000	0000
0531	1	7	1969	0401	0428
0623	1	1	1975	0401	0428
0915	2	5	1971	0401	0428
1213	2	0	1978	0401	0428

A Parent Case After Addition of Child Records

PERSONID	SEX	EDUCAT	BIRTHYR	MOTHER	FATHER
0401	1	16	1947	0000	0000

PERSONID	CHILDID
0401	0531
0401	0623
0401	0915
0401	1213

Summary Records Created from Retrievals

Retrieval 1 (TRIPS Database)

NCAR	NPEOP
2	3

Retrieval 2 (TRIPS Database)

PERSONID	OWNCAR	AGE	YEAR
01	02	44	1980
03	01	19	1976
04	02	16	1980

Retrieval 1 (PEOPLE Database)

FATHER	MOTHER	EDUCDAD	EDUCMOM
0428	0401	19	16
0428	0401	19	16
0428	0401	19	16
0428	0401	19	16

Retrieval 2 (PEOPLE Database)

YRMOM	YRCHILD	SEXCHILD	CHILDID	BIRTHAGE
1947	1978	2	1213	31

VOLUME TESTING OF STATISTICAL SOFTWARE -- THE STATISTICAL ANALYSIS SYSTEM (SAS)

Arnold W. Bragg
SAS Institute and North Carolina State University
PO Box 5847, Raleigh NC 27650

ABSTRACT

Solutions for two complex file management problems are proposed using the Statistical Analysis System (SAS). SAS is an integrated system for data management, statistical analysis, data reduction and summarization, color graphics, and report writing. Several fundamental concepts of SAS are reviewed and four methods of solution are suggested. Detailed descriptions of each of the problem solutions are presented, including the input/output volume at each stage (a reasonable performance metric for comparing similar systems). Comparisons among the four methods are discussed, and program listings for each solution are included.

KEYWORDS

Volume testing, SAS, the Statistical Analysis System, data management, complex data structures, statistical software, large data bases

INTRODUCTION

The Statistical Analysis System (SAS) is an integrated system for reading, editing, retrieving, transforming, manipulating, managing, and maintaining data. SAS has facilities for statistical analysis (including econometric and time series analysis), large-scale data reduction and summarization, report writing, and color graphics. SAS operates on IBM 360/370 and compatible hardware under OS, OS/VS, and CMS and runs in batch (OS) as well as interactive (OS/TSO, CMS) execution modes. Because SAS is a unified system for data management and analysis the SAS user can perform complex file management tasks, tabulation, statistical analysis, and graphical display without having to access other software packages or systems.

TWO PROBLEMS IN MANAGING DATA HAVING A COMPLEX STRUCTURE

Two complex file/data handling problems have been proposed (TEI81), hereinafter denoted 'problem 1' and 'problem 2'. Each problem consists of two tasks. The data structure of problem 1 (the TRIPS data collection) is a hierarchical representation of four record types:

HOUSES (House + other variables)

CARS (House, Car, Model, + other variables)

PERSONS (House, Person, Age, + others)

TRIPS (House, Person, Trip, Owncar, Duration + other variables)

Each HOUSES record contains zero or more CARS records and zero or more PERSONS records. Each PERSONS record contains zero or more TRIPS records. Task 1 requests a tabulation of households by the number of cars owned and by the number of persons over 16 years of age in the household. Task 2 requests a tabulation of trips of 3 days or longer duration taken in a car belonging to the household by the age of the person taking the trip and by the model year of the car.

The data structure of problem 2 (the PERSONS data collection) is based upon a single record type:

PEOPLE (Person, Sex, Education, Birthyear, Mother, Father + other variables)

Task 1 requests a tabulation of offspring by the educational level of the mother and by the educational level of the father. Task 2 requests a tabulation of mothers by the age of the mother at the birth of the last child and by the sex of the last child born. A more detailed specification of each problem with its associated data description can be found in (TEI81 appearing in this volume) and in (COU80).

SAS DATASETS, PROCEDURES, PROGRAMS, AND DATA TYPES

Before presenting a possible solution to the complex file problems, we must briefly review several of the fundamental concepts of SAS. A SAS dataset consists of a collection of data values arranged in a rectangular table. The rows in this table are SAS observations and the columns are SAS variables. A SAS dataset may have up to 4000 variables; there is no system imposed limit on the number of observations a SAS dataset may contain. A SAS observation usually corresponds to one record of raw data; each field in the record generally is represented by one SAS variable. A file of N homogeneous records each comprised of M fields might be portrayed as a SAS dataset of N observations and M variables:

	var 1	var 2	. . . .	var M
obs 1				
obs 2				
.				
.				
.				
obs N				

For example, the second problem in our volume test reads a collection of records which we can represent as a SAS dataset named PEOPLE:

	PERSON	SEX	EDUC	MOTHER	FATHER
obs 1	Sarah	F	8	0	0
.					
.					
obs 33	John	M	4	0	0
.					
.					
obs 3200	Ann	F	12	Sarah	John
.					
.					

Relationships among observations in this example are represented by the variables MOTHER and FATHER: Ann's (person 3200) mother is Sarah (person 1); Ann's father is John (person 33).

There are several major advantages in representing a file of data records as a SAS dataset: SAS datasets are the standard interface between raw data and the library of SAS analytical and statistical programs (called SAS procedures or PROCs); SAS datasets contain, in addition to actual data values, descriptive information about each of the variables in the dataset such as the name, type, length, format, and label of the variable as well as the source statements used to create the SAS dataset; SAS datasets can be edited, transformed and reshaped (observations added or deleted, variables added or deleted); two or more SAS datasets can be concatenated, interleaved observation by observation, merged left to right, and updated to yield new SAS datasets; SAS datasets can be saved on disk or tape storage devices for processing at a later time. There is no limit to the number of SAS datasets one can create and/or manipulate during a SAS job.

SAS procedures can be classified into several broad categories: descriptive statistics, analysis of designed experiments, regression analysis, multivariate analysis, econometric and time series analysis, graphics, interfaces with other systems, SAS utilities, OS utilities, and special purpose. The SAS PROCedure FREQ will be used to generate the required crosstabulations for problems 1 and 2.

SAS programs consist of collections of free-format SAS statements having a PL-1 like syntax. SAS supports arrays of variables (character and numeric), IF-THEN-ELSE, DO-END, DO-WHILE, DO-UNTIL, DO-OVER, and iterative DO constructs as well as the standard relational, Boolean, arithmetic, and string operators and functions. Excellent error diagnostics assist the SAS user in correcting syntax errors as well as detecting errors in the data.

SAS variables can be represented either as character or numeric data types. Character variable values can range from 1 to 200 bytes in length; numeric variable values can be stored in 2 to 8 bytes and are represented as real binary (floating point).

ASSUMPTIONS AND METHODS

1. Raw data can be read directly from the TRIPS (problem 1) and PERSONS (problem 2) data collections; it is not necessary to translate, decode, reformat, or restructure the data prior to processing.

2. Because SAS is a unified data management/data analysis system, it is not necessary to invoke other systems either for data management or for crosstabulations.

3. SAS requires no codebook, dictionary, or schema definitions. The SAS program statements in Figures 1 and 2 are the complete programs necessary for solving problems 1 and 2 respectively.

4. Because SAS supports character as well as numeric data types, a considerable savings in both input/output volume and CPU resources can be realized in this application by defining item values which will undergo no arithmetic processing as character variables. Table 1 illustrates the savings in CPU resources.

5. Further optimization is possible by KEEPing only those variables required for solution of each problem. Therefore, even though the original datasets contain all variables read, the working datasets which are sorted, merged, and crosstabulated might contain only a small subset of the total variable set. Table 1 illustrates this savings in CPU resources.

6. Variable placement in the problem datasets is immaterial.

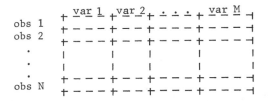

```
STAGE   PROGRAM STATEMENTS

        //ONE JOB . . .
        // EXEC SAS
        //IN1 DD UNIT=TAPE,VOL=SER=TEITEL,LABEL=(1,NL),DISP=OLD,DCB=(RECFM=FB,LRECL=120,BLKSIZE=2400)
        //SYSIN DD *

        *--READ TAPE FOR PROBLEM 1 AND BUILD WORKING DATASETS---------------------------------;

  1     DATA HOUSES(KEEP=HOUSE RECTYPE ONEA01-ONEA51 TWOA01-TWOA20 FOURA01-FOURA05)
             CARS(KEEP=HOUSE RECTYPE CAR YEAR ONEB01-ONEB51 TWOB01-TWOB20 FOURB01-FOURB04)
             PERSONS(KEEP=HOUSE RECTYPE PERSON AGE ONEC01-ONEC51 TWOC02-TWOC20 FOURC01-FOURC05)
             TRIPS(KEEP=HOUSE RECTYPE PERSON TRIP OWNCAR DAYS ONED02-ONED51 TWOD01-TWOD20 FOURD01-
                      FOURD04) ;
             INFILE IN1 ;
             RETAIN PERSON ;
             INPUT HOUSE $CHAR4. RECTYPE $ 5 @ ;
             IF RECTYPE='1' THEN DO ; INPUT @10 (ONEA01-ONEA51) ($1.) (TWOA01-TWOA20) ($2.)
                (FOURA01-FOURA05) ($4.) ; OUTPUT HOUSES ; END ;
             ELSE IF RECTYPE='2' THEN DO ; INPUT CAR $ 6-7 (TWOB01-TWOB20) ($2.) (FOURB01-FOURB05)
                ($4.) ; YEAR=FOURB05 ; OUTPUT CARS ; END ;
             ELSE IF RECTYPE='3' THEN DO ; INPUT @6 PERSON $CHAR2. @10 (ONEC01-ONEC51) ($1.)
                (TWOC01-TWOC20) ($CHAR2.) (FOURC01-FOURC05) ($4.) ; AGE=TWOC01 ; OUTPUT PERSONS ; END ;
             ELSE IF RECTYPE='4' THEN DO ; INPUT @8 TRIP $CHAR2. (ONED01-ONED51) ($1.)
                (TWOD01-TWOD20) ($CHAR2.) (FOURD01-FOURD05) ($4.) ; OWNCAR=ONED01 ; DAYS=TWOD20 ;
                OUTPUT TRIPS ; END ;
             ELSE PUT _INFILE_ ;

        *--COMPLEX FILE PROBLEM 1, TASK 1------------------------------------------------------;

  2     DATA ; MERGE HOUSES(IN=H) PERSONS CARS ; BY HOUSE ; RETAIN LCAR LPERSON ;
             CARCOUNT+(LCAR¬=CAR) ; PERCOUNT+((LPERSON¬=PERSON)*(AGE>'16')); LCAR=CAR ;
             LPERSON=PERSON ; IF LAST.HOUSE ; IF H THEN OUTPUT ; CARCOUNT=. ; PERCOUNT=. ;
             LCAR=' ' ; LPERSON=' ' ;

  3     PROC FREQ ; TABLES CARCOUNT*PERCOUNT / NOCOL NOROW NOPERCENT ;
             TITLE HOUSEHOLDS:  NUMBER OF CARS BY NUMBER OF PERSONS OVER 16 ;
             TITLE3 CARCOUNT DENOTES NUMBER OF CARS OWNED ;
             TITLE5 PERCOUNT DENOTES NUMBER OF PERSONS OVER 16 YEARS OF AGE IN HOUSEHOLD ;
             TITLE7 ' ' ;

        *--COMPLEX FILE PROBLEM 1, TASK 2------------------------------------------------------;

  4     PROC SORT DATA=PERSONS ; BY HOUSE PERSON ;

  5     DATA TRIPPERS ; MERGE TRIPS PERSONS(IN=P) ; BY HOUSE PERSON ; IF DAYS>=' 3' AND P AND
             OWNCAR>'0' ;

  6     PROC SORT DATA=TRIPPERS ; BY HOUSE OWNCAR ;

  7     DATA ; MERGE TRIPPERS(IN=T RENAME=(OWNCAR=CAR)) CARS(IN=C) ; BY HOUSE CAR ; IF T AND C ;

  8     PROC FREQ ; TABLES AGE*YEAR / NOCOL NOROW NOPERCENT ;
             TITLE TRIPS LONGER THAN 2 DAYS:  AGE BY YEAR OF CAR ;
             TITLE3 AGE DENOTES AGE OF PERSON TAKING TRIP ;
             TITLE5 YEAR DENOTES YEAR CAR WAS BUILT ;
             TITLE7 ' ' ;

        /*
        //
```

Figure 1.
SAS program statements for complex file problem 1, tasks 1 and 2.

```
STAGE   PROGRAM STATEMENTS

        //TWO JOB . . .
        // EXEC SAS
        //IN2 DD UNIT=TAPE,VOL=SER=TEITEL,LABEL=(2,NL),DISP=OLD,DCB=(RECFM=FB,LRECL=120,BLKSIZE=2400)
        //SYSIN DD *

        *--READ TAPE FOR PROBLEM 2 AND BUILD WORKING DATASET-------------------------------;

1       DATA PEOPLE(DROP=ONE01 TWO01) ;
        INFILE IN2 ;
        INPUT PERSON $CHAR5. (ONE01-ONE51) ($1.) (TWO01-TWO15) ($CHAR2.) @87 BIRTHYR $CHAR4.
             (FOUR02-FOUR06) ($CHAR4.) @111 MOTHER $CHAR5. FATHER $CHAR5. ;
             SEX=ONE01 ; EDUC=TWO01 ;

        *--COMPLEX FILE PROBLEM 2, TASK 1-------------------------------------------------;

2       PROC SORT DATA=PEOPLE ; BY FATHER ;

3       PROC SORT DATA=PEOPLE OUT=PARENT(RENAME=(FATHER=FTHR MOTHER=MTHR)) ; BY PERSON ;

4       DATA PA ; MERGE PEOPLE(IN=P) PARENT(RENAME=(PERSON=FATHER EDUC=FEDUC)) ; BY FATHER ; IF P ;

5       PROC SORT DATA=PA ; BY MOTHER ;

6       DATA MA ; MERGE PA(IN=P) PARENT(RENAME=(PERSON=MOTHER EDUC=MEDUC)) ; BY MOTHER ; IF P ;

7       PROC FREQ ; TABLES MEDUC*FEDUC / NOCOL NOROW NOPERCENT MISSING ;
             TITLE PERSONS:  MOTHER''S EDUCATION BY FATHER''S EDUCATION ;
             TITLE3 OFFSPRING COUNT ;
             TITLE5 MEDUC DENOTES MOTHER''S EDUCATION ;
             TITLE7 FEDUC DENOTES FATHER''S EDUCATION ;
             TITLE9 ' ' ;

        *--COMPLEX FILE PROBLEM 2, TASK 2-------------------------------------------------;

8       PROC FORMAT ; VALUE $SEXFMT        1=FEMALE
                                           2=MALE            ;

9       PROC SORT DATA=PEOPLE OUT=MA(RENAME=(MOTHER=MTHR)) ; BY PERSON ;

10      PROC SORT DATA=PEOPLE ; BY MOTHER ;

11      DATA AGES ; MERGE PEOPLE(IN=P) MA(IN=M RENAME=(PERSON=MOTHER BIRTHYR=MBIRTHYR)) ;
             BY MOTHER ; IF P AND M ; BIRTHAGE=BIRTHYR-MBIRTHYR ;

12      PROC SORT DATA=AGES ; BY MOTHER BIRTHAGE ;

13      DATA AGESEX ; SET AGES ; BY MOTHER ; IF LAST.MOTHER ;

14      PROC FREQ ; TABLES BIRTHAGE*SEX / NOROW NOCOL NOPERCENT ;
             TITLE MOTHERS:  AGE AT LAST CHILDBIRTH BY SEX OF LAST CHILD ;
             TITLE3 LAST BIRTH COUNT ;
             TITLE5 BIRTHAGE DENOTES MOTHER''S AGE ;
             TITLE7 SEX DENOTES SEX OF LAST OFFSPRING ;
             TITLE9 ' '  ;     FORMAT SEX $SEXFMT. ;

         /*
         //
```

Figure 2.
SAS program statements for complex file problem 2, tasks 1 and 2.

7. Any or all variables may function as sort keys.

8. The problem data specification indicates that both files (problem 1 and problem 2) are in sorted id order. This fact will be useful in at least one stage of both solutions.

9. For documentation, each DATAset operation and PROCedure invocation will be assigned a reference number. These numbers (or STAGES) will be noted in subsequent sections and in Figures 1 and 2.

10. Each problem has been solved in four slightly differing ways to illustrate the levels of performance associated with each assumption. Table 1 summarizes CPU time for each task in each of these methods. Method 1 uses the character data type for all variables and maintains all variables read as the width component of volume testing. The method 1 solution is documented in subsequent sections and is illustrated in Figures 1 and 2. Method 2 uses the numeric data type while maintaining all variables in all datasets. Method 3 uses the character data type but maintains only those variables needed at each stage. Method 4 uses the numeric data type for this subset of variables.

Having noted our working assumptions, we present method 1 solutions to problems 1 and 2. Each problem will be divided into two tasks, and each task further subdivided into STAGES consisting of a SAS DATA or PROC step.

SOLUTION TO PROBLEM 1, TASK 1

STAGE 1: Read the TRIPS tape file (7905 records, 120 bytes each, total bytes read=948,000) writing four SAS datasets: HOUSES (300 observations, 78 variables, record length=120, total bytes =36,000), CARS (1046 observations, 79 variables, record length=122, total bytes=127,612), PERSONS (1194 observations, 79 variables, record length=122, total bytes=145,668), and TRIPS (5365 observations, 80 variables, record length=122, total bytes=654,530). The four SAS datasets comprise 963,810 bytes.

STAGE 2: Merge datasets HOUSES (300 observations, 78 variables, record length=120, total bytes=36,000), PERSONS (1194 observations, 79 variables, record length=122 bytes, total bytes=145,668), and CARS (1046 observations, 79 variables, record length=122 bytes, total bytes=127,612) by the common variable HOUSE. The number of cars in the household (CARCOUNT) and the number of persons over sixteen years of age in the household (PERCOUNT) are counted as the observations are merged. The resulting dataset DATA1 (300 observations, 236 variables, record length=366, total bytes=109,800) consists of one observation for each household record.

STAGE 3: Invoke the SAS crosstabulation routine PROC FREQ to tabulate variables CARCOUNT (the number of cars in the household) vs. PERCOUNT (the number of persons over sixteen years of age in the household) appropriately titled. Dataset DATA1 (300 observations, 236 variables, record length=366, total bytes=109,800) is processed to conclude the task.

SOLUTION TO PROBLEM 1, TASK 2

STAGE 4: Sort the PERSONS dataset (1194 observations, 79 variables, record length=122, total bytes=145,668) by the variables HOUSE and PERSON.

STAGE 5: Merge datasets TRIPS (5365 observations, 80 variables, record length=122, total bytes=654,530) and PERSONS (1194 observations, 79 variables, record length=122, total bytes=145,668) by the common variables HOUSE and PERSON. The resulting dataset TRIPPERS (3403 observations, 156 variables, record length=233, total bytes= 792,899) consists of observations for persons taking trips of at least 3 days duration in cars owned by the household.

STAGE 6: Sort the TRIPPERS dataset (3403 observations, 156 variables, record length=233, total bytes=792,899) by the variables HOUSE and OWNCAR.

STAGE 7: Merge datasets TRIPPERS (3403 observations, 156 variables, record length=233, total bytes=792,899) and CARS (1046 observations, 79 variables, record length=122, total bytes=127,612) by the common variables HOUSE and CAR. The variable OWNCAR in the TRIPPERS dataset is renamed CAR for this merge. The resulting dataset DATA2 (3403 observations, 232 variables, record length= 344, total bytes=789,496) consists of observations matching each trip selected in STAGE 5 with the model year of the car in which the trip was taken.

STAGE 8: Invoke the SAS crosstabulation routine PROC FREQ to tabulate variables AGE (the age of the person taking the trip) vs. YEAR (the model year of the car) appropriately titled. Dataset DATA2 (3403 observations, 232 variables, record length=344, total bytes=789,496) is processed to conclude the task.

SOLUTION TO PROBLEM 2, TASK 1

STAGE 1: Read the PERSONS tape file (8000 records, 120 bytes each, total bytes read=960,000) writing one SAS dataset: PEOPLE (8000 observations, 75 variables, record length=124, total bytes= 992,000). The SAS dataset comprises 992,000 bytes.

STAGE 2: Sort the PEOPLE dataset (8000 observations, 75 variables, record length=124, total bytes=992,000) by the variable FATHER.

STAGE 3: Sort the PEOPLE dataset (8000 observations, 75 variables, record length=124, total bytes=992,000) by the variable PERSON. Name the sorted dataset PARENT (8000 observations, 75 variables, record length=124, total bytes=992,000) renaming variables FATHER and MOTHER to FTHR and MTHR respectively.

STAGE 4: Merge datasets PEOPLE (8000 observations, 75 variables, record length=124, total bytes=992,000) and PARENT (8000 observations, 75 variables, record length=124, total bytes=992,000) by the common variable FATHER. The variables PERSON and EDUC in the PARENT dataset are renamed FATHER and FEDUC respectively for this merge. The resulting dataset PA (8000 observations, 78 variables, record length=136, total bytes=1,088,000) consists of observations matching persons with their fathers.

STAGE 5: Sort the dataset PA (8000 observations, 78 variables, record length=136, total bytes =1,088,000) by the variable MOTHER.

STAGE 6: Merge datasets PA (8000 observations, 78 variables, record length=136, total bytes= 1,088,000) and PARENT (8000 observations, 75 variables, record length=124, total bytes=992,000) by the common variable MOTHER. The variables PERSON and EDUC in the PARENT dataset are renamed MOTHER and MEDUC respectively for this merge. The resulting dataset MA (8000 observations, 79 variables, record length=138, total bytes=1,104,000) consists of observations matching persons merged in STAGE 4 with their mothers.

STAGE 7: Invoke the SAS crosstabulation routine PROC FREQ to tabulate variables MEDUC (mother's education) vs. FEDUC (father's education) appropriately titled. Dataset MA (8000 observations, 79 variables, record length=138, total bytes =1,104,000) is processed to conclude the task.

SOLUTION TO PROBLEM 2, TASK 2

STAGE 8: Using PROC FORMAT, define a SAS format to translate encoded SEX variable values 1 and 2 to literal strings 'FEMALE' and 'MALE' respectively on output (see STAGE 14).

STAGE 9: Sort the PEOPLE dataset (8000 observations, 75 variables, record length=124, total bytes=992,000) by the variable PERSON. Name the sorted dataset MA (8000 observations, 75 variables, record length=124, total bytes=992,000) renaming the variable MOTHER to MTHR.

STAGE 10: Sort the PEOPLE dataset (8000 observations, 75 variables, record length=124, total bytes=992,000) by the variable MOTHER.

STAGE 11: Merge datasets PEOPLE (8000 observations, 75 variables, record length=124, total bytes=992,000) and MA (8000 observations, 75 variables, record length=124, total bytes=992,000) by the common variable MOTHER. The variables PERSON and BIRTHYR in the MA dataset are renamed MOTHER and MBIRTHYR respectively for this merge. The resulting dataset AGES (6000 observations, 78 variables, record length=141, total bytes= 846,000) consists of observations matching mothers with their children. The mother's age at birth of the child, BIRTHAGE, is calculated for each observation.

STAGE 12: Sort the AGES dataset (6000 observations, 78 variables, record length=141, total bytes=846,000) by the variables MOTHER and BIRTHAGE.

STAGE 13: Subset the AGES dataset (6000 observations, 78 variables, record length=141, total bytes=846,000). The resulting dataset AGESEX (2400 observations, 78 variables, record length=141, total bytes=338,400) consists of observations for the last child born to each mother.

STAGE 14: Invoke the SAS crosstabulation routine PROC FREQ to tabulate variables BIRTHAGE (the mother's age at the birth of her last child) vs. SEX (the child's sex) appropriately titled. The variable SEX is decoded on output based on value definitions specified in STAGE 8 using PROC FORMAT. Dataset AGESEX (2400 observations, 78 variables, record length=141, total bytes= 338,400) is processed to conclude the task.

METHOD COMPARISONS

Table 1 illustrates the wide differences in CPU time among the four methods investigated. Performance can be optimized by defining variables which undergo no arithmetic processing as character rather than numeric data types. For example, the string '7' stored as a single character (one byte) value rather than in its eight byte real binary numeric representation achieves a considerable economy in input/output volume; this economy is significant when large numbers of variables are involved (again, the width component of volume testing). Similarly, SAS allows users to specify the LENGTH of numeric variables allowing numeric variable values to be stored in fewer than eight bytes, although when these values are used for processing in later DATA or PROC steps their representations are extended to eight bytes by appending non-significant zeroes.

A comparison of method 1 vs. 2 indicates a savings of nearly 40% in CPU time for problems 1 and 2 can be achieved by using character rather than eight byte numeric variable representations. An equally dramatic reduction in input/output volume is likewise achieved. For smaller numbers of variables the magnitude of the savings is understandably less, though certainly no less significant; method 3 saves 6-10% over method 4's numeric representation.

Although contrary to the problem specification and perhaps to the entire notion of volume testing, reducing the number of variables maintained in working datasets at intermediate stages can also save resources. Methods 1 vs. 3 and 2 vs. 4 differ in that all non-essential variables have been DROPped from consideration in the latter cases, resulting in savings of up to 76%. Combining both notions of economy, we see that method 3 executes

METHOD	DATA TYPE	VARIABLES MAINTAINED	PROBLEM 1 INPUT	PROBLEM 1 TASK 1	PROBLEM 1 TASK 2	PROBLEM 1 TOTAL	PROBLEM 2 INPUT	PROBLEM 2 TASK 1	PROBLEM 2 TASK 2	PROBLEM 2 TOTAL
1	CHAR	ALL	7.13 sec	0.80 sec	4.94 sec	12.87 sec	5.84 sec	10.29 sec	7.83 sec	23.96 sec
2	NUMERIC	ALL	11.48	1.02	8.27	20.77	10.07	15.92	12.20	38.19
3	CHAR	ESSENTIAL ONLY	1.15	0.45	2.85	4.45	1.30	7.59	5.59	14.48
4	NUMERIC	ESSENTIAL ONLY	1.39	0.46	3.05	4.90	1.59	7.90	5.83	15.32

Table 1.
Varying the variables' data type and the number of variables maintained throughout the intermediate stages can have a dramatic effect on the CPU time utilization for each task.

in 18.93 CPU seconds vs. 58.96 CPU seconds for method 2, less than one-third the time for the worst case. All runs were made on an Amdahl 470/V8 running MVS and utilizing model 3330-1 direct access storage devices; SAS release 79.3B was the production version. These figures are intended only to illustrate the variation in performance of the four solution strategies and are essentially meaningless in other contexts.

In conclusion, we note that the current production version of SAS is capable of solving both complex file manipulation problems; furthermore, as an integrated data management/data analysis system, it is a simple matter to invoke a data reduction and summarization routine such as PROC FREQ to generate the required crosstabulations. A five-fold savings in input/output volume and a 40% reduction in CPU time can be achieved with a prudent selection of data variable representations; ignoring non-essential variables in intermediate stages can save more than 75% in CPU time and several orders of magnitude in total input/output volume for these specific problems in data management.

REFERENCES

COU80 Council, K. A. (editor). SAS Applications Guide - 1980 Edition. SAS Institute Inc. Cary NC. 1980. pp. 67-70.

TEI81 Teitel, Robert F. "Volume Testing of Statistical Software." Proc. Computer Sci. and Statistics: 13th Symp. on the Interface. Springer-Verlag. March 1981. Appearing in this volume.

SOLVING COMPLEX DATABASE PROBLEMS IN P-STAT

Roald and Shirrell Buhler, P-STAT, Inc.

ABSTRACT

At the 13th Interface, six different packages submitted their solutions to four problems designed by Robert F. Teitel -- two each for two different data bases. P-STAT proved to be very well suited for handling these problems. The tasks involving the first data base required only a single pass through the system data file even in the standard version of P-STAT. The problems involving the second data base were also easy to do in standard P-STAT even though they required several passes of the file. However, these problems could also be done in a single pass when the file was built using P-STAT's new P-RADE database enhancement.

Keywords: Complex data base; aggregation; direct access retrieval; hierarchical structure; P-STAT programming language.

1. THE PROBLEMS.

The first two problems involved a file with a hierarchical structure.

```
H       A household record, followed by
  C     0 to 9 car records, followed by
  P     0 or more person records.
        A person record may be followed
    T   by trip records.
  P
  P
    T
    T
  P
H       And then the next household is
        followed by its car, person,
            and trip records.
```

Each record type contains the household id number. Car records contain a car number and the model year of the car. Person records contain the person's age. Trip records contain the number of the car that was used if the trip was made in a family car.

The first problem involved aggregation across households to produce a table of household counts on persons over 16 by number of cars. The second problem involved moving infomation from car and person records to the trip records to produce a count of trips taken, showing person's age by car model year.

These problems can both be done directly by TABLES, the crosstabulation command, using the P-STAT programming language to do the aggregation and selection.

The second database contained records on individuals including their sex, education, birthdates, and pointers to their mother's record and their father's record. Zero in the mother id or father id fields indicated that that record was not present in the file.

ID	SEX	EDUCA TION	YEAR OF BIRTH	MOTHERS ID	FATHERS ID
1	1	16	1929	6	7
2	2	12	1935	6	7
3	1	12	1932	8	0
4	2	8	1905	15	17
5	2	10	1910	15	17
6	2	12	1900	33	44
7	1	12	1898	0	0
8	2	8	1910	0	56

etc.

The third problem involved locating the records for each person's mother and father and then producing a table of mother's education by father's education. Given the data above, this required looking at records 6 and 7 to get parent information while processing the first person in the file.

The fourth problem was to construct a table of mother's age by the sex of her last child. This required accessing the records in the file in birth within mother order, finding the last born child and then retrieving the mother's birth year from the mother record.

Both of these problems are easy to do in standard P-STAT even though it takes several commands: two commands to sort the file, once in father's id order and once in mother's id order, followed by two COLLATE commands to join the information back to the individual records before the tables can be made. In the P-RADE enhanced version of P-STAT, each table can be done directly from the P-RADE database in a single step.

All of the problems can be handled in any version of P-STAT 78. The solutions here were done in release 6.0 and may be somewhat different from solutions in earlier releases because of new enhancements to the P-STAT programming language.

2. BUILDING THE FIRST DATABASE.

The first complex data base contains four record types: 1) households; 2) cars; 3) people; and 4) trips. The raw data file is in household order. Each household may contain some number of cars and some number of persons. Each person may or may not have some associated trips.

The layout of the four records permitted the necessary variables to be cleanly accessed and named. This allowed a single P-STAT system file to be built rather than four separate files, eliminating the need for sorts or collates to move the data around. The one, obvious consequence of mixing record types within one file is that all references to original variables need to be preceeded by a test on type of record. This is, however, very simple within P-STAT and the single file can be processed quite naturally.

The original data file contained 80 variables. These were all retained when the problems were run but are omitted here because they add nothing to this paper and detract from clarity.

```
DATA = TRIPS,
  NV = 7,
  LENGTH = 120,
  UNIT = TRIPS.INPUT,
  DES = TD$
*LAB   ID, RECORD.TYPE, CAR.NUMBER,
*LAB   CAR.USED, AGE, TRIP.LENGTH,
*LAB   MODEL.YEAR
*FMT ( I4, I1, I2, T10, I1, T61, I2,
*FMT   T99, I2, T117, I4 )
*READ
```

2.1 Problem 1.

The first problem was number of cars by number of drivers in a household. This can be solved by using the TABLES command and doing the aggregation on-the-fly in P-STAT's programing language.

```
TABLES, IN = TRIPS
(IF RECORD.TYPE=1, SET P(1) TO 0,
                   SET P(2) TO 0 )
(IF RECORD.TYPE=2, SET P(1) TO P(1) + 1)
(IF RECORD.TYPE=3  .AND. AGE > 16,
                   SET P(2) TO P(2) + 1)
(IF LAST (ID), RETAIN )
(SETX CARS             TO P(1) )
(SETX PERSONS.OVER.16 TO P(2) )
(C CARS, PERSONS.OVER.16 )      $

T = ( HOUSEHOLD COUNT )
  CARS (0,9) BY PERSONS.OVER.16 (0,8) $
```

The two language elements of special interest here are the use of permanent variables and the LAST function. There are 20 permanent variables -- P(1) to P(20) -- which are available throughout the P-STAT run. They are set to missing at the beginning of the run and may be changed or used in the data modifications. Here the first two P variables are set to zero each time a new household record is read. P(1) is incremented each time a car record is read. P(2) is incremented each time a person record is read if the person's age is greater than 16.

The LAST function is tested to find the last record of a given household. Records which are not the last are dropped from further analysis. When the last record for a household is read the two P variables are placed in the new variables, CARS and PERSONS.OVER.16. These are then given to the TABLES command.

HOUSEHOLD COUNT

PERSONS.OVER.16

CARS	0	1-2	3+	ROW TOTALS
0	4	9	24	37
1-2	13	17	46	76
3+	22	48	117	187
TOTAL N	39	74	187	300

NOTE: There were 4 apparently vacant households which had neither car nor person records. These could have been omitted from the table if desired. This table was interactively post-processed within the TABLES command so that it would fit within the size constraints of this two column layout.

2.2 Problem 2.

The second table to be produced from the TRIPS file was age of person by year of car for each trip taken. This requires moving information from one record type to another. Again the use of P variables makes this easy.

```
TABLES, IN = TRIPS
  ( IF RECORD.TYPE = 2,
       SET P(CAR.NUMBER) TO MODEL.YEAR)
  ( IF RECORD.TYPE  = 3,
       SET P(20) TO AGE )
  ( IF RECORD.TYPE  = 4  .AND.
       TRIP.LENGTH >= 3  .AND.
       P(20) .INRANGE. (17, 99), RETAIN )
  ( SETX PERSONS.AGE TO P(20))
  ( SETX YEAR.OF.CAR TO P(CAR.USED))
  ( C PERSONS.AGE, YEAR.OF.CAR ),
 NO.PASS $

 T=( NUMBER OF TRIPS IN HOUSEHOLD CARS )
   ( BY FAMILY MEMBERS OVER 16 )
   ( LASTING 3 OR MORE DAYS )
 PERSONS.AGE (17,99) BY
 YEAR.OF.CAR (1960, 1981) $
```

In the car records, CAR.NUMBER contains the number of the car record within the household and MODEL.YEAR contains the year. When the first car record for a household is processed, CAR.NUMBER is 1 and, therefore, the model year of the first car is placed in P(1). When the third car record for a household is processed, CAR.NUMBER is 3 and, therefore, the model year for the third car is placed in P(3). These values will remain until the car records for the next household are read. When a person record is read, P(20) is set to that person's age.

As each person's trip records are read, they are examined to determine whether the trip was in a family car. CAR.USED is the variable which indicates which family car was used. If the trip was taken in the second family car (CAR.USED=2), then P(CAR.USED) is P(2) which contains the model year for that second family car.

NUMBER OF TRIPS IN HOUSEHOLD CARS
BY FAMILY MEMBERS OVER 16
LASTING 3 OR MORE DAYS

YEAR.OF.CAR

PERSONS AGE	1960-1969	1970-1979	1980-1981	ROW TOTALS
17-29	265	250	58	573
30-49	422	446	49	917
50 AND OVER	720	784	162	1666
TOTAL N	1407	1480	269	3156

NOTE: this table was also interactively post-processed within the P-STAT TABLES command. This was done to relabel and combine rows and columns so that they would fit within the limits of this paper.

3. BUILDING THE SECOND DATABASE.

The second data base is somewhat trickier than the first. Each case represents a person and contains background information. In addition it contains the ID field for their mother's and father's records. These fields are zero if the parent information is not in the file. Both problems require access to more than one record at the same time. The first step is to build a P-STAT system file. Again, extra variables which are not used in the problems are omitted from these examples.

```
DATA = PEOPLE,
   NV = 6,
   UNIT = PEOPLE.INPUT,
   DES = PD,
   LENGTH = 120$
*LAB   ID, SEX, EDUCATION,
*LAB   YEAR.OF.BIRTH,
*LAB   MOTHERS.ID, FATHERS.ID
*FMT ( I5, I1, T57, I2, T87, I4,
*FMT   T111, 2I5 )
*READ
```

Problems three and four require several commands in standard P-STAT but only one command in the P-RADE enhanced version. Both methods are shown here. To use P-RADE, a direct access file must first be made from the P-STAT file.

A P-RADE file has two parts. One, the file itself, can be thought of as a modified P-STAT file, stored on a direct access device (i.e., disk) so that each case can be independently accessed. The second part holds one or more keys.

A given case can be found very rapidly when its values on the key variables are provided. A key can also be used to read the entire file in the sort order of the variables that make up that key. A P-RADE file can, therefore, be read by any P-STAT command in any key order.

```
MAKE.PRADE.FILE,
     IN = PEOPLE,
     OUT = PEOPLE.PRADE,
     DES = PD,
     UNIT = PRADE,
     VAR = ID $
```

The primary key will be the person ID. A primary key must have a unique value for each case. A secondary key named SIBLINGS is also constructed using MOTHERS.ID and YEAR.OF.BIRTH as the key variables. This key facilitates the solution of the final problem. A P-RADE file may have up to 9 secondary keys. Secondary keys do not need to be unique for each case.

```
MAKE.PRADE.KEY,
     IN = PEOPLE.PRADE,
     KEY.NAME = SIBLINGS,
     VAR = MOTHERS.ID / YEAR.OF.BIRTH $
```

3.1.1 Problem 3, Using P-STAT Files.

The third problem is to produce a table of mother's education by father's education. The first step in standard P-STAT is to sort the file on the FATHERS.ID and create a second file in father order.

```
SORT,
  IN = PEOPLE,
  OUT = FATHER.ORDER,
  VAR = FATHERS.ID  $
```

The second step is to COLLATE the father's information back on to each child record. The file FATHER.ORDER contains all children sorted in the order of their fathers id numbers. The file PEOPLE contains the fathers in their own ID order. To make sure that father information is joined to each child we use the option REPEAT=RIGHT. Without this, the father information would be joined only to the first of his children. We also use FILL=LEFT to supply missing data for those people who did not have fathers in this file.

```
COLLATE,
  LEFT = FATHER.ORDER,
  RIGHT = PEOPLE
     ( SETX FATHERS.EDUC TO EDUCATION )
     ( C ID, FATHERS.EDUC ),
  REPEAT = RIGHT,
  LVAR = FATHERS.ID,
  RVAR = ID,
  DROP = RIGHT,
  FILL = LEFT,
  OUT = F1 $
```

The next step is to retrieve the mother information. This time we sort the file which has children and fathers into the order of the mother's id. We use YEAR.OF.BIRTH as a second sort variable so that the family file will be in the proper sort order for our final table.

```
SORT,
  IN = F1,
  OUT = MOTHER.ORDER,
  VAR = MOTHERS.ID / YEAR.OF.BIRTH $
```

The final step in making a family file is to collate the mother information in the same way that we collated the father information.

```
COLLATE,
 LEFT = MOTHER.ORDER,
 RIGHT = PEOPLE
  ( SETX MOTHERS.EDUC  TO EDUCATION )
  ( SETX MOTHERS.BIRTH TO YEAR.OF.BIRTH)
  ( C ID, MOTHERS.EDUC, MOTHERS.BIRTH ),
 REPEAT = RIGHT,
 LVAR = MOTHERS.ID,
 RVAR = ID,
 DROP = RIGHT,
 FILL = LEFT,
 OUT = FAMILIES $
```

The education data was originally coded as years of education. It is recoded into education groups on-the-fly, as the data is passed to the TABLES command.

```
TABLES, IN = FAMILIES
( C MOTHERS.EDUC, FATHERS.EDUC)
(FOR (1+) SET .X. TO NCOT(.X., 12, 16)),
  NO.PASS $

LABELS = (1) NO COLLEGE (2) COLLEGE
         (3) GRADUATE SCHOOL /,

T=MOTHERS.EDUC(1,3) BY FATHERS.EDUC(1,3)$
```

FATHERS.EDUC

MOTHERS EDUC	NO COLLEGE	COLLEGE	GRADUATE WORK	ROW TOTALS
NO COLLEGE	2843	947		3790
COLLEGE		316	945	1261
GRADUATE WORK	949			949
TOTAL N	3792	1263	945	6000

3.1.2 Problem 3, Using a P-RADE File.

When the input file is PEOPLE.PRADE, the table can be created directly by use of the LOOKUP function in the P-STAT programming language.

```
TABLES, IN = PEOPLE.PRADE
  ( SETX MOTHERS.EDUC TO NCOT
     ( LOOKUP ( EDUCATION, MOTHERS.ID ),
         12, 16 ))
  ( SETX FATHERS.EDUC TO NCOT
     ( LOOKUP ( EDUCATION, FATHERS.ID ),
         12, 16 ))
  ( C MOTHERS.EDUC, FATHERS.EDUC ) $

 LABELS = (1) NO COLLEGE (2) COLLEGE
          (3) GRADUATE SCHOOL /,

T=MOTHERS.EDUC(1,3) BY FATHERS.EDUC(1,3)$
```

LOOKUP is a function which looks up a value in another case. The arguments for the LOOKUP function are: 1) the name of the variable whose value is needed and; 2) the name of the variable in the current case which points to the "other" case. For each row in the file, MOTHERS.EDUC will be set to the variable EDUCATION as it is found in the record pointed to by the value in MOTHERS.ID. For example, if case 1034 has the value 378 in the MOTHERS.ID field, case 378 will be "looked up" and the value found there for EDUCATION will be given to the new variable MOTHERS.EDUC. FATHERS.EDUC will be the variable EDUCATION as it is found in the record pointed to by the value in FATHERS.ID.

It should be noted that using LOOKUP does increase the wall clock time of a pass, due to increased arm movement.

3.2 Problem 4.

The final table shows the mother's age when her last child was born by the sex of that child. This table can be directly produced from either the P-STAT file FAMILIES created for problem 3 or from the P-RADE file PEOPLE.PRADE.

When the FAMILIES file is used, MOTHERS.BIRTH is already in the file and it is only necessary to locate the youngest child and compute the mothers age at birth. The LAST function can be used because the file was already sorted on YEAR.OF.BIRTH within MOTHER.ID. This means that for each mother, the last child will be the youngest one. AGE.LAST.BIRTH is recoded on-the-fly into age groups.

When the table is created from the P-RADE file, it is necessary to use the LOOKUP function to retrieve the mothers year of birth. The age calculation and the recoding into age groups is done at the same time. Again, LAST can be used to locate the youngest child because the file is accessed using the SIBLINGS key created when the file was built.

```
TABLES, IN = FAMILIES
  ( IF LAST ( MOTHERS.ID ), RETAIN)
  ( C YEAR.OF.BIRTH, MOTHERS.BIRTH,
      MOTHERS.ID, SEX )
  ( SETX AGE.LAST.BIRTH TO YEAR.OF.BIRTH
        - MOTHERS.BIRTH )
  ( SET AGE.LAST.BIRTH TO NCOT
        (.., 18, 25, 45/5 )),
   NO.PASS $


 -- or if this is the P-RADE file ---

 TABLES, IN = PEOPLE.PRADE
   ( IF LAST ( MOTHERS.ID ), RETAIN )
   ( SETX AGE.LAST.BIRTH TO NCOT
     ( YEAR.OF.BIRTH - LOOKUP
           ( YEAR.OF.BIRTH, MOTHERS.ID ),
       18, 25, 45/5 )),
 KEY = SIBLINGS,
 NO.PASS $
```

The table definitions are identical whether the TABLES command references the P-STAT file or the P-RADE file.

```
 LABELS = SEX (1) MALE (2) FEMALE /
 AGE.LAST.BIRTH (1) 18 OR UNDER (2) 19-25
                (3) 26-30       (4) 31-35
                (5) 36-40       (6) 41-45
                (7) 46 + /,
```

```
T = ( MOTHER'S AGE AT LAST CHILDBIRTH )
    ( BY SEX OF LAST CHILD )
AGE.LAST.BIRTH BY SEX $
```

MOTHER'S AGE AT LAST CHILDBIRTH
BY SEX OF LAST CHILD

AGE LAST BIRTH	SEX: MALE	SEX: FEMALE	ROW TOTALS
18 OR UNDER	300		300
19-25	300	600	900
26-30	300		300
31-35	300	600	900
TOTAL N	1200	1200	2400

4. CONCLUSIONS.

These four problems are interesting because they serve to highlight the recent developments in package capabilities in the areas of language and file organization. Ten years ago, P-STAT had an integrated set of commands using a single type of file, enhanced by a programming language that did not, for example, have a LAST function. The problems, nonetheless, could have been done by P-STAT in 1971, using sequences of sorts and collates. In other words, the problems were solvable only by using explicit commands.

Today, we find that commands are but one of three components: language, command, and file organization, with which a package can solve problems like these four. Here the first two problems are solved using the P-STAT programming language. TABLES gets the data after all the work has been done.

The last two problems are solved using two components; the P-RADE file organization, which suits the problem nicely, and the P-STAT programming language, which makes the most of the file organization. Once again, each problem is solved as the file is read by the TABLES command. The CPU time benefit of single pass processing speaks for itself.

We see this trend accelerating. P-STAT will very likely have a third (transposed) file organization and additional language enhancements in time for the problems which surely will be presented at the next Interface.

VOLUME TESTING OF SPSS

Jonathan B. Fry, SPSS Inc.

ABSTRACT

Four complex file - handling problems proposed by Robert Teitel are posed in accompanying paper. A solution to one using the SPSS Batch System* is presented and the associated costs are detailed. Control language to solve all four problems using the SPSS-X system,* the follow-on system to SPSS presently under development, are also presented, along with estimates of the input/output volume required.

INTRODUCTION

Each of the participants has been invited to present the results, and especially the costs, of running the same set of complex file problems on the same data using his package. The purpose is to see how various approaches and implementations do on some typical problems of this class with substantial volumes of data.

It should come as no suprise to SPSS Batch System users that this author was able to solve only one of the four tabulation problems using SPSS alone. Since little would be gained by examination of the author's manipulations outside the package, only the one problem the author could solve will be presented.

However, the introduction of SPSS-X, the successor to SPSS, is in prospect, and that system is sufficiently far along that, while the problems cannot be run yet, the commands to run them can be layed out and the input/output volumes involved can be estimated with some precision. This is done here.

For the work done with SPSS, <u>SPSS</u>, Second Edition, serves as a reference. For the SPSS-X work, there are no public references yet: all the specifications are SPSS Inc. private documents.

SPSS RESULTS

The first step in the process was to convert the trips file to a SPSS system file. This was done because real users can be expected to want to save variable names, labels, and variables created or changed by writing a system file. The actual runs done here would be cheaper without this step.

The input/output volumes for this job are:

Item	Bytes
Read the trips data file 120 bytes per record * 7905 records	948,600
Write a work file 81 variables + 1 control word per case * 7905 cases * 4 bytes per word	2,592,840
Read that same work file	2,592,840
Write the SPSS system file: 2,592,840 bytes of data + 2400 bytes of dictionary	2,595,240
Total	8,729,520

The CPU time required for this job on the University of Chicago's Amdahl 470 V7 was 6.4 seconds, nearly all of which was input editing. All files were blocked as close as possible to 4,628 bytes per block, an efficient block size on 3350 disk drives. The total input/output operation count was 1900. 192K of main memory was required. The total cost, at daytime commercial rates and normal priority, was $3.82.

The first actual problem was to count households by the number of persons over 16 years of age and the number of cars. Neither of these figures is considered to be present in the household record.

This can be done in two SPSS runs: the first creates a file whose cases represent households and whose variables include the number of persons over 16 years of age and the number of cars, using the procedure AGGREGATE. The second SPSS run crosstabulates these two variables.

The first step read the system file of the trips data base once and wrote the aggregated file. There are 300 households on the file and each record in the aggregated file is 4 words long. The second step only reads the aggregated file. The input/output volumes are, then:

Item	Bytes
Read the system file	2,595,240
Write the aggregated file 16 bytes per record * 300 records	4,800
Read the aggregated file	4,800
Total	2,604,840

This run took 1.5 CPU seconds and cost $1.38.

*SPSS and SPSS-X are trademarks of SPSS Inc.

SPSS-X RESULTS.

SPSS-X is the successor system to SPSS, and will be on the market this year. All the test problems can be run using this system when it works. When this was written, none could actually be run.

However, a fairly accurate input/output analysis can be performed based on coded and specified functions.

SPSS-X differs from SPSS in dozens of ways; those pertinent to this analysis are:

.. System files and work files are maintained in double precision on the IBM systems.

.. A file compression feature is available for system files. At some cost in CPU time, this feature will encode integers from -99 to 155 in one byte each. The overhead in file space is about one byte per 16 variables in each case.

.. The SAVE FILE operation will not require that the data be written on a work file.

.. All of the limitations which have restricted the SPSS user to a single data file per run are gone.

.. The procedure AGGREGATE no longer produces a "raw" binary file. Instead, its output file can be a system file or can replace the current active file.

The effect of data compression can be seen by computing the length of a "trip" case.

Variable	Bytes
HOUSE (4 digits)	8
PERSON (2 digits)	1
TRIP (2 digits)	1
OWNCAR, TA2 to TA51 (1 digit each)	51
TB1 to TB19, DAYS (2 digits each)	20
TD1 to TD5 (4 digits each) 5*8=	40
Control words (overhead) (approx.)	5

Total	126

Using single precision rather than double, SPSS needs 328 bytes for the same case.

All the different files are similar enough so that little error will be introduced by using the 126 byte figure for all of them. Since compression is based on the actual value of each variable in each case, and the four digit fields often contain small integers, many cases will not be this long. Still, the figure will serve for this analysis.

Each SPSS or SPSS-X system file includes a dictionary, containing such things as the names, types, missing values, and labels of the variables on that file. The dictionaries in this analysis are estimated to occupy 24 bytes per variable. This is a very rough estimate, but the dictionary size will certainly be small compared to the data size in all of the files described here.

The SPSS-X user will prefer to build four homogeneous files for his trips data base. The input/output volumes involved are:

Item	Bytes
Read the data file. 4 passes	
* 7905 cases	
* 120 bytes	3,794,400
Write the system files (combined)	
7905 cases	
* 126 bytes per case	996,030
Dictionary: 24 bytes per entry	
* 80 entries per file	
* 4 files	7,680

Total	4,798,110

The sizes of the individual files are:

File	Cases	* 126 bytes/case	+ 1,920 bytes =
HOUSES	300	37,800	39,720
CARS	1,046	131,796	133,716
PERSONS	1,194	150,444	152,364
TRIPS	5,365	675,990	677,910

The first task, to count households by the number of persons over 16 and the number of cars, is done in the following steps:

.. Using the PERSONS file and the procedure AGGREGATE, form a file whose cases are households and whose variables are house ID and number of persons over 16.

.. Using the CARS file and AGGREGATE, form a file whose cases are households and whose variables are house ID and number of cars.

.. Using MATCH FILES, form the union of the households identified in the two aggregated files. The result file will contain the union of their sets of variables.

.. Count the resulting cases by number of persons over 16 years of age and number of cars.

The input/output volumes involved are:

Item	Bytes
Read PERSONS file	152,364
Write aggregated file	
data: 300 cases	
* 2 variables	
* 8 bytes/ variable/case	4,800
dictionary: 24 bytes per entry	
2 entries per file	48
Read CARS file	133,716
Write aggregated file (see above)	4,848
Read both aggregated files	
2 files * 4848 bytes per file	9,696
Total	305,472

The second task, again using the trips data base, is to count the trips of at least three days duration, in a car belonging to the household, by the age of the person making the trip and the age of the car. The steps involved are:

.. Use MATCH FILES and the TRIPS file; look up the age of the person making the trip from the PERSONS file.

.. Select the trips of at least three days duration in cars belonging to the household.

.. Sort the resulting trips file by house ID and car ID in order to use MATCH FILES with the CARS file to look up the age of the car.

.. Count.

There are 3403 trips of at least three days duration in a car belonging to the family in the file, so the input/output volumes are:

Item	Bytes
Read the TRIPS file	677,910
Read the PERSONS file	152,364
Write a work file for sorting	
3403 cases	
* 5 variables per case	
* 8 bytes/variable/case	136,120
Read same work file into sort	136,120
Sort. The file is almost in order,	
so no merge passes are expected.	
2 passes * 136,120 bytes	272,240
Write same work file after sort	136,120
Read work file after sort	136,120
Read CARS file	139,476
Total	1,797,990

The remaining tasks use the PEOPLE data base. Again, the user can be expected to prefer the system file form of the file. If data compression is used, each of the 8000 cases on the people file will require 145 bytes, as shown below.

Variable	Bytes
PERSON (5 digits)	8
SEX, A2 to A51 (1 digit each)	51
EDUC, B2 to B15 (2 digits each)	15
BIRTHYR, D2 to D6 (4 digits each) (6*8)	48
MOTHER, FATHER (5 digits each) (2*8)	16
AGE (2 digits)	1
Control codes	6
Total	145

The dictionary is estimated to be (81 variables * 24 bytes per variable =) 1944 bytes long, giving a total size of

Data - 145 bytes per case	
* 8000 cases	1,160,000
Dictionary	1,944
Total length	1,161,944 bytes

The input/output volume involved in creating this file is then

Item	Bytes
Read original data file	
120 bytes per record	
* 8000 records	960,000
Write PEOPLE system file	1,161,944
Total	2,121,994

The third tabulation task is to count persons by father's education and mother's education. Since neither of those variables is represented directly, the file itself must be used to look them up.

The steps involved are:

.. Extract a file to be used for looking up education in this task and age in the next. Note that this is entirely a cost-saving device - the PEOPLE system file itself could be used instead. The extract file's cases are 11 bytes each:

Variable	Bytes
PERSON (5 digits)	8
EDUC (2 digits)	1
AGE (2 digits)	1
Control byte	1
Total	11

It has 8000 cases, so 88,000 bytes of data. Its dictionary is about 3*24 = 72 bytes long, giving a total file length of 88,072.

.. Look up father's education by sorting the cases by father ID, then using MATCH files to look it up.

.. Look up mother's education in the same way.

.. Recode the missing education values to get them into the table, and tabulate.

The input/output volume involved in this breaks down as follows:

Item	Bytes
Read the PEOPLE system file	1,161,944
Write the extract file (PERTAB)	88,072
Write a work file to be sorted: 5 variables * 8000 cases * 8 bytes/variable/case	320,000
Sort that work file: (assuming no extra merge passes) 2 reads + 2 writes	1,280,000
Read sorted work file into MATCH FILES	320,000
Read the PERTAB extract file into MATCH FILES	88,072
Write 2nd work file: 2 variables * 8000 cases * 8 bytes/case/variable	128,000
Sort 2nd work file: 2 reads+2 writes	512,000
Read sorted 2nd work file into 2nd MATCH FILES	128,000
Read PERTAB into 2nd MATCH FILES	88,072
Total	4,114,160

The final task was to count mothers on the PEOPLE file by the age at which they gave birth to their youngest child and the sex of that child.

The steps involved are:

.. Extract a file of offspring, as identified by a valid mother ID number.

.. Sort that file so that offspring of the same mother are grouped together and in ascending age order (youngest first).

.. Using the AGGREGATE procedure, build a file whose cases are youngest children and whose variables are mother ID, age, and sex.

.. Using MATCH FILES and the PERTAB table set up for the previous task, look up the mother's age, compute the difference in ages, and tabulate. Formally, this relies on the one-to-one correspondence between mothers and youngest children.

The input/output volumes involved are:

Item	Bytes
Read the PEOPLE system file	1,161,944
Write a work file of offspring 3 variables * 6000 cases * 8 bytes/variable/case	144,000
Sort that work file, assuming no extra merge passes: 2 reads + 2 writes	576,000
Read the sorted work file into AGGREGATE	144,000
Write a work file of last offspring 3 variables * 2400 cases * 8 bytes/variable/case	57,600
Read that work file into MATCH FILES	57,600
Read the PERTAB extract file into MATCH FILES	88,072
Total	2,229,216

REFERENCES

Nie, N.H, et al.: Statistical Package for the Social Sciences, New York, McGraw-Hill, 1975

Hull, C. Hadlai; Nie, N. H.: SPSS Update for Releases 7 and 8, New York, McGraw-Hill, 1979

Appendix A - Job Stream to Create the Trips System File

```
//TRIPSYS JOB (3YC080,SSJF,E,SPSS),FRY,
           TERMINAL=Y,REGION=193K
        EXEC SPSS,PARM-20K
//FT02F001 DD DCB=BLKSIZE=4628
//FT04F001 DD DSN=$3YC080.SSJF.#$TTL.TRIPSYS,
//            DISP=(,CATLG),
//            UNIT=SYSDA,VOL=SER=SRES01,
//            SPACE=(4628,(565,50),RLSE),
//            DCB=BLKSIZE=4628
//FT08F001 DD DSN=$3YC080.SSJF.#$TTL.TRIPS,
//            DISP=SHR
RUN NAME       MAKE THE TRIPS INTO A SYSTEM FILE

INPUT MEDIUM   DISK
DATA LIST      FIXED / 1
                  HOUSE 1-4 (A)
                  RECTYPE 5 (A)
                  ID2 6-7 (A)
                  TRIP 8-9 (A)
                  A1 TO A51 10-60
                  B1 TO B20 61-100
                  C1 TO C5 101-120

PRINT FORMATS  HOUSE TO TRIP (A)

FILE NAME      TRIPSYS TRIPS DATA BASE
SAVE FILE
```

APPENDIX B - Job Stream for the First Tabulation

```
//PROB1 JOB (3YC080,SSJF,E,SPSS),FRY,TERMINAL=Y,
//            REGION=204K
//         EXEC SPSS,PARM=20K
//FT02F001 DD DCB=BLKSIZE=4628
//FT03F001 DD DSN=$3YC080.SSJF.#$TTL.TRIPSYS,
//            DISP=SHR,
//            UNIT=,VOL=
//FT09F001 DD DSN=$3YC080.SSJF.#$TTL.AGGREG,
              DISP=(,CATLG),
              UNIT=SYSDA,VOL=SER=SRES01,
              SPACE=(TRK,(3,3),RLSE),
              DCB=(RECFM=FB,LRECL=16,BLKSIZE=4624
RUN NAME       COUNT ADULTS AND CARS BY HOUSEHOLD

COMMENT        THE DATA BASE USED IS THE SPSS
               SYSTEM FILE FORM OF THE ORIGINAL
               FILE.

GET FILE       TRIPSYS

COMMENT        TRIPS RECORDS ARE NOT NEEDED HERE.

SELECT IF      (RECTYPE NE '4')

TASK NAME      CREATE FILE WITH COUNTS OF ADULTS
               AND CARS PER HOUSEHOLD

COMMENT        INCLUDE THE HOUSEHOLD RECORD TO
               INSURE WE GET A RECORD FOR EACH
               HOUSEHOLD.

*IF            (RECTYPE EQ '3' AND B1 GT 16)
               ADULT=1
*IF            (RECTYPE EQ '2') CAR=1
AGGREGATE      GROUPVARS=HOUSE/
               VARIABLES=ADULT CAR/ AGGSTATS=SUM
OPTIONS        1
STATISTICS     3

//         EXEC SPSS,PARM=20K
//FT02F001  DD  DCB=BLKSIZE=4628
//FT08F001  DD  DSN=$3YC080.SSJF.#$TTL.AGGREG,
               DISP=SHR

RUN NAME       COUNT HOUSEHOLDS FROM THE TRIPS
               DATA BASE
TASK NAME      BY NUMBER OF PERSONS OVER 16 AND
               NUMBER OF CARS
INPUT MEDIUM   DISK
DATA LIST      BINARY/ 1 ADULTS CARS 3-4
VAR LABELS     ADULTS NUMBER OF PERSONS OVER 16/
               CARS NUMBER OF CARS

CROSSTABS      TABLES= ADULTS BY CARS
OPTIONS        1 3 4 5
```

APPENDIX C - SPSS-X Commands to Create Trips and People System Files.

```
TITLE           CONSTRUCT THE 'TRIPS' DATA BASE

FILE INFO       INDATA .../  /* PRESORTED INPUT
                HOUSES .../  /* HOUSEHOLD FILE
                PERSONS .../ /* PERSONS FILE
                TRIPS   .../ /* TRIPS FILE
                CARS .../    /* CARS FILE

SUBTITLE        MAKE THE HOUSES FILE

FILE TYPE       SIMPLE, FILE INDATA,  RECORD 5

RECORD TYPE     1             /* HOUSEHOLD RECORD
DATA LIST       /
                   HOUSE 1-4
                   HA1 TO HA51 10-60
                   HB1 TO HB20 61-100
                   HD1 TO HD5  101-120

RECORD TYPE     OTHER SKIP

SAVE FILE       HOUSES/COMPRESSED
NEW FILE


SUBTITLE        MAKE THE PERSONS FILE

FILE TYPE       SIMPLE,  RECORD 5

RECORD TYPE     3             /* PERSONS RECORD
DATA LIST       /
                   HOUSE 1-4
                   PERSON 6-7
                   PA1 TO PA51 10-60
                   AGE PB2 TO PB20 61-100
                   PC1 TO PC5 101-120

RECORD TYPE     OTHER SKIP

SAVE FILE       PERSONS/ COMPRESSED
NEW FILE


SUBTITLE        MAKE CARS FILE
FILE TYPE       SIMPLE,  RECORD 5

RECORD TYPE     2             /* CARS RECORD
DATA LIST       /
                   HOUSE 1-4
                   CAR   6-7
                   CA1 TO CA51 10-60
                   CB1 TO CB20 61-100
                   CD1 TO CD4 CARYEAR 101-120
RECORD TYPE     OTHER SKIP

SAVE FILE       CARS/ COMPRESSED
NEW FILE

SUBTITLE        MAKE THE TRIPS FILE

FILE TYPE       SIMPLE,  RECORD 5

RECORD TYPE     4             /* TRIPS RECORD
DATA LIST       /
                   HOUSE 1-4
                   PERSON 6-7
                   TRIP  8-9
                   OWNCAR TA2 TO TA51 10-60
                   TB1 TO TB19 DAYS 61-100
                   TD1 TO TD5 101-120
RECORD TYPE     OTHER SKIP

SAVE FILE       TRIPS/COMPRESSED
NEW FILE

 /* *******************************************

TITLE           BUILD PEOPLE FILE
FILE INFO       PDATA.../    /* INPUT PEOPLE FILE
                PEOPLE.../   /* SPSS SYSTEM FILE

DATA LIST       FILE PFILE/
                   PERSON 1-5
                   SEX A2 TO A51 6-56
                   EDUC B2 TO B15 57-86
                   BIRTHYR D2 TO D6 87-110
                   MOTHER FATHER 111-120

COMPUTE         AGE = 1981-BIRTHYR

SAVE FILE       PEOPLE/ COMPRESSED
```

APPENDIX D - SPSS-X Commands to Perform Tabulation Tasks.

```
TITLE          BOB TEITEL'S COMPLEX FILE PROBLEM 1

COMMENT                   SORT ORDER:
FILE INFO      PERSONS .../ /*HOUSE, PERSON
               HOUSES  .../ /* HOUSE
               CARS    .../ /* HOUSE, CAR
               TRIPS   .../ /* HOUSE, PERSON, TRIP
               WPER    .../ /* PERSONS WORK FILE

SUBTITLE       TASK 1 - COUNT HOUSEHOLDS BY N OF
               CARS BY N OF ADULTS

GET FILE       PERSONS/ KEEP HOUSE AGE
COMPUTE        PER16 = AGE> 16
AGGREGATE      FILE WPER/ BREAK HOUSE/
               PER16 'NUMBER OF PERSONS OVER 16'
               = SUM(PER16)

GET FILE       CARS/ KEEP HOUSE
AGGREGATE      FILE */ BREAK HOUSE/
               NCARS 'NUMBER OF CARS' = N

MATCH FILES    FILE WPER/ FILE */ BY HOUSE/
               KEEP PER16 NCARS

RECODE         ALL (SYSMIS = 0)

CROSSTABS      PER16 BY NCARS

SUBTITLE       TASK 2

COMMENT        COUNT TRIPS AT LEAST 3 DAYS LONG
               IN CARS BELONGING TO HOUSEHOLDS BY
               THE AGE OF THE PERSON MAKING THE
               TRIP AND THE YEAR OF THE CAR

MATCH FILES    FILE=TRIPS/ TABLE=PERSONS/
               BY HOUSE PERSON/
               KEEP HOUSE, CAR, AGE, DAYS, OWNCAR
               (DAYS>=3 AND OWNCAR NE 0)

SORT CASES     HOUSE, CAR
MATCH FILES    FILE=*/ TABLE=CARS/ BY HOUSE, CAR/
               KEEP AGE, CARYEAR

CROSSTABS      AGE BY CARYEAR
```

```
TITLE          BOB TEITEL'S COMPLEX FILE PROBLEM 2

FILE INFO      PEOPLE .../      /* SORTED BY PERSON
               PERTAB .../      /* PERSON TABLE

COMMENT        PERTAB WILL BE USED IN THIS TASK
               AND THE NEXT

SUBTITLE       COUNT PERSONS BY FATHER'S AND
               MOTHER'S EDUCATION

GET FILE       PEOPLE/
               KEEP PERSON EDUC AGE FATHER MOTHER
SAVE FILE      PERTAB/ KEEP PERSON EDUC AGE/
               COMPRESSED

SORT CASES     FATHER
MATCH FILES    TABLE = PERTAB/
               RENAME (FATHER PAEDUC=PERSON EDUC)
               FILE = */
               BY FATHER
               KEEP MOTHER PAEDUC

SORT CASES     MOTHER
MATCH FILES    TABLE = PERTAB/
               RENAME (MOTHER MAEDUC=PERSON EDUC)/
               FILE=*/ BY MOTHER/
               KEEP PAEDUC MAEDUC/

RECODE         ALL (SYSMIS = 99)
MISSING VALUES ALL (99)

CROSSTABS      PAEDUC BY MAEDUC
OPTIONS        1

SUBTITLE       TASK 2

COMMENT        COUNT MOTHERS BY AGE AT WHICH THEY
               GAVE BIRTH TO THEIR YOUNGEST
               OFFSPRING AND BY THE SEX OF THAT
               CHILD.

GET FILE       PEOPLE/ KEEP MOTHER AGE SEX
SELECT IF      (MOTHER NE 0)
SORT CASES     MOTHER AGE
AGGREGATE      FILE */ BREAK MOTHER/
               AGE = FIRST(AGE)/ SEX = FIRST(SEX)

MATCH FILES    FILE */ TABLE PERTAB/
               RENAME MOTHER=PERSON MOMAGE=AGE/
               BY MOTHER/ KEEP AGE MOMMAGE SEX/

COMPUTE        BIRTHAGE = MOMAGE-AGE
VAR LABELS     BIRTHAGE
                    MOTHER'S AGE AT LAST BIRTH/
               SEX SEX OF LAST OFFSPRING/

FILE LABEL     MOTHERS
CROSSTABS      BIRTHAGE BY SEX
```

VOLUME TESTING OF STATISTICAL SYSTEMS

Pauline R. Nagara, Institute for Social Research
Michael A. Nolte, Institute for Social Research

ABSTRACT

This paper outlines the steps taken to solve data management and analysis problems based on two complex files. One file, TRIPS, contained four groups of variables in a three-level hierarchical structure resembling that of the 1972 National Travel Survey. The second file, PERSONS, contained records in which link variables defined relationships to other records in the file. Both problems were solved using the current implementation of OSIRIS at the University of Michigan. The solution to the problems based on the TRIPS dataset involved use of OSIRIS structured file creation and retrieval procedures. The solutions to the problems based on the PERSONS dataset were based on the use of OSIRIS sort and merge procedures. OSIRIS instructions and the cross-tabulations that resulted are presented for all four problem segments.

Keywords: statistical analysis package, data management, structured files, complex files, OSIRIS

Introduction.

The session on "Volume Testing of Statistical Systems" involved use of statistical packages to perform common data management and analysis tasks on complex files. We carried out these tasks using the version of OSIRIS available to users of the University of Michigan Amdahl 470V/8. Since the test data files were quite small (8000 cases, each case having a logical record length of 120 bytes) all data manipulations were carried out interactively, using disk files for input/output operations.1

As a first step, each data file was copied from the transmittal tape to disk. To describe the data contained in each file, OSIRIS dictionaries were created using &DICT (see Figures 1.1 and 2.2). The test data were then ready for use by OSIRIS.

Overview of Problem 1. The dictionary summarized in Figure 1.1 and the TRIPS data were first processed by SBUILD to create an OSIRIS hierarchical dataset since the problem could be easily resolved with such a dataset. In contrast to the usual rectangular dataset, an OSIRIS hierarchical dataset permits the storage of different record types(with different lengths and composition) in one dataset. It further permits their storage in a hierarchical relationship. The schema for the TRIPS dataset is seen in Figure 1.2. The instructions for creation of a structured file are listed in Figure 1.3. The format of the input records was retained for the output records except for the sort or identification fields prefixing each output data record. These sort fields are used by OSIRIS to maintain the data records in the hierarchical structure created by SBUILD. Once created, an hierarchical dataset can be retrieved in a number of ways depending on the underlying structure of the database. Occurrences of groups in the data structure are passed to the calling program as temporary data records called entries. The solutions to problems 1.1 and 1.2 illustrate how entry definitions may be changed as appropriate for analysis purposes.

Problem 1.1 This problem called for a count of households by the number of cars in the household by the number of persons older than sixteen. Since these figures did not exist in the dataset they had to be generated from a household's car and person records. The unit of analysis was household and the necessary information was derived from person and car records. Therefore, we defined an entry definition that returned one record for every household and included car IDs from the car records and age from person records. Since there was a variable number of car records (0-7) and person records (0-8) per household, this definition was designed to permit the retrieval of a maximum of seven car records and eight person records. If less than the maximum number of records existed, the corresponding variable in the entry was set to missing data. The format of this entry is shown in Figure 1.4.

The entries were then processed by RECODE (the data transformation program) ,which created two temporary variables : 1) a count of the number of cars and 2) a count of the number of persons older than 16 years. These temporary variables were transmitted to TABLES (the cross tabulation program) which generated the solution to Problem 1.1 (See Figures 1.5 and 1.6)

Problem 1.2 The second problem called for a count of trips by trip taker's age and model year of the trip car. The count was further restricted to trips taken in a household car (0=non-household car) which were of at least 3 days duration. Thus the unit of analysis for problem 1.2 was trips

Figure 1.1: OSIRIS Dictionary for TRIPS data records

Variable Number	Variable Name	Type	Location	Width	MD1 Value	MD2 Value
1	HOUSEHOLD ID	C	1	4		
2	RECORD TYPE	C	5	1		
203	CAR/HOUSEHOLD ID	C	6	2	99	
204	MODEL YEAR	C	117	4		
301	PERSON/HOUSEHOLD ID	C	6	2		
302	AGE	C	61	2	0	
401	TRIP/PERSON ID	C	8	1		
402	OWNCAR	C	10	1		
403	TRIP DURATION	C	99	2		

instead of households. The same structured file was used, but with a different entry definition. This definition retrieved entries at the trip level, gathering information from car records, one person record, and one trip record. The format of the entry is shown in Figure 1.7.

The entries were again processed by RECODE (see Figure 1.8) to match the trip car's ID with one of the household cars and to save its model year in a temporary variable. The temporary variable and trip taker's age were then transmitted to TABLES, which produced the results shown in Figure 1.9. In this TABLES setup, a filter was used to screen out the trips taken in non-household cars and trips of less than three days. For the sake of simplicity, the age of the trip taker was bracketed into eight categories, while the car model year was bracketed into three.

Figure 1.2: TRIPS File after &SBUILD

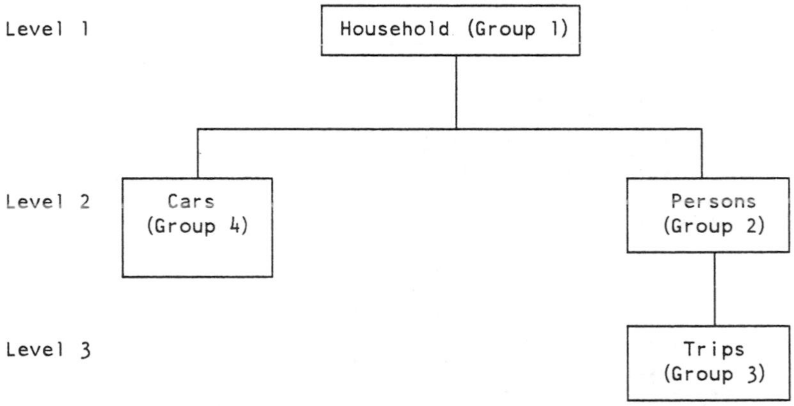

Figure 1.3: Creating the TRIPS dataset

```
&SBUILD DICTIN=-DI DATAIN=-TRIPS DICTOUT=-SDI DATAOUT=-SDA DATATEMP=-T
1:INCLUDE V2=1
2:INCLUDE V2=2
3:INCLUDE V2=3
4:INCLUDE V2=4
   CREATING TRIPS DATASET WITH 4 RECORD TYPES AT 3 LEVELS
PRINT=OUTD
GNUM=1 NAME='HOUSEHOLD RECORD' LEVEL=1 VARS=1-80 FILTER=1
GNUM=2 NAME='PERSON RECORD' LEVEL=2  VARS=1-2,301,4-55,302,57-80  -
   LINK=(G1.V1:G2.V1) FILTER=3
GNUM=3 NAME='TRIP RECORD' LEVEL=3 VARS=1-2,301,401-402,6-74,403,76-80  -
   LINK=(G1.V1:G3.V1,G2.V1:G3.V1,G2.V301:G3.V301) ID=1,301,401    FILTER=4
GNUM=4 NAME='CAR RECORD' LEVEL=2 VARS=1-2,203,4-79,204  -
   FILTER=2 ID=203 LINK=(G1.V1:G4.V1)
&END
```

Figure 1.4: Entry for Problem 1.1

Car 1 ID	....	Car 7 ID	Person 1 Age	....	Person 8 Age

Figure 1.5: Setup for Problem 1.1

```
&ENTRY
UNIT=1
G4+G2
G=4 LEVEL=1 OPT MAXOCC=7 VINCR=1 REN=(203:501)
G=2 LEVEL=1 OPT MAXOCC=8 VINCR=1 REN=(302:601)
&END
&RECODE
    R1=COUNT(99,V501-V507)
    R1=7-R1
    R3=1
    R2=0
LOOP IF SELECT(FROM=V601-V608,BY=R3) GT 16 THEN R2=R2+1
    IF R3 LT 8 THEN R3=R3+1 AND GO TO LOOP
    NAME R1'NUMBER OF CARS/HSHLD',R2'COUNT OF AGE>16'
&END
&TABLES DICTIN=-SDI DATAIN=-SDA
    PROBLEM 1, TABLE 1
RECODE=1 ENTRY=1
VARS=R2 STRATA=R1 TITLE='HOUSEHOLD COUNT BY # CARS AND # ADULTS>16'
```

Figure 1.6: Solution to Problem 1.1

VARIABLE R2 COUNT OF AGE>16
STRATA(ROW) R1 NUMBER OF CARS/HSHLD

HOUSEHOLD COUNT BY # CARS AND # ADULTS>16

	1	2	3	4	TOTALS
0	9	9	7	8	33
1	15	8	9	6	38
2	15	8	8	7	38
3	14	11	6	7	38
4	14	10	7	7	38
5	15	7	8	7	37
6	13	9	8	7	37
7	14	8	7	8	37
TOTALS	109	70	60	57	296

Figure 1.7: Entry for Problem 1.2

Car 1 ID	...	Car 7 ID	Car 1 Year	...	Car 7 Year	Person n Age	Trip Car	Trip Days

Figure 1.8: Setup for Problem 1.2

```
&ENTRY
UNIT=3 ENTRY=2
G4+G2+G3
G=4 L=1 MAXOCC=7 VINCR=1 REN=(203,204:501,511)
&END
&RECODE
RECODE=2
    R2=1
LOOP IF SELECT(FROM=V501-V507,BY=R2) EQ V402 THEN -
    R1=SELECT(FROM=V511-V517 BY=R2) AND GO TO END
    IF R2 LT 7 THEN R2=R2+1 AND GO TO LOOP
    REJECT
    NAME R1'OWNCAR/MODEL YEAR'
END  CONTINUE
&END
&TABLES
EXCLUDE V402=0 OR V403=<3
   PROBLEM 1, TABLE 2
ENTRY=2 RECODE=2
STRATA=302 VARS=R1 -
TITLE='TRIP COUNT BY TRIP TAKER'S AGE AND YEAR OF CAR'
&END
```

Figure 1.9: Solution to Problem 1.2

```
VARIABLE       R1       OWNCAR/MODEL YEAR
STRATA(ROW)  V302     AGE

TRIP COUNT BY TRIP TAKER'S AGE AND YEAR OF CAR
```

	1960-69	1970-79	1980-81	TOTALS
12-19	224	114	32	370
20-29	185	217	48	450
30-39	240	223	15	478
40-49	182	223	34	439
50-59	181	230	56	467
60-69	204	239	23	466
70-79	240	171	53	464
80+	95	144	30	269
TOTALS	1,551	1,561	291	3,403

Overview of Problem 2. This problem presupposed the existence of a large database consisting of person-level information. This database contained information on inter-generational relationships; that is, for each person record in the database there could be another person record containing data on a parent or child. Any person record could be linked to another via the FATHER or MOTHER identification fields. If there were no such link, then the parent identification fields contained missing data. The storage sequence of the records in the database was not related to family relationships. Each record contained 73 variables, of which we were interested in SEX, EDUCATION and BIRTHDATE. Other variables were retained throughout the various data manipulations, but were not used in the analysis. The OSIRIS solutions to the two sub-problems based on this dataset involved the use of straightforward sorting and merging techniques. Both output and temporary datasets remained in rectangular format. The structure of the input data file is summarized in Figure 2.1. The dictionary records for the variables used in the test are reproduced in Figure 2.2.

Problem 2.1 The first portion of this problem involved production of a cross-tabulation of parent characteristics (i.e., educational attainment) of offspring. In producing an OSIRIS solution to this problem, we assumed that each person record represented a child, whether or not parental data were actually present. Given this assumption, the logical solution to the problem was to merge the entire file to itself twice -- once in order to merge in potential mother records, and again to merge in potential father records. This process was embodied in the OSIRIS instructions listed in Figure 2.3.

A brief description of the OSIRIS commands in Figure 2.3 is in order. The first step (&COPYSORT) sorts all person records on the mother identification number (V72). The sorted records are then merged (&UPDATE) with the original person records. Since a mother can have more than one child, a mother record may be linked to more than one person record. The first &COPYSORT/&UPDATE combination produced the data structure summarized in Figure 2.4. The sort/merge (&COPYSORT/&UPDATE) procedure was then repeated for potential father records. The data structure that resulted is summarized in Figure 2.5. Once the offspring-parent dataset had been constructed, it was a simple task to produce the required cross-tabulation (See Figures 2.6 and 2.7).

Problem 2.2 The second portion of problem 2 involved production of a cross-tabulation in which the unit of analysis was youngest offspring. Required was a table in which the age of each mother at birth of youngest child (bracketed) was crossed with the sex of that offspring. The solution to this problem was based on a sort of the merged person/mother/father file on person-mother id (V72) and person-birthdate (V68). The primary sort field of each person record was now mother-id, the secondary sort field being birth-year. This enabled us to use a &RECODE procedure to select the last offspring of each mother (See Figure 2.8).

A &RECODE procedure was then used in conjunction with the &TABLES command to produce the solution (see Figure 2.9). Note the use of a filter to exclude person records with no mother data.

The reader will note that we did not choose to convert the PEOPLE data file into an OSIRIS hierarchical file. The reason for this is that a hierarchical file would have required the creation of a common variable linking father and mother records. It would have been possible to create such a variable and then use &SBUILD to create a structured file, but retention of the rectangular format proved to be cheaper and faster. If the test data file had been somewhat larger (i.e., 37,000+ records and 3,000,000+ bytes) it might have been necessary to use magnetic tape as the medium for intermediate storage as well as for input and output files. Since OSIRIS provides the user with tools for easy manipulation of magnetic tape files, this circumstance would not have affected the viability of the rectangular format.

Figure 2.1: Entry for Problem 2.1

Person Id	Sex	V3-V53	Education	V54-V67	Year of Birth	Mother Id	Father Id

Figure 2.2: OSIRIS Dictionary for PEOPLE data records

Variable Number	Variable Name	Type	Location	Width	MD1 Value	MD2 Value
1	PERSON-ID	C	1	5	0	
2	SEX	C	6	1	0	9
53	EDUCATION	C	57	2	0	99
68	YEAR-OF-BIRTH	C	87	4	0	9999
72	MOTHER-ID	C	111	5	0	99999
73	FATHER-ID	C	116	5	0	99999

Figure 2.3: Building the PEOPLE dataset

```
$RUN ISR:OSIRIS.IV
&MTS CRE -V72 SIZE=270P
&MTS CRE -DATEMP SIZE=560P
&MTS COPY MN.DIPR TO -DIM
&MTS CRE -V73 SIZE=270P
&MTS CRE -DAOUT SIZE=800P
&COPYSORT SORTIN=-PEOPLE SORTOUT=-V72 DICTIN=MN.DIPR SORT=V72
&UPDATE DICTIN1=MN.DIPR DATAIN1=-V72 DICTIN2=-DIM DATAIN2=-PEOPLE -
      DICTOUT=-DITEMP DATAOUT=-DATEMP
COMPLEX FILE PROCESSING -- PROBLEM 2.1 <STEP 1>
MAXB=0
INFI=IN1 VARS=ALL ID=V72 DUPS=ALL
INFI=IN2 VARS=ALL ID=V1 RENUM=201 HOLD OPT
END
&COPYSORT SORTIN=-DATEMP SORTOUT=-V73 DICTIN=-DITEMP SORT=V73
&UPDATE DICTIN1=-DITEMP DATAIN1=-V73 DICTIN2=MN.DIPR DATAIN2=-PEOPLE -
      DICTOUT=MN.DIPR2.1 DATAOUT=-DAOUT
COMPLEX FILE PROCESSING -- PROBLEM 2.1 <STEP 2>
MAXB=0
INFI=IN1 VARS=ALL ID=V73 DUPS=ALL
INFI=IN2 VARS=ALL ID=V1 RENUM=401 HOLD OPT
END
```

Figure 2.4: Merging Mother Records

Person record (V1-V73)	Mother record (V201-V273)

Figure 2.5: Merging Father Records

Person record (V1-V73)	Mother record (V201-V273)	Father record (V401-V473)

Figure 2.6: Setup for Problem 2.1

```
&RECODE
   RECODE=1
&  **** BRACKET EDUCATIONAL VARIABLES ****
   V253=BRAC(V253,TAB=1,ELSE=0,1-8=1,9-11=2,12-19=3)
   V453=BRAC(V453,TAB=1)
&TABLES DICTIN=MN.DIPR2.1 DATAIN=-DAOUT
COMPLEX FILE PROCESSING -- PROBLEM 2.1 <STEP 3>
RECODE=1
STRATA=V253 VARS=V453 TITLE='EDUCATIONAL ATTAINMENT OF RESPONDENT PARENTS'
&END
```

Figure 2.8: Setup for Problem 2.2

```
&RECODE
 RECODE=2
&   *** SAVE VARIABLES:  MOTHER ID (R72), FILE START COUNTER (R999),
&                        SEX AND BIRTHDATE OF OFFSPRING (R2, R68),
&                        MOTHER BIRTHDATE (R268)
 CARRY (R72,R999,R2,R68,R268)
&   *** IF FIRST RECORD OF FILE, LOAD CARRY VARIABLES WITH COMPARISON INFO
IF R999 EQ 0 THEN R72=V72 AND R999=1 AND R68=V68 AND R268=V268
&   *** CHECK TO SEE IF NEW MOTHER ID FOR CURRENT OFFSPRING RECORD:
&       IF A NEW MOTHER THEN OUTPUT SAVE VARIABLE DATA AND THEN
&          SAVE NEW OFFSPRING INFORMATION (AND MOTHER DATA); ELSE,
&          SAVE NEW MOTHER ID AND CARRY INFO AND REJECT CURRENT RECORD
&
           IF (V72 GT R72 OR EOF) THEN R500 = R68 - R268 AND             -
           R500=BRAC(R500,ELSE=9,0-18=1,19-24=2,25-30=3,31-40=4,41-50=5  -
               AND R600=R2 AND -
               R68=V68 AND R268=V268 AND R2=V2 AND R72=V72 -
           ELSE -
               R68=V68 AND R268=V268 AND R2=V2 AND REJECT
&   *** NAME R-VARIABLES
           NAME R500 'AGE OF MOTHER (BRACKET)'
           NAME R600 'SEX OF OFFSPRING'
&TABLES DICTIN=MN.DIPR2.1 DATAIN=-DAOUT2
EXCLUDE V72=0
COMPLEX FILE PROCESSING -- PROBLEM 2.2
RECODE=2
STRATA=R500 VARS=R600 TITLE='LAST BIRTH COUNT'
&END
```

Figure 2.7: Solution to Problem 2.1

VARIABLE V453 EDUCATION OF FATHER
STRATA(ROW) V253 EDUCATION OF MOTHER

EDUCATIONAL ATTAINMENT OF RESPONDENT PARENTS

	0	1	2	3	TOTALS
0	2,000	0	0	0	2,000
1	0	1,579	949	0	2,528
2	0	0	0	947	947
3	0	949	0	1,576	2,525
TOTALS	2,000	2,528	949	2,523	8,000

Note that 2,000 of the person records do not have a father or mother record. For these data records, the parent variables were padded with missing data -- in this case, zero (0).

Figure 2.9: Solution to Problem 2.2

VARIABLE R600 SEX OF OFFSPRING
STRATA(ROW) R500 AGE OF MOTHER (BRACKET)

LAST BIRTH COUNT

	1	2	TOTALS
1	300	0	300
2	300	600	900
3	300	0	300
4	300	600	900
TOTALS	1,200	1,200	2,400

Conclusions.

The Volume Testing exercise highlighted some important advantages of OSIRIS in building and using complex databases, whether in structured of rectangular format. With respect to structured files, the OSIRIS user need only specify complex relationships among data records once, during the build process. The OSIRIS system will maintain these relationships until they are changed by the user, no matter what storage medium is used to contain the database. When structured files are not appropriate, rectangular datasets are easy to manipulate using OSIRIS data management procedures. No matter what file structure is chosen, the user will be free from size constraints, both in number of cases and number of variables. OSIRIS processes records sequentially, thus the portion of the database accessed during analysis is not limited by core constraints.

Although cost comparisons were not specifically a part of this exercise, we gathered cost and CPU time figures for each processing step. Figure 3 summarizes the steps and the costs involved in each. Both cost and CPU data reflect the fact that both problem sets in this exercise were processed during normal working hours. The data represent MTS University/Government rates at NORMAL priority and do not include connect time, peripheral charges or indirect and miscellaneous costs.

---

[1] In the MTS operating system, each raw data file occupied approximately 260 pages of disk space. For those who are unfamiliar with computing in the MTS environment, MTS disk storage is measured in pages, where each page is equal to 4096 bytes. An MTS disk page is also roughly equivalent to 2 disk tracks in an OS/JCL environment.

Figure 3: Volume Testing Processing Summary

Step Number	Step Name	Step Cost	CPU Seconds
1.1	SBUILD	$2.61	14.26
1.1	ENTRY	.004	.02
1.1	RECODE	.02	.15
1.1	TABLES	.65	1.53
1.2	ENTRY	.003	.03
1.2	RECODE	.03	.16
1.2	TABLES	.97	2.26
	Problem 1 Totals:	$4.29	18.41
2.1	COPYSORT	$2.04	4.34
2.1	UPDATE	3.35	7.47
2.1	COPYSORT	2.40	5.05
2.1	UPDATE	3.56	7.87
2.1	RECODE	0.06	0.15
2.1	TABLES	1.07	2.45
2.2	COPYSORT	2.93	5.78
2.2	RECODE	0.08	0.17
2.2	TABLES	0.99	2.39
	Problem 2 Totals:	$17.47	38.06

Workshop 7

Random Number Generation

Organizer: James E. Gentle, IMSL, Inc.

Chair: William J. Kennedy, Iowa State University

Invited Presentations:

In Search of Correlation in Multiplicative Congruential Generators with Modulus $2^{31} - 1$, George S. Fishman, University of North Carolina at Chapel Hill, and Louis R. Moore, III, University of North Carolina at Chapel Hill

Portability Considerations for Random Number Generators, James E. Gentle, IMSL, Inc.

IN SEARCH OF CORRELATION IN MULTIPLICATIVE CONGRUENTIAL GENERATORS WITH MODULUS 2^{31} -1

George S. Fishman and Louis R. Moore, III
University of North Carolina at Chapel Hill

Abstract

This paper describes an empirical search for correlation in sample sequences produced by 16 multiplicative congruential random number generators with modulus 2^{31} - 1. Each generator has a distinct multiplier. One multiplier is in common use in the LLRANDOM and IMSL random generation packages as well as in APL and SIMPL/1. A second is used in SIMSCRIPT II. Six multipliers were taken from a recent study that showed them to have the best spectral and lattice test properties among 50 multipliers considered. The last eight multipliers had the poorest spectral and lattice test properties for 2-tuples among the 50. A well known poor generator, RANDU, with modulus 2^{31}, was also tested to provide a benchmark for evaluating the empirical testing procedure.

A comprehensive analysis based on test statistics derived from cumulative periodograms computed for each multiplier for each of 512 independent replications of 16384 observations each showed evidence of excess high frequency variation in two multipliers and excess midrange frequency variation in three others, including RANDU. Also evidence exists for a bimodal spectral density function for yet another multiplier. An examination of the test results showed that the empirical evidence of a departure from independence did not significantly favor the eight poorest multipliers. This observation is in agreement with a similar observation made by the authors in an earlier study of these multipliers that principally concentrated on their distributional properties in one, two and three dimensions. This consistency raises some doubt as to how one should interpret the results of the spectral and lattice tests for a multiplier. Also, the three multipliers considered superior in the earlier study maintain that position in the current study.

Key Words

lattice test; periodogram; random number generation; simulation; spectral test.

Introduction

This paper presents a comprehensive account of an empirical search for correlation in multiplicative congruential generators with modulus 2^{31}- 1. These generators are members of the class of generators

$$Z = \{Z_i \equiv AZ_{i-1} \pmod M;\ i = 1,2,\ldots\} \qquad (1)$$

where, provided M is a prime number, one chooses the multiplier A in a way that guarantees a period of $M-1$ and where one assumes that the normalized quantities $X_i = Z_i/M; i = 1,2,\ldots$ behave as independent uniform deviates. The number $M = 2^{31} - 1$ is prime and provides a convenient modulus for 32 bit computers. The relative accuracy of the uniformity and independence assumptions is a function of A, and a theoretical literature exists for measuring the extent to which *overlapping* k-tuples $(Z_i, Z_{i+1}, \ldots, Z_{i+k-1}), (Z_{i+1}, Z_{i+2}, \ldots, Z_{i+k}), (Z_{i+2}, Z_{i+3}, \ldots, Z_{i+k+1}), \ldots$ are distributed uniformly in the k-dimensional lattice with sides of length M-1.

The spectral test (Coveyou and MacPherson,1967) and the lattice test (Beyer, Roof and Williamson, 1970 and Marsaglia,1972) describe two methods of evaluating the extent of nonuniformity. These tests evaluate *global* nonuniformity. A recent empirical study (Fishman and Moore, 1978), designed to examine local uniformity of nonoverlapping 1-tuples, 2-tuples and 3-tuples for 16 multipliers with $M = 2^{31} - 1$, revealed significant nonuniformity for several of them. Moreover, no clear-cut association was found between these empirically discovered aberrations and the theoretical measures of nonuniformity as given by the spectral and lattice tests. At least two explanations for this failure are possible. Firstly, the theoretical tests focus on overlapping k-tuples, whereas the empirical work used nonoverlapping k-tuples. Secondly, the existence of serious local nonuniformity does not preclude the absence of serious global nonuniformity.

This earlier empirical study also used a comprehensive form of the runs up and down test to detect departures from independence. Because the runs procedure relies only on the signs of the differences between successive observations, one has the feeling that ignoring information on magnitude limits the power of the test. Indeed, this suspected limitation may be responsible for the lack of a definitive conclusion there when testing exclusively for independence. To overcome this inadequacy in the old study, the present work examines in detail the correlation structure of sample sequences produced by the 17 multipliers (A) in the original study. Multiplier I, A=16807, is in common use in LLRANDOM (Learmonth and Lewis, 1973), APL (Katzan, 1971), SIMPL/1 (1972) and IMSL (1977). Multiplier II, A=630360016, is implemented in SIMSCRIPT II (Kiviat, Villaneuva and Markowitz, 1969). Multipliers III through VIII come from a study of 50 multipliers by Hoaglin (1976), who evaluated them as best using the spectral and lattice tests. Multipliers IX through XVI are the poorest performers among Hoaglin's 50 multipliers on the spectral and lattice tests with regard to 2-tuples and multiplier XVII is RANDU, the random number generator used for many years in IBM's Scientific Subroutine Library. Marsaglia (1972) has demonstrated the notably poor global behavior of this generator on the lattice test for 3-tuples, 4-tuples and 5 tuples, but showed no aberration for 2-tuples. Note that RANDU, which is included as a benchmark, is not a prime modulus generator. Table 1 lists the multipliers.

Table 1
Multipliers

	A
I.	16807
II.	630360016
III.	1078318381
IV.	1203248318
V.	397204094
VI.	2027812808
VII.	1323257245
VIII.	764261123
IX.	1214170817
X.	190576451
XI.	1865416386
XII.	507435369
XIII.	2139391393
XIV.	1277850838
XV.	163252367
XVI.	55279518
XVII.	65539

A comprehensive periodogram analysis was used to search for correlation. For each multiplier, 512 independent replications each with 16384 observations were collected. A periodogram was computed for each realization and empirical distribution functions (e.d.f.'s) were computed for 11 test statistics, based on the 512 periodograms, whose distributions are known under the null hypothesis (H_0) of independence. Four alternative hypotheses were considered, each designed to consider the existence of distinct forms of variation in the sample sequences.

Graphical analysis of the e.d.f.'s of the test statistics together with the examination of tabulated test values failed to provide consistent evidence of correlation for multipliers II, V through XIII and XV. Multipliers IV and XVI showed evidence of high frequency variation, multipliers I, III and XVII showed evidence of variation with period closer to 4 than to 2 and multiplier XIV provided evidence consistent with variation at two frequencies, one with period less than 4 and the other with period greater than 4.

Since a comparison of empirical test results and corresponding lattice test measures failed to reveal a relationship across multipliers, one is led to the conclusion that empirical testing provides local information not discernible via the spectral and lattice tests. Combining the empirical evidence here with the empirical evidence in Fishman and Moore (1978) and the theoretical evaluation in Hoaglin (1976), we note that multipliers II, V and VIII are the least suspect among the 17 multipliers. The earlier empirical study contains a discussion of the implications of alternative algorithms on execution time with regard to these multipliers, and Appendix A of Fishman (1978) contains seeds for multipliers II and V that are spaced 100,000 numbers apart.

Periodogram Analysis

Independence Setting

Consider a sequence of observations $X_1,\ldots,X_n$ and the corresponding periodogram

$$P_k = \frac{2}{n}\left|\sum_{j=1}^{n} X_j e^{-i2\pi jk/n}\right|^2 \qquad i \equiv \sqrt{-1} \tag{2}$$

$$m = \lfloor (n-1)/2 \rfloor - 1, \qquad k = 1,\ldots,m+1$$

which leads to the *cumulative normalized periodogram* (c.n.p)

$$S_k = \frac{\sum_{j=1}^{k} p_j}{\sum_{j=1}^{m+1} p_j} \qquad k = 1,\ldots,m\ . \tag{3}$$

If $X_1,\ldots,X_n$ are i.i.d. and normal, then $S_1,\ldots,S_m$ have the distribution of the order statistics of m independent uniform random variables on (0,1) and

$$F_m(s) = \frac{1}{m}\sum_{k=1}^{m} I_{[S_k,1]}(s) \qquad 0 \le s \le 1\ , \tag{4}$$

where

$$I_{[a,b]}(x) = 1 \qquad a \le x \le 1$$
$$= 0 \qquad \text{otherwise,}$$

is an *empirical distribution function* (e.d.f.). If $X_1,\ldots,X_n$ have finite variance but are not normal, then the order statistics property holds asymptotically ($n\to\infty$) for $S_1,\ldots,S_m$ and $\{F_m(s),\ 0 \le s \le 1\}$ is asymptotically an e.d.f. Therefore, a test of independence of $X_1,\ldots,X_n$ is equivalent to a test of the hypothesis H_0: $S_1,\ldots,S_m$ are order statistics of m independent uniform random variables on (0,1). See Bartlett (1955) and Durbin (1969).

Conclusions and Recommendations

With regard to the search for correlation, our analysis suggests its presence in finite sequences from multipliers I, III, IV, XIV, XVI and XVII. With regard to an association between our findings and the spectral and lattice test values (see Fishman and Moore, 1978) for 2-tuples, no connection is apparent. That is, two of the best six multipliers are suspect here whereas two of the worst eight multipliers are suspect; hardly an overwhelming affirmation of the discriminating properties of the spectral and lattice tests for correlation.

To recommend an appropriate multiplier, one needs to combine the empirical results of this study with lattice and spectral test measurements and the empirical findings in Fishman and Moore (1978). Because of their poor showing on the spectral and lattice tests, we are inclined to put aside multipliers IX through XVI. RANDU (multiplier XVII) is also omitted. Among the first eight, the results of the current study cast no suspicion on multipliers II and V through VIII. The earlier Fishman and Moore study aroused principal concern about multipliers I, III and VII, with secondary concern about IV and VI. The earlier study encourages a choice of multipliers II, V or VIII, and the present study in no way contradicts these suggestions. Seeds spaced 100,000 numbers apart for multipliers II and V appear in Appendix A of Fishman (1978). As mentioned in the earlier study, computational efficiency is best for multiplier V.

References

1. Anderson, T. W. and D. A. Darling (1952). "Asymptotic Theory of Certain Goodness of Fit Criteria Based on Stochastic Processes," *The Annals of Mathematical Statistics*, 23, 193-212.

2. _____(1954). "A Test of Goodness of Fit," *Journal of the American Statistical Association*, 49, 765-9.

3. Bartlett, Maurice S. (1955). *An Introduction to Stochastic Processes*. Cambridge, England: Cambridge University Press.

4. Beyer, W. A., R. B. Roof and Dorothy Williamson (1971). "The Lattice Structure of Multiplicative Congruential Pseudo-Random Vectors," *Mathematics of Computation*, 25, 345-63.

5. Birnbaum, Z. W. and R. Pyke (1958). "On Some Distributions Related to the Statistic D_n^+ Statistic," *The Annals of Mathematical Statistics*, 179-87.

6. Coveyou, R. R. and R. D. MacPherson (1967). "Fourier Analysis of Uniform Random Number Generators," *Journal of the Association for Computing Machinery*, 14, 100-19.

7. Cox, David R. and Peter A. W. Lewis (1966). *The Analysis of Series of Events*. New York City: Methuen Publishing Co,

8. Durbin, James (1961). "Some Methods of Constructing Exact Tests," *Biometrika*, 48, 41-55.

9. _____(1969). "Tests of Serial Independence Based on the Cumulated Periodogram," *Bulletin of the International Statistical Institute*, 42, 1039-48.

10. Durbin, James (1973). *Distribution Theory for Tests Based on the Sample Distribution Function*. Philadelphia, Pennsylvania: Society for Industrial and Applied Mathematics.

11. Dwass, Meyer (1958). "On Several Statistics Related to Empirical Distribution Functions," *The Annals of Mathematical Statistics*, 29, 188-91.

12. Fishman, George S. (1978). *Principles of Discrete Event Simulation*. New York: John Wiley and Sons.

13. Fishman, George S. and Louis R. Moore, III (1978). "A Statistical Evaluation of Multiplicative Congruential Generators with Modulus 2^{31} -1," Technical Report 78-11, Curriculum in Operations Research and Systems Analysis, University of North Carolina at Chapel Hill.

14. Gnedenko, B. V. and V. S. Mialevic (1952). "Two Theorems on the Behavior of Empirical Distribution Functions," *Doklady Nauk. SSSR (N.S.)*, 85, 25-7.

15. Hoaglin, David (1976). "Theoretical Properties of Congruential Random-Number Generators: An Empirical View," Memorandum NS-340, Department of Statistics, Harvard Universtiy.

16. IBM (1972). *SIMPL/1 Program Reference Manual*. SH19-5060-0.

17. International Mathematical and Statistical Libraries, Inc. (1977). *IMSL Library*, Houston, Texas.

18. Katzan, H., Jr. (1971). *APL User's Guide*. New York: Van Nostrand Reinhold.

19. Learmonth, G. and P. A. W. Lewis (1973). "Naval Postgraduate School Random Number Generator Package LLRANDOM, " Monterey, California: Naval Postgraduate School.

20. Lewis, Peter A. W. (1960). "Distribution of the Anderson-Darling Statistic," *The Annals of Mathematical Statistics*, 1118-24.

21. Lewis, Peter A. W., A. S. Goodman and J. M. Miller (1969). "A Pseudo-Random Number Generator for the System 350," *IBM Systems Journal*, 8(2), 136-45.

22. Marsaglia, George (1972). "The Structure of Linear Congruential Sequences," in *Applications of Number Theory to Numerical Analysis*, ed. S. K. Zaremba, New York: Academic Press.

23. Pyke, R. (1959). "The Supremum and Infimum of the Poisson Process," *The Annals of Mathematical Statistics*, 568-76.

24. Smirnov, N. V. (1939). "Sur les ecarts de la courbe de distribution empirique," (French summary), *Rec.Math.*, 6, 3-26.

PORTABILITY CONSIDERATIONS FOR RANDOM NUMBER GENERATORS

James E. Gentle, IMSL, Inc.

ABSTRACT

With increasing use being made of computer-generated random samples, questions concerning the quality of the random number generators become increasingly important. Various studies have been made of the quality of basic generators and some generators have been identified as fairly reliable. As with any computer software, however, it is necessary to know more than that the method or "program" is good. It is also necessary to know that the program will perform well in each computer-compiler environment in which it is to be executed. Portability of a program within a class of environments means that the program will run correctly in each environment. A portable program for random number generation that has been found to produce high quality results in one computer-compiler environment will be just as reliable in other environments, and hence portability allows for efficiency in testing of random number generators. Another desirable aspect of portability is that Monte Carlo studies performed with portable random number generators are reproducible elsewhere, and results of studies by one researcher are more easily extensible by another researcher.

Various computer-compiler characteristics relevant to the question of portability of U(0,1) pseudorandom generators and the necessary programming considerations are discussed.

Keywords: Statistical computing; Pseudorandom numbers; Portability; Monte Carlo

INTRODUCTION

The technological advances in computing machinery that have resulted in much cheaper computing have had a concomitant salutary effect on the attitudes of most users. Of the two traditional criteria for quality of software, fidelity (correctness, accuracy, etc.) and efficiency, the current attitude requires fidelity before efficiency can even be a consideration. This change in attitudes is quite evident in the methods of computer generation of pseudo-random numbers that are considered acceptable. For random number generation, efficiency has long been an important consideration because of the large numbers of computations required in many of the Monte Carlo applications. Selection of random number generation methods based on criteria which attempt to strike a balance between fidelity and efficiency often resulted in use of approximations and shortcuts. An extreme example is the use of the sum of twelve uniform deviates to approximate a normal deviate.

When efficiency is a primary consideration in the construction of computer programs, even if shortcut approximations are eschewed, the programs will very likely capitalize on special characteristics of the computer/compiler environment in which the programs are to be used. The programs may execute efficiently in a given environment but very inefficiently or not at all in a different environment. For the freer exchange of scientific methods and information, however, ability to move programs from one environment to another is highly desirable. Portability of a program within a set of environ ments means that the program will run correctly in each environment.

Two separate aspects of the quality of computer programs are the quality of the underlying algorithms and the validity of the implementation. For portable software, evaluation of the validity of the implementation is simpler, because results from one environment carry over to others. In the case of random number generation, in which the assessment of the quality of the underlying algorithm is not straightforward, portability of the programs become even more important, because then the accumulating evidence, much of which is subjective, on the quality of the underlying algorithm is relevant in any environment in the set over which the random number generator is portable.

The increased speed of computers and the consequent deemphasis of efficiency allow meaningful efforts to be expended on the development of portable programs for random number generation. This situation is to be contrasted with the state of affairs twelve years ago, when a portable generator was first proposed in the literature (Kruskal, 1969). The cycle length of that early generator was only 2048. That generator represents an extreme sacrifice of fidelity to achieve portability, because of the high premium on efficiency. (That generator did, however, include in its set of environments sixteen-bit-word computers, which some later proponents of portability ignore.)

PORTABILITY

"Portability" refers to a source program unit. Both hardware and system software contribute to nonportability; hence in speaking of portability it is necessary to refer to computer/compiler environments (in which "compiler" implicitly includes other system software). Portability of a program unit across a set of computer/compiler environments is taken to mean that the program unit, without modification, can be compiled and executed correctly in all environments in the set. (See Aird, Battiste and Gregory, 1977, for further discussion of portability and related concepts.)

Hardware characteristics which most directly affect portability are the codes in which the numeric quantities are represented and the manner in which arithmetic operations are performed. Almost all codes for numeric representation employ an underlying binary scheme and almost all employ at least two basic representations: integer (i.e., no scaling) and floating point (with scaling), but differ in whether a binary base is used or some other base such as octal or hexadecimal is used; how many bits are available for each numeric unit; and, in floating point, how these bits are allocated between the scaled value and the scaling factor. Arithmetic operations may differ depending on how many extra bits, if any, are available for numeric operand.

Software characteristics which affect portability include such things as extensions to the basic language (Fortran, in this paper), extent of optimization in compilation, differences in intrinsic and basic external functions, and so on. Whether or not rounding occurs in floating point computations, which may be determined by hardware, system software, or firmware, also affects portability of software, especially programs for random number generation, as we shall see.

A key component in any attempt to develop portable software is restriction of the language used to a subset of the language family which is uniform over the intersection of dialects in the family. The PFORT subset of Fortran is useful for this purpose, particularly because of the availability of a verifier to insure that programs conform to its standards (Ryder, 1974).

Programs coded in the PFORT standard, however, may yield different results on different computers due to different word lengths and different rounding schemes. Use of integer arithmetic has been advocated by some (e.g. Greenwood, 1976, 1977) in an attempt to avoid such problems. In the widely-used computers that have 32 bits for ordinary representation of numeric quantities, it turns out that just the difference in having a full set of 31 bits, as in integer mode, and having fewer than 31 bits, as in the fraction portion of the floating point mode, can be quite important for quantities of the magnitude occuring in some commonly used random number generators. The integer mode must be avoided in many computations, however if a program is to be portable across a set of environments which includes any of the currently popular 16 bit minicomputers; and so a method formerly advocated to improve portability may actually decrease it.

Portability in its fullest sense is not attainable across any set of current computer/compiler environments that is at all comprehensive. Successful solutions to the portability problem have involved use of preprocessors or converters to tailor a source program for a given environment. In the converter portable approach described by Aird et al. (1977), the source program in its basis form will compile and execute correctly on at least one environment; and a converter program is available to make appropriate changes to the basis form to generate a version for each other environment.

The definitions of portability and various qualifications of portability involve 1) movement of the program unit from one environment to another, 2) compilation of the program, and 3) "correct" execution--meaning the output is correct. The qualifications of portability, such as converter portability, relate to consideration of what is required during the "movement" to satisfy the second and third points above. These other two aspects of the definition of portability relate to the effect of the movement. The second point--compilation--is not moot; but the third point--"correctness" of output--is not quite straightforward. If the movement from one environment to another is to have no effect, the output would not change; and "correctness" of output might imply that there exists a single standard by which the output can be measured; but "correctness" in this restrictive sense is not as useful a concept as correctness of solution. In this sense, however, "correctness" is problem oriented rather than program oriented; an underlying problem with an associated set of correct solutions must be assumed. Consideration of portability must involve definition of the problem and its set of correct solutions. The solution is generally an n-tuple, some of whose elements may vary and yet the solution may remain within the correct solution set (as in a linear programming problem with multiple solutions). Verification of portability, therefore, is not a trivial task.

In the case of programs for problems whose correct solution sets consist of single points for any single input, correctness may be assessed by comparing the output to a single standard; and ('qualifier') portable (i.e., portable, converter portable, etc.) programs for these problems produce output that is invariant to the environment (within rounding of the output). The general problem of random number generation does not fall within this class of problems. (How could something "random" produce the same output every time whether in the same environment or not?) Pseudorandom uniform number generation as a practical problem, however, relies on algorithms which do yield a single correct output for a given input, and so it is possible to assess portability with regard to a standard output. ('Qualifier') portable programs for uniform random numbers, implementing the commonly used algorithms, should produce output unaffected by the environment (within rounding). Greenwood (1981) has referred to this invariance of output to the environment as reproducibility, and has shown that certain algorithms for generation of variates from discrete distributions also may be implemented in programs with this property. Programs for random number generation implementing an acceptance-rejection algorithm by means of Fortran IF statements on REAL expressions cannot be invariant to environment over a reasonable set of computer/compiler types. The sequence generated by such programs may be grossly different in different environments, yet each sequence may be "correct" and the program itself may be considered "('qualifier') portable".

PORTABLE RANDOM NUMBER GENERATORS

The extremely large amount of computation in many applications employing random generators caused most early developers to employ whatever computer/compiler specific method that was available to enhance efficiency in that environment. Except for the paper of Kruskal (1969) there seemed to be little interest in portable generators until the latter 1970's. Two papers by Greenwood (1976, 1977) described portable composite linear congruential generators employing integer arithmetic. The environment set for portability of the Greenwood generators included only computers having at least 31 bits for integer representation. Schrage (1979) described two generators for the simple, but good, multiplicative congruential generator using a multiplier of 16807 and modulus of $2^{31}-1$ (Lewis et al., 1969). One of Schrage's generators, RAND, used integer mode in which 31 bits were assumed; and the other one, DRAND, used double precision assuming no more than 31 bits for the fraction. The former assumption seems not to be restrictive enough, and the latter seems overly restrictive for the computers currently in widespread use. Schrage's generators, within the limits indicated, seem fairly portable. They would occasionally yield a 1.000000 on Xerox, Dec 10, Dec 11, VAX, and Honeywell computers, however. This could be very distressing if the output of RAND or DRAND were to be used to generate a random variate from a non-uniform distribution with infinite range using an inverse CDF technique. Because a 1 is incorrect output for a uniform random number generator, RAND and DRAND are not portable to those computers on which it would result.

The basic IMSL random number generator, GGUBS, (see Exhibit 1) is converter portable across a set of environments that includes twelve different computer types. (The number of computer/compiler environments is considerably larger than twelve. Numbers such as this can be inflated considerably by counting certain models of a basic computer type, or by counting Amdahl, Itel, and IBM look-alikes as different.) The computer on which the basis form program will compile and execute correctly without processing by the converter is an IBM 360 or 370. The command statements for the converter have "C" in column 1 and "$" in column 2. The converter will produce clean listings for a given target computer by selectively removing statements and/or removing the "C" from comment statements (See Exhibit 2). For GGUBS, the resulting programs are very similar except that the CDC version uses single precision while all other versions use double. The fundamental requirement is that the exact product of 16807 and any permissible value of DSEED be representable in the precision of the computer. Since this may require as many as 46 bits, double precision is required for all computers in the set except CDC. The question of precision for use in the CDC must involve a compromise between the two desiderata, uniformity across types and efficiency. Since efficiency differences (despite the remarks of the introduction!) are rather compelling in this case, the CDC version has entirely different code from the other versions.

Because some computers round and others truncate following floating point operations (and some computers allow user choice), it is necessary to use different values for scaling the integral values of DSEED into the open interval (0,1), if the number of bits of single precision fraction is less than 31.

The routine GGUBS implements the same multiplicative congruential generator as proposed by Lewis et al. (1969) and Schrage (1979). Because of some empirical test results of Fishman and Moore (1978), another converter portable random number generator, GGUBT, with a different multiplier was added to the IMSL Library. This multiplier, 397204094, would cause spillover even in double precision fractions, so the modular reduction was implemented in three statements:

```
DSEED1 = DMOD(32768.DO*DSEED,D2P31M)
DSEED2 = DMOD(23166.DO*DSEED,D2P31M)
DSEED  = DMOD(12121.DO*DSEED1 + DSEED2, D2P31M)
```

GGUBT executes almost three times as slowly as GGUBS.

Both GGUBS and GGUBT are invariant across the set of computer/compiler environments supported; that is, either GGUBS or GGUBT yield exactly the same sequence of seeds or DSEED values on all environments supported. In addition, the first 23 bits of the mantissa of the U(0,1) numbers will be exactly the same.

CONCLUDING REMARKS

Although computing efficiency is not the overriding concern as in former years, certain differences in computer architecture suggest slight differences in programs to avoid deleterious efficiency loss. Moreover, differences in arithmetic operations, such as whether or not rounding is done, prevent the use of a single source program in random number generation. Converter portability seems to be a useful compromise between full portability and the realities of various computer types.

The use of random number generators for simulation studies has become very important in the development of statistical tools. Hoaglin and Andrews (1975) noted that approximately 19% of the articles in the 1973 volume of the Journal of the American Statistical Association contained reports of Monte Carlo simulation studies. In 1978 this figure for the same journal was 29% and in 1979 it was 35%. The reader of the reports of these simulation studies does not care how many CPU seconds were burned up, but (s)he does care about the fidelity of the random number generators used.

The primary concern for random number generators is fidelity. Portability and reproducibility of generators allow assessments of fidelity performed in one environment to be relevant in a different environment.

REFERENCES

Aird, T. J., E. L. Battiste, and W. C. Gregory (1977). Portability of mathematical software coded in Fortran. ACM TOMS 3, 113-127.

Fishman, G. S., and L. R. Moore (1978). A statistical evaluation of multiplicative congruential generators with modulus $2^{31}-1$. Tech Rep. 78-11, Operations Research and Systems Analysis, University of North Carolina.

Fox, P. A. (1976). PORT Mathematical Subroutine Library. Bell Laboratories, Murray Hill.

Greenwood, J. A. (1976). A fast machine-independent long-period generator for 31-bit pseudo-random numbers. Compstat 76. J. Gordesch and P. Naeve, eds. Physica Verlag, Wien, 30-36.

Greenwood, J. A. (1977). Portable generators for the random variables usual in reliability simulation. Recent Developments in Statistics. J. R. Barra et al, eds. North Holland Publishing Company. New York, 677-688.

Greenwood, J. A. (1981). A portable formulation of the alias method for random numbers with discrete distributions. Communications in Statistics B10.

Hoaglin, D. C., and D. F. Andrews (1975). The reporting of computation-based results in statistics. American Statistician 29, 122-126.

IMSL (1980). The IMSL Library Reference Manual, Edition 8. IMSL, Inc., Houston.

Kennedy, W. J., and J. E. Gentle (180). Statistical Computing. Marcel Dekker, New York.

Kruskal, J. B. (1969). Extremely portable random number generators. CACM 12, 93-94.

Lewis, P. A. W., A. S. Goodman, and J. M. Miller (1969). A pseudo-random number generator for the system 360. IBM Systems Journal 8, 136-146.

Ryder, B. G. (1974). The PFORT verifier. Software Practice and Experience 4, 359-377.

Schrage, L. (1979). A more portable Fortran random number generator. ACM TOMS 5, 132-138.

```
C   IMSL ROUTINE NAME   - GGUBS                                         GGUS0010
C
C-----------------------------------------------------------------------
C$    /  BASIS FOR ALL ENVIRONMENTS
C$    OPTIONS, PRECISION=SINGLE
C
C   COMPUTER            - IBM/SINGLE
C
C   LATEST REVISION     - JUNE 1, 1980
C
C   PURPOSE             - BASIC UNIFORM (0,1) PSEUDO-RANDOM NUMBER
C                           GENERATOR
C
C   USAGE               - CALL GGUBS (DSEED,NR,R)
C
C   ARGUMENTS    DSEED  - INPUT/OUTPUT DOUBLE PRECISION VARIABLE
C                           ASSIGNED AN INTEGER VALUE IN THE
C                           EXCLUSIVE RANGE (1.D0, 2147483647.D0).
C                           DSEED IS REPLACED BY A NEW VALUE TO BE
C                           USED IN A SUBSEQUENT CALL.
C                NR     - INPUT NUMBER OF DEVIATES TO BE GENERATED.
C                R      - OUTPUT VECTOR OF LENGTH NR CONTAINING THE
C                           PSEUDO-RANDOM UNIFORM (0,1) DEVIATES
C
C   PRECISION/HARDWARE  - SINGLE/ALL
C
C   REQD. IMSL ROUTINES - NONE REQUIRED
C
C   NOTATION            - INFORMATION ON SPECIAL NOTATION AND
C                           CONVENTIONS IS AVAILABLE IN THE MANUAL
C                           INTRODUCTION OR THROUGH IMSL ROUTINE UHELP
C
C   COPYRIGHT           - 1980 BY IMSL, INC. ALL RIGHTS RESERVED.
C
C   WARRANTY            - IMSL WARRANTS ONLY THAT IMSL TESTING HAS BEEN
C                           APPLIED TO THIS CODE. NO OTHER WARRANTY,
C                           EXPRESSED OR IMPLIED, IS APPLICABLE.
C
C-----------------------------------------------------------------------
C
      SUBROUTINE GGUBS (DSEED,NR,R)
C$    CONSTANTS
C        D2P31(IBM)    = 2147483648.D0
C        D2P31(XEROX)  = 2147483711.D0
C        D2P31(DGC)    = 2147483648.D0
C        D2P31(DEC11)  = 2147483711.D0
C        D2P31(VAX)    = 2147483711.D0
C        D2P31(PRIME)  = 2147483711.D0
C        D2P31(HP3000) = 2147483648.D0
C        D2P31(UNIVAC) = 2147483648.D0
C        D2P31(HIS)    = 2147483655.D0
C        D2P31(DEC10)  = 2147483655.D0
C        D2P31(BGH)    = 2147483648.D0
C$
C$    IF(H.LT.60) HLT60,HGE60
CC                                  SPECIFICATIONS FOR ARGUMENTS
      INTEGER            NR
      REAL               R(NR)
      DOUBLE PRECISION   DSEED
CC                                  SPECIFICATIONS FOR LOCAL VARIABLES
      INTEGER            I
C$    IF(HP3000) 1 LINE, 1 LINE
C     DOUBLE PRECISION   D2P31M,D2P31,DMOD
      DOUBLE PRECISION   D2P31M,D2P31
CC                                  D2P31M=(2**31) - 1
CC                                  D2P31 =(2**31)(OR AN ADJUSTED VALUE)
      DATA               D2P31M/2147483647.D0/
C$    DATA               D2P31/D2P31/
```

```
      DATA              D2P31/2147483648.D0/
CC                                FIRST EXECUTABLE STATEMENT
      DO 5 I=1,NR
         DSEED = DMOD(16807.D0*DSEED,D2P31M)
    5 R(I) = DSEED / D2P31
C$LBL HLT60
CC                                SPECIFICATIONS FOR ARGUMENTS
C     INTEGER           NR
C     REAL              R(NR)
C     DOUBLE PRECISION  DSEED
CC                                SPECIFICATIONS FOR LOCAL VARIABLES
C     INTEGER           I
C     REAL              S2P31,S2P31M,SEED
CC                                S2P31M = (2**31) - 1
CC                                S2P31 = (2**31)
C     DATA              S2P31M/2147483647.E0/,S2P31/2147483648.E0/
CC                                FIRST EXECUTABLE STATEMENT
C     SEED = DSEED
C     DO 5 I=1,NR
C        SEED = AMOD(16807.E0*SEED,S2P31M)
C   5 R(I) = SEED / S2P31
C     DSEED = SEED
C$LBL HGE60
      RETURN
      END
```

EXHIBIT 1

```
C   IMSL ROUTINE NAME   - GGUBS                                         GGUS0010
C
C-----------------------------------------------------------------------
C
C   COMPUTER            - IBM/SINGLE
C
C   LATEST REVISION     - JUNE 1, 1980
C
C   PURPOSE             - BASIC UNIFORM (0,1) PSEUDO-RANDOM NUMBER
C                           GENERATOR
C
C   USAGE               - CALL GGUBS (DSEED,NR,R)
C
C   ARGUMENTS    DSEED  - INPUT/OUTPUT DOUBLE PRECISION VARIABLE
C                           ASSIGNED AN INTEGER VALUE IN THE
C                           EXCLUSIVE RANGE (1.D0, 2147483647.D0).
C                           DSEED IS REPLACED BY A NEW VALUE TO BE
C                           USED IN A SUBSEQUENT CALL.
C                NR     - INPUT NUMBER OF DEVIATES TO BE GENERATED.
C                R      - OUTPUT VECTOR OF LENGTH NR CONTAINING THE
C                           PSEUDO-RANDOM UNIFORM (0,1) DEVIATES
C
C   PRECISION/HARDWARE  - SINGLE/ALL
C
C   REQD. IMSL ROUTINES - NONE REQUIRED
C
C   NOTATION            - INFORMATION ON SPECIAL NOTATION AND
C                           CONVENTIONS IS AVAILABLE IN THE MANUAL
C                           INTRODUCTION OR THROUGH IMSL ROUTINE UHELP
C
C   COPYRIGHT           - 1980 BY IMSL, INC. ALL RIGHTS RESERVED.
C
C   WARRANTY            - IMSL WARRANTS ONLY THAT IMSL TESTING HAS BEEN
C                           APPLIED TO THIS CODE. NO OTHER WARRANTY,
C                           EXPRESSED OR IMPLIED, IS APPLICABLE.
C
C-----------------------------------------------------------------------
C
      SUBROUTINE GGUBS (DSEED,NR,R)
C                                  SPECIFICATIONS FOR ARGUMENTS
      INTEGER            NR
      REAL               R(NR)
      DOUBLE PRECISION   DSEED
C                                  SPECIFICATIONS FOR LOCAL VARIABLES
      INTEGER            I
      DOUBLE PRECISION   D2P31M,D2P31
C                                  D2P31M=(2**31) - 1
C                                  D2P31 =(2**31)(OR AN ADJUSTED VALUE)
      DATA               D2P31M/2147483647.D0/
      DATA               D2P31/2147483648.D0/
C                                  FIRST EXECUTABLE STATEMENT
      DO 5 I=1,NR
         DSEED = DMOD(16807.D0*DSEED,D2P31M)
    5 R(I) = DSEED / D2P31
      RETURN
      END
```

EXHIBIT 2

Workshop 8

Understanding Time Series Analysis

Organizer: Emanuel Parzen, Texas A&M University

Chair: Emanuel Parzen, Texas A&M University

Invited Presentations:

Recent Developments in Spectrum and Harmonic Analysis, David J. Thomson, Bell Laboratories

On Some Numerical Properties of ARMA Parameter Estimation Procedures, H. Joseph Newton, Texas A&M University

Time Series Recursions and Self-Tuning Control, Victor Solo, Harvard University

Some Recent Developments in Spectrum and Harmonic Analysis

David J. Thomson

Bell Laboratories
Whippany, New Jersey 07981

ABSTRACT

In estimating the spectrum of a stationary time series from a finite sample of the process two problems have traditionally been dominant: first, what algorithm should be used so that the resulting estimate is not severely biased; and second, how should one "smooth" the estimate so that the results are consistent and statistically significant.

Within the class of spectrum estimation procedures that have been found successful in the various engineering problems considered, bias control is achieved by iterative model formation and prewhitening combined with robust procedures (the "robust filter"), while "smoothing" is done by an adaptive nonlinear method.

Recently a method has been found which, by using a "local" principal components expansion to estimate the spectrum, provides new solutions to both the bias and smoothing problems and also permits a unification of the differences between windowed and unwindowed philosophies. This estimate, which is an approximate solution of an integral equation, consists of a weighted average of a series of direct spectrum estimates made using discrete prolate spheroidal sequences as orthogonal data windows.

Keywords
Spectrum estimation, prolate spheroidal wave functions, data windows, mixed spectra, harmonic analysis, analysis of variance, smoothing.

1. Introduction

This paper reconsiders the old problem in time series analysis of reliably estimating the spectral density function from a finite sample of an approximately Gaussian stationary process. In particular we consider three related topics: *first,* What general form should the estimation procedure take? *second,* Should one use a data window , and if so which one, and *third;* How should one "smooth" the resulting estimate?

The fact that these questions may be considered distinct is historical; in the following we attempt to summarize some of the developments which led to them and some recent developments towards their unification. While this theory is somewhat orthogonal to the current emphasis on rational spectrum estimates, *i.e.* autoregressive, moving average, or combined (ARMA) estimates plus their Burg and "maximum entropy" variants, it has significant impact on these problems, in particular on the "super resolution" estimation question.

For several years following the 1946 Royal Statistical Society conference[1] most of the work in the field was directed either towards solving computational problems or to finding better "lag windows" so that the perceived conflict between resolution and variance could be minimized. Advances in computer technology and the discovery of the fast Fourier transform eventually led to more difficult problems being considered but fifteen years ago papers on the "smoothing" problem and the trade between resolution and variance were common and most practitioners used some variation of the Blackman-Tukey method. A few, notably Tukey[1967] and Welch[1967], had advanced to use of direct estimates based on the recently discovered fast Fourier transform. Even though prewhitening had been recommended in Blackman and Tukey in 1957 and earlier, the major bias problems were still largely ignored.

Ten years ago it was beginning to be recognized that the fundamental conflict in spectrum estimation was not between resolution and variance but between resolution and bias. This realization prompted the use of prolate spheroidal wave functions for data windows and led to the use of autoregressive models for prewhitening filters. While the use of these techniques resulted in better estimates, the use of data windows (not to be confused with the older lag windows used for smoothing) also posed a dilemma: first, they clearly could result in vastly superior estimates to unwindowed estimates; second, they effectively gave different weighting to observations which were equally valid. This conflict is difficult to resolve.

Again, five years ago, the spectrum estimation procedures in common use had changed significantly from their predecessors: *robust* procedures were coming into use, the old smoothers were finally being replaced by adaptive techniques; and the structure of the estimation process was becoming clearer. Details of these procedures are contained in Thomson[1977a,b] and Kleiner *et al* [1979]. The increasing power of these techniques, however, also made it clear that spectrum estimation, as a subject, was an art consisting of a collection of heuristic methods only loosely bound by theory. Worse, much of the existing theory, particularily the parts depending on asymptotics, was inadaquate.

Three years ago Slepian[1978] described *discrete* prolate spheroidal wave functions and sequences. Using these functions as a basis does much to unify the theory of spectrum estimation: A simple ANOVA-like test for line components makes spectrum estimation and harmonic analysis distinct problems, rather than two names for nearly the same one; The "smoothing" problem disappears; The dilemma between different windowing philosophies can be peacefully resolved; and, not least, the method provides considerable insight into some of the parametric modelling problems as well.

2. Orthogonal Window Estimates

In this section we present a brief outline of a new and somewhat more unified theory which has a number of interesting features: first, it is an explicit small sample theory with the sample size entering explicitly into the methods and performance bounds; second, it explains the use of data windows and provides a resolution of the dilemma posed earlier; third, smoothing in the conventional manner is unnecessary as the estimate is consistent; fourth, the procedure is data-adaptive; fifth, it provides an analysis of variance test for line components including the process mean (this also distinguishes harmonic analysis from spectrum estimation); and sixth, in the case of multivariate data it results in new *classes* of estimates. As a particular example of the latter, the technique results in *two* distinct estimates of coherence, one for line components, one for the continuum.

We assume that the available data consists of N samples, $x_0, x_1, \cdots, x_{N-1}$, representing a stationary, real, ergodic, near Gaussian, time series. N is assumed to be finite. We initially assume that the observations are centered, that is that $\mathbf{E}\{y_k\}=0$, but later will mention a new estimate of the mean.

We assume that frequency, f, and radian frequency, $\omega=2\pi f$, are defined on their principal domains $(-½, ½)$ and $(-\pi, \pi)$ respectively. When estimates are computed on a frequency mesh we assume that the mesh spacing is less than the equivalent Nyquist frequency, $1/2N$.

We begin with the *time centered*† Cramer representation (see Doob[1953])

$$x_n = \int_{-1/2}^{1/2} e^{i2\pi\nu(n-\frac{N-1}{2})}\, dZ(\nu)$$

In simple cases where there are no line components (which in the line-test procedure becomes the null hypothesis) we have, as usual,

$$\mathbf{E}\{dZ(\nu)\}=0$$

but when there are line components we assume the extended representation‡

$$\mathbf{E}\{dZ(\nu)\}=a\,\delta(\nu-f_0)$$

for the simplest case of a single line component. The case for multiple lines is similar but algebraically more complex.

† The centered representation assumes that the center of the observation epoch is at the origin. It is used simply for notational convenience.

‡ Note that this definition results in a considerable extension of the usual definition of stationarity.

This representation permits a distinction between harmonic and spectrum analysis: *harmonic analysis* is concerned with the *first* moments of $dZ(\nu)$, while *spectrum analysis* is the problem of estimating the *second*, and higher, moments of $dZ(\nu)$, and in particular the second *central* moment. To quote Parzen[1981]

> "Spectral analysis has as its aim the determination of the properties of the function $Z(\nu)$."

For notational simplicity it is convenient to also define the Fourier transform of the observations, $\{x\}$, in time centered form:

$$\tilde{x}(f) = \sum_{n=0}^{N-1} e^{-i2\pi f(n-\frac{N-1}{2})} x_n$$

so that, using the spectral representation for the data in this formula, we have

$$\tilde{x}(f) = \int_{-1/2}^{1/2} \frac{\sin N\pi(f-\nu)}{\sin\pi(f-\nu)}\, dZ(\nu)$$

which is the convolution of the Cramer process, $dZ(\nu)$, with a Dirichlet kernel. The normal interpretation of this equation is that the observed power at frequency f is a complex weighted average of the power at all frequencies with, for N sufficiently large, most of the weight being at frequencies "close" to f.

An alternative viewpoint is to consider equation (5) as a linear Fredholm integral equation of the first kind for $dZ(\nu)$. Clearly, since we are dealing with a projection operator, it is impossible to obtain exact or unique solutions: what we desire is a solution which is both statistically and numerically plausible. We thus contemplate "solving" the integral equation in some local interval about f, say $(f-W, f+W)$ with the condition that we want the statistics of the "solution" and not the "solution" itself. From this viewpoint the formal solution process for the integral equation can be used as a guide to the statistical approach. Investigation of the solution of an integral equation over an interval naturally suggests an eigenfunction approach and, fortunately, a very detailed description of the eigenfunctions and eigenvalues of the Dirichlet kernel has recently been given by Slepian[1978].

2.1. Diversion: Discrete Prolate Spheroidal Wave Functions and Sequences

In this section we give a short list of formulae and properties of these eigenfunctions from Slepian's paper. The eigenfunctions, denoted by $U_k(N, W; f)$, $k=0, 1, \cdots N-1$ are known as *discrete prolate spheroidal wave functions* and are solutions of the equation:

$$\int_{-W}^{W} \frac{\sin N\pi(f-f')}{\sin\pi(f-f')} U_k(N,W;f')\,df' = \lambda_k(N,W)\cdot U_k(N,W;f)$$

where W, $0<W<½$ is the bandwidth defining "local" and is normally of the order $1/N$. The functions are ordered by their eigenvalues

$$1 > \lambda_0(N, W) > \lambda_1(N, W) > \cdots > \lambda_{N-1}(N, W) > 0.$$

The first $2NW$ eigenvalues are extremely close to 1 and, of particular relevence here, *of all functions which are the Fourier transform of an indexlimited sequence the discrete prolate spheroidal wave function,* $U_0(N, W; f)$ *has the greatest fractional energy concentration in* $(-W, W)$. For small k and large N the degree of this concentration is given by Slepian's asymptotic expression (in slightly different notation)

$$1-\lambda_k(N,W) \sim \frac{4(\pi N \sin\pi W)^{1/2}}{k!\cos\pi W}\left[\frac{8N\sin\pi W}{\cos^2\pi W}\right]^k\left[\frac{1-\sin\pi W}{1+\sin\pi W}\right]^N$$

or, for larger N with $N\pi W = c$,

$$1-\lambda_k(N,W) \sim 4\sqrt{\pi c}\,\frac{(8c)^k}{k!}e^{-2c}$$

From a conventional spectrum estimation viewpoint this expression gives the fraction of the total energy of the spectral window outside the main lobe *(i.e.* outside $(-W, W)$).

The eigenfunctions, $U_k(N, W; f)$, $k = 0, 1, \cdots N-1$ are *doubly orthogonal,* that is they are orthogonal over $(-W, W)$

$$\frac{1}{\lambda_k(N,W)}\int_{-W}^{W} U_j(N,W;f)\cdot U_k(N,W;f)\,df = \delta_{j,k}$$

and orthonormal over $(-\tfrac{1}{2}, \tfrac{1}{2})$.

$$\int_{-1/2}^{1/2} U_j(N,W;f)\cdot U_k(N,W;f)\,df = \delta_{j,k}$$

The Fourier transforms of the discrete prolate spheroidal wave functions are known as *discrete prolate spheroidal sequences*

$$v_n^{(k)}(N,W) = \frac{1}{\epsilon_k\lambda_k(N,W)}\int_{-W}^{W} U_k(N,W;f)\,e^{-i2\pi f\left(n-\frac{N-1}{2}\right)}df$$

valid for $k = 0, 1, \cdots, N-1$ and all n. ϵ_k is 1 for k even and i for k odd. Because of the double orthogonality there is a second Fourier transform

$$v_n^{(k)}(N,W) = \frac{1}{\epsilon_k}\int_{-1/2}^{1/2} U_k(N,W;f)\,e^{-i2\pi f\left(n-\frac{N-1}{2}\right)}df$$

valid for both $n, k = 0, 1, \cdots, N-1$. As one would expect, the finite discrete Fourier transform of the prolate spheroidal sequence results in the discrete prolate spheroidal wave functions

$$U_k(N,W;f) = \epsilon_k\sum_{n=0}^{N-1} v_n^{(k)}(N,W)\,e^{i2\pi f\left(n-\frac{N-1}{2}\right)}$$

It should also be noted that the discrete prolate spheroidal sequences satisfy a Toeplitz matrix eigenvalue equation

$$\sum_{m=0}^{N-1}\frac{\sin 2\pi W(n-m)}{\pi(n-m)}\cdot v_m^{(k)}(N,W) = \lambda_k(N,W)\cdot v_n^{(k)}(N,W)$$

and so are easily computed. Like the wave functions the $\{v_n^k\}$ are doubly orthogonal, being orthogonal on $(-\infty, \infty)$ and orthonormal on $[0, N-1]$.

2.2. Orthogonal Window Estimates, Continued.

Returning to the main spectrum estimation problem and using standard methods of solving Fredholm integral equations as a guide to the statistical approach we are led to consider expansions of the relevant functions in the proper basis, that is in the basis set consisting of the eigenfunctions of the kernel. In this case the basis functions are the discrete prolate spheroidal wave functions and we first consider the expansion coefficients of dZ in $(f-W, f+W)$

$$a_k(f) = \frac{1}{\sqrt{\lambda_k(N,W)}}\int_{-W}^{W} U_k(N,W;\nu)\,dZ(f+\nu)$$

The normalization implicit in this definition results in $\mathrm{E}\{|a_k(f)|^2\} = S$ when the spectrum, S, is white.

Now consider an *estimate*† of these coefficients obtained by expanding the Fourier transform, $\tilde{x}(f)$ over the interval $(f-W, f+W)$ on the same set of basis functions

$$\hat{a}_k(f) = \frac{1}{\lambda_k(N,W)}\int_{-W}^{W} U_k(N,W;\nu)\,\tilde{x}(f+\nu)\,d\nu$$

By using the basic integral equation which expresses the projection operation from dZ unto $\tilde{x}$ $\hat{a}_k(f)$ may be expressed in terms of dZ as

$$\hat{a}_k(f) = \int_{-1/2}^{1/2} U_k(N,W;\zeta)\,dZ(f+\zeta)$$

from which it is apparent that the estimated coefficient, $\hat{a}_k(f)$, also is defined so that $\mathrm{E}\{|\hat{a}_k(f)|^2\} = S$ for a white spectrum.

An alternative form of this estimate may be obtained by writing the Fourier transform, $\tilde{x}(f)$, directly in terms of the data and using the definition of the discrete prolate spheroidal sequences to obtain

$$\hat{a}_k(f) = \sum_{n=0}^{N-1} x_n\cdot\frac{v_n^{(k)}(N,W)}{\epsilon_k}\,e^{-i2\pi f\left(n-\frac{N-1}{2}\right)}$$

This expression is worthy of careful study as it is simply the discrete Fourier transform of the *data multiplied by a data window.* We note in particular $|\hat{a}_0(f)|^2$ is the best known direct estimate of spectrum for a given W, (Thomson[1971], Eberhard[1973], Kaiser[1974]) However, when used by itself as it always has been in the past, this estimate has had to be smoothed to produce a consistent result. In

† Note that in this case "estimate" is being used more in a numerical than a statistical sense. The estimated coefficient is conditioned on the particular realization and is not meant in an ensemble sense.

addition, in previous usage, the other coefficients, $a_1(f), \cdots$ were never present. To preview the approximate approach, one can show, using standard methods and assuming that $\mathbf{E}\{dZ\}=0$,

$$\mathbf{E}\{|\hat{a}_k(f)|^2\} = |U_k(N,W;f)|^2 * S(f)$$

Thus the different coefficients squared all produce estimates of the spectrum whose expected values are convolutions of the true spectrum, $S(f)$, with the kth spectral window, $|U_k(N,W;f)|^2$. Moreover, since $\hat{a}_k(f)$ and $\hat{a}_j(f)$ are approximately uncorrelated for j, k small the average of these different estimates can be better than the individual estimates. For the same reasons that weighting is used both in Wiener filtering and in the numerical solution of integral equations, see Lawson & Hanson[1974], it is necessary to use a weighted average in this case also. Specifically window leakage, expressed via the smaller eigenvalues associated with the higher functions, causes the corresponding coefficients to be both biased and correlated.‡

We thus introduce a sequence of weight functions, $d_k(f)$, which, like the coefficients they modify, are functions of frequency and defined so that the mean square error between $a_k(f)$ and $\hat{a}_k(f)\cdot d_k(f)$ is minimized.

$$a_k(f)-\hat{a}_k(f)d_k(f)=\frac{1}{\sqrt{\lambda_k(N,W)}}\int_{-W}^{W} U_k(N,W;\zeta)dZ(f+\zeta)$$

$$-\, d_k(f)\int_{-½}^{½} U_k(N,W;\zeta)\,dZ(f+\zeta)$$

or, collecting regions of integration,

$$=\left[\frac{1}{\sqrt{\lambda_k(N,W)}} - d_k(f)\right]\cdot\int_{-W}^{W} U_k(N,W;\zeta)\,dZ(f+\zeta)$$

$$-\, d_k(f)\,⨍ U_k(N,W;\zeta)\,dZ(f+\zeta)$$

where the cut integral is defined as

$$⨍ = \int_{-½}^{½} - \int_{-W}^{W}$$

From this expression it can be seen that the error consists of the sum of two terms; one defined on $(-W, W)$, the other on the remainder of the principal domain. Because both of these integrals are with respect to the random orthogonal measure dZ they are independent and consequently the mean square error is simply the sum of the squares of the two terms. Using the orthogonal increment properties of dZ again *and assuming that the spectrum varies slowly over* $(-W, W)$ the mean square value of the first integral is well approximated by

$$\mathbf{E}\{|\int_{-W}^{W} U_k(N,W;\zeta)dZ(f+\zeta)|^2\} \approx \lambda_k(N,W)\,S(f)$$

The second integral is more difficult and depends on the detailed shape of the spectrum over the external domain. However, by considering its average value over all frequencies, a value adequate for the determination of weights may be obtained

$$\underset{f}{ave}\,\mathbf{E}\{\left|⨍ U_k(N,W;\zeta)dZ(f+\zeta)\right|^2\} = \underset{f}{ave}\,⨍ U_k(N,W;\zeta)^2 dS(f+\zeta)$$

$$= \sigma^2(1-\lambda_k(N,W))$$

where σ^2 is the process variance.

Combining these two integrals and minimizing the mean square error with respect to $d_k(f)$ gives the approximate optimum weight

$$d_k(f) \approx \frac{\sqrt{\lambda_k(N,W)}\,S(f)}{\lambda_k(N,W)\,S(f) + (1-\lambda_k(N,W))\,\sigma^2}$$

and the corresponding average of the spectral density function

$$\hat{S}(f) = \frac{\sum_{k=0}^{[2NW]} |d_k(f)\,\hat{a}_k(f)|^2}{\sum_{k=0}^{[2NW]} |d_k(f)|^2}$$

In the preceding few expressions there is one problem; notably that the spectrum, S, occurs in the expression for the weight. In practice it is necessary to use the estimated spectrum to determine the weights, which then determine the spectrum estimate and so on iteratively. The iteration is stable, and converges rapidly if one begins by using the two lowest coefficients to define the initial estimate of spectrum.

There are several features of this estimate which are strikingly different from conventional spectrum estimates. First, there are no *arbitrary* windows, the windows which appear being the natural eigenfunctions of the Fourier transform. Second, the estimate is consistent: for a fixed value of W the number of "large" eigenvalues will be approximately $2WN$ so that the estimate will have $4WN$ degrees of freedom. Thus W is the only arbitrary parameter and defines the trade between resolution, bias, and variance. Third, the estimate is data adaptive and the weighting has the effect that, in situations where the dynamic range of the spectrum is very large and the sample size small relative to the required resolution, regions where the spectrum is large will have the expected stability whereas regions where the spectrum is low may have fewer degrees of freedom.

In addition to these features we note that, by allowing both sides of the argument, the dilemma

‡ The formulation used here can also be used to assign reasonable distributions to the coefficients. For the zero mean spectrum estimation case the distribution is conditionally *non-central* chi-square two. The non-centrality parameter increases with the coefficient order, is correlated between coefficients, and depends on the spectrum *outside* $(-W, W)$.

posed earlier has been resolved. In this estimate we began with the *unwindowed* Fourier transform, $\tilde{x}(f)$ and obtained an estimate which is the result of a sequence of windowed estimates.

3. Harmonic Analysis

In engineering and physical problems, and one suspects also in economics, *harmonic analysis* is frequently of greater interest than *spectrum analysis*. The reason for this is obviously the vast number of periodic or almost periodic phenonema which occur everywhere. In such cases the process is usually described as having a non-zero mean value function consisting of a number of sinusoidal terms at various frequencies plus perhaps a polynomial trend plus a stationary random process of the type we have been considering. In terms of the spectral representation this, in practice, amounts to having $E\{dZ(f)\} = \sum \mu_m \delta(f-f_m)$, the extended representation introduced earlier. For algebraic simplicity we consider the case where the lines are distinct and separated by at least $2W$. When this condition is not true the algebra required becomes considerably more tedious.

Assuming a line component at frequency f_0

$$E\{\hat{a}_k(f)\} = \mu\, U_k(N,W; f-f_0)$$

so that, at f_0,

$$E\{\hat{a}_k(f_0)\} = \mu\, U_k(N,W; 0)$$

In addition, again on the assumption that the continuous component of the spectrum in the vicinity of f_0 is slowly varying or "locally white"

$$cov\{\hat{a}_k(f), \hat{a}_j^*(f)\} \approx S(f) \cdot \delta_{j,k}$$

Thus one can estimate the mean by standard methods

$$\hat{\mu}(f) = \frac{\sum_{k=0}^{[2NW]} U_k(0)\,\hat{a}_k(f)}{\sum_{k=0}^{[2NW]} U_k^2(0)}$$

and compute an F statistic with 2 and $[2NW]-2$ degrees of freedom for significance

$$F(f) = \frac{([2NW]-1)\,|\hat{\mu}(f)|^2}{\sum_{k=0}^{[2NW]} |\hat{a}_k(f) - \hat{\mu}(f)U_k(N,W;0)|^2}$$

A particularily interesting example is that of estimating the process mean. It is common experience that, when one estimates and subtracts the sample average prior to estimating the spectrum, the periodogram estimate of $S(0)$ is 0 and better estimates are typically biased low. Here one obtains a slightly better estimate for the mean while the estimated spectrum near the origin is much more accurate.

References

[1] 1946, J. Royal Statist. Soc., Supplement, **VIII,** pp 27-97.

Blackman,R.B. & Tukey,J.W. [1958], **The Measurement of Power Spectra,** Dover, New York.

Doob,J.L. [1953], **Stochastic Processes,** J. Wiley & Sons

Eberhard,A. [1973], *An Optimal Discrete Window for the Calculation of Power Spectra,* IEEE Trans. **AU-15,** pp 37-43.

Kaiser,J.F. [1974], *Nonrecursive Digital Filter Design Using the I_0 − sinh Window,* IEEE Inter. Symp. Circuits & Systems Proc., pp 20-23.

Kleiner,B., Martin,R.D., & Thomson,D.J.[1979], *Robust Estimation of Power Spectra (with discussion),* J. Royal Statist. Soc. **B-41** pp 313-351.

Lawson,C.L. & Hanson,R.J. [1974], **Solving Least Squares Problems,** Prentice-Hall.

Parzen,E. [1980], *Modern Empirical Statistical Spectral Analysis,* Texas A & M University, Tech Rep. **N-12,** May, 1980.

Slepian, D. [1978] *Prolate Spheroidal Wave Functions, Fourier Analysis, and Uncertainty- V: The Discrete Case,* Bell System Tech. J. **57** pp 1371-1429.

Thomson,D.J. [1971], **Spectral Analysis of Short Series,** *Thesis,* Polytechnic Institute of Brooklyn. (available from University Microfilms)

Thomson,D.J. [1977a], *Spectrum Estimation Techniques for Characterization and Development of WT4 Waveguide-I,* Bell System Tech. J., **56** pp 1769-1815.

Thomson,D.J. [1977b], *Spectrum Estimation Techniques for Characterization and Development of WT4 Waveguide-II,* Bell System Tech. J., **56** pp 1983-2005.

Tukey,J.W. [1967], *An Introduction to the Calculations of Numerical Spectrum Analysis,* **Spectral Analysis of Time Series,** B.Harris, Ed., Wiley & Sons.

Welch,P.D. [1967], *The Use of the Fast Fourier Transform for Estimation of Spectra: A Method Based on Time Averaging over Short, Modified Periodograms,* IEEE Trans. Audio Electroacoust. **AU-15,** pp 70-74.

ON SOME NUMERICAL PROPERTIES OF ARMA PARAMETER ESTIMATION PROCEDURES

H. Joseph Newton
Institute of Statistics,
Texas A&M University

Abstract

This paper reviews the algorithms used by statisticians for obtaining efficient estimators of the parameters of a univariate autoregressive moving average (ARMA) time series. The connection of the estimation problem with the problem of prediction is investigated with particular emphasis on the Kalman filter and modified Cholesky decomposition algorithms. A result from prediction theory is given which provides a significant reduction in the computations needed in Ansley's (1979) estimation procedure. Finally it is pointed out that there are many useful facts in the literature of control theory that need to be investigated by statisticians interested in estimation and prediction problems in linear time series models.

Key Words

ARMA models, Maximum likelihood estimation, Minimum mean square error prediction, Kalman filter algorithm, Modified Cholesky decomposition algorithm

1. INTRODUCTION

Let $\{\varepsilon(t), t\varepsilon Z\}$, Z the set of integers, be a white noise time series of uncorrelated zero mean random variables having common variance σ^2. Then the time series $\{Y(t), t\varepsilon Z\}$ satisfying

$$Y(t) = -\sum_{j=1}^{p} \alpha(j)Y(t-j)+\varepsilon(t)+ \sum_{k=1}^{q} \beta(k)\varepsilon(t-k) \quad (1.1)$$

is called an autoregressive moving average process of order p and q (ARMA(p,q)). If p=0 then $Y(\cdot)$ is a moving average process of order q, MA(q), while if q=0 $Y(\cdot)$ is an autoregressive process of order p, AR(p).

Defining $\alpha(0)=\beta(0)=1$ and the complex valued polynomials

$$g(z)= \sum_{j=0}^{p} \alpha(j)z^j \ , \ h(z)= \sum_{k=0}^{q} \beta(k)z^k \ ,$$

we can write (1.1) as

$$\sum_{j=0}^{p} \alpha(j)Y(t-j)= \sum_{k=0}^{q} \beta(k)\varepsilon(t-k)$$

or

$$g(L)Y(t)=h(L)\varepsilon(t),$$

where L is the lag or back shift operator, $L^jY(t)=Y(t-j)$, $j\varepsilon Z$. If the zeros of $g(\cdot)$ are outside the unit circle then $Y(\cdot)$ is stationary, i.e. it can be written as an infinite order moving average process, while if the zeros of $h(\cdot)$ are outside the unit circle then $Y(\cdot)$ is invertible, i.e. it can be written as an infinite order autoregressive process.

The ARMA model has been very useful in analyzing time series data. Given a sample realization $\underset{\sim}{Y}_T^T=(Y(1),\ldots,Y(T))$ from $Y(\cdot)$ one seeks estimators $\hat{\underset{\sim}{\alpha}}=(\hat{\alpha}(1),\ldots,\hat{\alpha}(p))^T$, $\hat{\underset{\sim}{\beta}}=(\hat{\beta}(1),\ldots \hat{\beta}(q))^T$, and $\hat{\sigma}^2$ of the parameters $\underset{\sim}{\alpha},\underset{\sim}{\beta}$, and σ^2, as well as memory-t, horizon-h, minimum mean square error linear predictors and prediction variances $Y(t+h|t)$, $\sigma^2_{t,h}=E\{Y(t+h)-Y(t+h|t)\}^2$ of $Y(t+h)$ given $Y(1),\ldots,Y(t)$ for a variety of memories and horizons $t=t_1,\ldots,t_2$, $h=h_1,\ldots,h_2$. Thus $Y(t+h|t)=\underset{\sim}{\lambda}^T_{t,h}\underset{\sim}{Y}_t$ where $\underset{\sim}{\lambda}_{t,h}$ minimizes

$$S(\underset{\sim}{\ell})=E\{Y(t+h)-\underset{\sim}{\ell}^T\underset{\sim}{Y}_t\}^2 \ ,$$

and $\sigma^2_{t,h}=S(\underset{\sim}{\lambda}_{t,h})$. This gives (see Whittle (1963), p.47) that $\underset{\sim}{\lambda}_{t,h}$ satisfies the normal equations

$$\Gamma_{Y,t}\underset{\sim}{\lambda}_{t,h}=\underset{\sim}{r}_{t,h} \ ,$$

while

$$\sigma^2_{t,h}=R_Y(0)-\underset{\sim}{r}^T_{t,h}\ \Gamma^{-1}_{Y,t}\underset{\sim}{r}_{t,h},$$

where $\underset{\sim}{r}^T_{t,h}=(R_Y(t+h-1),\ldots,R_Y(h))$ with $R_Y(v)=E(Y(t)Y(t+v)), v\varepsilon Z$, and $\Gamma_{Y,t}$ is the $(t\times t)$ Toeplitz covariance matrix of $\underset{\sim}{Y}_t$, i.e. $\Gamma_{Y,t}$ has (j,k)th element $R_Y(k-j)$. We can thus define $\Gamma_{Y,t}$ by its first row; $\Gamma_{Y,t}=\mathrm{TOEPL}(R_Y(0),\ldots,R_Y(t-1))$.

The usual assumption made for doing estimation and prediction in ARMA models is that $\varepsilon(\cdot)$ (and thus $Y(\cdot)$) is a Gaussian process. Thus the maximum likelihood estimators $\hat{\underset{\sim}{\alpha}}, \hat{\underset{\sim}{\beta}}$, and $\hat{\sigma}^2$ maximize the Gaussian likelihood

$$L(\underset{\sim}{\alpha}, \underset{\sim}{\beta}, \sigma^2/\underset{\sim}{Y}_T) = (2\pi)^{-T/2} |\Gamma_{Y,T}|^{-\frac{1}{2}} \exp(-\tfrac{1}{2}\underset{\sim}{Y}_T^T \Gamma_{Y,T}^{-1} \underset{\sim}{Y}_T),$$

while $Y(t+h|t)$ and $\sigma^2_{t,h}$ are the conditional mean and variance of $Y(t+h)$ given $Y(1),\ldots,Y(t)$.

In this paper we 1) review attempts by statisticians to obtain $\hat{\underset{\sim}{\alpha}}, \hat{\underset{\sim}{\beta}}$, and $\hat{\sigma}^2$ (or estimators asymptotically equivalent to them), 2) show how recent methods are closely connected with finding predictors, and 3) propose an improvement of Ansley's (1979) estimation procedure. This procedure is currently regarded, at least in many situations, as the most numerically efficient available. Finally we will be able to compare the two most popular methods available.

2. APPROXIMATE METHODS

There have been three basic types of procedures used to estimate the parameters of the ARMA model: 1) estimators that are derived heuristically and then shown to be asymptotically equivalent to the MLE, 2) estimators obtained by maximizing a function asymptotically equivalent to the likelihood L, and 3) directly maximizing L. The procedures have evolved as both computing software and hardware have improved.

As an example of a type 1 procedure we consider Hannan's (1969) method. The method consists of 1) finding consistent initial estimators $\hat{\underset{\sim}{\alpha}}^{(0)}$ and $\hat{\underset{\sim}{\beta}}^{(0)}$, 2) performing an alternating procedure of the form $\hat{\underset{\sim}{\alpha}}^{(0)} \to \hat{\underset{\sim}{\beta}}^{(1)}, \hat{\underset{\sim}{\beta}}^{(1)} \to \hat{\underset{\sim}{\alpha}}^{(1)}, \hat{\underset{\sim}{\alpha}}^{(1)} \to \hat{\underset{\sim}{\beta}}^{(2)}$, 3) combining $\hat{\underset{\sim}{\beta}}^{(1)}, \hat{\underset{\sim}{\beta}}^{(2)}$ to obtain asymptotically efficient $\tilde{\underset{\sim}{\beta}}$, 4) $\tilde{\underset{\sim}{\beta}} \to \tilde{\underset{\sim}{\alpha}}$, and 5) possibly iterating the process by returning to (2) with $\tilde{\underset{\sim}{\alpha}}$ replacing $\hat{\underset{\sim}{\alpha}}^{(0)}$.

The initial estimators $\hat{\underset{\sim}{\alpha}}^{(0)}$ and $\hat{\underset{\sim}{\beta}}^{(0)}$ are obtained by noting the following facts about the ARMA model:

1) $$\sum_{j=0}^{p} \alpha(j) R_Y(j-v) = 0 \ , \quad v = q+1,\ldots,q+p \tag{2.1}$$

2) By writing

$$g(L)Y(t) = h(L)\varepsilon(t) = X(t), \tag{2.2}$$

we note that $X(\cdot) \sim MA(q, \underset{\sim}{\beta}, \sigma^2)$, <u>i.e.</u> $X(\cdot)$ is an $MA(q)$ process with parameters $\underset{\sim}{\beta}$ and σ^2. Thus by the first equality in (2.2) we have

$$R_X(v) = \sum_{j,k=0}^{p} \alpha(j)\alpha(k) R_Y(j+v-k), \quad v \in Z, \tag{2.3}$$

while from the second equality we have

$$R_X(v) = \begin{cases} \sigma^2 \sum_{k=0}^{q-|v|} \beta(k)\beta(k+|v|) \ , & |v| \le q \\ 0 \ , & |v| > q \ . \end{cases} \tag{2.4}$$

Then defining the consistent sample autocovariances

$$R_T(v) = \frac{1}{T} \sum_{t=1}^{T-|v|} Y(t)Y(t+|v|) \ , \quad |v| < T \ ,$$

$\hat{\underset{\sim}{\alpha}}^{(0)}$ is found by solving (2.1) with $R_T(v)$ replacing $R_Y(v)$, while $\hat{\underset{\sim}{\beta}}^{(0)}$ is obtained by first using $\hat{\underset{\sim}{\alpha}}^{(0)}$ and $R_T(\cdot)$ in (2.3) to obtain consistent estimators $R_X^{(0)}(0),\ldots,R_X^{(0)}(q)$ which are then used in (2.4) to get $\hat{\underset{\sim}{\beta}}^{(0)}$ via Wilson's (1969) or Bauer's (1955) algorithm.

Steps 2 and 4 of Hannan's method are performed in the frequency domain by noting from (2.2) that the spectral density functions $f_Y(\cdot)$, $f_X(\cdot)$, and $f_\varepsilon(\cdot)$ of $Y(\cdot)$, $X(\cdot)$, and $\varepsilon(\cdot)$ are related by (since $f_\varepsilon(w) = \sigma^2/2\pi$, $w \in [0, 2\pi]$)

$$|g(e^{iw})|^2 f_Y(w) = \frac{\sigma^2}{2\pi} |h(e^{iw})|^2 = f_X(w), \tag{2.5}$$

Then we can rewrite (2.5) as, dropping arguments for convenience,

$$\frac{|g|^2 f_Y}{f_X^2} = \frac{4\pi^2/\sigma^2}{2\pi} \frac{1}{|h|^2} \tag{2.6}$$

$$\frac{f_Y}{|h|^2} = \frac{\sigma^2}{2\pi} \frac{1}{|g|^2} \tag{2.7}$$

$$\frac{(2\pi/\sigma^2)^2 |g|^2 f_Y}{|h|^4} = \frac{4\pi^2/\sigma^2}{2\pi} \frac{1}{|h|^2} \ . \tag{2.8}$$

Thus the right hand sides of (2.6), (2.7), and (2.8) are in the form of an autoregressive spectral density, and thus for $v \ge 0$,

$$\sum_{j=0}^{q} \beta(j) \int_0^{2\pi} \frac{|g|^2 f_Y}{f_X^2} e^{i(j-v)w} dw = \delta_v \frac{4\pi^2}{\sigma^2} \ , \tag{2.9}$$

$$\sum_{j=0}^{p} \alpha(j) \int_0^{2\pi} \frac{f_Y}{|h|^2} e^{i(j-v)w} dw = \delta_v \sigma^2 \ , \tag{2.10}$$

$$\sum_{j=0}^{q} \beta(j) \int_0^{2\pi} \frac{|g|^2 f_Y}{|h|^4} e^{i(j-v)w} dw = \delta_v \sigma^2 \ . \tag{2.11}$$

Now f_Y can be estimated by the periodogram

$$f_T(w) = \frac{1}{2\pi} \sum_{|v|<T} R_T(v) e^{-ivw} \ ,$$

and since $X(\cdot) \sim MA(q, \underset{\sim}{\beta}, \sigma^2)$ we have that

$$f_X(w) = \frac{1}{2\pi} \sum_{|v| \le q} R_X(v) e^{-ivw} \ .$$

Then for example, in an obvious notation, the step $\hat{\alpha}^{(0)} \to \hat{\beta}^{(1)}$ is done by solving (2.9) with $g^{(0)}$, f_T, and $f_X^{(0)}$ replacing g, f_Y, and f_X and the fast Fourier transform used to calculate a rectangular sum approximation to the integrals needed. Then (2.10) is used for $\hat{\beta}^{(1)} \to \hat{\alpha}^{(1)}$ and $\beta \to \alpha$ while (2.11) is used for $\hat{\alpha}^{(1)} \to \hat{\beta}^{(2)}$. Finally the combination of $\hat{\beta}^{(1)}$ and $\hat{\beta}^{(2)}$ is done using an analogy with two-stage least squares.

In a pivotal article, Akaike (1973) noted that Hannan's method was the same as one step in directly maximizing L using the Newton-Raphson procedure with an approximation to the Hessian matrix of L. This observation led many researchers to turn their attention to other possible approaches to directly maximize L. We will consider two of these in the next section.

As an example of a type 2 estimation procedure we consider the method of Box and Jenkins (1970). In large samples the term $\exp(-\frac{1}{2}\underset{\sim}{Y}_T^T\Gamma_{Y,T}^{-1}\underset{\sim}{Y}_T)$ dominates the term $|\Gamma_{Y,T}|^{-\frac{1}{2}}$. Thus Box and Jenkins suggest maximizing

$$L'(\underset{\sim}{\alpha}, \underset{\sim}{\beta}, \sigma^2|\underset{\sim}{Y}_T) = \exp(-\underset{\sim}{Y}_T^T\Gamma_{Y,T}^{-1}\underset{\sim}{Y}_T)$$

which can be done in a nonlinear regression framework. Thus it can be shown that

$$\underset{\sim}{Y}_T^T\Gamma_{Y,T}^{-1}\underset{\sim}{Y}_T = \sum_{t=-\infty}^{T} [\varepsilon(t)]^2 \tag{2.12}$$

where $[\varepsilon(t)] = E(\varepsilon(t)|\underset{\sim}{Y}_T, \underset{\sim}{\alpha}, \underset{\sim}{\beta})$. Thus the Box-Jenkins procedure consists of replacing $-\infty$ by $-Q$ in (2.12) and approximating $[\varepsilon(t)]$ by backforecasting.

It is generally agreed that in the case of large T and/or zeros of h(z) far from the unit circle that estimation methods of type 1 and type 2 give very reasonable results. However if one is faced with a situation where the above conditions are not satisfied, then the numerical properties of the procedures nullify the benefit of their simplicity. Thus iterations often fail to converge in a reasonable time and systems of equations that must be solved become highly ill-conditioned. Thus there has been a need for more numerically stable, exact maximum likelihood methods.

3. EXACT MAXIMUM LIKELIHOOD PROCEDURES

With the advent of iterative nonlinear optimization procedures which require only a starting value and evaluation of the function to be optimized for given values of its arguments (see Dennis and More (1977), for example), the most recent procedures for ARMA parameter estimation have centered on evaluating L for given values of $\underset{\sim}{\alpha}$, $\underset{\sim}{\beta}$, and σ^2.

In this section we consider two such methods for evaluating L: 1) using the Kalman filter algorithm and 2) using covariance matrix decomposition methods, particularly the Cholesky decomposition algorithm. The basic purpose of these algorithms is to obtain the one step ahead prediction errors $e(t) = Y(t) - Y(t|t-1)$ and prediction variances $\sigma_t^2 = \sigma_{t-1,1}^2$, $t = 1,\ldots, T$ since one can easily show that

$$|\Gamma_{Y,T}| = \prod_{t=1}^{T} \sigma_t^2, \quad \underset{\sim}{Y}_T^T\Gamma_{Y,T}^{-1}\underset{\sim}{Y}_T = \sum_{t=1}^{T} \frac{e^2(t)}{\sigma_t^2}.$$

To summarize, the two methods currently regarded as most numerically efficient for maximizing L consist of two stages:

1) Find initial estimators as described above for Hannan's method.

2) Use an iterative, derivative free nonlinear optimization algorithm to find the maximum likelihood estimators, using either the Kalman filter algorithm (Akaike (1974), Harvey and Phillips (1979), Jones (1980), Pearlman (1980), or Gardner, Harvey, and Phillips (1980), for example) or the Cholesky decomposition algorithm (see Pagano and Parzen (1973), Pagano (1976), Phadke and Kedem (1978), Ansley (1979), Newton (1980), Newton and Pagano (1981), for example) to find the e(t) and σ_t^2 needed to evaluate L.

Kalman Filter Algorithm

Consider the following two equations:

$$\underset{\sim}{Y}_{t+1} = H_t\underset{\sim}{Y}_t + G_t\underset{\sim}{U}_t \quad \text{(State Equation)}$$

$$\underset{\sim}{X}_{t+1} = S_{t+1}\underset{\sim}{Y}_{t+1} + \underset{\sim}{V}_{t+1} \quad \text{(observation equation)}$$

where the $\underset{\sim}{Y}$'s are unobservable N-vectors, the $\underset{\sim}{X}$'s are observable M-vectors, H_t, G_t, and S_t are known (NxN), (NxL), and (MxN) matrices, and $\underset{\sim}{U}_t$, $\underset{\sim}{V}_t$ are independent zero mean L and N dimensional white noise series with known covariance matrices Q_t and R_t.

Then given initial values $\underset{\sim}{Y}_0$ and $P_0 = \text{var}(\underset{\sim}{Y}_0)$ one finds the $\underset{\sim}{Y}_t$'s and $P_t = \text{cov}(\underset{\sim}{Y}_t)$ by the Kalman filter algorithm

$$\underset{\sim}{Y}_{t+1} = \underset{\sim}{Z}_{t+1} - K_{t+1}(S_{t+1}\underset{\sim}{Z}_{t+1} - \underset{\sim}{X}_{t+1})$$

$$P_{t+1} = W_{t+1} - K_{t+1}S_{t+1}W_{t+1}$$

$$\underset{\sim}{Z}_{t+1} = H_t\underset{\sim}{Y}_t, \quad K_{t+1} = W_{t+1}S_{t+1}^T \times (S_{t+1}W_{t+1}S_{t+1}^T + R_{t+1})^{-1}$$

$$W_{t+1} = H_tP_tH_t^T + G_tQ_tG_t^T. \tag{3.1}$$

A number of authors in the statistical literature have pointed out that a simple application of this algorithm to ARMA models can be made to obtain the e(t) and σ_t^2. Thus since

$$Y(t+j)|t+1) = \begin{cases} Y(t+j|t) + g_j\varepsilon(t+1), & j = 1, \ldots, m-1 \\ \sum_{k=1}^{p} \alpha(k)Y(t+m-k|t) + g_m\varepsilon(t+1), & j = m \end{cases}$$

where $g_1 = 1$ and $g_j = \beta(j-1) + \sum_{k=1}^{j-1} \alpha(k)g_{j-k}, j \geq 2$

and m = max(p, q+1), we can write the equation of state

$$\underset{\sim}{Y}_{t+1} = H\,\underset{\sim}{Y}_t + \underset{\sim}{g}\,\varepsilon(t+1)\ ,$$

where $\underset{\sim}{g}^T = (g_1, g_2, \ldots, g_m)$, and

$$H = \begin{bmatrix} 0 & 1 & 0 & \ldots & 0 \\ 0 & 0 & 1 & \ldots & 0 \\ \cdot & \cdot & \cdot & & \cdot \\ \cdot & \cdot & \cdot & & \cdot \\ \cdot & \cdot & \cdot & & \cdot \\ 0 & 0 & 0 & \ldots & 1 \\ \alpha(m) & \cdot & \cdot & \ldots & \alpha(1) \end{bmatrix},$$

where $\alpha(j) = 0$ if $j > p$.

Further, since $Y(t+1|t+1) = Y(t+1)$, we can write the observational equation

$$\underset{\sim}{X}_{t+1} = \underset{\sim}{S}^T\,\underset{\sim}{Y}_{t+1} + \underset{\sim}{V}_{t+1}\ ,$$

where $\underset{\sim}{X}_{t+1} = Y(t+1)$, $\underset{\sim}{S}^T = (1, 0, \ldots, 0)$, and $\underset{\sim}{V}_{t+1}$ is an observational error random variable.

Cholesky Decomposition Algorithm

An (nxn) symmetric matrix A_n is positive definite if and only if it can be written

$$A_n = L_{A,n}D_{A,n}L_{A,n}^T\ , \tag{3.2}$$

where $L_{A,n}$ is a unit lower triangular matrix and $D_{A,n}$ is a diagonal matrix with positive diagonal elements. This is the modified Cholesky decomposition of A_n. We note that we can also write the Cholesky decomposition $A_n = M_{A,n}M_{A,n}^T$ where $M_{A,n} = L_{A,n}D_{A,n}^{\frac{1}{2}}$ but we will consider exclusively the modified decomposition.

The decomposition (3.2) is unique and nested, i.e. $L_{A,j}$, $D_{A,j}$, $L_{A,j}^{-1}$, $D_{A,j}^{-1}$ are the (jxj) principal minors of $L_{A,n}$, $D_{A,n}$, $L_{A,n}^{-1}$, and $D_{A,n}^{-1}$ respectively for $j \leq n$. Thus we can denote the (j,k)th element of $L_{Y,t}$ and $D_{Y,t}$ in the decomposition of the (KxK) covariance matrix $\Gamma_{Y,K}$ of a time series Y(·) as $L_{Y,j,k}$ and $D_{Y,j,k}$ respectively for $t \leq K$. If Y(·) is a purely nondeterministic covariance stationary time series with spectral density function f(·), i.e.

$$\sigma_\infty^2 = 2\pi \exp\{\frac{1}{2\pi}\int_0^{2\pi} \log f(w)dw\} > 0\ ,$$

then e(1) = Y(1), while

$$e(t) = Y(t) - \sum_{k=1}^{t-1} L_{Y,t,t-k}e(t-k),\ t \geq 2$$

$$\sigma_t^2 = D_{Y,t,t}\ ,$$

and in fact (Newton and Pagano (1981))

$$Y(t+h|t) = \sum_{k=h}^{t+h-1} L_{Y,t+h,t+h-k}e(t+h-k) \tag{3.3}$$

$$\sigma_{t,h}^2 = \sum_{k=0}^{h-1} L_{Y,t+h,t+h-k}^2 D_{Y,t+h-k,t+h-k}\ . \tag{3.4}$$

Further,

$$\lim_{K\to\infty} L_{Y,K,K-j} = \beta_\infty(j),\ j \geq 0 \tag{3.5}$$

$$\lim_{K\to\infty} D_{Y,K,K} = \sigma_\infty^2\ , \tag{3.6}$$

where the $\beta_\infty(\cdot)$ are the coefficients in the MA(∞) representation of Y(·). While these facts appear to be known in the literature of control theory they do not appear to exist in the statistical literature, at least in the general form of equations (3.3) through (3.6).

Now major simplifications of these equations occur when Y(·) is an ARMA process. Thus (Pagano and Parzen (1973), Pagano (1976), Newton (1980)) if Y(·) is an MA(q) we have that $\Gamma_{Y,j,k} = 0$ if $|j-k| > 0$ and thus $L_{Y,j,k} = 0$ if $j - k > 0$. Further, Pagano and Parzen (1973) state in an attempt to find $Y(t+h|t)$ and $\sigma_{t,h}^2$ that if $Y(\cdot) \sim ARMA(p, q, \underset{\sim}{\alpha}, \underset{\sim}{\beta}, \sigma^2)$, and one forms the series $X(t) = \sum_{j=0}^{p} \alpha(j)Y(t-j)$, $t > p$, then X(p+1), ..., X(T) is a realization from an MA(q, $\underset{\sim}{\beta}$, σ^2) process and one can combine the MA(q) prediction algorithm with an AR(p) prediction algorithm. Ansley (1979) makes this same transformation of Y(·) to X(·) to get the e(t) and σ_t^2 necessary to evaluate L.

The methods of Pagano and Parzen (1973) and Ansley (1979) can be summarized theoretically by combining the equations (3.3) through (3.6) above with the following theorem due to Newton and Pagano (1981).

Theorem

Let $Y(\cdot) \sim ARMA(p, q, \underset{\sim}{\alpha}, \underset{\sim}{\beta}, \sigma^2)$ and $Z(\cdot) \sim AR(p, \underset{\sim}{\alpha}, \sigma^2)$ with associated covariance matrix sequences $\Gamma_{Y,t} = L_{Y,t}D_{Y,t}L_{Y,t}^T$ and $\Gamma_{Z,t} = L_{Z,t}D_{Z,t}L_{Z,t}^T$. Then

$$L_{Y,t} = L_{Z,t}L_{X,t} ,$$

$$D_{Y,t} = D_{X,t} ,$$

where $\underset{\sim}{X}_t = (X(1), \ldots, X(t))^T = L^{-1}_{Z,t}\underset{\sim}{Y}_t$ has jth element

$$X(j) = \begin{cases} Y(j) & , j = 1 \\ \sum_{k=0}^{j-1} \alpha_{j-1}(k)Y(j-k) & , j = 2, \ldots, p \\ \sum_{k=0}^{p} \alpha(k)Y(j-k) & , j > p \end{cases}$$

where the $\alpha_j(\cdot)$ are easily obtained by performing the Levinson (1974) recursion for j = p-1, ..., 1, with $\alpha_{p+1}(k) = \alpha(k)$:

$$\alpha_j(i) = \frac{\alpha_{j+1}(i) - \alpha_{j+1}(j+1)\alpha_{j+1}(j+1-i)}{1 - \alpha^2_{j+1}(j+1)} ,$$

$$i \le j < p.$$

Thus

$$\Gamma_{X,i,j} = \begin{cases} \sum_{m=\max(1,j-p)}^{j} \alpha_{j-1}(j-m) \\ \quad \times \sum_{\ell=\max(1,i-p)}^{i} \alpha_{i-1}(i-\ell)R_Y(\ell-m), i,j \ge 1 \\ R_X(|i-j|) = \sigma^2 \sum_{k=0}^{q-|i-j|} \beta(k)\beta(k+|i-j|), \\ \quad |i-j| \le q \quad i,j > p \\ 0 \quad \text{if } 1 \le j \le p,\ i > p, \text{ and } i - j > q \\ \quad \text{or if } i,j > p \text{ and } |i-j| > q \end{cases}$$

and

$$\lim_{K\to\infty} L_{X,K,K-j} = \beta(j),\ j = 1, \ldots, q \tag{3.7}$$

$$\lim_{K\to\infty} D_{X,K,K} = \sigma^2. \tag{3.8}$$

Thus e(1) = X(1), while

$$e(t) = X(t) - \sum_{k=1}^{\min(q,j-1)} L_{X,t,t-k}e(t-k),\ t \ge 1$$

$$\sigma^2_t = D_{X,t,t} ,$$

and in fact

$$Y(t+h|t) = X(t+h|t) - \sum_{j=1}^{p} \alpha(j)Y(t+h-j|t) ,$$

$$X(t+h|t) = \begin{cases} \sum_{k=h}^{q} L_{X,t+h,t+h-k}e(t+h-k), \\ \qquad h = 1, \ldots, q \\ 0 \qquad h > q \end{cases}$$

$$\sigma^2_{t,h} = \sum_{k=0}^{h-1} D_{X,t+h-k,t+h-k} \times \left\{ \sum_{\ell=t+h-k}^{t+h} L_{Z,t+h,\ell}L_{X,\ell,t+h-k} \right\}^2 .$$

Thus to find the one step ahead prediction errors and variances one need only compute the q nonzero elements of the successive rows of L_X until the convergence properties (3.7) and (3.8) take effect. Further, one can then also find the Y(t+h|t) from these same quantities and the e(·). Note however that to find $\sigma^2_{t,h}$ one also needs to find the elements of $L_{Z,t}$. These can be found simply by noting that the upper (pxp) principal minor $L_{Z,p}$ can be found by inverting the lower triangular matrix $L^{-1}_{Z,p}$, while

$$L_{Z,j,k} = \gamma(j-k),\ j \ge k \ge p ,$$

where $\gamma(0) = 1, \gamma(1), \gamma(2), \ldots$ are the coefficients in the MA(∞) representation of Z(·). Further, the elements in rows p+1, ... in the first (p-1) columns of L_Z are obtained by

$$L_{Z,p+j,k} = - \sum_{r=1}^{p} \alpha(r)L_{Z,p+j-r,k} ,\ 1 \le k < p,\quad j \ge 1,$$

and

$$\lim_{j\to\infty} L_{Z,p+j,k} = 0 , \qquad 1 \le k < p .$$

Thus the predictors and prediction variances Y(t+h|t) and $\sigma^2_{t,h}$ can be obtained using either the Kalman filter algorithm or the Cholesky decomposition algorithm. Also the convergence properties described in the Cholesky algorithm can be incorporated into the Kalman filter algorithm.

4. DISCUSSION

Thus we have seen that algorithms developed recently for finding maximum likelihood estimators of ARMA process parameters have essentially consisted of applying established algorithms for finding the minimum mean square error linear one step ahead predictors and prediction variances. We have shown how the Cholesky algorithm can be used to find more than one step ahead predictors and variances as well. We note that both the Kalman filter and Cholesky decomposition algorithm can be extended easily to the multivariate ARMA case. We also note that we have not attempted to survey all the recent work in the ARMA estimation area, but rather have emphasized those algorithms that appear to be most widely used in the literature.

We consider next the relative speed and stability of the Kalman filter and Cholesky decomposition algorithms. Pearlman (1980) shows that the number of operations needed to find the e(t) and σ_t^2 via the Kalman filter algorithm is approximately T(2p + 3m + 3), with m = max(p,q+1), while for the Cholesky algorithm it is T(p + ½(q+1)(q+4)). Thus if $q \geq 5$ the Kalman algorithm is faster.

Now the fact that the Kalman filter algorithm performs a number of "matrix squaring" opertions (for example $H_t P_t H_t^T$ in (3.1) has caused many investigators to question its stability. However a number of authors have suggested methods for avoiding these operations (see Paige (1976) for example). The Cholesky decomposition is well known for its numerical stability (see Wilkinson (1967) for example). However more work is needed before a final conclusion can be made about which of the two methods is to be preferred.

Finally we point out that there are a variety of numerical methods for analyzing linear time series models in the work of control theorists particularly in a series of papers of Kailath (see the references in Aasnaes and Kailath (1973)) that need to be incorporated into the statistical literature.

5. REFERENCES

Aasnaes, H. B. and Kailath, T. (1973). An innovations approach to least squares estimation - part VII: some applications of vector autoregressive moving average models, IEEE trans on Automatic Control, AC-18, 601-607.

Akaike, H. (1973). Maximum likelihood identification of Gaussian autoregressive moving average models, Biometrika, 60, 255-265.

Akaike, H. (1974). Markovian representation of stochastic processes and its applications to the analysis of autoregressive moving average processes. Annals of the Institute of Statistical Mathematics, 26, 363-387.

Ansley, C. (1979). An algorithm for the exact likelihood of a mixed autoregressive-moving average process. Biometrika, 66, 59-65.

Bauer, F. L. (1955). Ein direktes iterationsverfahren zur Hurwitz-Zerlegung eines polynoms, Archiv Elekt. Ubertr., 9, 285-290.

Box, G. E. P. and Jenkins, G. M. (1970). Time Series Analysis forecasting and control. Holden-Day, San Francisco.

Dennis, J. E. and More, J. J. (1977). Quasi-Newton methods, motivation and theory. SIAM Rev., 19, 46-89.

Gardner, G., Harvey, A. C. and Phillips, G. D. (1980). An algorithm for exact maximum likelihood estimation of autoregressive-moving average models by means of Kalman filtering, JRSS C, 29, 311-322.

Hannan, E. J. (1969). The estimation of mixed moving average autoregressive systems. Biometrika, 56, 579-593.

Harvey, A. C. and Phillips, G. D. (1979). Maximum likelihood estimation of regression models with autoregressive-moving average disturbances. Biometrika, 66, 49-58.

Jones, R. H. (1980). Maximum likelihood fitting of ARMA models to time series with missing observations. Technometrics, 22, 389-395.

Levinson, N. (1974). The Wiener RMS error criterion in filter design and prediction. Journal of Mathematical Physics, 15, 261-278.

Newton, H. J. (1980). Efficient estimation of multivariate moving average autocovariances. Biometrika, 67, 227-231.

Newton, H. J. and Pagano, M. (1981). The finite memory prediction of covariance stationary time series. Texas A&M Statistics Technical Report N-21.

Pagano, M. and Parzen, E. (1973). Timesboard; A time series package. Proc. of Comp. Sci. and Stat.: 7th Ann. Symp. on the Interface, ed. by W. J. Kennedy, Statistical Laboratory, Iowa State Univ., Ames, Iowa.

Pagano, M. (1976). On the linear convergence of a covariance factorization algorithm. Journ. Assoc. Comp. Mach., 23, 310-316.

Paige. (1976). Numerical computations for some estimation problems in Engineering, Proc. of Comp. Sci. and Stat.: 9th Ann. Symp. on the Interface, Harvard University.

Pearlman, J. G. (1980). An algorithm for the exact likelihood of high-order autoregressive-moving average process. Biometrika, 67, 232-233.

Phadke, M. S. and Kedem, G. (1978). Computation of the exact likelihood function of multivariate moving average models. Biometrika, 65, 511-519.

Whittle, P. (1963). Prediction and Regulation by linear least squares methods. English Universities Press, London.

Wilkinson, J. H. (1967). The solution of ill-conditioned linear equations. In Mathematical Methods for Digital Computers II (A. Ralston and H. S. Wilf, eds.) 65-93.

Wilson, G. (1969). Factorization of the covariance generating function of a pure moving average process. SIAM J. Numer. Anal., 6, 1-7.

TIME SERIES RECURSIONS AND SELF-TUNING CONTROL

Victor Solo, Harvard University

ABSTRACT

Recursive estimates are estimates (of parameters in a time series model) that are computed in a sequential fashion (i.e. updated quickly as new observations become available). The uses of such "real" time parameter estimators include real-time forecasting and self-tuning control. Here it is shown how "real" time parameter estimators can be constructed for time series models; also an heuristic discussion of their convergence behavior is given. The analysis and synthesis of self-tuning controllers is also discussed.

KEYWORDS: Time series, control, real time, self-tuning control, adaptive control, convergence.

1. Introduction

Especially since the invention of the Kalman filter [1] a problem of great interest to engineers has been the real time (or sequential) estimation of parameters in time series models. An algorithm that allows rapid updating of parameter estimates (in a time series model) with each new observation will be called a real time parameter estimator.

The main uses of real time parameter estimators include (i) forming part of a real time forecasting scheme viz. electric power demand is sometimes forecast and measured on an hourly basis; (ii) forming part of a "self-tuning" or adaptive control scheme. In this scheme the parameters of the system or the control law are updated with each new observation. This example is discussed further in Section 4 below. (iii) More recently a use in econometrics and engineering for the detection and/or tracking of change (Pagan [3]).

The remainder of this article is organized as follows. Section 2 discusses the construction of real time parameter estimators from an intuitive point of view. Section 3 discusses briefly how they may be heuristically analyzed. Finally, Section 4 discusses the self-tuning control problem.

2. Construction of Real Time Parameter Estimators

(a) A Regression Model

The construction is illustrated for a simple regression model and the intuition gained used to suggest forms for time series model. Consider the regression

$$y_k = \underset{\sim}{x}_k'\underset{\sim}{\beta}_0 + \varepsilon_k\,, \quad k = 1, 2, \ldots$$

where y_k is an observed sequence, ε_k is a white noise (i.e. iid $(0, \sigma^2)$) sequence and $\underset{\sim}{x}_k$ is a p-dimensional vector of regressors. In an autoregressive time series model we should have $\underset{\sim}{x}_k' = (-y_{k-1} \ldots y_{k-p})$. Finally, $\underset{\sim}{\beta}_0$ is a vector of unknown parameters.

Consider the estimation of $\underset{\sim}{\beta}_0$ by sequential minimization of a sum of squares function

$$L_N(\underset{\sim}{\beta}) = \frac{1}{2}\sum_1^N e_k^2(\underset{\sim}{\beta})$$

where

$$e_k(\underset{\sim}{\beta}) = y_k - \underset{\sim}{x}_k'\underset{\sim}{\beta}$$

is the residual sequence based on β. The idea is to sequentially determine $\hat{\underset{\sim}{\beta}}_N$ the solution to the equation

$$\frac{dL_N}{d\hat{\underset{\sim}{\beta}}_N} = \underset{\sim}{0}\,.$$

There are a number of ways to do this but the following method is chosen for the intuition it provides in more complicated cases. Observe that

$$L_N(\underset{\sim}{\beta}) = L_{N-1}(\underset{\sim}{\beta}) + \frac{1}{2}e_N^2(\underset{\sim}{\beta})\,.$$

So that, noting $de_N/d\underset{\sim}{\beta} = -\underset{\sim}{x}_N$, gives

$$dL_N/d\underset{\sim}{\beta} = dL_{N-1}/d\underset{\sim}{\beta} + e_N de_N/d\underset{\sim}{\beta} \tag{1a}$$

$$= dL_{N-1}/d\underset{\sim}{\beta} - \underset{\sim}{x}_N e_N\,. \tag{1b}$$

Also call

$$\underset{\sim}{P}_N^{-1} = d^2L_N/d\underset{\sim}{\beta}d\underset{\sim}{\beta}'$$

and observe that

$$\underset{\sim}{P}_N^{-1} = \underset{\sim}{P}_{N-1}^{-1} + \underset{\sim}{x}_N\underset{\sim}{x}_N'\,. \tag{2}$$

Now consider the following Taylor series about $\hat{\underset{\sim}{\beta}}_N$

$$\underset{\sim}{0} = \frac{dL_N}{d\hat{\underset{\sim}{\beta}}_N} = \frac{dL_N}{d\hat{\underset{\sim}{\beta}}_{N-1}} + \frac{d^2L_N}{d\underset{\sim}{\beta}_N d\underset{\sim}{\beta}_N'}(\hat{\underset{\sim}{\beta}}_N - \hat{\underset{\sim}{\beta}}_{N-1})$$

so that

$$\hat{\underset{\sim}{\beta}}_N = \hat{\underset{\sim}{\beta}}_{N-1} - \left(\frac{d^2L_N}{d\beta_N d\beta_N'}\right)^{-1}\frac{dL_N}{d\hat{\underset{\sim}{\beta}}_{N-1}}\,. \tag{3}$$

This does provide a solution to the problem of sequentially estimating β_0 but as it stands it is computationally burdensome. We can evade this problem by returning to (1a), (1b) evaluated at $\hat{\beta}_{N-1}$ to see

$$\frac{dL_N}{d\hat{\beta}_{N-1}} = \frac{de_N}{d\hat{\beta}_{N-1}} e_N = -\underset{\sim}{x}_N e_N .$$

Thus we find in (3)

$$\hat{\beta}_N = \hat{\beta}_{N-1} - \underset{\sim}{P}_N \frac{de_N}{d\hat{\beta}_{N-1}} e_N \qquad (4a)$$

or

$$\hat{\beta}_N = \hat{\beta}_{N-1} + \underset{\sim}{P}_N \underset{\sim}{x}_N e_N \qquad (4b)$$

$$e_N = y_N - \underset{\sim}{x}_N' \hat{\beta}_{N-1} . \qquad (5)$$

If we invert equation (2) to

$$\underset{\sim}{P}_N = \underset{\sim}{P}_{N-1} - \frac{\underset{\sim}{P}_{N-1} \underset{\sim}{x}_N \underset{\sim}{x}_N' \underset{\sim}{P}_{N-1}}{1 + \underset{\sim}{x}_N' \underset{\sim}{P}_{N-1} \underset{\sim}{x}_N} , \qquad (6)$$

then equations (4b), (5), (6) provide a fully sequential procedure for estimating $\underset{\sim}{\beta}$. Notice the stochastic interpretations (in this linear case)

$$\text{var} \, (\hat{\beta}_N) = \sigma^2 \underset{\sim}{P}_N$$

$$\text{var} \, (e_N) = \sigma^2 (1 + \underset{\sim}{x}_N' \underset{\sim}{P}_{N-1} \underset{\sim}{x}_N) .$$

It is worth dwelling on the structure of the estimator in (4). It has the form

$$\underset{\sim}{\beta}_{new} = \underset{\sim}{\beta}_{old} + \text{correction term}$$

and

$$\text{correction term} = [\text{gain matrix}] \times [\text{gradient vector}]$$
$$\times [\text{innovation}] .$$

For convergence of $\hat{\beta}_N$ to β_0 it is required that

$$[\text{gain matrix}] \to 0 .$$

If this is the case the estimator is called a self-tuning estimator. If the gain matrix does not converge to zero the estimator is called an adaptive estimator. (This terminology seems implicit in the literature but is not standard.) The significance of this is that adaptive estimators are capable of tracking time varying parameters while self-tuning ones are not. The innovation term contains the new information in the observation y_N beyond what could be predicted based on $\underset{\sim}{x}_N$ and $\hat{\beta}_{N-1}$.

(b) Time Series Models

There are basically two approaches. In the first it is observed that the time series model may be written as a pseudo-linear regression. Consider the ARMAX model (ε_n is a white noise)

$$y_n + a_1 y_{n-1} + \ldots + a_{n_a} y_{n-n_a}$$

$$= b_1 u_{n-1} + \ldots + b_{n_b} u_{n-n_b} + \varepsilon_n + c_1 \varepsilon_{n-1} + \ldots + c_{n_c} \varepsilon_{n-c}$$

which can be written

$$y_n = \underset{\sim}{\phi}_n' \underset{\sim}{\theta} + \varepsilon_n$$

where

$$\underset{\sim}{\theta}' = (a_1 \ldots a_{n_a} b_1 \ldots b_{n_b} c_1 \ldots c_{n_c})$$

and

$$\underset{\sim}{\phi}_n' = (-y_{n-1} \ldots -y_{n-n_a} u_{n-1} \ldots u_{n-n_b} \varepsilon_{n-1} \ldots \varepsilon_{n-n_c}).$$

This suggests the following scheme by analogy with the solution (4b), (5), (6)

$$\hat{\underset{\sim}{\theta}}_n = \hat{\underset{\sim}{\theta}}_{n-1} + \underset{\sim}{P}_n \hat{\underset{\sim}{\phi}}_n \hat{e}_n$$

$$\hat{e}_n = y_n - \hat{\underset{\sim}{\phi}}_n' \hat{\underset{\sim}{\theta}}_{n-1}$$

$$\underset{\sim}{P}_n = \underset{\sim}{P}_{n-1} - \underset{\sim}{P}_{n-1} \hat{\underset{\sim}{\phi}}_n \hat{\underset{\sim}{\phi}}_n \underset{\sim}{P}_{n-1} / (1 + \hat{\underset{\sim}{\phi}}_n' \underset{\sim}{P}_{n-1} \hat{\underset{\sim}{\phi}}_n)$$

$$\hat{\underset{\sim}{\phi}}_n' = (-y_{n-1} \ldots -y_{n-n_a} u_{n-1} \ldots u_{n-n_b} \hat{e}_{n-1} \ldots \hat{e}_{n-n_c}).$$

In the second approach the real time parameter estimator is constructed as a first order solution to the problem of sequentially minimizing

$$\sum_1^n e_k^2(\underset{\sim}{\theta})$$

where

$$e_k(\underset{\sim}{\theta}) = (1 + C(L))^{-1}((1 + A(L))y_k - B(L)u_k)$$

where

$$L\xi_k = \xi_{k-1}$$

and

$$A(L) = \sum_1^{n_a} a_i L^i \quad \text{etc.}$$

First observe that

$$\frac{de_k}{d\underset{\sim}{\theta}'} = \left(\frac{de_k}{d\underset{\sim}{a}'} \frac{de_k}{d\underset{\sim}{b}'} \frac{de_k}{d\underset{\sim}{c}'} \right)$$

$$= -(1 + C(L))^{-1} \underset{\sim}{\phi}_n(\underset{\sim}{\theta})$$

where

$\underset{\sim}{\phi}_n(\underset{\sim}{\theta})$

$= (-y_{n-1} \cdots -y_{n-n_a} u_{n-1} \cdots u_{n-n_b} e_{n-1}(\underset{\sim}{\theta}) \cdots e_{n-n_c}(\underset{\sim}{\theta}))$.

With this in mind the second type of recursion is found by analogy with (4a), (5), (6) as

$$\hat{\underset{\sim}{\theta}}_n = \hat{\underset{\sim}{\theta}}_{n-1} - \tilde{\underset{\sim}{P}}_n \tilde{\underset{\sim}{\phi}}_n \hat{e}_n$$

$$\hat{e}_n = y_n - \hat{\underset{\sim}{\phi}}_n \hat{\underset{\sim}{\theta}}_{n-1}$$

$$\tilde{\underset{\sim}{P}}_n = \tilde{\underset{\sim}{P}}_{n-1} - \frac{\tilde{\underset{\sim}{P}}_{n-1}\tilde{\underset{\sim}{\phi}}_n \tilde{\underset{\sim}{\phi}}'_n \tilde{\underset{\sim}{P}}_{n-1}}{1 + \tilde{\underset{\sim}{\phi}}'_n \tilde{\underset{\sim}{P}}_{n-1} \tilde{\underset{\sim}{\phi}}_{n-1}}$$

together with

$$\tilde{\underset{\sim}{\phi}}_n = -\hat{c}_{n-1,1}\tilde{\underset{\sim}{\phi}}_{n-1} \cdots -\hat{c}_{n-1,n_c}\tilde{\underset{\sim}{\phi}}_{n-n_c} - \hat{\underset{\sim}{\phi}}_n$$

where

$$\hat{\underset{\sim}{\phi}}'_n = (-y_{n-1} \cdots -y_{n-n_a} u_{n-1} \cdots u_{n-n_b} \hat{e}_{n-1} \cdots \hat{e}_{n-n_c}).$$

Notice that $\tilde{\phi}_n$ is a sequential approximation to $de_n/d\hat{\underset{\sim}{\theta}}_{n-1}$.

These two basic approaches apply to very general time series model (see Solo [2]).

3. Convergence Analysis of Self-Tuning Parameter Estimators

Here it will be indicated how the convergence behavior of a self-tuning parameter estimator can be related to a pair of nonlinear ordinary differential equations (due to Ljung [4]).

Consider the ARMAX scheme

$$e_n(\underset{\sim}{\theta}) = (1 + C(L))^{-1}((1 + A(L))y_n - B(L)u_n)$$

$$= y_n - \underset{\sim}{\phi}'_n(\theta)\underset{\sim}{\theta}$$

where

$$\underset{\sim}{\theta}' = (a_1 \cdots a_{n_a} b_1 \cdots b_{n_b} c_1 \cdots c_{n_c})$$

($\underset{\sim}{\theta}_0$ is the true value)

$$\underset{\sim}{\phi}_n(\underset{\sim}{\theta}) = (-y_{n-1} \cdots -y_{n-n_a} u_{n-1} \cdots u_{n-n_b} e_{n-1}(\underset{\sim}{\theta}) \cdots e_{n-n_c}(\underset{\sim}{\theta}))$$

with recursion

$$\hat{\underset{\sim}{\theta}}_n = \hat{\underset{\sim}{\theta}}_{n-1} + \underset{\sim}{P}_n \underset{\sim}{\phi}_n e_n$$

$$\underset{\sim}{P}_n^{-1} = \underset{\sim}{P}_{n-1}^{-1} + \underset{\sim}{\phi}_n \underset{\sim}{\phi}'_n$$

$$\underset{\sim}{\phi}'_n = (-y_{n-1} \cdots -y_{n-n_a} u_{n-1} \cdots u_{n-n_b} e_{n-1} \cdots e_{n-n_c}).$$

Now call

$$\underset{\sim}{R}_n = n^{-1}\underset{\sim}{P}_n^{-1} \text{ so } \underset{\sim}{R}_n^{-1} = n\underset{\sim}{P}_n$$

by analogy with the least squares we expect

$$\underset{\sim}{R}_n \to E(\underset{\sim}{\phi}_n(\underset{\sim}{\theta}_0)\underset{\sim}{\phi}'_n(\underset{\sim}{\theta}_0)) \text{ with probability 1.}$$

Now reorganize the algorithm as

$$\hat{\underset{\sim}{\theta}}_n - \hat{\underset{\sim}{\theta}}_{n-1} = \frac{1}{n}\underset{\sim}{R}_n^{-1}\underset{\sim}{\phi}_n e_n$$

$$\underset{\sim}{R}_n - \underset{\sim}{R}_{n-1} = \frac{1}{n}(\underset{\sim}{\phi}_n \underset{\sim}{\phi}'_n - \underset{\sim}{R}_{n-1}).$$

Now it is suggested that this pair of stochastic difference equations may be analyzed by looking at the behavior of their first moments (viz. $\underset{\sim}{\theta}_n$ = "$E(\hat{\underset{\sim}{\theta}}_{n-1})$") which will obey (roughly)

$$\underset{\sim}{\theta}_n - \underset{\sim}{\theta}_{n-1} = \frac{1}{n}\underset{\sim}{R}_n^{-1}E(\underset{\sim}{\phi}_n(\underset{\sim}{\theta})e_n(\underset{\sim}{\theta}))$$

$$\underset{\sim}{R}_n - \underset{\sim}{R}_{n-1} = \frac{1}{n}\{E(\underset{\sim}{\phi}_n(\underset{\sim}{\theta})\underset{\sim}{\phi}'_n(\underset{\sim}{\theta})) - \underset{\sim}{R}_{n-1}\}.$$

Further introduce the time scale

$$\tau_n = \sum_1^n \frac{1}{k}$$

which has increments

$$d\tau_n = \tau_n - \tau_{n-1} = \frac{1}{n}$$

which tend to zero as $n \to \infty$ so that asymptotically τ_n measures a continuous time scale. The equations become

$$\frac{d\underset{\sim}{\theta}(\tau)}{d\tau} = \underset{\sim}{R}_\tau^{-1}E(\underset{\sim}{\phi}_n(\underset{\sim}{\theta})e_n(\underset{\sim}{\theta}))$$

$$\frac{d\underset{\sim}{R}_\tau}{d\tau} = \underset{\sim}{H}(\underset{\sim}{\theta}(\tau)) - \underset{\sim}{R}_\tau$$

where $\underset{\sim}{H}(\underset{\sim}{\theta}(\tau)) = \underset{\sim}{H}_\tau = E(\underset{\sim}{\phi}_n(\underset{\sim}{\theta})\underset{\sim}{\phi}'_n(\underset{\sim}{\theta}))$.

Now the idea is that the convergence of the estimator should be related to the stability of this pair of ordinary differential equations (ode's) (at least if this pair is not stable then the original algorithm cannot be expected to converge). The stability of ordinary differential equations is often analyzed by means of a Lyapunov function which is usually a quadratic form. A suitable such function here is

$$T_\tau = (\underset{\sim}{\theta}(\tau) - \underset{\sim}{\theta}_0)' \underset{\sim}{R}_\tau (\underset{\sim}{\theta}(\tau) - \underset{\sim}{\theta}_0).$$

The idea is to compute $dT/d\tau$ and show it to be negative. First however the first equation is reexpressed as follows. The following pseudo-Taylor series holds (as the reader can verify)

$$e_n(\underset{\sim}{\theta}) = e_n(\underset{\sim}{\theta}_0) + \underset{\sim}{\tilde{\phi}}_n'(\underset{\sim}{\theta}, \underset{\sim}{\theta}_0)(\underset{\sim}{\theta} - \underset{\sim}{\theta}_0)$$

where

$$\underset{\sim}{\tilde{\phi}}_n(\underset{\sim}{\theta}, \underset{\sim}{\theta}_0) = (1 + C_0(L))^{-1} \underset{\sim}{\phi}_n(\underset{\sim}{\theta}).$$

(The subscript zero denotes a true value.) So calling

$$\underset{\sim}{M}(\underset{\sim}{\theta}) = E\left[\underset{\sim}{\phi}_n(\theta) \frac{\underset{\sim}{\phi}_n'(\theta)}{1 + C_0(L)}\right]$$

and observing that $e_n(\underset{\sim}{\theta}_0) = \varepsilon_n$ is an iid sequence uncorrelated with $\underset{\sim}{\phi}_n(\underset{\sim}{\theta})$ (which depends only on data up to time $n-1$) we find

$$E(\underset{\sim}{\phi}_n(\underset{\sim}{\theta}) e_n(\underset{\sim}{\theta})) = \underset{\sim}{M}(\underset{\sim}{\theta})(\underset{\sim}{\theta} - \underset{\sim}{\theta}_0).$$

Thus the pair of ode's is

$$\frac{d\underset{\sim}{\theta}(\tau)}{d\tau} = \underset{\sim}{R}_\tau^{-1} \underset{\sim}{M}(\underset{\sim}{\theta}(\tau))(\underset{\sim}{\theta}(\tau) - \underset{\sim}{\theta}_0)$$

$$\frac{d\underset{\sim}{R}_\tau}{d\tau} = \underset{\sim}{H}(\underset{\sim}{\theta}(\tau)) - \underset{\sim}{R}_\tau.$$

Now consider if $T_\tau = (\underset{\sim}{\theta}(\tau) - \underset{\sim}{\theta}_0)' \underset{\sim}{R}_\tau (\underset{\sim}{\theta}(\tau) - \underset{\sim}{\theta}_0)$

$$\frac{dT}{d\tau} = (\underset{\sim}{\theta} - \underset{\sim}{\theta}_0)' \left(\underset{\sim}{M}(\underset{\sim}{\theta}(\tau)) + \underset{\sim}{M}'(\underset{\sim}{\theta}(\tau)) + \underset{\sim}{H}(\underset{\sim}{\theta}(\tau)) - \underset{\sim}{R}_\tau\right)$$
$$(\underset{\sim}{\theta}(\tau) - \underset{\sim}{\theta}_0)$$

$$\leq -T$$

if, for all $\underset{\sim}{\theta}$ (in a stability region)

$$\underset{\sim}{M}(\underset{\sim}{\theta}) + \underset{\sim}{M}'(\underset{\sim}{\theta}) + \underset{\sim}{H}(\underset{\sim}{\theta}) = 2E\left[\left(\frac{1}{1 + C_0(L)} - \frac{1}{2}\right)\underset{\sim}{\phi}_n(\underset{\sim}{\theta})\underset{\sim}{\phi}_n'(\underset{\sim}{\theta})\right]$$

is negative semidefinite (nsd). To see the condition required here call

$$H(L) = \frac{1}{1 + C_0(L)} - \frac{1}{2}$$

and let α be an arbitrary fixed p-dimensional vector. Then

$$\underset{\sim}{\alpha}'(\underset{\sim}{M}(\underset{\sim}{\theta}) + \underset{\sim}{M}'(\underset{\sim}{\theta}) + \underset{\sim}{H}(\underset{\sim}{\theta}))\underset{\sim}{\alpha} = 2E[H(L)(\underset{\sim}{\alpha}'\underset{\sim}{\phi}_n(\underset{\sim}{\theta}))^2]$$

$$= \frac{2}{2\pi} \int_{-\pi}^{\pi} H(e^{i\omega}) f_{\phi\phi}(\omega) d\omega$$

where $f_{\phi\phi}(\omega)$ is the spectrum of $\underset{\sim}{\alpha}'\underset{\sim}{\phi}_n(\underset{\sim}{\theta})$.

However, the left-hand side is real so this is

$$= \frac{1}{\pi} \int_{-\pi}^{\pi} \text{Re}\{H(e^{i\omega})\} f_{\phi\phi}(\omega) d\omega \geq 0$$

if $\text{Re}\{H(e^{i\omega})\} \geq 0$. That is

$$H(L) = \frac{1}{1 + C_0(L)} - \frac{1}{2}$$

is required to be __positive real__. For the positive real condition expressed in terms of the coefficients of $C_0(L)$ (see Ljung [4]).

The results of a typical real time parameter estimation are shown in Figure 1. The data was simulated from the model

$$(1 - 1.5L + .7L^2)y_t = (1 + .5L^2)u_t + (1 - L + .2L^2)\varepsilon_t$$

(L is the lag or backshift operator) where the u_t sequence was a pseudo random binary noise sequence. The estimation method used was the pseudo-regression one discussed above. (The example is taken from Söderström et al. [10]).

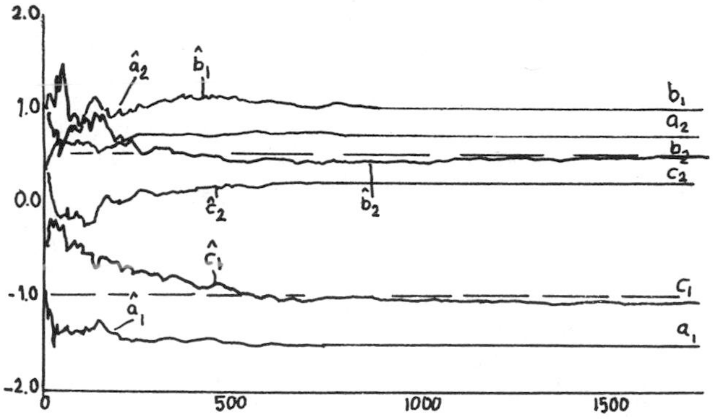

__Figure 1.__ Realization of a Real Time Parameter Estimator for an ARMAX model

This analysis described here can be made rigorous (see Ljung [4] and Hannan [5]) with one complication: namely the estimator has to be monitored so that if it should become unstable it can be projected back into a stable region. Recently an algorithm that does not require monitoring has been given in Solo [6].

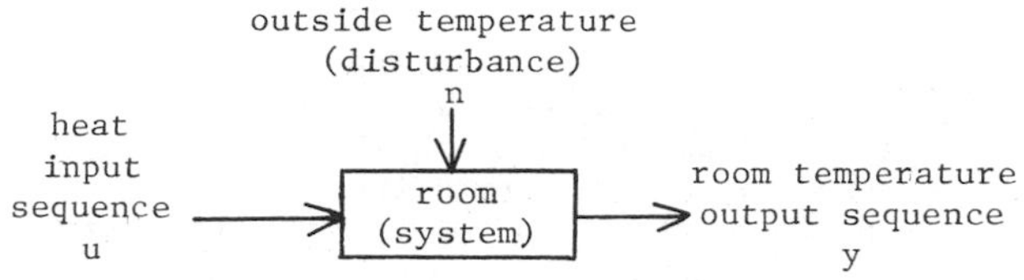

__Figure 2.__ A Control Problem: Warming a Room

4. Self-Tuning Controllers

We begin with a brief summary of the linear control problem. A simple example is the central-heating or air-conditioning of a room. Figure 2 shows the basic situation. There is a dynamic system, a room, subject to an input sequence u (heat), a disturbance sequence n (outside temperature) and an output sequence y (the room temperature). If the outside air temperature (the disturbance) changes so does the room temperature (the output y) unless an adjustment is applied by the input or control sequence u (more or less heat). A discrete time transfer function description is

$$y_k = T_s(L)u_k + n_k, \quad k = 0, 1, \ldots$$

where L is the lag or backwards operator and $T_s(L)$ is a rational function e.g.,

$$T_s(L) = \frac{b_1 L + b_2 L^2}{1 - a_1 L - a_2 L^2}.$$

The simplest automatic control problem consists in using (some form of thermostat as) a control scheme to make the system output y follow a desired or reference (temperature) y^* whatever n does. If y^* = constant we have the regulator problem. If y^* is time varying it is the servo problem.

If the disturbance sequence n cannot be measured the basic solution is to use feedback control. A simple feedback scheme is shown in Figure 3.

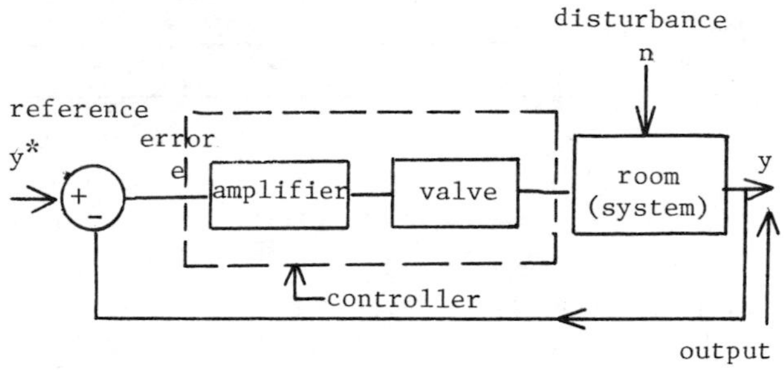

Figure 3. Unity Feedback Control

The idea is to measure the error $e = y^* - y$ and feed it back through a controller (consisting typically of an amplifier to increase the energy in the error signal to a level that allows it to drive a physical device (a valve) that actually alters u). We have now a closed-loop system. A more general feedback scheme allows the controller to appear in the feedback loop as in Figure 4.

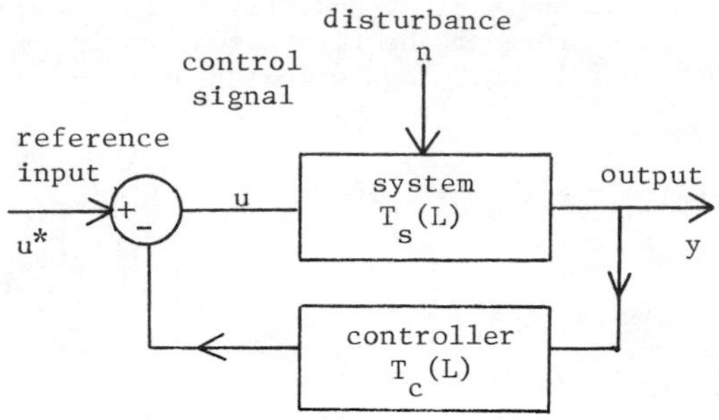

Figure 4. Controller in Feedback Loop

In this configuration u^* is a desired input that produces a desired output y^* through a designer specified transfer function $T^*(L)$ as $y_k^* = T^*(L)u_k^*$. A mathematical description for this scheme is as follows

Forward loop
$$y_k = T_s(L)u_k + n_k$$
$$= (1 + A(L))^{-1}(B(L)u_k + (1 + C(L))\varepsilon_k)$$

Feedback loop
$$u_k = u_k^* - T_c(L)y_k^*$$
$$= u_k^* - (1 + A_c(L))^{-1}B_c(L)y_k$$

where $A(L) = \sum_1^{n_a} a_i L^i$ etc.

Actually it is useful to employ an even more general description as

Forward loop
$$(1 + A(L))y_k = B(L)u_k + (1 + C(L))\varepsilon_k$$

Feedback loop
$$R(L)u_k = (1 + T(L))u_k^* - S(L)y_k$$

where $R(L) = \sum_1^{n_R} r_i L^i$ etc.

The control design problem is to choose R, T, S to ensure y_k is close to $y_k^* = (1 + A^*(L))^{-1}B^*(L)u_k^* = T^*(L)u_k^*$ whatever $n_k = (1 + A(L))^{-1}(1 + C(L))\varepsilon_k$ does. (Here the disturbance has been modelled by a stationary process; other types of models will be useful too.) The design transfer relation $T^*(L)$ incorporates features such as speed of response, no sudden production of large energy surges).

A first solution to this stochastic control problem is by minimum variance control (see Box and Jenkins [7], Astrom [8]). Choose R, T, S so that

$$E(y_k - y_k^*)^2 = \text{minimum}.$$

This solution often produces controllers that require large control effort (i.e. $E(u_k^2)$ large). Thus we can introduce a cost of control and find R, T, S so that

$$E(y_k - y_k^*)^2 + \lambda E(u_k^2) = \text{minimum}.$$

This is the so called LQG problem (linear quadratic Gaussian) much studied in the control engineering literature). For the self-tuning control scheme only the minimum variance controller will be discussed. To find the minimum variance controller reorganize the forward equation

$$(1+A(L))y_k = B(L)u_k + (1+C(L))\varepsilon_k$$

as

$$(1+C(L))(y_k - y_k^* - \varepsilon_k)$$

$$= (C(L)-A(L))y_k + B(L)u_k - y_k^* - C(L)y_k^* = \text{rhs}.$$

If $b_1 \neq 0$ (i.e. unit delay in B(L)) this suggests choosing u_k so that the rhs is zero i.e. yielding the control law

$$B(L)u_k = (1+C(L))y_k^* - (C(L)-A(L))y_k.$$

This can be compared with the feedback loop equation

$$R(L)u_k = (1+T(L))y_k^* - S(L)y_k$$

giving R = B, T = C, S = C − A.

In any case with rhs = 0 we find

$$(1+C(L))(y_k - y_k^* - \varepsilon_k) = 0$$

so that if C(L) is a stable polynomial

$$y_k - y_k^* - \varepsilon_k \to 0 \quad \text{as} \quad k \to \infty.$$

The interested reader can find a simple illustration of a minimum variance controller on pages 178-179 of Astrom's book [8]. If the delay in B(L) is greater than unity the minimum variance controller is more complicated (see Astrom [8]).

To implement minimum variance control recursively we write the control law

$$B(L)u_k = (1+C(L))y_k^* - (C(L)-A(L)y_k$$

as

$$\underset{\sim}{\phi}_{k-1}\underset{\sim}{\theta} = y_k$$

where

$$\underset{\sim}{\theta}' = (b_1 \ldots b_{n_b}\; c_1 - a_1 \ldots c_{n_a} - a_{n_a}\; -c_1 \ldots -c_{n_c}$$
$$(c_i = 0,\; n_c < i \leq n_a)$$

$$\underset{\sim}{\phi}'_{k-1} = (u_{k-1} \ldots u_{k-n_b}\; y_{k-1} \ldots y_{k-n_a}\; -y_{k-1}^* \ldots -y_{k-n_c}^*).$$

Then we use the recursion

$$\hat{\underset{\sim}{\theta}}_k = \hat{\underset{\sim}{\theta}}_{k-1} + r_{k-1}^{-1}\underset{\sim}{\phi}_{k-1}e_k$$

$$e_n = Y_k - \underset{\sim}{\phi}'_{k-1}\hat{\underset{\sim}{\theta}}_{k-1}$$

$$r_{k-1} = r_{k-2} + \underset{\sim}{\phi}'_{k-1}\underset{\sim}{\phi}_{k-1}$$

$$\underset{\sim}{\phi}'_k\hat{\underset{\sim}{\theta}}_k = y_{k+1}^*.$$

This last equation is the control law and produces u_{k+1} from $\hat{\underset{\sim}{\theta}}_k$ and previous y_k's and u_k's. Now we show not $\hat{\underset{\sim}{\theta}}_k \to \underset{\sim}{\theta}$ w.p.1. (since it does not) rather

$$y_k - y_k^* - \varepsilon_k \to 0 \quad \text{w.p.1.}$$

which is all that is of interest for the control scheme. It turns out that $C(L) + \frac{1}{2}$ must be positive real: see Goodwin et al. [9].

References

[1] Kalman, R. E. (1960). A new approach to linear filtering and prediction problems, Trans. ASME Ser. D, 82, 35-44.

[2] Solo, V. (1980). Some aspects of recursive parameter estimation, Int. Jour. Control, 32, 395-410.

[3] Pagan, A. (1980). Some identification and estimation results for regression models with stochastically varying coefficients, Jour. Econometrics, 13, 341-365.

[4] Ljung, L. (1977). On positive real transfer functions and the convergence of some recursive schemes, IEEE Trans. Autom. Control, 22, 539-551.

[5] Hannan, E. J. (1976). The convergence of some recursions, Ann. Statist., 4, 1258-1270.

[6] Solo, V. (1979). The convergence of AML, IEEE Trans. Autom. Control, 24, 958-963.

[7] Box, G. E. P. and Jenkins, G. M. (1976). Time Series Analysis, Forecasting and Control, Holden Day.

[8] Astrom, K. J. (1970). Introduction to Stochastic Control Theory, Academic Press.

[9] Goodwin, G. C., Ramadge, P. J., Caines, P. E. (1979). Discrete time stochastic adaptive control, in Proceedings of the 18th IEEE Conference on Decision and Control, also to appear in SIAM Jour. Control.

[10] Söderström, T., Ljung, L., and Gustaussan, J. (1974). A Comparative study of recursive identification methods, Report 7427, Division of Automatic Control, Lund Inst. Technol, Lund, Sweden.

Workshop 9

Measurement and Evaluation of Software

Organizer: Herman P. Friedman, IBM Systems Research Institute

Chair: Herman P. Friedman, IBM Systems Research Institute

Invited Presentations:

On Measurement and Evaluation of Software: A View from the Chair, Herman P. Friedman, IBM Systems Research Institute

Can Statistical Methods Help Solve Problems in Software Measurement?, Frederick G. Sayward, Yale University

Measuring the Performance of Computer Software -- A Dual Role for Statisticians, Ivor Francis, Cornell University

Software Metrics: Paradigms and Processes, Marvin Denicoff, Office of Naval Research, and Robert Grafton, Office of Naval Research

A Proposal for Structural Models of Software Systems, J.C. Browne, University of Texas at Austin

Software Metrics: A Key to Improved Software Development Management, J.E. Gaffney, Jr., IBM Corporation

ON MEASUREMENT AND EVALUATION OF SOFTWARE: A VIEW FROM THE CHAIR

Herman P. Friedman, IBM Systems Research Institute

Keywords: modeling, software methodology, software metrics, software systems, software evaluation, science and statistics

The purpose of this session was to identify problem areas in the measurement and evaluation of software that require close collaboration between computer professionals and statisticians.

Four of the five speakers in this session were members of a panel commissioned to analyze and evaluate problems in the area of "software metrics." These are efforts to develop quantitative indices of merit that would be useful in the management of the process of producing software as well as to evaluate the program products themselves. There are a number of recurring "needs" expressed in the report of this panel (11) as well as in special volumes of collected papers surveying the field (1) (2). Among these are the need for more and better collected data, a need for developing theories and methods for generalizing experimental results from one domain to another, and a need for critical appraisal of the hundred or more indices already proposed for use. Many of the authors are expressing an existential dilemma of theorization vs. experimentation, A. Kaplan (10). Experimental methods are inseparable from the development and application of hypothesis and theories. These are needed to tell us when we can generalize from the experimental situation, what controls must be imposed, what adjustments must be made to the data afterwards. Without such conceptual guides, the physical operations themselves remain haphazard or problematical in significance. (This process proves something, but I don't know what.) This latter lament is reflected in the paper by F. Sayward in (11). What is being asked for is a richer, broader and more explicit conceptual frame for experimentation in software. A conceptual framework is an embodiment of all the ideas and assumptions that enter into the design and interpretation of the experiment. Even exploratory experiments call upon theory to make sense of what the exploration has discovered.

At present this conceptual frame is fragmented. There is no unified framework for studying the creation and management of programs and for studying the performance of programs. Human beings are essentially part of the process of producing computer programs as well as the ultimate user of programs. Thus the computer professional and statistician have to understand the role of theory and experimentation in the behavioral (10), (12) as well as the physical sciences (4).

The paper by R. Grafton (14), defines the importance of the area for the government and asks that greater effort be expanded to put the field of software measurement and evaluation on a more "scientific" basis. A major goal (for Grafton) is to define generally accepted quantitive figures of merit in software so that decisions with regard to design and evaluationand selection, can be made on a more "objective" basis. See (5) for related issues.

The paper by F. Sayward (16) summarizes the panel report described in (11). In this paper an important distinction is made between software and algorithms. Algorithms are precise and can be examined studied and shaped like diamonds, they are forever. Software is amorphous, constantly changing it must be continually evaluated, maintained, and eventually replaced. Seven key stages in the life cycle of software are defined and questions and problems for investigation are discussed. In addition F. Sayward comments critically on the statistical experiments reported in the literature involving people. He asserts that experiments reported in the literature do not take into account the dynamics of change of the software over time. In addition almost all reported experiments did not allow for comparison between different language classes. Fred Sayward also describes issues of software reliability and testing. He concludes that computer scientists should learn more about statistical methods. He also asks for assistance from statisticians in the design of experiments and in validating the proposed software "metrics".

In his paper J. C. Browne (15), comments on the current use of models in managing the development of software. He asserts that most models in use are "empirically based" phenomological models. They are of some use in prediction, but provide no explanation of how? In addition he asserts that the current models in use are not invariant across the hierarchical levels of design; system, program, module and code. He proposes a "structural" or explanatory model based on concepts and tools of existing data models in the data base literature.

Many of the "metrics" that have been proposed attempt to measure the effort or complexity in developing and understanding software. These "metrics" can save serve many purposes. They can be used to evaluate the software development program or the software products. They can be used to estimate the cost of producing the product and the quality of the product. If the measures are proven consistent across the different levelsof design, they can be used to monitor the stability and quality of the product during its life cycle.

The software science "metrics" of M. Halstead have been among those widely investigated. The metrics of Halstead are based on analogies with information theory and thermodynamics. The measurements are based on simple counts directly available from the language used to express the algorithm. The claim has been made, but not fully substantiated, that relationships developed from these measurements represent "law like" properties of algorithms that are invariant under translation from one language to another.

A question currently under study by many people is whether these metrics "actually" measure effort and complexity. Of course this begs the question of what is meant by effort and complexity. Effort has been defined as the resources required to program a job, and complexity as the number of errors. Measurements have been taken in different organizations by different people on different programs and modules in an attempt to investigate the relationship between Halstead metrics, effort and errors for the purpose of developing prediction equations to be used in the software development process. Usually the analysis involves making scatter plots and doing regression and correlation studies to investigate the strength of the relationship between the measures. The paper by J. Gaffney (17) is typical of this type of investigation. He discusses the difficulties with collection of data in an environment with limited amounts of control. There are also technical issues of instrumentation for automatic data collection. J. Browne would not be happy with this type of study, because the resulting models although predictive are not explanatory. Many statisticians would want to investigate with "robust" procedures, the sensitivity of the estimated correlations to one or two possibly influential measurements.

The last paper in the session by Ivor Francis (18), introduces methods of classification to the area of software evaluation. In this paper I. Francis discusses the role of classification in science and uses modern methods of cluster analysis to classify statistical software packages.

This important paper describes the many roles of a statistician in relation to software measurement and evaluation and broadens the view of statistics and statisticians to the computer professional.

There are many other problems in the design and analysis of software systems that could benefit from classification methods. For example, early on in the design process the system is divided into component parts and their connections. The art of such partitioning is to absorb as many relations between elements in a cluster as possible and thus leave few inter-cluster connections. L. Belady and C. Evangelisti (3), describe an algorithm for clustering a large number of program modules and data structures. Also, a measure of complexity for comparing the quality of partitioning is given. The procedure uses graph-theory and the measure of complexity is based on the number of nodes and edges in each cluster of a partition.

A similar problem is dealt with by A. Wong in (13). In his research effort, a systematic approach has been proposed for the decomposition of an overall set of functional requirements for a complex design, into sub-problems that will exhibit the key characteristics of good design: strong coupling within sub-problems and weak coupling between them. As before graph-theory is the descriptive tool. However A. Wong has created and used innovative and powerful new methods of cluster analysis for graph decomposition. This procedure solves some of the technical problems inherent in previous methods and produces clusters useful to the systems designer.

A key issue in understanding soft measurements is variability. What D. Knuth said in (1971) is very relavant today.

"There has been widespread realization that more data about language use is needed; we cannot really compare two different compiler algorithms until we understand the input data they deal with. Of course, the great difficultyis that there is no such thing as a 'typical programmer'; there is a tremendous variation among programs written by different people with different backgrounds and sympathies and indeed there is considerable variation even in different programs written by the same person. Therefore we cannot trust any measurements to be very accurate, although we can measure the degree of variation in an attempt to determine how significant it is."

To some extent the new disciplines of "software engineering" are attempting to reduce this variability. How successful they are is a question for study. When it comes to measurements of large running programs, issues of data complexity and data quality will have to be faced by the statistician and computer scientist.

"Data Complexity - depends on the number of explanatory and response variables, the form imposed by the method of data collection and in addition on such complications as missing values.

"Data Quantity - can range from amounts for which each data value can be known as individual amounts so vast that only a small sample can ever be analyzed."

Similar issues arise in computer performance evaluation and are discussed in (7).

The performance that the users of a software product experience depends on thespecifics of the computer system. Thus models in the development of software have to interface with models of computer systems performance in order to predict the performance of software products.

The techniques of multidimensional scaling and related techniques developed by statisticians and psychometricians should prove useful in understanding the users perception of complexity, quality, and usability. This could be a fruitful area for collaboration. Statisticians and computer scientists could define experiments in order to "derive" dimensions directly rated to people's perception. These methods can also cope with individual differences. A good reference is (19).

As A. Kaplan (10) describes:

"The meeting place between theory and experiment is one of those dangerous intersections at which neither vehicle is allowed to proceed till the other has gone by; the remarkable thing is that traffic moves most freely only when both roads are well-traveled."

In the area of software measurement statisticians can play a much more important role than traffic cop. In fact as the following quote from R. A. Fisher (6) indicates the computer scientists are capable of policing themselves.

"When any scientific conclusion is supposed to be proved on experimental evidence, critics who still refuse to accept the conclusion are accustomed to take one of two lines of attack. They may claim that the interpretation of the experiment is faulty, that the results reported are not in fact those which should have been expected had the conclusion drawn been justified, or that they might equally well have arisen had the conclusions drawn been false. Such criticisms of interpretation are usually treated as falling within the domain of statistics. They are often made by professed statisticians against the work of others whom they regard as ignorant of or incompetent in statistical technique; and, since the interpretation of any considerable body of data is likely to involve computations, it is natural enough that questions involving the logical implications of the results of the arithmetical processes employed, should be relegated to the statistician. At least I make no complaint of this convention. The statistician cannot evade the responsibility for understanding the processes he applies or recommends. My immediate point is that the questions involved can be dissociated from all that is strictly technical in the statistician's craft, and, when so detached, are questions only of the right use of human reasoning powers, with which all intelligent people, who hope to be intelligible, are equally concerned, and on which the statistician, as such, speaks with no special authority. The statistician cannot excuse himself from the duty of getting his head clear on the principles of scientific inference, but equally no other thinking man can avoid a like obligation."

In summary, computer professionals and statisticians require an understanding of the nature of science in the behavioral as well as physical sciences. Statisticians and statistics should be used at beginning stages of investigation to aid in the development of conceptual models and to provide insight and understanding into the nature of software processes and the structure of the measurements. The prevalent view in the computer science literature of statistics appears to be one of support, confirmation and validation, rather than illumination.

Finally, I recommend to those interested in this area to read references (1), (2), (8), (9), (11), and (12) to learn more of the problems and issues in this area of emerging importance.

REFERENCES

1. Basili, V. R., Tutorial On Models and Metrics For Software Management and Engineering. Computer Society Press, IEEE, N. Y., N. Y. 1980.

2. Belady, L., Editor. Proceedings of the IEEE, Special Issue on Software Engineering, Sept. 1980, Vol. 68, #9. IEEE, N. Y., N. Y.

3. Belady, L. and Evangelisti, C., "System Partitioning and Its Measure, IBM Research Report, T. J. Watson Research Center, Yorktown Heights, N. Y.

4. Campbell, N. Foundations of Science, Dover Publications, Inc., New York.

5. Elkana, Y., et al. Editors. Toward a Metric of Science: The Advent of Science Indicators. John Wiley, 1980.

6. Fisher, R. A. Design of Experiments, Hafner, 1966.

7. Friedman, H. P., "Statistical Methods in Computer Performance Evaluation," in Experimental Computer Performance Evaluation. Edited by D. Ferrai and M. Spadoni. North Holland Publishing Co. 1980.

8. Gilb, T., Software Metrics, Winthrop, 1977.

9. Halstead, M. Elements of Software Science. Elsevier, North Holland, 1977.

10. Kaplan, A., The Conduct of Inquiry. Chandler Publishing Co., San Francisco, Ca., 1964.

11. Perlis, A., Sayward, F, Shaw, M. Editors. Software Metrics: An Analysis and Evaluation. MIT Press, Cambridge, Mass. 1981.

12. Schneiderman, B., Software Psychology. Winthrop, 1980.

13. Wong, M. A., "A Graph Decomposition Technique Based on a High-Density Cluster Model on Graphs." Technical Report #14, July 1980, Center for Information for System Research. MIT Sloan School.

14. Grafton, R. B., "Software Metrics: The Need for Paradigms and Processes," Proceedings Computer Science and Statistics Interface Meeting, Held at Carnegie Mellon University, Sprinter Verlang Pub. 1981.

15. Browne, J. C., "Role of Models and Abstractions in Software Evaluation," Proceedings Computer Science and Statistics Interface Meeting, Held at Carnegie Mellon University, Sprinter Verlang Pub. 1981.

16. Sayward, F. G., "Can Statistical Methods Help Solve Problems in Software Measurement," Proceedings Computer Science and Statistics Interface Meeting, Held at Carnegie Mellon University, Springer Verlang Pub. 1981.

17. Gaffney, J. E., "Software Metrics: A Key to Improved Software Development Management," Proceedings Computer Science and Statistics Interface Meeting Held at Carnegie Mellon University, Springer Verlang Pub. 1981.

18. Francis, I., "Measurement of Software Performance: A Dual Role for the Statistician," Proceedings Computer Science and Statistics Interface Meeting Held at Carnegie Mellon University, Springer Verlang Pub. 1981.

19. Carrol, J. D. and J. Kruskal, "Multidimensional Scaling," International Encyclopedia of Statistics, Volumes 1 and 2. The Free Press, 1978.

20. Knuth, D., "An Empirical Study of FORTRAN Programs," Software Practice and Experience, Vol. 1, pp. 105-133 (1971).

CAN STATISTICAL METHODS HELP SOLVE PROBLEMS IN SOFTWARE MEASUREMENT?

Frederick G. Sayward
Department of Computer Science
Yale University

Abstract

The study of software metrics involves the creation and analysis of quantitative indices of merit which can be assigned to software, either existing or proposed. These measurements of software provide informational aids to be used in making software lifecycle decisions.

A panel commissioned to analyze and evaluate the problems in the emerging field of software metrics has recently issued a report. An overview of the panel's findings, including how statisticians might be of help in solving some problems in software measurement, is presented.

Keywords

Cost estimation, human factors, manpower loading, program complexity measures, single subject experiment, software experiments, software lifecycle, software metrics, software reliability estimation.

1. Introduction

During the spring of 1979 Marvin Denikoff of the Office of Naval Research asked Alan Perlis, Mary Shaw, and myself for our opinion on the merit of the research which was being conducted on software metrics. After a quick review of the literature it became apparent that the central issues revolved around the following question:

> Can there be assigned to software and the processes associated with its design, development, use, maintenance, and evolution, indices of merit that can support quantitative comparisons and evaluation of software?

We decided that an in-depth study was needed to identify useful areas in software metrics and software experimentation, to analyze the research being done in these areas, and to recommend directions which these areas should follow in the near term. Eventually this work would lead to the development of a scientific basis for analyzing and evaluating software which is so desperately needed by those involved with software management.

We recognized early that a large fraction of the computer science research community held little hope for software metrics. Consequently we decided that part of our effort should be in focusing the attention of the research community on the problems and methods in software metrics. It was also recognized that statisticians should play an important role in the development of a science of software metrics. It was decided that part of our goal would be to define what that role should be.

We decided that the software metrics issues could be best studied by a panel consisting of computer scientists and statisticians involved in software metrics. The panel members were:

Vic Basili, University of Maryland
Les Belady, IBM Corporation
Jim Browne, University of Texas
Bill Curtis, ITT Corporation
Rich DeMillo, Georgia Institute of Technology
Ivor Francis, Cornell University
Richard Lipton, Princeton University
Bill Lynch, Xerox Corporation
Merv Muller, World Bank
Alan Perlis, Yale University
Jean Sammet, IBM Corporation
Fred Sayward, Yale University
Mary Shaw, Carnegie-Mellon University

Alan Perlis served as chairman of the Study Panel with Fred Sayward and Mary Shaw serving as assistants.

An organizational meeting was held at Yale University

This research was supported in part by the Office of Naval Research under Research Contract N00014-79-0672.

September 10-12, 1979 at which time an initial state of the art presentation was given and the software metrics issues were discussed. Members were initially asked to study the current status of a variety of software metric areas. A second meeting was held on January 31-February 1, 1980 at Las Vegas. State of the art evaluations were presented and final topics were assigned. Each panel member wrote at least one paper consisting of a state of the art evaluation and directional recommendations for an area of software metrics. The papers were presented at a meeting held in Washington, DC on June 30, 1980 at the National Academy of Science.

The proceedings of the Washington meeting will appear in a book entitled *Software Metrics: An Analysis and Evaluation* to be published by MIT Press early in 1981 [5]. The book also contains an extensive annotated bibliography on software metrics and an edited version of the discussion which took place at the meeting follows each paper, where applicable.

In this paper the study framework adopted by the panel will be presented along with some overview material on how statistical methods are being used by computer scientists. In particular, a review of three software metrics areas studied by the panel, software cost estimation, human factors experiments, and software reliability estimation, will be given. Also, some observations and recommendations will be given on how and where statisticians can help computer scientists working on problems in software metrics.

2. *What is Software?*

Software is more than source code. During the study we used the term software as a generic for all of the stages gone through in tailoring a computer system to solve some particular problem. Software is a non-terminating process and the product, software, is evolutionary and shaped both by the nature of its use and the intent of its design. Specification of software is generally incomplete and arrives at a satisfactory state through evolution and use.

Software is subject to a perpetual tension. Being purely symbolic, it can be perfected, guaranteed, arbitrarily extended, and reproduced at almost no cost. Being purely symbolic it can be easily changed, adapted, mishandled, corrupted, generalized, altered by use and discarded. Far from being reproducible at no cost, the replication of software introduces significant extra cost in maintenance and replacement no matter who is responsible for these two activities.

If by software we mean program, every piece of software is (a representation of) an algorithm. Hence the results of the study of algorithms in computer science should suggest methods for controlling and improving software. To a limited degree, computer science has already helped. But the behavior of software is different from that of algorithms. We observe the following comparisons:

1. Software is rarely as precisely specified as algorithms. It is often pointless to speak of software as a map from its input data to its output data.
2. Unlike algorithms, software changes its intent while under specification, design, development, use and replacement.
3. Software is generally huge while algorithms are described as being precise.
4. Software is managed (or mismanaged), while algorithms are created, perfected and proven correct.

Hence we conclude that software is not the same as algorithms, that the study of software is not the same as the study of algorithms, and that software problems are not the same as algorithm problems.

2.1. *Software Problems and The Lifecycle Model*

Software problems are problems in managing, controlling, predicting and optimizing the success, limitations, and cost of the evolution of software during its lifecycle. The software metrics study panel decided to fix upon a model for software development in which the dynamics of software play a significant role. The model chosen was the lifecycle model in which software is seen as passing through seven stages:

1. Requirements Analysis
2. Specification
3. Design
4. Implementation
5. Testing and Integration
6. Maintenance and Enhancement
7. Replacement or Retirement

In isolation, these stages occur at progressively later times but backtracks from one stage to an earlier one may occur at any time. Revision of specification, alteration of requirements, change in environment and erroneous implementation may interrupt the flow of normal development or spawn sub-processes having their own lifecycles.

The software metrics panel focused on the questions that naturally arise in guiding software through the lifecycle model. Questions were formulated early in the study and a set of them were selected for detailed analysis. Among the questions listed in the formulation were two questions common to all the stages:

Is it time to go onto the next stage?

Is a backtrack to an earlier stage needed?

At each stage there is also a set of questions about the software and the project whose precise answers are needed to dictate an optimal lifecycle control flow. Software metrics addresses such questions but it is not sufficiently developed yet to provide precise answers to most of them.

2.1.1. Requirements Analysis

While this area of system development has an enormous influence on system software, the questions of concern here must lie outside the domain of software metrics -- for it is not until the requirements are fixed that the structure of the software can begin to take shape. What is needed is a link from requirements analysis to software specification, but this link must necessarily not be software-oriented, just as a link from an informal to a formal model cannot be formal.

2.1.2. Specification

At this stage an informal statement of the problem and its proposed solution has been prepared. Questions to be answered include:

- What is the cost of production?
- What are the memory requirements of the software?
- What are the speed requirements of the software?
- How long will it take to produce?
- When will it have to be replaced?
- What manpower loading should be used?
- Is the project feasible? That is, does the expected production time exceed the time when the software will be of use?

2.1.3. Design

At this stage a detailed formal statement of the problem and its proposed solution have been prepared. This includes a development plan for all future stages of the lifecycle. Questions to be answered include:

- What machine configuration to use?
- What language to use?
- Is it possible to incorporate the work of others or must everything be built in house?
- How will the availability of tool X affect factor Y?
- How close to its limits is the system expected to run?
- What are the potential future enhancements?
- Should the system be all encompassing from which subsystems are carved out or should the system be primitive on which specific systems are built?

2.1.4. Implementation

Some questions need to be answered before implementation begins and others arise during implementation. They include:

- What developmental technology should be used? Should the system be built all at once or should it be constructed through a sequence of executable prototypes?
- What programming discipline should be used? Chief programmer? Cottage industry?
- Is the project on schedule?
- Is the project on the budget?
- Is the implemented code correct? If not, how close is it to meeting the specification?
- What is the quality of the implemented code? Is it understandable? Is it maintainable? Is it enhanceable?

2.1.5. Testing and Integration

At this stage the chief question is: Does the implementation meet the specification? This usually reduces to questions concerning what the implementation actually does, what resources it uses, and how easy it is to use. Such questions can be asked about individual modules and about integrated modules working in concert. The decisions to be made include:

- Should testing be done top down or bottom up?
- Which of the available testing methodologies should be used?
- What levels of satisfactory testing are sufficient?
- How well does the testing environment approximate the execution environment?
- How will subsequent error reports be handled?

2.1.6. Maintenance and Enhancement

These two very different activities are often linked because both result in a re-release of the system. Maintenance is similar to testing except that the software execution environment has changed from a controlled world of testing to the hurly-burly of actual use. Every repair and update must be tested, so the questions generated by maintenance are similar to the ones above in the testing stage. Enhancement, on the other hand, is a post-release augmentation of the system specifications to meet unforeseen demands. Enhancement may cause a backtrack all the way to the requirements analysis stage.

The questions we ask about enhancement include:

- What is the cost of the enhancement? Is it worthwhile?
- Will the enhancement speed up or delay replacement?
- What is the re-release strategy?

Once it has been decided that an enhancement should take place, there is an automatic backtrack at least to the specification stage of the lifecycle.

2.1.7. Replacement and Retirement

Among the questions asked when considering replacement or retirement of a system are:

- Has the problem outgrown the program?
- Has technology moved beyond the program?
- Has a critical support resource for the system become unavailable?
- Would it cost less to re-build the system than to maintain and enhance the system?
- How should the system be phased out?
- Should there be a change in the language in which the system is written? In the machine on which the system runs?

2.2. Software Metrics

Software metrics are quantitative indices of merit which are assigned to software, either existing of proposed. Their stated goal is to be an aid in making lifecycle decisions. Many, but not all of them are based on measurements made on programs. Others are based on measuring time; errors, programming effort, and programmer performance.

There is no lack of proposed software metrics. What lacks is a scientific discipline for studying software metrics. The types of studies needed include 1) ways of providing an interpretation for a particular software metric in the context of the lifecycle, 2) ways of evaluating competing software metrics, 3) ways of determining when and where to use which software metric, and 4) ways of determining the cost of applying particular software metrics.

3. Software Cost and Manpower Loading

During the requirements and specification stages of the lifecycle two important questions which arise are:

- How much will the software cost?
- How many people should be used at a given time?

From historical data collected on the software lifecycle one sees that

- 60% of lifecycle costs are for the post-development stages.
- Manpower load builds up gradually, peaks at development completion, and then gradually diminishes.

One model which attempts to provide an answer to these questions is the Putnam model [6] which was analyzed by Vic Basili during the study. The model is based on a hardware development model which noted that there are regular patterns of manpower buildup and phase out which are independent of the type of work being done. That is, it is related to the way people solve problems.

The Putnam model is based on an assumption that in software projects the manpower load over time follows a Rayleigh curve. The input to the model are two parameters:

- t_d: an estimate of the software development time.
- K: an estimate of total manpower effort needed for the software.

This results in the so-called lifecycle equation:

$$y = K\ (1 - \exp\{-at^2\})$$

where

- y is the manpower used at time t.
- a is a parameter determined by the time at which y′ reaches its maximum value (shape parameter). It turns out that

$$a = 1\ /\ (2\ t_d^2)$$

From the lifecycle equation it is a simple exercise to derive an estimate for the optimal manpower load at time t:

$$y' = 2\ K\ a\ t \exp\{-a\ t^2\}$$

Other such simple formulae can be derived for estimating lifecycle costs and development costs.

It would be very nice if software were governed by the Putnam model because of its simplicity. However, there is little evidence to support the model. It would seem that in this area statisticians could be of great help in analyzing lifecycle data in an effort to see if the assumptions of the Putnam model hold.

Among the analysis and recommendations coming out of the panel were those of Demillo and Lipton claiming that the present searches for simple formulae for predicting the cost of large scale software efforts are very likely to fail. They explain how measurement theory rejects most of the formulae that have been suggested. On a positive side, they feel that the other popular method of predicting cost from historical data is more likely to produce useful results.

They recommend that predicting from historical data receive continued support and that more effort and thought go into establishing a better data base for these studies. They also recommend that an analogy from weather forecasting suggests a refinement which should be explored. That is, the development of micro theories of software costing (which don't necessarily scale up) and the development of large scale computational techniques (such as clustering) which integrate the micro theories to make a cost prediction for large systems.

4. Human Factors

In the design, implementation, testing, and maintenance stages, a question which constantly arises involves which programming tools and programming methodologies should be used in order to increase programmer performance.

Several researchers have been conducting experiments aimed at understanding the effect of programming tools and programming methodologies on programmer performance. During the study Curtis and Sayward analysed this area. To date the designs used in these "software experiments" have followed the traditional approach in which subjects are randomly drawn from a population and then randomly assigned to two or more groups. The effect of the hypothesized improvement factor is then evaluated by comparing the mean intergroup change observed on some measured factor.

Popular topics of study for programming tools in these experiments have included programming language features, time sharing versus batch, debugging aids, and program testing tools. For programming methodologies people have studied structured programming, code reading methods, program complexity measures, flowcharting, source code commenting, variable nomenclature, debugging strategies, and testing strategies. Popular measures of programmer performance have been code comprehension, the time needed to do a programming task, and the accuracy achieved in doing a programming task.

Rarely in software experiments have hypotheses been formally stated at the outset. Rather, the paradigm is an informal introduction to the factors under consideration, the experiment design, the data collected, the application of statistical inference tools, and conclusions. The reported internal validity considerations are rarely satisfactory. Each paper usually ends with a note that the findings suggest further investigation should be done, indicating that even the authors have little faith, as yet, in the external validity of their software experiments.

In [1], Brooks gives a detailed account of some internal validity flaws found in current software experiments. He cites several examples of subject and material selection, summarized below, which make many experiments suspect. With respect to measurement selection, he states that new measures of the effort required for program construction and program understanding based on cognitive models of program-programmer interaction are needed.

Brooks' criticism of subject selection is based on the methods used to circumvent the problems of cost, non-availability and wide ability differences of using experienced professional programmers. Most software experiments have used beginning or intermediate level student programmers as subjects. There is little proof of the necessary internal validation issue that experienced programmers use at a faster rate the same problem solving procedures as do beginners. Also, guaranteeing that groups of begining programmers have equal ability is not trivial.

The problem with material selection is not so much internal as external. For internal validity, the programming tasks selected must be comparable across the uncontrolled variables. Program complexity measures have been somewhat helpful in this respect. Factorization designs can also be used.

For external validity, Brooks states that the programs used in software experiments are not representative of the programs being developed in the real world because of their small size. Since it is universally accepted that developing a large system is not just a matter of scaling up from developing a small system, small scale material will be a potential source of external invalidity until an accepted model of the effects of program size is developed.

Brooks states that all of the experiments done to date need to be replicated on larger programs before any generality of their results can be accepted. Although not stated, presumably experienced professional programmers would also have to be used in attempting to generalize the current software experiments by replicating them on large scale software.

It was the opinion of the study panel that Brooks' suggestion is not only economically infeasible but also premature. This was based on the following question:

> Given that the software lifecycle is so dynamic over time, should we be content in our software experiments to draw conclusions based on a *single* ex post facto measurement?

Designing software experiments based on this principle would seem to be an area where statisticians could be of invaluable help to computer scientists.

The panel recommended continued support for the type of

small scale many subject experiments which have been done since they will lead to interesting new hypotheses and they will produce a gradual refinement of design techniques for strengthening internal validity. Also recommended was that, given the dynamic nature of the software lifecycle model, cross-over and single subject designs [4] might be more natural for software experiments because of their heavy dependence on a time-series of data.

5. *Software Reliability Estimation*

Two questions which arise during the testing and integration lifecycle stages are, "Should the software be released?" and "What is the mean time to failure of the software?" This is the problem of software reliability estimation which was analysed by DeMillo and Sayward during the software metrics study.

To date researchers who have worked on estimating software reliability have followed the traditional methods of reliability theory which has more or less been successfully applied to hardware [3]. The paradigm for software has been to estimate an initial software error content at the end of the software development stage (at time T). Also at this point in time a statistical model is determined, usually by controlled observations of system errors (the "failures" of the traditional theory). During the testing and integration stage (the time interval [T,T+t]) enough data is gathered to determine a failure rate r(t). This, coupled with a variety of subjective evidence concerning the behavior of the system in operation, then determines an estimate of the underlying failure distribution. The predictive model is then applied in the interval [T+t,T+t+k], where k is a "regenerative" time less than the operational lifetime of the system, to predict the operational failure behavior of the software.

The panel's major observation on software reliability was that attempts at applying hardware reliability estimation theory to software reliability estimation have led to unnatural and often contradictory assumptions. Among the assumptions found in these models are [2]:

1. The number of initial program errors can be reliably estimated.
2. Error detection rate (failure rate) is proportional to the number of remaining errors.
3. Errors are discovered one at a time.
4. Once an error is detected, it can be found and removed immediately.
5. The rate at which errors occur is constant.
6. Removing an error reduces the total number of errors by one.
7. Errors are statistically independent.
8. The distribution of program inputs is known.
9. The rate at which errors are detected is proportional to the amount of time spent debugging.
10. The size of a program is constant over its lifetime.

It would seem that statisticians could help computer scientists by developing methods for accepting or, more likely, rejecting most of these assumptions. This would involve a detailed categorization of lifecycle data gathered from large-scale software projects.

Among other difficulties in software reliability estimation is the fact that software is not a fixed object and hence the stochastic requirements of hardware reliability theory cannot be satisfied. The panel recommended that, rather than striving for ways to assign a probability of correct operation to software, attempts to assign a probability to the processes used to validate software might be a more fruitful path to follow. These validation processes, mainly software testing tools, are fixed over time and thus distributions for them can be studied empirically. Then a Baysian based reliability estimate for software validated by the process can be derived.

6. *Summary*

In addition to the detailed studies on particular topics in software metrics, the panel made several observations of a general nature on the current activities in software metrics. These include:

- Software metrics are being widely *used* and *misused* by computer scientists.
- Computer scientists are poorly trained in statistical techniques.
- There is a lack of agreed on definitions and measures.
- There is a lack of properly collected and categorized lifecycle data.
- There is a lack of knowledge on how to design, conduct, and validate software experiments.

It was felt by the panel that there is a need for much more interaction between computer scientists and statisticians to:

- Confirm the proper use of software metrics.
- Detect the misuse of software metrics.
- Suggest proper experimental methods and evaluation techniques.
- Suggest techniques for validating or rejecting the assumptions being made by computer scientists in models of software and its processes.

- Suggest methods of collecting, validating, and categorizing software lifecycle data.

It is hoped that this paper will make statisticians aware of the software metrics problems facing computer scientists and that it will encourage statisticians to work with computer scientists in a combined effort aimed at resolving some of these problems.

References

[1] R. Brooks.
Studying programmer behavior experimentally: The problems of proper methodology.
Communications of the ACM 23(4), April, 1980.

[2] *DACS Software Engineering Research Review.*
Quantitative Software Models edition, Data and Analysis Center for Software, IIT Research Institute, Rome, NY, March 1979.
Order Number SRR-1.

[3] K. C. Kapur and L. R. Lambertson.
Reliability in Engineering Design.
Wiley, 1977.

[4] T. Kratochwill, editor.
Single Subject Research: Strategies for Evaluating Change.
Academic Press, New York, 1978.

[5] A. Perlis, F. Sayward, M. Shaw, editors.
Software Metrics: An Analysis and Evaluation.
MIT Press, Cambridge, Mass, 1981.

[6] L. Putnam.
A General Empirical Solution to the Macro Software Sizing and Estimating Problem.
IEEE Transactions on Software Engineering SE-1(2), 1975.

MEASURING THE PERFORMANCE OF COMPUTER SOFTWARE - A DUAL ROLE FOR STATISTICIANS

Ivor Francis, Cornell University

Abstract

In describing and modelling the properties of any system, science traditionally uses quantitative measures. The emerging science of software is no exception. Statisticians, traditionally the arbiters of evidential inference in science, can play an important role in defining quantitative measures - metrics - of software performance and the equally important quantitative measures of the computational difficulty of problems to be solved. Computer Science, in embracing empirical research in software, must now consult the discipline of experimental design.

When the software is statistical in application, statisticians are also in the role of beneficiaries of this scientific study of software. As users they should insist that evaluations of performance measure not only the usual completion times of workloads on certain machines, but also the accuracy of computed solution and the usefulness of the output.

This paper describes a classification system for statistical software which is based on quantitative measures drawn from the "life-cycle" of a complete statistical analysis: file building, editing, data display, exploration, model building.

Keywords: software performance, quantitative measures, classification, software science, evaluation, experimental design, statistical software.

This paper is concerned with determining the quality of software which has been written to be used by someone other than the programmer. In an article entitled "Home-Computer Software Lags" in the New York Times, November 26, 1980, Peter J. Schuyten wrote,

"For those who think they might like to own a personal computer but are unwilling to go through the trials of learning to program one, off-the-shelf, or "canned," software is available to make a system function. ...

"While it would be nice to report that much of what is available is useful, entertaining or just plain functional, the fact is that the personal computer industry has not yet done a good job in providing users with adequate software.

"In fact, with some noteable exceptions, software intended for the consumer market is either poorly executed, flawed by programming errors, or inexorably complicated to master."

This could be a description of the state of scientific software, although it might be more accurate to say that little is known about the quality of scientific software. A recent GAO report states that Federal software conversion costs run at more than $450 million a year, and that the Social Security Administration made unauthorized payments of approximately $1 billion between 1974 and 1976 largely because of "incomplete, untested or erroneous computer programs". Other Federal agencies rely on the accuracy of computer programs. For example the F. D. A. must rely on the accuracy of a drug company's software used in analyzing tests of a new drug.

As far as software with statistical applications is concerned, software which is available for wide distribution has proliferated in recent years. Packaged computer software has become an important and permanent feature of statistical practice as an increasing number of researchers rely on packaged programs for their analyses. Yet little has been known of the number of these programs, their detailed characteristics, and which ones were dangerous. Under these circumstances it is not surprising that no standards exist for either the development or use of such programs: some program developers have given insufficient attention to accuracy and to methods of protecting the user against his misusing the programs; users, on the other hand, in publishing the results of analyses in which computers have been used, typically fail to identify precisely the program and computer used. In addition, with minimal modifications and management some programs could have been transplanted to similar habitats, thereby eliminating the need for developing new programs which essentially duplicated others already in existence.

In the mid-1960's individuals began to report shortcomings of available statistical software, for example Longley (1967). Later professional societies, beginning with the American Statistical Association in 1974, began to show concern in this growing use by statisticians and non-statisticians alike of "statistical" software, the quality of which was unknown. Even non-statistical societies have become concerned, such as the American Society for Testing Materials which in 1980 established a task force to investigate the accuracy of statistical computer programs.

For thousands of years man has used science to impose some simplicity on the complex world around him so that he might begin to comprehend his environment and perhaps control it.

"Harken to the miseries that beset mankind. They were witless erst and I made them to have sense and be endowed with reason. Though they had eyes to see, they saw in vain, they had ears but heard not, but, like to shapes in dreams throughout their length of days without purpose they wrought things in confusion... They had no sign either of winter or flowery spring or fruitful summer, whereon they could depend, but in everything they wrought without judgement, until such time as I taught them to discern the rising of the stars and their settings, aye, and numbers, too, chiefest of sciences, I invented for them, and the combining of letters, creative mother of the muses' arts, wherewith to hold all things in memory." [Prometheus Bound, Aeschylus, c.470 B.C.]

This quotation suggests a positive answer to the question asked by Alan Perlis (See Perlis, Sayward, and Shaw, 1980):

"Can there be assigned to software and the processes associated with its design, development, use, maintenance, and evolution, indices of merit that can support quantitative comparisons and evaluation of software?"

In this paper we will demonstrate that at least as far as the use of statistical software is concerned, indices of merit can be assigned, indeed have been assigned, to software to support quantitative comparisons and evaluations.

The first step in any new scientific study is a linguistic system and secondly (in fact simultaneously) a classification scheme. Science is "a mental construct, by means of which a collection of objective data is arranged in a model and expressed linguistically for certain ends." (Dingle, 1938). The field of Statistics, a tool of Science, has been described as "a method of research which tends to clarify the systematic collection of quantative observations by the comparison of numerous typical groups." (Kaufmann, 1913).

Statisticians have two roles to play in the science of software. First, in their traditional role of arbiters of evidential reasoning in science, they can assist in developing appropriate quantitative measures, in the collection and exploratory analysis of data, in the building of mathematical models, and in the design and analysis of experiments to test the reliability of the software. Second, in a less accustomed role of being directly interested in the specific content and results of the research, in that they are either potential users of the results or have a professional concern in the implications of the results, statisticians should insist on evaluation criteria for statistical software appropriate to their priorities. The term "computer performance evaluation" has been used to measure completion time of workloads in certain machines, which is of paramount importance to computer scientists. But to users, "performance" means more: the user is also interested in the nature of the output from a software system in response to the practical problem facing the user. He is interested in the content and the form of the output, that is, the accuracy and usability.

Measurement and Classification

To illustrate the use of simple quantitative measures to describe and classify software, consider a classification system for statistical software founded on a "life-cycle" model of a complete statistical analysis. The stages in this life-cycle are listed in Table 1.

Table 1: Stages in the Life-Cycle of a Complete Statistical Analysis

1. Data acquisition
2. File building
3. Editing
4. Data description
5. Data exploration
6. Model building

In order to investigate the state of existing statistical software, a questionnaire was prepared consisting of fifty-five questions to probe the capabilities of a statistical program in each of these stages. The answers to each question were on a simple four-point scale reflecting the usefulness of the program in the particular area of that question. The rating scheme is described in Table 2. Examples of what was meant by "complete coverage" of three statistical characteristics are shown in Table 3. Each question on the questionnaire had similarly detailed instructions. For the more qualitative questions regarding portability, ease of use, and maintenance, the usefulness ratings in the first column of Table 2 were explicated: for example, question 55 and 56 regarding the desirable statistical and computer training a user needs, a "3" rating indicated that very little training was required (and therefore the program has a high "usefulness" for many users), while a "0" rating indicated that at least a specialized bachelor's degree was needed.

This questionnaire was sent to the developers and users of over a hundred statistical programs that are widely distributed for general use in statistical analysis. The complete results of this survey are published in Francis and Wood (1981). A portion of the results for twenty-one programs rated on twenty of the questions is shown in Figure 1. On this figure appears the developer's rating of his own program on each of the twenty items, as well as the average rating of several users of that program on each of these items.

For each program the developer's rating is plotted, the height of each point from the lower edge of each box representing a number from 0 to 3, and each of these developer's points is connected by a continuous line to assist the reader's eye. For example, for SPSS the developer's first two ratings on questions 14 and 11 are 2 and 1, followed by ratings of 2, 1, 2, 1, 2, 0, 3, 0, etc. In addition the user's average rating is plotted on the same grid. If the users' rating differs from the corresponding developer's rating, a vertical line is drawn connecting the two points: if the users' rating is higher than the developer's rating, the line is thin; if the users' rating is lower than the developer's rating, the line is thick. Thus a thick line indicates that the developer has over-rated his package compared with the users. It is therefore easy to see, for example, that developers and users of SPSS agree rather well, except that the developer over-rated on question 55: the developer feels that no training in statistics is needed for a user to be an effective user, but all users agree that a bachelor's degree in statistics is needed. Likewise it is easy to see that the developer of EASYTRIEVE has a higher opinion of this program than do the users.

Table 2: Numerical Usefulness Ratings For Characteristics of Software

Rating	Meaning for Statistical Characteristics
0 -- Low	No facilities in this area, or not an intended purpose.
1 -- Modest	A few functions in this area, or a minor purpose or byproduct.
2 -- Moderate	Moderate capabilities, or a significant purpose.
3 -- High	Complete coverage of all aspects in this area, or one of the principal purposes.

Table 3: Examples of Characteristics of Statistical Software

11 Ability to Define and File Complex Data Structures: Hierarchies, matrices, vectors, tables, variable by case or case by variable.

24 Power and Flexibility in Computing Tables: Ability to recode variables, produce multi-way hierarchical tables, weighted summary tables, permit user control of arithmetic operations on the table.

30 Multiple Regression: Stepwise regression, all possible regressions, ridge regression, wide variety of residual plots, summary statistics (Durbin-Watson, etc.), regression through the origin.

The users in this survey were suggested by the developers: each developer was asked to supply a list of three users who might be willing to give their assessment of the package's capabilities. These users should (1) not be at the institution that supports the package, (2) not have been connected with the development of the package, and (3) be familiar with at least one other statistical package so that they would have some basis for comparison.

Since it was possible that users who were suggested by a developer might be slightly biassed in favor of the developer, an experiment was conducted to compare the ratings of these users with those of users who were selected independently. No systematic bias was found, and in fact these developer-suggested users were seen to be more reliable.

It is not possible to tell whether the discrepancies in Figure 1 between the two sets of ratings are due to the users' being unaware of the capabilities, which may possibly exist as unpublished options, or whether the developers are misrepresenting these capabilities. Also it is possible that the wording of some questions was not entirely clear or, even if it was clear, that it was not understood.

However, redundant questions were included in the questionnaire, and those questions of doubtful reliability were excluded from the analysis. In addition, each user was asked to attach to each rating a "confidence" score (0 = don't know, 1 = low, 2 = medium, 3 = high) reflecting the user's confidence that this rating was accurate. Only those ratings with confidence scores of 2 or better were used. Finally, users' responses to availability items, such as 54 and 50, were not recorded, the reason being that few users would have this information.

The Classification System

Apart from the questions related to the user interface -- portability, usability, maintenance-- which appeared at the end of the questionnaire, all other questions up to question 48 concerned statistical capabilities, and were in the order reflecting the "life-cycle" in Table 1. Using the developers' ratings on fourteen of these capability items, we ordered the programs and re-ordered the items so that the high ratings would cluster down the diagonal of this 21 by 14 matrix (programs by items).

Thus the first two programs in Figure 1, namely ADABAS and SIR, are rated (by their developers) very high on the first two questions, numbered 14 and 11. The next four programs are rated high on the next two questions numbered 24 and 25. The next six programs rate quite high on the first four questions, but also have some capabilities on the next six questions. The next five questions rate uniformly high on the six questions numbered 30 through 33 and also on some of the last four questions 35 through 47. Finally,

Figure 1: STATISTICAL SOFTWARE: A COMPARISON OF DEVELOPERS' AND USERS' RATINGS

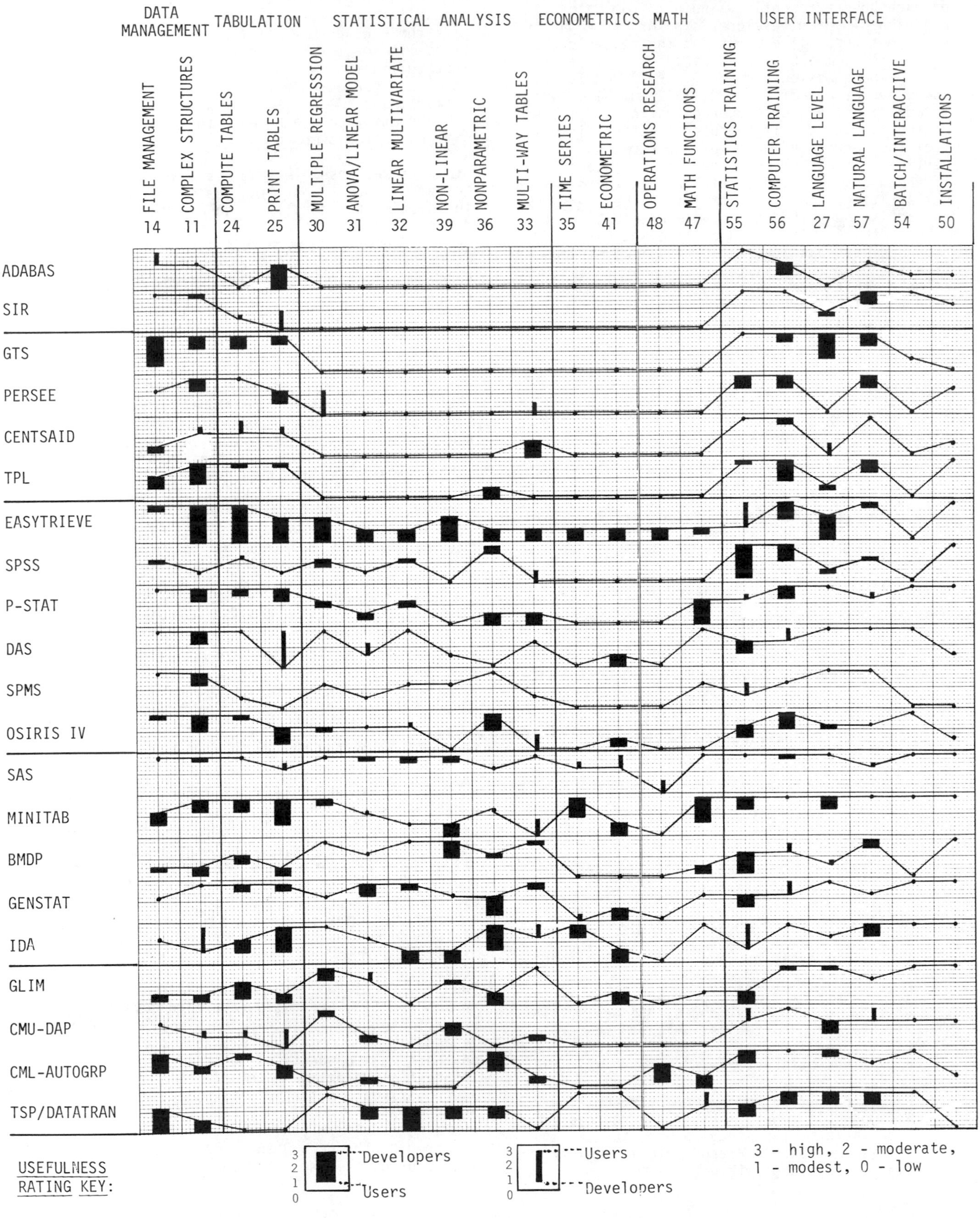

the last four programs are relatively weak on the first four questions and have occasional high ratings on the next ten questions, from 30 to 47.

Given the ordering of the questions according to the "life-cycle" of Table 2, and given the clustering of high capability ratings down the diagonal, it follows that the resultant ordering of the programs also reflects the life-cycle model: the first two programs are for data management, the next four are primarily for tabulation, the next six, since they are strong in filing and tabulation, but also have some statistical analysis capabilities, are broad survey analysis packages. The next five are general-purpose statistical analysis systems, some of which also have filing and tabulation capabilities as well as mathematical functions available. The last four are special-purpose statistical programs.

In Figure 1 we can see that even the course quantitative measures defined in Tables 2 and 3 can enable a potential user to begin to comprehend the usefulness of this statistical software.

Evaluation and Improvement

Figure 1, and its larger counterpart in Francis and Wood (1981), sets the stage for more in-depth evaluations. Some evaluations which compare certain statistical aspects of a few programs have been conducted: Chambers (1973) -- least squares algorithms; Francis and Sedransk (1979) -- analyzing survey data; Heiberger (1976) -- analysis of variance; Ling (1974) -- algorithms for means and variances; Searle (1979) -- unbalanced analysis of variance; Velleman, Seaman and Allen (1979) -- multiple regression; and many others listed in the bibliography of Francis (1979).

Quantitative measures of convenience and usefulness have also been used in comparing statistical systems. For example, Bryce and Hilton (1975) conducted an experiment to compare the difficulty of installation of a number of statistical packages. They measured the effort of three systems programmers in installing each of three packages, using a Latin square experimental design. Francis and Valliant (1976) describe an experiment to compare the ease of use of two packages for novices, using a simple crossover design. Thisted (1976) performed a further experiment to assess the adequacy of user documentation and control languages of three packages for an audience with some computer experience and statistical training. Again, he used a crossover design.

These evaluations are difficult and often expensive, and so they can examine only a few features of the software; furthermore, the software is constantly changing and so reviews are soon out of date. One of the reasons for changes in the software is that evaluations and reviews frequently disclose errors or shortcomings.

To calibrate the accuracy of an instrument a scientist needs:

a.) to characterize quantitatively the input signals over some specified range,

and

b.) to measure quantitatively the response for signals in that range.

In terms of the use of software, these translate into:

a.) to specify and characterize quantitatively the difficulty of the problem to be solved,

and

b.) to measure quantitatively the response of the software, that is,

i) to measure quantitatively the accuracy of computed solutions, and

ii) to assess usability of the output.

A series of papers on the accuracy of programs for multiple regression provides examples of these components of a calibration of software. Longley (1967) compared the performance of several, widely-used multiple regression programs by submitting a single test problem to each program, and qualitatively comparing their computed solutions with an extremely precise solution which he obtained using special subroutines on an extended precision machine. Chambers (1973), in comparing the accuracy of various algorithms for regression calculations, used the "minimum number of significant figures" in regression coefficients that were in agreement with those of a "best" program as a quantitative measure of accuracy of an algorithm.

Beaton, Rubin, and Barone (1976), however, showed that Longley's precise solution was not the best solution for that test problem, and they defined a "perturbation index" as a quantitative measure of the difficulty of the problem to be solved, which incorporated the variance of the data and the word length of the machine. Velleman and Francis (1975) argued that Chambers' measure was inadequate and proposed other quantitative measures of accuracy.

Velleman, Seaman and Allen (1977) used these measures of difficulty of problems and these measures of accuracy of solutions in an experiment which measured the accuracy of several programs as a function of the difficulty of a sequence of problems. Thus it was possible to plot the "response curve" for a program, the accuracy of its computations for problems of increasing difficulty.

We have seen, therefore, that this experiment did not just happen. It evolved over ten years during which time different quantitative measures were tried and found wanting.

From such response curves it is possible to

propose an index of merit to evaluate performance of computing systems, incorporating difficulty, accuracy, cost and convenience.

There is a need for well-designed experiments to evaluate certain features of software, particularly experiments which compare at least two software systems. There is a need for these experiments to be replicated under different environments.

To capture the many dimensions of problem difficulty, accuracy of solutions, and usefulness of output by a few quantitative measures presents challenges both applied and theoretical. To carry out the many and continuing experiments necessary to monitor the quality of software will require the contributions of many researchers.

Finally, there is a need for standards of software performance expressed in quantitative terms. How these standards should be formulated is debatable, but meanwhile, in the absence of articulated standards, the results of published comparative experiments using quantitative measures as advocated in this paper do provide some *de facto* standards, since the best performances give users some indication of what they can expect from new programs. These *de facto* standards will evolve over time as the state of the art improves.

REFERENCES

Beaton, A. E., D. B. Rubin and J. L. Barone. (1976) "The Acceptability of Regression Solutions: Another Look at Computational Accuracy," *Journal of the American Statistical Association*, 71, 353, 158-168.

Bryce, G. R. and H. G. Hilton. (1975) "Local Installation of Packages," *Proceedings of the Statistical Computing Section*, American Statistical Association, 13-15.

Chambers, J. M. (1973) "Linear Regression Computations: Some Numerical Statistical Aspects," *Bulletin of the International Statistical Institute*, 45, 4, 255-267.

Dingle, H. (1938) "The Rational and Empirical Elements in Physics," *Philosophy*, 13, 148-165.

Francis, I., ed. (1979) *A Comparative Review of Statistical Software*. Voorburg, The Netherlands: International Association for Statistical Computing.

Francis, E. and J. Sedransk. (1979) "A Comparison of Software for Processing and Analyzing Surveys," *Bulletin of the International Statistical Institute*, 48.

Francis, I. and R. Valliant. (1976) "The Novice With a Statistical Package: Performance without Competence," *Proceedings of Computer Science and Statistics, 8th Annual Symposium on the Interface*, 110-114.

Francis, I. and L. Wood. (1981) *Statistical Software: A Comparative Review for Developers and Users*. New York: Elseiver-North Holland. (to appear)

Heiberger, R. M. (1976) "A Conceptualization of Experimental Designs and their Specification and Computation with ANOVA Programs," *Proceedings of Statistical Computing Section*, American Statistical Association, 13-14.

Kaufmann, A. (1913) *Theorie und Methoden der Statistik*, Tubingen: J.C.B. Mohr.

Ling, R. F. (1974) "Comparison of Several Algorithms for Computing Sample Means and Variances," *Journal of American Statistical Association*, 69, 348, 859-866.

Longley, J. (1967) "An Appraisal of Least Squares Programs," *Journal of American Statistical Association*, 62, 319, 819-841.

Perlis, A., F. Sayward, M. Shaw, eds. (1980) *Software Metrics: An Analysis and Evaluation*, MIT Press.

Searle, S. (1979) "Deciphering the Output of ANOVA Programs for Unequal-Subclass-Numbers Data Using Benchmark Data Sets," *Proceedings of Computer Science and Statistics, 12th Annual Annual Symposium on the Interface*.

Thisted, R. A. (1976) "User Documentation and Control Language: Evaluation and Comparison of Statistical Computer Packages," *Proceedings of Statistical Computing Section*, American Statistical Association.

Velleman, P. F. and I. Francis. (1975) "Measuring Statistical Accuracy of Statistical Regression Problems," *Proceedings of Computer Science and Statistics, 8th Annual Symposium on the Inferface*, 122-127.

Velleman, P. F., J. R. Seaman, and I. E. Allen. (1977) "Evaluating Package Regression Routines," *Proceedings of Statistical Computing Section*, American Statistical Association, 82-83.

SOFTWARE METRICS: PARADIGMS AND PROCESSES

Marvin Denicoff
Robert Grafton
Office of Naval Research

ABSTRACT

Computer science and software engineering have no precise, well understood, standardized and accepted metrics. In this paper problems in obtaining software metrics will be discussed, and the relationship to statistics will be emphasized. Important advances in software metrics could come about thru the synergystic relationship between statistics and software engineering. A significant challenge to software metrics researchers is to stay within the traditional scientific paradigm of hypothesis, evaluation, criticism and review in the face of intense demands for software metrics.

KEYWORDS: Metrics for software, computer language measurement, software experiments, paradigms for software measurement, comparison of software systems, software data.

"Count what is countable, measure what is measurable, and what is not measurable, make measurable."

Galileo Galilei

The purpose of this paper is to argue for the rational development of indices of merit (i.e. metrics) for use in software engineering. In analogy traditional engineering disciplines are all marked by the availability of precise and well understood parameters of measurement. Numerous examples come readily to mind: pressure, temperature, volts, length, area, and volume. Less well known to the layman are metrics for viscosity and illumination, and parameters of fluid flow such as the Reynolds as Mach numbers. There are literally thousands of metrics available for workers in every scientific and engineering field. These all have elements in common: they are precise, well understood, standardized, and accepted by the scientific communities that use them. Moreover, they have evolved to their present useful state thru a long term process of scientific examination and investigation. Indeed, the process continues as science and engineering advance.

Software engineering and computer science need metrics. By analogy, they would evolve thru a similar process of scientific evaluation and investigation; however, the analogy is incomplete. Metrics for the established science and engineering disciplines are based in the physical sciences while those of computer science, like software itself, will be derived from human ingenuity. Already there is no lack of proposed metrics, and many are being used, out of necessity, without the benefit of a deep understanding of them. Time is not on the side of software metrics either, since the demand is high, thereby inhibiting detailed examination, gradual acceptance and eventual standardization.

SOFTWARE CONCERNS

It is not the purpose here to discuss the crisis in software illustrated by a very recent data point that software costs are growing at the rate of 15% per year while productivity is increasing at less than 3%. As great a concern is the accusation that the growth of the field is proceeding without scientific foundations or underpinnings. The absence of experimentation and measurement has led to doubts about the fundamental designs of software packages. It has raised questions on cost-effectiveness tradeoffs over the whole software life cycle process. Most serious, this lack of a history of metrics and experimentation caters to the accusation that the technological growth of computer software is both undirected and unstructured.

Further evidence of this universal concern with the lack of measurement comes from a GAO report of June 1978 on managing weapons systems. It stated that "there exists no DOD performance criteria to measure software quality and to establish a basis for its acceptance or rejection." The Secretary of Defense's response to this reported deficiency was brief and candid. "We concur. We regret and underscore the importance of the need. The Department of Defense will quickly embrace such measures when they are available."

SOME METRIC CONCERNS

The difficulties inherent in developing software metrics are illustrated by one example. In today's state of the art no one knows an objective measure for comparing two complex software designs or choosing between two or more competing programming languages. The current DOD effort to develop and implement a standard programming language for the tactical arena is an example of this problem. That is, choosing among competing languages. The field lacks a formalism of metrics and experiments; thus such decisions as the choice between declaring an extent tactical language as the standard or designing a completely new language, let alone making evaluations across competing new designs, are resolved by exclusive reliance on "expert judgment." We do not want to deprecate the value of experts in the case of selecting the standard DOD tactical and computer language, now called ADA. In fact, a particular purpose of ONR's interest in metrics is to permit the supplementing of opinion with empirical data.

Continuing the example of evaluating programming languages, one can see a need to develop hypotheses, metrics, and experiments for comparing languages on such features as: portability, maintainability, extensibility, capability for interfacing with other language code, efficiency of object code, ease of learning and training, etc. More difficult than identifying the explicit features upon which programming language comparisons can be made is the need to address such issues as: How one assigns weights to the importance of the individual features; varying the weights in consonance with the requirements of different programming environments; providing an answer to the question of whether one can or should develop a figure of merit scoring approach over all features; assessing the tradeoffs across language properties. Inherent to the language comparison situation just described are some nontrivial research issues: For example; reaching agreement on an appropriate set of metrics, establishing the feasibility of collecting relevant data, exploring methodologies for comparing different measures of the same gross property, coping with a multitude of competing, if not conflicting metrics, and defining an objective function that is specific to the task and that epitomizes the properties of their weights or priorities for the task.

Among the many research needs in software metrics is to determine if the experimental constructs of psychology and statistics are appropriate to software measurement, and if so, to what extent. A useful beginning would be the identification of software hypotheses. The further investigation might show sufficient grounds for establishing a scientific approach to substantiate or disclaim the hypotheses via experiments involving the collection and analysis of software data. Some examples of software hypotheses include: Determining whether programming language X is better than Y for a given set of programming tasks for a designated programming environment; and determining whether software test method X is stronger than Y for a given evaluation objective.

Essential to progress in software metrics is the capability for distinguishing those elements of software which can be measured from those which defy measurement. Few disciplines are able to make precise measurements of large, complex, hetrogeneous systems, consequently we must avoid the trap of insisting on and pursuing only that objective for software.

In addition to the internal concerns of the computer science and software communities with building mechanisms for objective evaluations and comparisons of software designs, software metrics should address an external concern. This is the comparison and evaluation of alternative software systems. Successful metrics work in this area will have a very strong impact on the navy and military decision process by which software is acquired. In a very real sense, we must be able to convince ourselves and others that investments in new software technology really constitutes a significant, measurable improvement.

EXPERIMENTATION

The matter of experimentation is, of course, crucial if we are interested in substantiating or refuting various hypotheses. Factors relevant to this concern with "experimental design" were noted in the discussion of making decisions on the choice across competing programming languages. Several other issues should be, but almost never are taken into account in the kind of primitive evaluation that sometimes passes for formal experiment in the software field. We will take up just a few of them.

To ensure validity of results in an experiment where two or more programming teams are coding up the same application, for example, where the goal is to choose between competing languages, it is necessary to certify that the teams are balanced in terms of the quality and qualifications of their members. It is altogether too easy for an experimenter to let bias creep in thru mismatched groups of programmers.

The matter of choosing "real world applications" on which to conduct software experiments is another issue for research attention. Too often in the past, we have arbitrarily chosen a single application for our experiment without addressing the question of whether that choice accidentally, unconsciously, or worse, deliberately, exactly satisfied our apriori biases on the outcome of the competition. Again, the complex research issue here is building a methodology for choosing that application or set of applications which ideally and fairly exercises all features across all competitors.

Another factor is the degree to which experience and learning contribute to the result when one is conducting experiments which involve assessing a new language. Particularly difficult in the case where a previously programmed application, is to be redone in a new language.

There are, of course many other important concerns; for example the number and duration of the tests, and the costs of the experiment in terms of money and people. This last factor may require a radically new approach to designing large scale software experiments. Statistics could contribute much to the design and analysis of software experiments where data points are severely limited--perhaps to just one.

CONCLUSION

The approach to software metrics must be made in a careful, scientific way marked by the traditional paradigm of hypothesis, evaluation, scientific criticism and review. Progress will most likely be incremental. The search for exact truth may never end and we need to value approximations to truth. An apparent need is a survey of extant measurement methodologies with an evaluation both of domains in which they apply and some insights into their utility for the field of software metrics.

Finally, contributions from statistics, especially in the design and analysis of experiments will aid substantially the long term research effort that will define those metrics and experimental methodologies that will become the basis for a science of computer software. We continue to believe that progress in all the areas embraced by software metrics will inspire the evolution of computer software from an art or engineering form to a science.

A PROPOSAL FOR STRUCTURAL MODELS OF SOFTWARE SYSTEMS

J. C. Browne, The University of Texas at Austin

Abstract

A software system is a collection of information. A structural model for software systems relates the information content of the system to perhaps abstract metrics characterizing the software system. It is proposed that one basis for structural models of software systems is the data models of data base management technology. This paper describes software systems from the viewpoint of defining a structural model as a schema in a data model. A proposed representation for some aspects of software systems is sketched.

Keywords: modeling, software methodology, software metrics, software systems, software evaluation

Models for Software Systems

A science of software systems must eventually develop structural models [BRO81] of software systems and the processes by which they are developed and maintained. This paper proposes that structural models of some aspects of software systems can be based upon the data models of modern data base principles. A software system can certainly be viewed as a structured collection of information. Data models represent the relationships between the entities of a structured collection of information.

This paper briefly discusses models for software systems. A more detailed discussion is given in Browne and Shaw [BRO81]. The representation of the static structure and execution behavior of software systems as data models is proposed.

Models and Abstractions for Software Systems

The first question which must be answered before one can develop models and abstractions of software systems is "what is a software system?" For the purpose of this exposition we include in a software system the requirements analysis, the specifications, the code, the documentation, specification of the execution environment and measurements of the system's execution behavior. It is our proposal that models of software systems should include all of these characteristics and should develop relationships between them and between the metrics chosen for system evaluation.

Figure 1 defines a classical representation of model systems in the physical and behavioral sciences. Let us look for a minute at this Figure. It is typically the case that the development of a model or a theory begins with observations from experiments. These experiments often suggest invariants between the properties or characteristics of the system. An invariant is an abstract relation or constraint involving significant variables at some level of abstraction. The invariants are guidelines toward the development of models which can be evaluated to produce predictions for the results of further experiments. A model must incorporate metrics whose values reflect the properties of interest. A metric for evaluation of software systems is simply a scale for the measurement of some property of interest. A model, therefore, reflects relationships between observables which transform across levels of abstraction to give metrics which are observable in measurement behaviors.

The problem in the application of this paradigm is to find invariants between the observed properties or characteristics of software systems and, having obtained these invariants, to construct models which allow the prediction of significant properties.

Let us apply the paradigm to modeling of the static properties of software systems. Models for static properties will have as input information specifications, code and documentation for the system. The relationships which can be directly specified are structural properties of the code and relationships between the code and its specifications and documentation. From these properties one would like to derive measures for such properties as verifiability, testability, maintainability, etc. It is clear that the difference in level of abstraction between these code relationships and say verifiability is too large to be bridged directly. Therefore intermediate concepts such as "complexity" must be invoked. This discussion highlights properties of a structural model necessary for successful application to static properties of software systems.

° It must support hierarchies of abstraction.

° It must admit of automatable evaluation since the volume of information defining a software system is very large.

The information defining the dynamic behavior of a software system includes the flow of control between program units, the resources consumed to execute a given functionality and specification of the execution environment. It is clear that, for example, the frequency with which a given unit is visited during execution must have an influence on the maintainability of a system. Therefore dynamic behavior information must be a part of the definition of a model which predicts high level qualities such as maintainability. This adds breadth of representational capability for different types of information to the list of requirements for a model representation for software systems.

Current Status of Software System Models

The current status of software system models and the prediction of the development cost and behavior of software systems is largely founded on phenomenological models of the development process and on complexity models of code [WAL77, PUT78]. These phenomenological models, while they have significant agreement among themselves, offer little insight into their structure or their operation. One phenomenological model of the development process [WAL77] predicts a power law behavior relating the effort required in development to the lines of code in the system. Several independent studies [BAS80] have found that the exponent in this equation is close to one. The fundamental question "Why is this the case?" has not, however, been answered.

Complexity models of code structures have also received a great deal of study. These models develop complexity measures based upon control flow, upon operation count and other properties of the program [MCC76, BEL80]. These several models often have substantial agreement among themselves. The several metrics show normally a substantial degree of correlation. However, the same procedure applied at a different level of abstraction often produces totally different evaluations of the same system. Finally, and perhaps most significantly, the complexity models focus only on code, they do not use specification and documentation information which are supposed to be the principal tools for the management of complexity. Curtis [CUR81] and Basili [BAS81] give valuable analyses of the current status of phenomenological models of software systems. The interested reader should consult these reviews.

It is the thesis of this paper that the essential need is to be able to evaluate software systems and to have <u>structural</u> models for software systems. Structural models aid in the development of insights, they provide understanding of <u>why</u> something happens and they are guidelines for the design of significant experiments.

Software Systems as Data Bases

It is clear that software systems can be viewed as information objects. It is therefore appropriate to consider modeling software systems with the principles for structuring general information systems; the tools and techniques of data modeling. Familiarity with data modeling principles must be assumed. There are several standard texts to which reference can be taken [DAT81, MAR77, ULL80].

Let us examine the general characteristics of data modeling by taking a typical business data management application, a personnel data base. The basic entities in the data base include the names of employees. The attributes of the employees include salaries, degrees, performance ratings, etc. There are abstractions relating the entities, departments, projects, etc. There are further relationships between entities and abstractions, and abstractions and abstractions. For example, "manages", or "is managed by".

These concepts, data, attributes, abstractions and relationships are defined in a schema defined in a data model. The questions which are to be answered by this data base are the metrics of the data base. In a personnel system a metric might be average salary for a given department or average salary for degrees and years of experience. The abstractions and relationships are defined so as to support answers to the questions (evaluation of the metrics).

Let us know look at a program as a data object. A static program consists of specifications, code and documentation. The requirement is to define entities, attributes, abstractions and relationships among the objects of a program which will lead to the ability to formulate invariants and to evaluate metrics against this data base.

Entities may appear at several levels of abstraction. They may include operations, lines of code, procedures or modules. The attributes may include size, control flow, numbers of operands, etc. Abstractions are built upon the entities just as in the personnel data base. Procedures, modules, algorithms and programs may also represent abstractions. The relationships among program units include "in", "uses", "invoked by", etc. Queries upon or metrics of this data base might include connectivity, volume of parameter flow, etc. These metrics must then be related to general characteristics such as maintainability and understandability.

Invariants would be the relationships between metrics such as complexity and lower level objects such as operand counts, etc. The invariants are probably defined with respect to error ranges.

The key question to be asked is "Is there a data model which expresses all of the needed relationships between the objects of software systems?" Experience suggests that a directed graph data model will provide an adequate representation of a program as an information structure.

A number of measurements are needed to help resolve this context. For example, it is important to determine the attributes appropriately by levels of abstraction, i.e., to resolve some inconsistencies found between complexity measurements at different levels of abstraction in previous studies. It is important also to determine empirically the relationships between entities and abstractions.

It is worth noting that the graph model data base as a system structure meets some of the important requirements for model structures. They are hierarchic with respect to abstraction definition and they support automatable evaluation.

The graph oriented data model includes entity-attribute sets as nodes and relationships as nodes or arcs. Metrics are correlations between the entities and abstractions.

We thus suggest that a data base approach to programs as information objects allows the definition of structured models, suggests experiments and simplifies and automates routine analysis.

Dynamic Behavior Models and an Integrated Model Structure for Software Systems

The requirement for models of execution behavior is for representation of the flow of control, the flow of data and the mapping of logical functionality to physical resources. Most common models of program execution, directed graphs, queueing network models and Petri-Nets, can all be formulated in the same data model proposed for static program properties.

Figure 2 illustrates the relationships between the several component models of software systems. Data model technology appears to provide a basis for a unified representation of static properties and dynamic behavior of software systems. This promise needs exploration. The challenging problems of mapping the intermediate level metrics of complexity, etc. remains almost untouched.

References

[BAS81] V. R. Basili, "Resource Models", in Software Metrics and Evaluation, edited by A. Perlis, F. Sayward and M. Shaw (MIT Press, Cambridge, 1981).

[BEL80] L. A. Belady, "Complexity of Programming: A Brief Summary", Proc. Workshop on Quantitative Models of Software, Reliability, Complexity and Cost (New York, 1980).

[BRO81] J. C. Browne and M. Shaw, "Towards a Scientific Basis for Software Evaluation", in Software Metrics and Evaluation, edited by A. Perlis, F. Sayward and M. Shaw (MIT Press, Cambridge, 1981).

[CUR81] W. Curtis, "The Measurement of Software Quality and Complexity", in Software Metrics and Evaluation, edited by A. Perlis, F. Sayward and M. Shaw (MIT Press, Cambridge, 1981).

[DAT81] C. J. Date, An Introduction to Database Systems (3rd Ed.), Addison-Wesley, Reading, 1981.

[MAR77] J. Martin, Computer Data Base Organization (2nd Ed.), Prentice-Hall, Englewood Cliffs, 1977.

[MCC76] J. J. McCall, "A Complexity Measure", IEEETSE 2 (1976).

[PUT78] L. H. Putnam, "A General Empirical Solution to the Macro Software Sizing and Estimating Problem", IEEETSE 4, 345 (1978).

[ULL80] J. D. Ullman, Principles of Database Systems, (Computer Science Press, Potomac, 1980).

[WAL77] C. E. Walston and C. P. Felix, "A Method of Programming Measurement and Evaluation", IBM Systems Journal 16, 1 (1977).

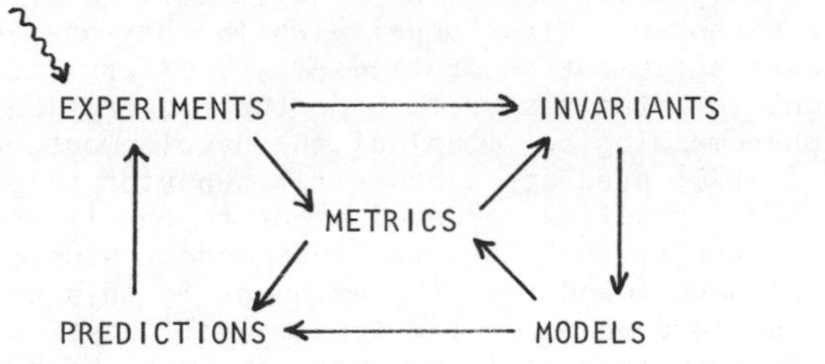

Figure 1: A SCHEMATIC OF THE SCIENTIFIC METHOD

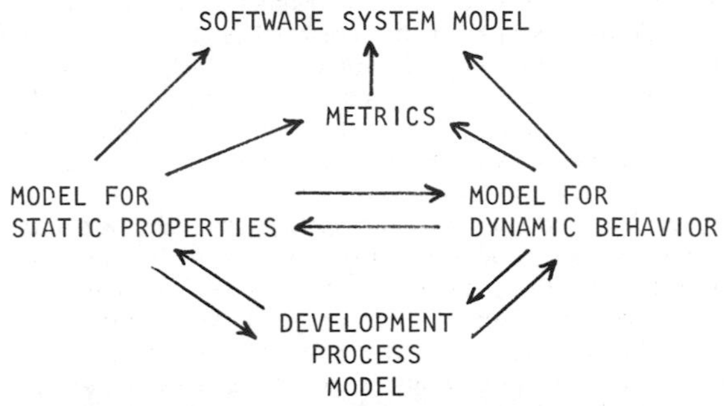

Figure 2: A SCHEMATIC OF COMPONENT MODEL RELATIONSHIPS FOR SOFTWARE SYSTEMS

SOFTWARE METRICS: A KEY TO IMPROVED SOFTWARE DEVELOPMENT MANAGEMENT

J. E. Gaffney, Jr., IBM, Federal Systems Division
Manassas, Virginia

ABSTRACT

This paper describes some of the potential for applying software metrics to the management of the software development process. It also considers some of the practical difficulties one typically faces in evolving and validating a software metric. One difficulty is the collection of baseline data in the real world of software production in which controlled experiments typically are not possible. The results of some recent quantitative 'metrics' investigations are presented and their practical implications for software estimation and control are cited. These investigations are thought to be representative of the process of evaluating software data not obtained under 'controlled' conditions such as is typically the situation in the natural science laboratory.

KEYWORDS

1. Software Metric
2. Software Volume
3. Software Potential Volume
4. Software Size Estimation
5. Conditional Jumps

INTRODUCTION

This paper describes two sets of experimental investigations in software science. They are illustrative of the process occurring at a number of universities and industrial laboratories to better understand the nature of the software development process and the software product it produces. The objective of an increased understanding is severalfold: to be aware of limitations inherent in the process and to be better able to predict and control the nature of that process and the resultant product in terms of: producing a software product that meets the user's requirements within prescribed budgetary and schedule objectives. Basically, quantitative measures of software, or 'metrics' are being developed in response to the observation, "if you can't measure it, you can't manage it."

The first investigation describes the potential application of the Halstead Software size equation(1) as the basis for estimating the amount of code to be written (program code size) based on several parameters discernable relatively early in the design process. A software size estimation algorithm using actual data about the function to be programmed, even if preliminary, has the potential for greater accuracy than an 'educated guess' as is frequently used at present. The second investigation considers the degree to which the number of conditional jumps in code is a function of the nature of the requirement the code is to satisfy and therefore is not controllable by the implementor. The number of conditional jumps is a significant metric because it has been found to relate to software development productivity(2),(3) and to the amount of testing to which a software product should be subjected(4). Hence, having some idea of the degree to which this factor may be controllable by the software developer is important. Both of these investigations used data from several major software products developed for the AN/UYK-7 military computer.

SOFTWARE CODE SIZE ANALYSIS AND ESTIMATION

This section describes the application of the Halstead program length equation(1) to modules from three software products. It is observed that this equation, although originally derived to apply to the implementation of small algorithms only, applies quite well to much larger aggregations of code which contain a multiplicity of algorithms. A procedure for estimating the amount of code required to implement a stated requirement is then derived from this equation. This procedure would satisfy an important need, that of accurately estimating the amount of code required to implement a stated requirement.

Halstead states that the number of tokens or symbols, N, constituting a program is a function of n_1, the operator vocabulary size and n_2, the operand vocabulary size. The software length equation is:

$$N = n_1 \log_2 n_1 + n_2 \log_2 n_2.$$

Actually, this formula is meant to apply to a small program (or one procedure of a large program) or function. It was derived to apply to the program expression of an algorithm. Thus, the number of tokens in a program consisting of a multiplicity of functions or procedures would be best found by applying the size equation to each function or procedure individually, and then summing up the results.

The Halstead length equation was applied to data from 14 modules of AN/UYK-7 code, using the interpretation of n_1 equals the number of unique instruction types (out of 144 possible) and n_2 equals the number of labels used.* Note that a

---

* The counting rules for obtaining n_1 and n_2 used here differed slightly from those suggested by some examples given by Halstead. Halstead counted each unique conditional jump-plus address combination as a unique operator. The figures for n_1 and n_2 used in the present paper, however, consider each unique conditional jump type as an operator type. The use of the same conditional jump

type plus a different address did not count as a unique operator type.

---

slightly more accurate approximation of n_2 would be to include the effect of fixed (or hardware) addressable registers and index registers. A regression was made on 14 modules for I, the number of instructions, against N, the number of tokens, given by the software length equation and then a value for the coefficient 'b' in the equation 'I = bN' was calculated. The coefficient value determined was b = .478. The correlation between $\hat{I}$ and I was found to be .91557. The data used is given in Table 1 and the corresponding plot of $\hat{I}$ versus I is shown in Figure 1. $\hat{I}$ and I were individually correlated with percent relative error $(\frac{\hat{I}-I}{I} \times 100)$. The correlation (r) in the first case was -.3764 and -.3163 in the second. These figures indicate a slight correlation between larger module size and smaller relative error.

The figure of .478 for multiplying N, the number of tokens, to obtain I, the number of instructions, is rather close to what one might expect for the ratio of the number of instructions to tokens for single address object code. Excluding instruction modifiers, index register indicators, etc., that ratio is .5 to .1.

The data above suggests a very important finding, that the Halstead sizing equation is extremely 'robust.' Also, it may apply to assemblages of code that are considerably more complex than 'procedures' that implement single 'algorithms' as was originally intended.

The Halstead size equation appears promising as the basis for a procedure for estimating the number of instructions, I, required to implement a procedure, or a module, and hence a software system consisting of a number (perhaps large) of such procedures or modules. Consider the size equation (multiplied by .478) to convert N to I = $.478\,(n_1 \log_2 n_1 + n_2 \log_2 n_2)$. An approach to code size estimation based on the work already done would be to:

° Use a 'standard' estimate for n_1, the size of the instruction repertoire. This was found to be 50.53 for the software about which data is provided here. Values for n_1 corresponding to the software type (e.g., operational software, fault location software, etc.) would probably be used in a practical estimation procedure.

° Estimate, n_2, the operand (label) vocabulary size based on estimating the number of conceptually unique inputs and outputs (n_2*) from the top level design and then estimating n_2 from this figure according to the equation, $n_2 = bn_2^*$. The factor b was estimated for the software products considered here to be 6.85.

Thus, an equation for estimating the number of instructions required to implement a procedure might be:

$$\hat{I} = .478(50.53 \log_2 50.53 + 6.85n_2^* \log_2 6.85n_2^*)$$

or

$$\hat{I} = 136.68 + 3.27n_2^* \log_2 6.85n_2^*,$$

where n_2* is the conceptually unique number of inputs and outputs for the procedure of interest.

As estimate for the overall software system would be obtained by summing the estimates for the procedures or modules constituting it. The figures for n_2* should be relatively easy to determine fairly early in the design cycle. For example, one could decompose an overall functional requirement into smaller elements. Then, represent the elements graphically as boxes and link the boxes with 'data lines' representative of the information that would be passed among them. The figures for n_2* would be obtained by counting these lines. Figure 2 depicts the process of hierarchically decomposing an overall requirement into smaller elements. The figure also shows how these elements might be presented in terms of the data flow among them. The decomposition methodology illustrated is an application of structured programming. Structured programming may be defined as the "... application of a basic problem decomposition method to establish a manageable hierarchical problem structure."[5] In practice, the n_2* figures might be obtained with the aid of a tool such as PSL/PSA ("Problem Statement Language/Problem Statement Analysis"), developed at the University of Michigan.[6]

An approach similar to estimating n_2*, as described above, in which n_2 would be estimated (equal to the number of internal variables in the procedure), has been suggested by Fitsos and Smith.[7,8]

CONTROLLABILITY OF CONDITIONAL JUMPS

Overall software development productivity has been found to be inversely proportional to the number of conditional jumps in a program.[2] Hence, an important question is the degree to which the number of conditional jumps is predictable and/or controllable. This prompted the investigation, described in this section, of the degree to which the number of conditional jumps in a body of software is a function of the nature of the implementation done by the programmer as distinguished from the requirement that the software is to effect. The data suggests that the number of conditional jumps is more a function of the requirement than the implementation.

A number of investigators have found relationships between the number of conditional jumps and productivity and the amount of testing to be done on a program. Chen determined the existence of a measure he calls the 'Control Structural Entropy' which is a function of the number of "IF's" in the program as well as probabilities of certain serial "IF" - structures.[2] He found that productivity is related to this measure. Gaffney found a strong negative correlation (-.77) between software development productivity and the proportion of conditional jumps in the code[3]. McCabe demonstrated that the number of basic paths of a program is related to the number of conditional jumps in its control structure[4]. The 'basic paths' are those paths from which any path through the programs can be structured. The size of the 'basic set' is called the 'cyclomatic number,' C. Paige[9] states that in structured programs, C is equal to the number of binary decisions (or conditional jumps) plus one. The number of basic paths is a very important figure because the amount of testing required for a program is related to number of independent paths in its structure.

The investigation involved the calculation of two metrics that represent the sizes of the requirement the program is to satisfy and of its implementation, respectively. The investigation included correlating the values of these metrics with the number of conditional jumps as a way of estimating the degree to which these metrics and the number of conditional jumps are related. The two metrics, 'potential volume' and 'volume,' were created by the late M. Halstead[1]. The 'potential volume' is a measure of the smallest size a software implementation of an algorithm could require (or the requirement). The 'volume' is a measure of the actual size of the program.

If the number of conceptually unique inputs and outputs is n_2^*, then, the 'potential volume' is:

$$V^* = (2+n_2^*) \log_2 (2+n_2^*), \text{ bits.}$$

Assume that a program consists of a sequence of characters or 'tokens.' For example, the instruction "Add A" is comprised by two tokens, "Add" and "A", where "Add" is an 'operator' and "A", is an "operand." Suppose there are 'N' tokens (or operators and operands) in the program. Let each one of them be selected from a set of N operators and operands. Then, the 'volume' is:

$$V = N \log_2 n, \text{ bits.}$$

Then, V measures the number of binary decisions made by the programmer in creating the stream of tokens which is his program, whereas V*, the 'potential volume', is a measure of the least number of such decisions that could be required in effecting the algorithm of interest and may be taken as a measure of the basic requirement that the program is to be designed to satisfy.

V* can be estimated as L*V, where L, is the program level. An approximation to L is $\frac{2}{n_1} \times \frac{n_2}{N_1}$, where n_1 is the operator vocabulary size, n_2 is the operand vocabulary size, and N_2 is the total number of operands used. V* was computed as equal to L*V for the operations described in the remainder of this section, principally because of the ease of obtaining n_1, n_2, and N_2, as compared to n_2^*.

The 'volume' and 'potential volume' are not orthogonal (independent) measures. Logically, the 'volume' must be a function of both the 'nature of the requirement' and of the 'implementation.' Thus, the 'volume' must be a function of the 'potential volume.' The 'potential volume' and the 'volume' were calculated for 25 modules from three large software products which are implemented on the AN/UYK-7 military computer. Two of these products are for operational sonar systems and the third is for a sonar trainer. The 'volume' and 'potential volume' measures were found to have a correlation of .7098 (or 50.4% of the variability of 'volume' is 'explainable' by the variability of 'potential volume'). Clearly, the two volume measures are not orthogonal (independent). Thus, the software engineer does not have complete latitude in the nature of the implementation of a requirement. The numerical results confirm what many 'know' qualitatively.

Regressions were made of the conditional jumps against 'potential volume' and 'volume,' jointly and independently. The corresponding correlations, multiple and single, were determined. The correlations are given in Table 2. The number of conditional jumps and the calculated values of 'potential volume' and 'volume' for which the correlations were computed are given in Table 3. Figures 3 and 4, respectively, provide plots of potential volume and volume versus number of conditional jumps.

Several important inferences can be made from these data, which of course, may not apply in all environments. They are:

1. The number of conditional jumps (and hence of the attributes such as amount of testing required) is more dependent on the nature of the problem (which is represented by the potential volume metric) than on the implementation approach. According to the data in Table 2, the measures representative of both the requirement and the implementation account for 58.8% of the variability experienced in the number of conditional jumps, and the measure representative of the requirement alone accounts for 39.9% of the variability. Hence, roughly 18.9% (58.8-39.9) of the variability in the number of conditional jumps is attributable to the implementation. This gives some idea of the degree of discretion available to the software engineer (at least with respect to the cases reported on here) in his use of conditional jumps, given a certain requirement to implement.

2. The potential volume and volume measures jointly account for a significant amount (about 59%) of the variation of the number of conditional jumps and hence of the amount of testing (at least development or unit testing) to be expected. However, some 41% is not accounted for by them. This indicates that the potential volume and volume are imperfect measures of requirements and implementation for a piece of code.

3. An additional question of dependency was addressed: What is the correlation between the number of conditional jumps and the volume, with potential volume held fixed? The partial correlation between the number of conditional jumps and volume, given potential volume, was calculated.

This figure was determined to be .5601. This indicates that with the requirement (as represented by the 'potential volume') held constant, 31.37% of the variability of the number of jumps is 'explainable' by the volume (or vice versa). This result is not completely 'pure,' because the volume metric reflects the nature of the requrement and the requirement/implementation interaction to some degree as well as the implementation. The 31.37% figure further supports the assertion of result number 2, above, that the requirement (for which the potential volume is the surrogate) has a considerably greater impact on the number of conditional jumps than does the implementation.

SOME COMMENTS ABOUT SOFTWARE SCIENCE EXPERIMENTS

The studies reported on here illustrate the fact that the 'ultimate' metrics have yet to be identified. For example, the 'potential volume' and 'volume' metrics do not fully measure the nature of either the program requirement (or minimum representation of the algorithm(s) it is to effect) or the program implementation.

It is very difficult to collect enough data that is representative of the software product in terms of accuracy. The industrial software development environment does not allow the software science investigator the luxury of controlled experiments. Often, the experimentor must be satisfied with many fewer data points than he would expect to find, say in a medical investigation situation. Yet, the need is there to be able to make quantitative statements about the nature of the software product and the process by which it is developed.

SUMMARY

The application of the Halstead software length equation to code written for the AN/UYK-7 military computer was presented. The application of the Halstead equation as the basis for a software size estimation procedure was described. The relationship between the number of conditional jumps in a program and a measure of the extent of the requirement the code is to satisfy was provided. It was indicated that the number of conditional jumps is more a function of the functional requirement than the variability of the implementation. Some of the difficulties in software metrics experimentation were described.

Bibliography

1. Halstead, M.H., "Elements of Software Science", Elsevier, 1977.
2. Chen, E.T., "Program Complexity and Programmer Productivity," IEEE Transactions on Software Engineering, May, 1978, Pg. 187.
3. Gaffney, J.E., Jr., "Program Control Complexity and Productivity"; IEEE Workshop on Quantitative Software Models for Reliability, Complexity, and Cost; October, 1979; IEEE Catalog #TH0067-9.
4. McCabe, T.H., "A Complexity Measure," IEEE Transactions on Software Engineering, December 1976; pg. 308.
5. Jones, R.W., "Structured Programming", IEEE "Computer", March, 1981; pg. 31.
6. Teichroew, D. and Hershey, E.A., "PSL/PSA: A Computer-Aided Technique for Structured Documentation and Analysis of Information Processing Systems;" IEEE Transactions on Software Engineering, January 1977.
7. Fitsos, G., "Vocabulary Effects in Software Science," Computer Software & Applications Conference; October, 1980; pg. 751; IEEE Catalog #80 CH1607-1.
8. Smith, C.P., "A Software Science Analysis of IBM Programming Products," TR 3.081; January, 1980; IBM, Santa Teresa Laboratory.
9. Paige, M., "A Metric For Software Test Planning," Computer Software and Applications Conference; op. cit., pg. 499.

Table 1. Halstead Size Approximation to I

Product/ Module Number	Actual No. of Instructions (I)	Approximation to I, $\hat{I} = .478(n_1 \log_2 n_1 + n_2 \log_2 n_2)$	Percent Relative Error $= \frac{\hat{I}-I}{I} \times 100$
<u>Product 1</u>			
Module 1.	1188	1017	-14.39
Module 2.	1200	641	-46.58
Module 3.	513	484	- 5.65
Module 4.	813	660	-18.82
Module 5.	1988	1212	-39.03
Module 6.	1143	1736	+51.88
<u>Product 2</u>			
Module 1.	4399	4196	- 4.61
Module 2.	1772	1420	-19.86
Module 3.	1593	1908	+19.77
<u>Product 3</u>			
Module 1.	454	577	+27.09
Module 2.	372	534	+43.55
Module 3.	755	365	-51.66
Module 4.	2566	2047	-20.23
Module 5.	2158	2961	+37.21
	Sum = 20914	Sum = 19758	Avg. = -6.04%

$$\text{OVERALL RELATIVE ERROR} = \frac{19758 - 20914}{20914} \times 100 = 5.53\%$$

Table 2. Conditional Jumps/Potential Volume/Volume Regression

Case	Regression Variable(s)	Correlation Coefficient Multiple/Single	Percent of Variation of Number of Cond. Jumps 'Explained' by Variable(s)
1	Potential Volume and Volume	.76646	58.75
2	Potential Volume	.63154	39.88
3	Volume	.75016	56.27

Table 3. Conditional Jumps/Potential Volume/Volume

Product/ Module Number	Number of Conditional Jumps	Potential Volume	Volume
Product 1			
Module 1.	125	211.0	16664
Module 2.	144	127.4	14108
Module 3.	052	116.5	05602
Module 4.	069	099.2	09322
Module 5.	242	278.2	25519
Module 6.	024	089.3	02211
Module 7.	062	089.7	07513
Product 2			
Module 1.	356	708.5	64353
Module 2.	240	270.5	22908
Module 3.	177	353.5	23376
Module 4.	104	103.7	11358
Module 5.	424	424.7	55377
Module 6.	030	116.4	01449
Product 3			
Module 1.	013	084.2	08189
Module 2.	030	066.4	04835
Module 3.	076	131.5	08650
Module 4.	038	294.0	57930
Module 5.	208	367.0	36531
Module 6.	055	078.9	04934
Module 7.	034	093.1	06066
Module 8.	063	095.9	05112
Module 9.	074	168.8	18364
Module 10.	009	623.1	05967
Module 11.	004	069.6	04436
Module 12.	000	127.4	07914

FIGURE 1

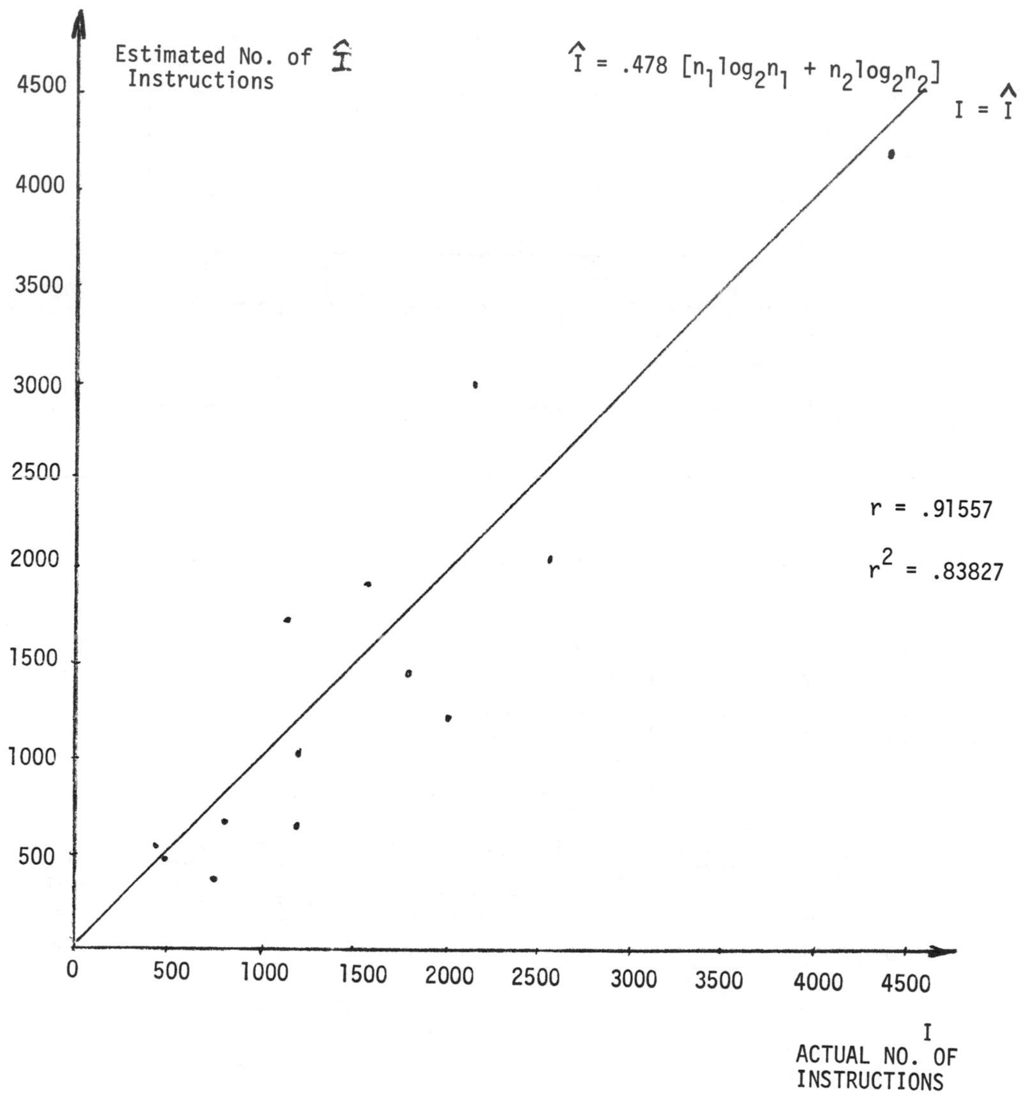
Estimated No. of Instructions Î
Î = .478 [n1 log2 n1 + n2 log2 n2]
I = Î
r = .91557
r2 = .83827
4500
4000
3500
3000
2500
2000
1500
1000
500
0
500
1000
1500
2000
2500
3000
3500
4000
4500
I
ACTUAL NO. OF INSTRUCTIONS

Figure 2

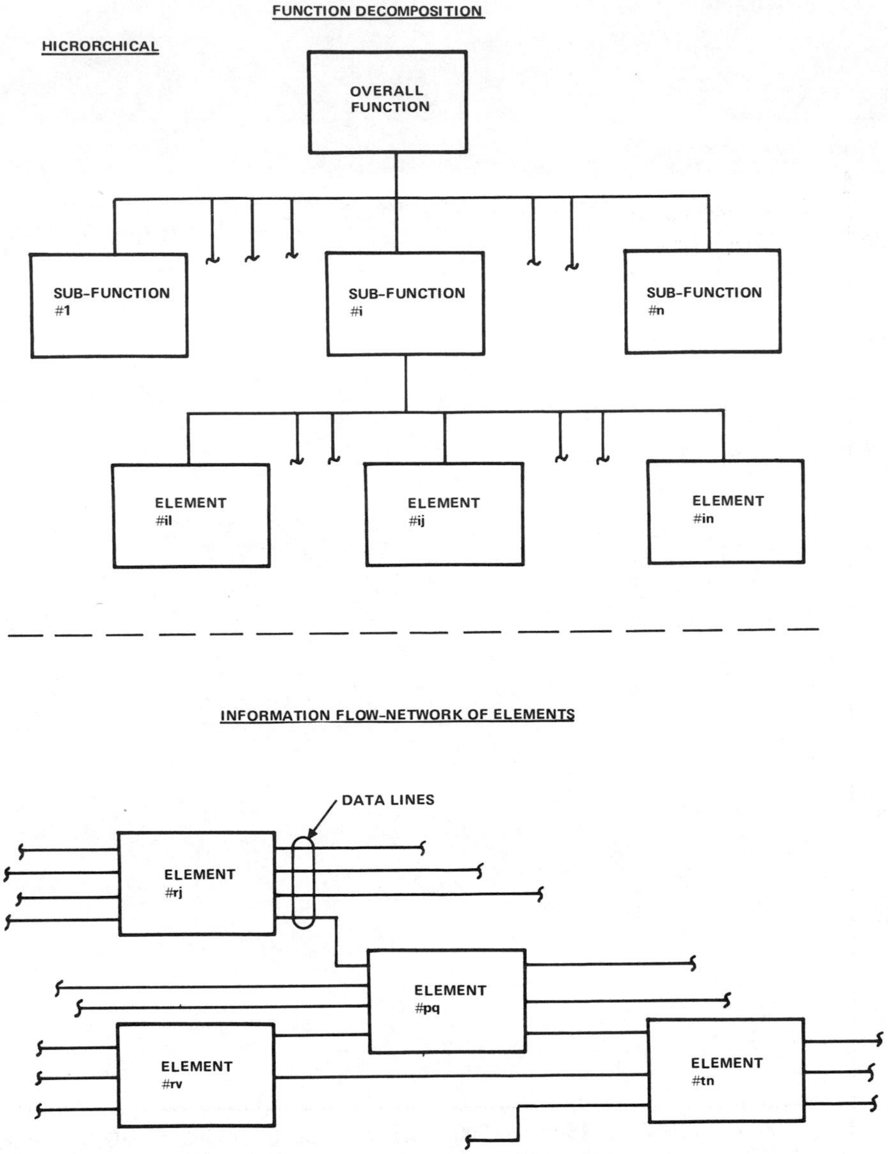

FUNCTION DECOMPOSITION
HICRORCHICAL
OVERALL FUNCTION
SUB-FUNCTION #1
SUB-FUNCTION #i
SUB-FUNCTION #n
ELEMENT #il
ELEMENT #ij
ELEMENT #in
INFORMATION FLOW-NETWORK OF ELEMENTS
DATA LINES
ELEMENT #rj
ELEMENT #pq
ELEMENT #rv
ELEMENT #tn

FIGURE 3

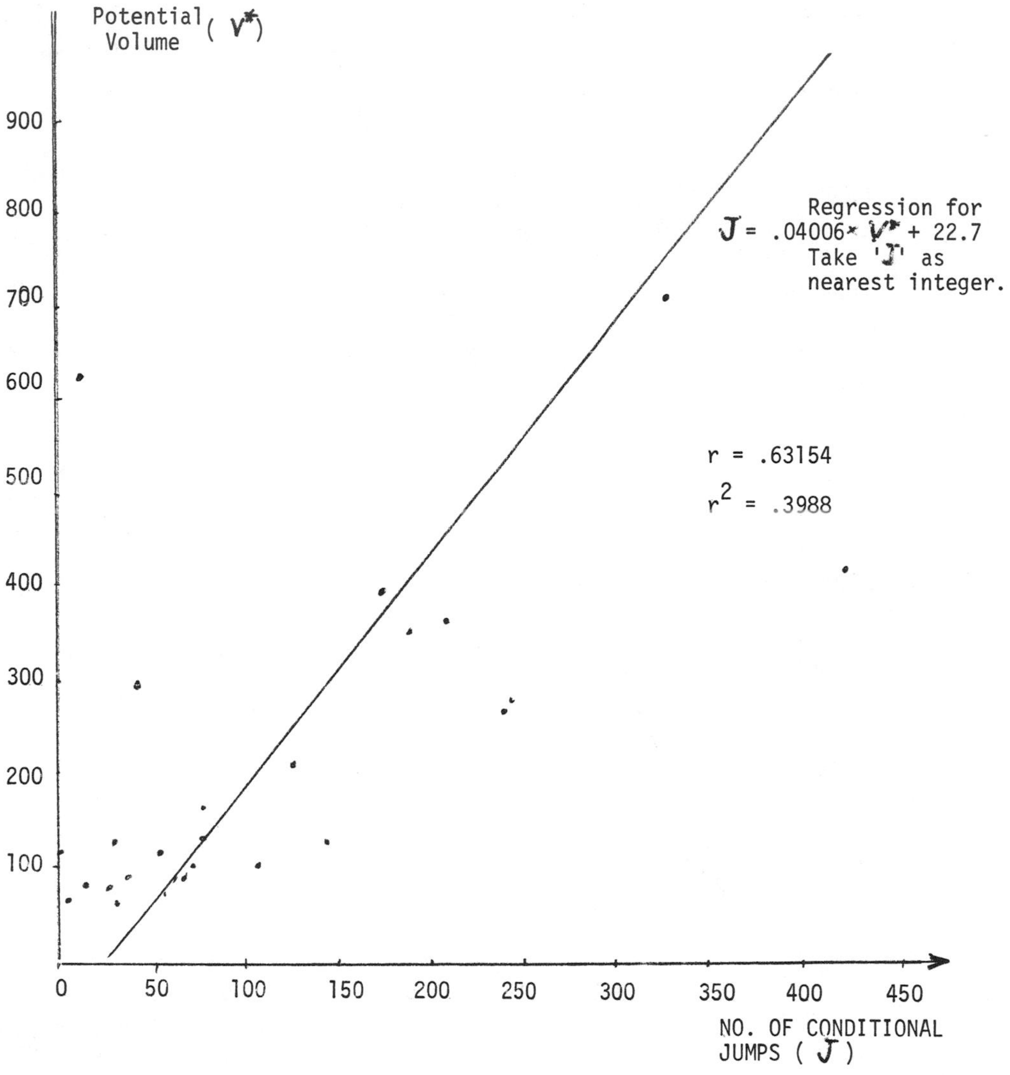

Potential Volume (V*)
900
800
700
600
500
400
300
200
100
0
50
100
150
200
250
300
350
400
450
NO. OF CONDITIONAL JUMPS (J)
Regression for
J = .04006 * V* + 22.7
Take 'J' as
nearest integer.
r = .63154
r2 = .3988

FIGURE 4

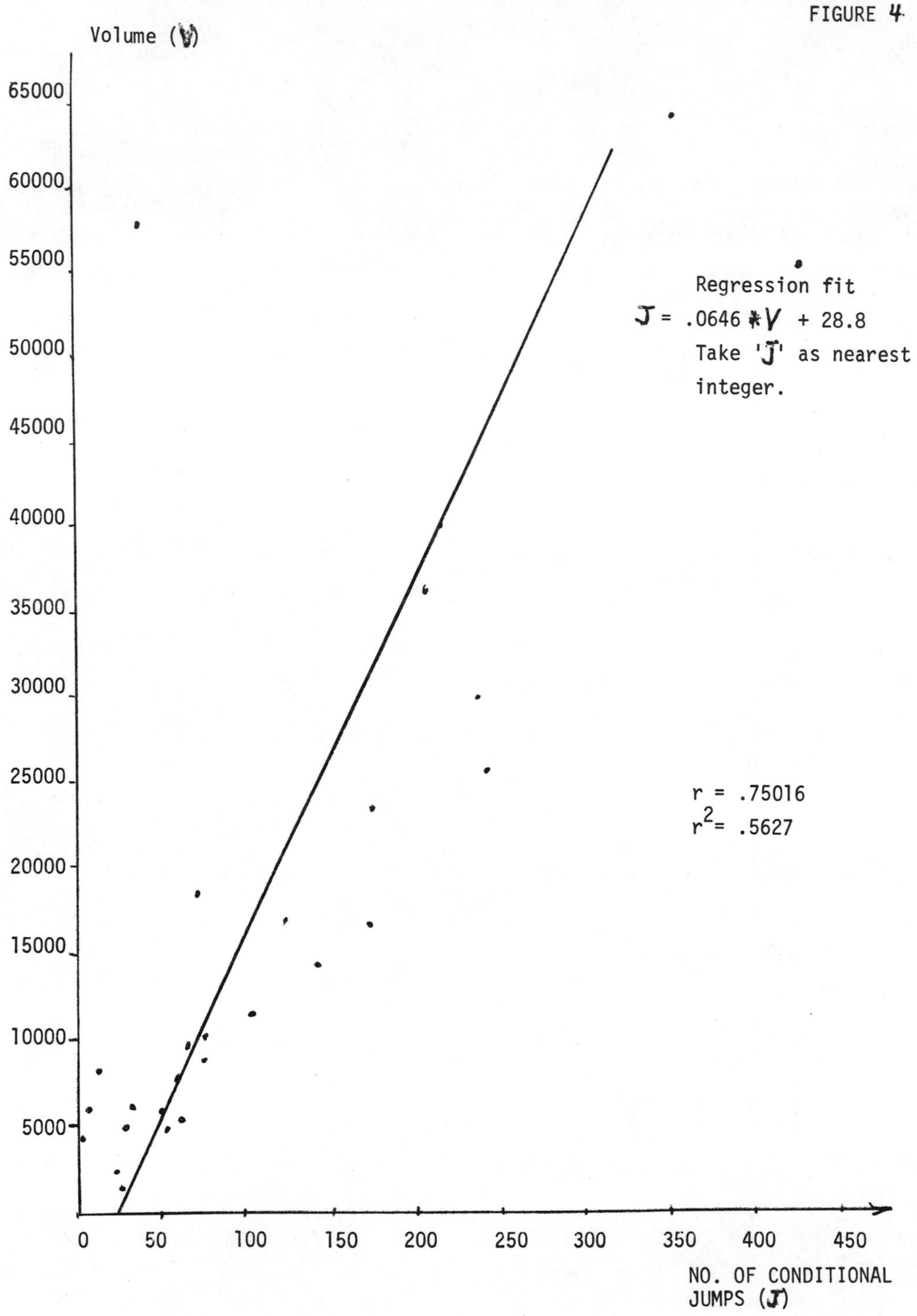
Volume (V)
65000
60000
55000
50000
45000
40000
35000
30000
25000
20000
15000
10000
5000
Regression fit
J = .0646 * V + 28.8
Take 'J' as nearest integer.
r = .75016
r2= .5627
0
50
100
150
200
250
300
350
400
450
NO. OF CONDITIONAL JUMPS (J)

Workshop 10

ORTHOGONALIZATION: AN ALTERNATIVE TO SWEEP

Organizer: David M. Allen, University of Kentucky

Chair: David C. Johnson, University of Kentucky

Invited Presentations:

Orthgonalization-Triangularization Methods in Statistical Computations, Del T. Scott, Brigham Young University, and G. Rex Bryce, Brigham Young University, and David M. Allen, University of Kentucky

ORTHOGONALIZATION-TRIANGULARIZATION METHODS IN STATISTICAL COMPUTATIONS

Del T. Scott, Brigham Young University

G. Rex Bryce, Brigham Young University

David M. Allen, University of Kentucky

ABSTRACT

Procedures for reducing a data matrix to triangular form using orthogonal transformations are presented, e.g., Householder, Givens, and Modified Gram-Schmidt. Using small numerical examples these procedures are compared to procedures operating on normal equations. We show how an analysis of variance can be constructed from the triangular reduction of the data matrix. Procedures for calculating sums of squares, degrees of freedom, and expected mean squares are presented. These procedures apply even with mixed models and missing data. It is demonstrated that all statistics needed for inference on linear combinations of parameters of a linear model may be calculated from the triangular reduction of the data matrix. Also included is a test for estimability. We also demonstrate that if the computations are done properly some inference is warranted even when the X matrix is ill-conditioned.

Keywords: *Orthogonalization, triangular reduction, Householder, Givens, Gram-Schmidt, Cholesky, Analysis of Variance, Expected Mean Squares, Linear Combinations, triangular reduction, estimability.*

1.0 INTRODUCTION

It has long been recognized that the triangularization of the matrix of coefficients of a set of simultaneous equations is a simple yet efficient means of obtaining a solution. Indeed, the author of the Chui-chang suan-shu or Nine Chapters on the Mathematical Art written in 250 BC in China, used the technique although a general systematic method of approach is not given (Boyer, 1968). The first systemization of the technique is apparently due to Gauss [1873].

Further systemization came through those who were using the technique in the practical setting of Geodesy. M. H. Doolittle [1878] published in the U. S. Coast and Geodesic report a method which after some slight procedural changes (Dwyer, 1941) became the abbreviated Doolittle method. Benoit [1924] attributes another method to a Commander Choleski, a commander of artillery in the Geographic Service of the French army who was killed early in the first world war. Variants of this method which has come to be known as the Choleski factorization or the square root method have been independently developed by several others (e.g., Schur 1917 , Banachiewicz 1937, Dwyer 1941). Dwyer [1941] seems to be the first to have supplied a formal proof of the validity of the techniques and indicated their usefulness beyond the solution of normal equations.

The next major developments seem to have been by numerical analysts. Recognizing the numerical inaccuracies inherent in sums of squares and crossproducts they proposed methods of obtaining solutions directly from the original observational equations rather than the more compact normal equations. This required that an orthogonal basis for the matrix of coefficients be found (perhaps only implicitly), along with an upper triangular matrix. Three basic ways of computing this orthogonal-triangular decomposition resulted: elimination (the Gram-Schmidt method), reflection (the Householder transformation; Householder, 1953) and rotation (the Givens transformation; Givens, 1954; Gentleman, 1972). Stewart [1973] gives a complete discussion of these methods.

However, it was up to the statistician to really plumb the depths of the triangulation and discover its many and varied uses. Dwyer [1941,

1944, 1945, 1951] discusses its use in correlation and regression analysis, Lucas [1950] and Gaylor, Lucas and Anderson [1970] discuss applications to expected mean squares and Rohde and Harvey [1965] present methods for estimating linear combinations of parameters along with their variances. In addition Graybill [1969, 1976] and Seber [1977] have made use of these techniques in texts dealing with the theory of linear models. The purpose of this paper is to present some of the statistical uses to which the triangular decomposition can be put. The uses discussed do not exhaust the possibilities but represent those the authors have found most beneficial in their own work.

1.1 Preliminaries and Notation

Define the orthogonal-triangular decomposi-tion of an n x p matrix X with rank(X) = $r \leq p$ as

$$X = QT \tag{1.1}$$

where Q is an orthonormal basis for X, i.e., $Q'Q = I$ an identity matrix with p - r zero's on the diagonal and T is a p x p upper triangular matrix. We shall consider the matrix Q as a triangularization operator since

$$Q'X = Q'QT = T. \tag{1.2}$$

We shall assume that the Q-operator as typified in (1.2) is any one of the class of algorithms which operate directly on the columns of X, i.e., Gram-Schmidt, Householder or Givens.
The definition of the Q-operator can be broadened to include procedures which yield an upper triangular matrix from the normal equations, e.g., Choleski or square root method, the Doolittle method, etc. If we define

$$Q = XR' \tag{1.3}$$

then

$$Q'X = R[\,X'X\,] = T \tag{1.4}$$

where R is defined as an "inverse" of T. Thus in this paper references to a Q-operator will refer to any of the algorithms which yield an upper triangular matrix whether from a rectangular system (1.2) or from a square symmetric system (1.4).

Since the matrix T will in general be less than full rank, it will be useful to define a generalized inverse of T. Without loss of generality, assume that the null rows of T are the last p-r rows and write

$$T = \begin{bmatrix} T_1 & T_{12} \\ 0 & 0 \end{bmatrix} . \tag{1.5}$$

Following the suggestion of Allen [1974] define

$$T^g = R' = \begin{bmatrix} T_1^{-1} & T_1^{-1}T_{12} \\ 0 & 0 \end{bmatrix} \tag{1.6}$$

so that $TR' = I$ and

$$R'T = \begin{bmatrix} I & T_1^{-1}T_{12} \\ 0 & 0 \end{bmatrix} . \tag{1.7}$$

It follows from (1.1) that $X'X = T'T$ and thus, that $R'R$ is a generalized inverse of $X'X$ from which it follows that $H = (X'X)^g(X'X) = R'T$. Of course, for X of full rank $R' = T^{-1}$.

We note that R' contains explicit information on the nature of the redundancy in X. Suppose $X = [\,X_1 \mid X_2\,]$ where $X_2 = X_1L$ for some matrix L. It follows that $Q = [\,Q_1 \mid 0\,]$ and from (1.3) we have $T_1 = Q_1'X_1$ and

$$T_{12} = Q_1'X_2 = Q_1'X_1'L = T_1L . \tag{1.8}$$

Since T_1 is of full rank $L = T_1^{-1}T_{12}$. But L is just a partition of (1.7) which can be computed by applying the backward solution to the columns of T_{12}. If the orthogonal basis Q is explicitly required, it may be found from the square symmetric case using (1.6) and (1.3).

1.2 Linear Models

We define the general linear model as

$$\underline{y} = X\underline{\beta} + \underline{\epsilon} \tag{1.9}$$

where $\underline{y}$ is an n x 1 vector of responses, X an n x p known coefficient matrix with rank(X) = r $\leq$ p, $\underline{\beta}$ is a p x 1 vector of unknown fixed parameters and/or random variables and $\underline{\epsilon}$ is a vector of error components with null means and variance-covariance matrix $\sigma^2 I$. The response vector could also be defined as a multivariate response matrix, and with the appropriate modifications, the following development would be applicable to MANOVA.

Applying the Q-operator to the model (1.9), we have

$$\begin{aligned} Q'\underline{y} &= Q'X\underline{\beta} + Q'\underline{\varepsilon} \\ \underline{t} &= T\underline{\beta} + \underline{\varepsilon}^* \end{aligned} \tag{1.10}$$

where $\underline{\varepsilon}^*$ is a vector of random variables with null means and variance-covariance matrix $\sigma^2 I$. It is important to note that $\underline{t}$ is sufficient in a statistical sense while T is "sufficient" in the sense that it carries all of the basic design information carried by the original X matrix.

A solution to (1.10) is obtained through a backward elimination algorithm in a straight forward manner since T is upper triangular. Clearly this is equivalent to pre-multiplying (1.10) by R' to obtain

$$\underline{\tilde{\beta}} = R'\underline{t} = H\beta + R'\underline{\epsilon}^* \tag{1.11}$$

where

$$E(\underline{\tilde{\beta}}) = H\beta = \begin{bmatrix} \underline{\beta_1} + T_1^{-1} T_2 \underline{\beta_2} \\ \underline{0} \end{bmatrix} = \begin{bmatrix} \underline{\beta_2} + L\, \underline{\beta_2} \\ \underline{0} \end{bmatrix}$$

and

$$V(\tilde{\beta}) = \sigma^2 R'R = \sigma^2 (X'X)^g,$$

with $\underline{\beta_2}$ being the set of redundant parameters in $\underline{\beta}$.

2.0 SOME SIMPLE ILLUSTRATIONS

The most often used decomposition algorithms are a systematic way of performing matrix multiplication to yield T. The remainder of this section is devoted to the Q-operator and some of its uses, later sections give the error analysis and demonstrate how to compute quantities needed in a "linear model" analysis using a Q-operator.

Table 2.1 shows some Q-operators with the type of decomposition generated. More specifics about the algorithms can be found in Dywer [1951], Graybill [1969, 1976], Lawson and Hanson [1974], Seber [1977], Chambers [1977] and Kennedy and Gentle [1980].

TABLE 2.1

SOME Q-OPERATORS, CATEGORIZED BY THE TYPE OF MATRIX USED AND THE RESULTING DECOMPOSITION

Matrix to be decomposed	A'B	T'T
X'X	Gauss-Doolittle	Cholesky
$X_{columns}$	Gram-Schmidt	Householder
X_{rows}	Givens	Givens

2.1 Example of Decomposition

Let

$$X = \begin{bmatrix} 1 & 1 & 1 \\ 1 & 3 & 9 \\ 1 & 4 & 16 \\ 1 & 6 & 36 \\ 1 & 7 & 49 \end{bmatrix} . \qquad (2.1)$$

Then using (1.1)

$$QT = \qquad (2.2)$$

$$\begin{bmatrix} .447 & -.670 & .555 \\ .447 & -.251 & -.407 \\ .447 & -.042 & -.536 \\ .447 & .377 & -.092 \\ .447 & .586 & .481 \end{bmatrix} \begin{bmatrix} 2.236 & 9.391 & 49.641 \\ \cdot & 4.775 & 38.702 \\ \cdot & \cdot & 8.541 \end{bmatrix}$$

where T is a triangular matrix; the dot, "•", or lack of any symbols represents 0. Simple multiplication demonstrates that QT = X. Now if X from (2.1) is used we have

$$X'X = \begin{bmatrix} 5 & 21 & 111 \\ 21 & 111 & 651 \\ 111 & 651 & 4035 \end{bmatrix}$$

and from (1.4)

$$T'T = \qquad (2.3)$$

$$\begin{bmatrix} 2.236 & \cdot & \cdot \\ 9.391 & 4.775 & \cdot \\ 49.641 & 38.702 & 8.541 \end{bmatrix} \begin{bmatrix} 2.236 & 9.391 & 49.641 \\ \cdot & 4.775 & 38.702 \\ \cdot & \cdot & 8.541 \end{bmatrix}$$

which equals X'X. Because T is "identical" in (2.2) and (2.3) other computations needed in an analysis are invariant to the Q-operator used in the original decomposition.

2.2 Partitioned Matrices

It is not uncommon to partition the matrix X as

$$X = [\, X_1 \mid X_2 \,]. \qquad (2.4)$$

X_2 could be used to represent the response vector for univariate analysis or response vectors for multivariate analysis or a partitioning of the independent variables. What ever the definition of X_2, T is also partitioned comformable to (2.4), i.e.,

$$T = \begin{bmatrix} T_1 & T_{12} \\ \cdot & T_2 \end{bmatrix}$$

where T_1, T_2 are triangular matrices and T_{12} is a rectangular matrix.

From (1.11) we see how a triangular system of equations is "easy" to solve by the "backward" operator (Graybill 1976). This backward operator can also be used to find the inverse of T,

$$T^{-1} = R' = \begin{bmatrix} T_1^{-1} & -T_1^{-1} T_{12} T_2^{-1} \\ & T_2^{-1} \end{bmatrix} . \qquad (2.5)$$

With T^{-1} calculated it is easy to find $(X'X)^{-1}$ because

$$(X'X)^{-1} = T^{-1} T^{-1\,\prime}. \qquad (2.6)$$

It is also possible to modify the backward operator to yield $(X'X)^{-1}$ directly. Obviously X must be full column rank for (2.5) and (2.6) to be valid. If X is not full rank, i.e., $X_2 = X_1 L$ (we mean that this is the actual relationship and is not based only on machine accuracy) refer to (1.5)-(1.8) to see T^g.

2.3 Matrix Multiplication

Orthogonalization algorithms do not explicitly find Q but systematically perform the matrix multiplications between Q' and X or Q' and X'X to generate T. Operating on X from (2.1), i.e.,

$$Q'\begin{bmatrix}1&1&1\\1&3&9\\1&4&16\\1&6&36\\1&7&49\end{bmatrix} = \begin{bmatrix}2.236&9.391&49.641\\ \cdot&4.775&38.702\\ \cdot&\cdot&8.541\end{bmatrix}$$

As previously stated since the end result of the orthogonalization procedure on X is the same, there is no need to differentiate between Gram-Schmidt, Householder, etc. unless the specific Q-operator has properties of interest. For example, the Cholesky essentially multiplies a matrix by R (Graybill 1976) thus

$$Q_C' \,[\, X'X \mid K' \,] = [\, T \mid T_K' \,].$$

So suppose $A'M^gA$ needs to be computed where M is symmetric and positive semi-definite. Then

$$Q_C'[\, M \mid A \,] = [\, T \mid A_t]$$

with the desired result $A_t'A_t$.

Now suppose $A'M^gB(B'M^gB)^{-1}B'M^gA$ is needed (a quantity used to construct hypothesis test Searle 1971). First

$$Q_C[\, M \mid A \mid B \,] = [\, T \mid A_t \mid B_t] \,.$$

Now another Q-operator, e.g., Givens, is used such that

$$Q_G[\, B_t \mid A_t] = [\, T_B \mid T_{AB}] \,.$$

$T_{AB}'T_{AB}$ is the desired computation. Performing matrix multiplication in this manner is one of the most accurate methods available.

The inherent properties of Q-operators can also be used to find orthogonal polynomials for equally or unequally spaced values, (X in (2.1) is an example). Thus

$$\begin{aligned} Q_C'[\, X'X \mid X' \,] &= [\, T \mid T_X \,] \\ &= [\, T \mid RX' \,] \\ &= [\, T \mid Q \,] \,. \end{aligned}$$

Now using X from (2.1)

$$\begin{bmatrix}1&1&1\\1&3&9\\1&4&16\\1&6&36\\1&7&49\end{bmatrix} =$$

$$\begin{bmatrix}.447&-.670&.555\\.447&-.251&-.407\\.447&-.042&-.536\\.447&.377&-.092\\.447&.586&.481\end{bmatrix}\begin{bmatrix}2.236&9.391&49.641\\ \cdot&4.775&38.702\\ \cdot&\cdot&8.541\end{bmatrix}$$

To get the integer values usually recognized as orthogonal polynomials Q is scaled, i.e.,

$$\begin{bmatrix}.447&-.670&.555\\.447&-.251&-.407\\.447&-.042&-.536\\.447&.377&-.092\\.447&.586&.481\end{bmatrix}\begin{bmatrix}2.236&0.0&0.0\\0.0&23.87&0.0\\0.0&0.0&54.09\end{bmatrix}$$

$$= \begin{vmatrix}1&-16&30\\1&-6&-22\\1&-1&-29\\1&9&-5\\1&14&26\end{vmatrix}$$

The previous demonstrates some of the uses of Q-operators to multiply a matrix by T^{-1} or T^g.

In addition to using the Q-operators to explicitly multiply matrices it is of interest to note the structure of the elements in T, i.e.,

$$\begin{aligned} T_{12}'T_{12} &= X_2'T_1^{-1}T_1^{-1\prime}X_2 \\ &= X_2'X_1(X_1'X_1)^{-1}X_1'X_2. \end{aligned}$$

Also because

$$T_2'T_2 + T_{12}'T_{12} = X_2'X_2$$

then clearly

$$T_2'T_2 = X_2'(I - X_1(X_1'X_1)^{-1}X_1')X_2 .$$

Another useful quantity used in the analysis of linear models is the "Hat-Matrix" (Belsley, Kuh, Welsh 1980) which is simply QQ'.

2.4 Reordering

Often in computing the various quantities of interest we need to change the order of the columns of X and re-orthogonalize. Because T contains all the necessary information if T is available from a previous computation it is not necessary to use the original X or X'X . All that is necessary is to reorder the columns of T and use an appropriate Q-operator. To demonstrate this concept using T from (2.2)

$$Q'\begin{bmatrix} 2.236 & 49.641 & 9.391 \\ \cdot & 38.702 & 4.775 \\ \cdot & 8.541 & . \end{bmatrix} = \begin{bmatrix} 2.236 & 49.641 & 9.391 \\ \cdot & 39.633 & 4.662 \\ \cdot & \cdot & 1.029 \end{bmatrix}$$

which is identical to reordering the original X matrix and using a Q-operator, i.e.,

$$Q'\begin{bmatrix} 1 & 1 & 1 \\ 1 & 9 & 3 \\ 1 & 16 & 4 \\ 1 & 36 & 6 \\ 1 & 49 & 7 \end{bmatrix} = \begin{bmatrix} 2.236 & 49.641 & 9.391 \\ \cdot & 39.633 & 4.662 \\ \cdot & \cdot & 1.029 \end{bmatrix}$$

2.5 Updating a Solution

Often a function of the matrix $(X'X + \underline{vv}')$ is needed. The vector $\underline{v}$ is the "new data" that is to be included in the analysis. The computations are easily performed using the Givens operator. Suppose T from operating on X is available, then

$$Q_G'\begin{bmatrix} T \\ \underline{v}' \end{bmatrix} = T^*$$

where

$$T^{*\prime}T^* = (X'X + \underline{vv}'). \qquad (2.7)$$

This concept of adding new data points to the original solution has application in ridge, robust, etc. analysis. The structure of the variance-covariance matrix of the observations in a mixed model analysis of variance is (2.7).

3.0 ARMCHAIR ERROR ANALYSIS

The accuracy of an algorithm is of primary consideration. A number of articles, presentations and books have compared the algorithms and application programs. Any discussion about the accuracy of algorithms assumes that the algorithms are coded in the best possible way. This means taking care with coding the order of the different operations.

It is also well known that simple operations on the columns of X, like "centering", improve the numerical stability of algorithms using normal equations. There is no real gain in accuracy if the centering or scaling operations are preformed on X'X. "Centering" is of most impostance in interpretation of results. The first step of a Q-operator that operates on columns of X is a "centering" type of operation.

For our purposes it is sufficient to show some of the results in Lawson and Hanson [1974]. Suppose

$$X = \begin{bmatrix} 1 & 1 \\ 1 & 1 \\ 1 & 1-\varepsilon \end{bmatrix},$$

then

$$X'X = \begin{bmatrix} 3 & 3 - \varepsilon \\ 3 - \varepsilon & 3 - 2\varepsilon + \varepsilon^2 \end{bmatrix} . \quad (3.1)$$

The computer we are using has η precision where $\eta > \varepsilon^2$ the because machine accuracy the quantity $3 - 2\varepsilon + \varepsilon^2$ becomes $3 - 2\varepsilon$ which causes X'X to be replaced by $(X'X)^*$, i.e.,

$$(X'X)^* = \begin{bmatrix} 3 & 3 - \varepsilon \\ 3 - \varepsilon & 3 - 2\varepsilon \end{bmatrix}$$

If a Q-operator, e.g., Cholesky, that operates on the normal equation is used then

$$T^* = \begin{bmatrix} \sqrt{3} & (3 - \varepsilon)/\sqrt{3} \\ 0 & 0 \end{bmatrix} .$$

If however a Q-operator, e.g., Householder, that operates on X is used then

$$T = \begin{bmatrix} \sqrt{3} & (3 - \varepsilon)/\sqrt{3} \\ 0 & 2\varepsilon/\sqrt{3} \end{bmatrix} .$$

Clearly $T \neq T^*$. As stated by Lawson and Hansen [1974].

> This difficulty is not a defect of the Cholesky decomposition algorithm but rather reflects the fact that the column vectors of X'X are so nearly parallel (linearly dependent) that the fact that they are not parallel cannot be established using arithmetic of relative precision η. ($\varepsilon^2 < \eta$)

Obviously an ε could be chosen so that even the Householder was not able to compute the "correct" results.

3.1 Comparisions and Summary

Table 3.1 contains comparisions of some well known Q-operators. Some of the differences observed in Table 3.1 are "not significantly" different but provide relative differences. Table 3.2 contains some of the advantages for each algorithm. The information in Table 3.1 and Table 3.2 indicate that one algorithm is not uniformly best. Thus the "best" algorithm will depend on a number of machine and data dependent factors.

There is another algorithm not discussed in this paper that is important, namely the Singular Value Decomposition-SVD (Lawson and Hansen 1974). The SVD can be thought of as a Q-operator. Finally the iterative algorithms proposed by Hemmerlee [1974, 1976] need to be considered whenever a computatonal method is selected.

TABLE 3.1

COMPARISON OF ALGORITHMS

BY ACCURACY, SPEED/COST AND STORAGE

HH = Householder

GS = Gram-Schmidt

GV = Given -- operating on rows

CH = Cholesky

GD = Gauss-Doolittle

Accuracy:

```
           G               C       G    G H
           D               H       V    S H

+=========+=========+=========+=========+
poor                                excellent
```

Speed/Cost:

```
        G    H  G    C            G
        V    H  S    H            D

+=========+=========+=========+=========+
slow                                 fast
```

Storage:

```
     G    H   G                G C
     S    H   V                D H

+=========+=========+=========+=========+
large                                  small
 O[np+p^2]   O[np]             O[p^2]
```

TABLE 3.2

ADVANTAGES OF RESPECTIVE ALGORITHMS

Householder
- High accuracy
- Requires X in core
- Stable
- Double/Single precision

Gram-Schmidt
- High accuracy
- Returns Q explicitly
- Requires X and T in core
- Stable
- Double/Single precision

Givens
- High accuracy
- Exploits presence of zeros
- Requires T and a row of X
- Fairly stable
- Double/Single precision

Cholesky
- Fairly accurate
- Requires normal equations
- Less stable
- Double precision required
- Computationally convenient

Gauss-Doolittle
- Less accurate
- Requires normal equations
- Less stable
- Double precision required

4.0 COMPUTATIONS FOR THE ANALYSIS OF VARIANCE

The general linear model presented in Section 1.3 is sufficiently general to cover a wide range of applications. In this section, we discuss its application to analysis of variance problems. Although the general formulation is applicable to any balanced or unbalanced design, its application to balanced designs would be computationally inefficient. Kennedy and Gentle [1980] review algorithms for the analysis of balanced designs that are computationally efficient and reduce computer storage requirements from the more general approach. We shall focus our attention on the unbalanced case.

In the analysis of variance, the X matrix of (1.9) could be the simple, full rank, orthogonal matrix of the cell means model (Hocking and Speed 1975), the "design matrix" of the usual overparameterized model (Searle 1971), or a full rank reparameterization of either of these (Bryce, Scott and Carter 1980, Graybill 1976). The parameter vector could be fixed unknown constants, e.g., the population cell means or linear combinations thereof or a combination of fixed constants and random variables. In the discussion that follows, we consider both cases.

A principal objective of this paper is to present computationally efficient procedures. In this context, the use of the usual overparameterized model with its singular X matrix is an inefficient use of computer storage. For this reason, we recommend a full rank reparameterization approach for most applications. However, the results presented here are developed in the general case of a singular matrix.

4.1 The Fixed Effects Model

In this subsection, the linear model (1.9) is hypothesized. It is assumed that a logical partitioning of $\underline{\beta}$ exists such that each partition corresponds to a meaningful subset of the parameters in an analysis of variance model, e.g., main effects, two-way interactions, etc. Denote this partition by

$$\underline{\beta}' = (\beta_0, \underline{\beta}_1', \ldots, \underline{\beta}_f'). \tag{4.1}$$

It is also useful to define a notation for the first i+1 factors in the model as

$$\underline{\beta}_{(i)}' = (\beta_0, \underline{\beta}_1, \ldots, \underline{\beta}_i), \quad i \leq f. \tag{4.2}$$

Identical partitions of $\underline{t}$ and T are also needed. However, since T is a pxp matrix, we define three partitions, i.e.,

$T = (T_{ij})$ (by both rows and columns)
$T = (T_i)$ (by rows only)

and $T_{(i)}$ which is analogous to (4.2). Note that $T_{(f)} = T$.

Formally, an analysis of variance is a decomposition of $\underline{y}'\underline{y}$ into a sum of independent quadratic forms. This decomposition corresponds to an ordering of a monotone increasing sequence of submatrices of X. For a set of integers $1 \leq k_1 < k_2 < \ldots < k_f = p$ let X_i be an $n \times k_i$ matrix of rank r_i consisting of the first k_i columns of X. It is assumed that $1 \leq r_1 < r_2 < \ldots < r_f = r$. Also let Q_i be an orthonormal basis for X_i as defined in (1.1) and (1.3). Let $P_i = X_i(X_i'X_i)^-X_i' = Q_iQ_i'$ be the projection matrix that projects on to the space of X_i. The expression

$$\underline{y} = P_1\underline{y} + (P_2-P_1)\underline{y} + \ldots + (P_f-P_{f-1})\underline{y} + (I-P_f)\underline{y}$$

represents an orthogonal partitioning of $\underline{y}$. Since the squared norm of a sum of orthogonal vectors is the sum of the squared norms of those vectors, the squared norm of $\underline{y}$ is

$\underline{y}'\underline{y}$

$= \| P_1 \underline{y} \|^2 + \| (P_2 - P_1)\underline{y}\|^2 + \ldots + \| (P_f - P_{f-1})\underline{y}\|^2 + \| (I - P_f)\underline{y}\|^2$

$= SS_1 \quad + \quad SS_2 \quad + \ldots + \quad SS_f \quad + \quad SSE$

If a Q-operator is applied to the augmented matrix $[\, X \mid \underline{y} \,]$, i.e.,

$$Q'[\, X \mid \underline{y} \,] = [\, T \mid \underline{t} \,] \qquad (4.3)$$

and if $\underline{t}_i$ is a vector partition of $\underline{t}$ containing from the $(k_{i-1}+1)^{th}$ to the $k_i{}^{th}$ elements of $\underline{t}$ as defined above, we have

$$SS_i = \underline{t}_i'\underline{t}_i, \quad i=0,1,\ldots,f \qquad (4.4)$$

$$SSE = \underline{y}'\underline{y} - \underline{t}'\underline{t} \qquad (4.5)$$

and r_i = the number of positive elements among the first k_i diagonal elements of T. (4.6)

To prove these statements, Let X be partitioned as $X = [\, X_i \mid X_j \,]$ where X_i is $n \times k_i$ and X_j is the remaining $p-k_i$ columns of X. Let Q be partitioned conformably and write

$$Q'[\, X \mid \underline{y} \,] = \left[\begin{array}{cc|c} Q_i'X_i & Q_i'X_j & Q_i'\,\underline{y} \\ 0 & Q_j'X_j & Q_j'\,\underline{y} \end{array}\right] =$$

$$= \left[\begin{array}{cc|c} T_{(i)} & T_{i,j} & \underline{t}_{(i)} \\ 0 & T_j & \underline{t}_j \end{array}\right] = [T|\underline{t}]$$

First note from (1.6) that $QQ' = X(X'X)^-X'$ and thus

$$\underline{t}'\underline{t} = \underline{y}'X(X'X)^-X'\underline{y} = \| P_f\underline{y}\|^2 \qquad (4.7)$$

and (4.5) is established. Now note that

$$\underline{t}_{(i)}'\underline{t}_{(i)} = \underline{y}'Q_iQ_i'\underline{y}$$

$$= \underline{y}'X_i(X_i'X_i)^-X_i'\underline{y} = \| P_i\underline{y}\|^2 .$$

and thus

$$\underline{t}_i'\underline{t}_i = \underline{t}_{(i)}'\underline{t}_{(i)} - \underline{t}_{(i-1)}'\underline{t}_{(i-1)}$$

$$= \| P_i\underline{y}\|^2 - \| P_{i-1}\underline{y}\|^2$$

$$= \| (P_i - P_{i-1})\underline{y}\|^2$$

$$= SS_i$$

which establishes (4.4). The last statement is verified by noting that since $T_{(i)} = Q_i'X_i$, the $\text{rank}(T_{(i)}) = \text{rank}(X_i) = r_i$ and since $T_{(i)}$ is triangular it's rank is given by (4.6).

The sums of squares developed in (4.4) are commonly referred to as sequential sums of squares since a sequence of models is being fit. The sequence increases monotonically from the simplest (one factor, generally the overall mean) to the full model (f-factors). The parameter estimates (partial regression coefficients) are available for any of this set of hierarchal submodels by backward solution of

$$T_{(i)}\hat{\underline{\beta}}_{(i)} = \underline{t}_{(i)}$$

i.e.,

$$\hat{\underline{\beta}}_{(i)} = R_{(i)}'\underline{t}_{(i)} \,. \qquad (4.8)$$

where $R_{(i)}$ is the inverse of $T_{(i)}$. These solutions will be BLUE with variance

$$V(\hat{\underline{\beta}}_{(i)}) = \sigma^2 R_{(i)}'R_{(i)} \,,$$

under the assumption that the model fit is appropriate. Otherwise, the expected value takes the form of (1.11). These solutions can be used to obtain estimated marginal means (Searle, Milliken and Speed, 1979), predicted values of individual cell means, linear combinations of the cell means (Bryce, et al., 1980), etc.

The expected values of the sums of squares (4.4) can be obtained by using (1.10), the partition

(4.1) and a well-known theorem on quadratic forms (Searle, 1971). Thus,

$$E(SS_i) = (r_i - r_{i-1})\sigma^2 + \underline{\beta}'T_i'T_i\underline{\beta}$$

$$= d_i\sigma^2 + \sum_{j=i}^{f} \sum_{k=i}^{f} \underline{\beta}_j'T_{ij}'T_{ik}\underline{\beta}_k \ . \qquad (4.9)$$

The expectation involves only the ith factor and those factors following it in the model. In a general unbalanced design, the partition cross products of (4.9) will generally be non-null. It would be convenient to have a simple numerical representation for (4.9) which would provide some insight into the nature of the contamination from the other model factors. Unfortunately, (4.9) can involve a large number of cross product terms with unknown parameters. One suggestion that has been found useful (Bryce, Scott and Carter, 1980) is to record

$$tr(T_{ij}'T_{ij})/d_i \qquad i,j = 1,2,\ldots,f \qquad (4.10)$$

for all i and j as an f x f matrix. Since the indicated matrix trace is just the sum of squares of the elements of T_{ij} scaled by the degrees of freedom, it gives an idea of the magnitude of the contamination present in the ith sum of squares from the jth model term.

All of the foregoing development is correct irrespective of the definition of X and $\underline{\beta}$. With the usual assumption of homogeneity of variance, it can be shown that the observed cell means and the total sum of squares of the observations constitute a set of sufficient statistics for the analysis of variance. Thus assuming s cells or treatment combinations in a designed experiment, we can reduce an n-dimensional problem to an s-dimensional problem. However, there will generally be some constraints, e.g., no interaction, that can be used to further reduce the number of unknowns. The constraints referred to above are known relations among the population cell means rather than restrictions among the parameters of the usual overparameterized model. Restrictions are generally imposed as a computational aid while constraintes have meaning in terms of the experiment being analyzed.

Additional simplifications and insights are possible if the assumption of a full rank X is appended. We assume that the linear model (1.9) is the result of a full rank reparameterization of the cell means model as outlined in Bryce, et al. [1980]. In addition, it is assumed that s-p constraints have been imposed upon the system to reduce X to an n x p matrix of full rank. Thus, the partitions of $\underline{\beta}$ correspond to linear combinations of the cell means which reflect quantities of interest in the analysis of variance.

In this case, the estimates (4.8) will be estimated linear combinations of the cell means under various levels of constraint. It is possible to use these estimates to solve directly for the individual cell means given that each level of constraint is justifiable.

It can be shown that

$$tr(T_{ij}'T_{ij}) = 0$$

if and only if the ith and j^{th} factor are orthogonal. Thus, the matrix (4.10) can be used to examine the confounding structure in an experimental design as well as to determine the contaminants in the sums of squares. Caution should be used in interpreting the magnitude of the elements of (4.10). Since in practice the noncentrality of (4.9) involves cross products with linear combinations of unknown means, the actual effect of any contaminant will be unknown.

Thus far, we have confined ourselves to a single order of the terms in the model (1.9). This limits us to a set of f hypotheses associated with the particular ordering of the f factors in the model. Some of these hypotheses may not be of interest, while hypotheses associated with other orders of the model may be of interest. This is easily accomplished directly from (1.10) by permuting the columns of T and rows of $\underline{\beta}$. Since the permuted T matrix will no longer be upper triangular, we perform an additional Q-operation which yields the required information for an analysis of variance of the reordered model.

Using this technique one can obtain the sums of squares to test all of hypotheses of interest which have been discussed in the recent literature (Kutner 1974, Speed, Hocking and Hackney, 1978). As an indication of the flexibility inherent in this approach, we present Table 4.1. A two-way model is assumed for simplicity. The reorderings necessary to obtain the appropriate sums of squares are given in the last column of the table. The symbols used are R = row main effects, C = column main effects and RC = row by column interaction.

The hypotheses H_1, H_4 and H_7 correspond to those tested in a balanced design. They have been recommended by several authors (Francis 1973, Kutner 1974, Speed et al., 1978). If these are the hypotheses of interest, one need not perform the indicated reorderings of the model. Bryce [1979] has shown that we merely partition R by columns conformable to (4.1), i.e.,

$$R = [\, R_0 \mid R_1 \mid \dots \mid R_f \,]$$

then

$$Q'[\, R_i \mid \underline{t} \,] = [\, T_R \mid \underline{t}_{R_i} \,]$$

and

$$\underline{t}_{R_i}' \underline{t}_{R_i} = SS(i) \qquad (4.11)$$

with

$$E(SS(i)) = d_i\sigma^2 + \underline{\beta}_i' T_R' T_R \underline{\beta}_i .$$

The same computational strategy is readily extended to the case where some cells contain no observations. In this case, one must either restrict himself to the observed cells and formulate hypotheses and estimates based upon only those cells (Hocking, Hackney and Speed, 1978) or make simplifying assumptions about higher order interactions (Bryce, Carter and Scott, 1980). The paper by Bryce et al. [1980] discusses the problem in the context of the Q-operator. Other recent papers dealing with the missing subclass problem are Henderson and McAllister [1978] and Hemmerle [1979].

4.1.1 Example -- Fixed Model

We now demonstrate the foregoing development through a sample data set taken from Chakravarti [1967]. The data are in a two-way layout with each factor having three levels and cell sizes $n_{11} = n_{13} = n_{22} = n_{32} = n_{33} = 2$ and $n_{12} = n_{21} = n_{23} = n_{31} = 3$. Using a full rank reparameterization of the cell means model which reflects row and column main effects and row by column interactions, the results of the Q-operator (4.3) are given in Table 4.2, where the dotted lines indicate the following partitions

$$[\, T \mid \underline{t} \,] = \left[\begin{array}{cccc|c} T_{00} & T_{01} & T_{02} & T_{03} & \underline{t}_0 \\ & T_{11} & T_{12} & T_{13} & \underline{t}_1 \\ & & T_{22} & T_{23} & \underline{t}_2 \\ & & & T_{33} & \underline{t}_3 \end{array}\right].$$

For example, the sum of squares for interactions is given by

$$SS_{0S} = \underline{t}_3'\underline{t}_3 = 18.74^2 + 2.14^2 + 2.03^2 + 18.71^2 = 709.91$$

TABLE 4.1

HYPOTHESES FOR A TWO-WAY MODEL WITH REQUIRED MODEL REORDERINGS

Hypothesis	Form	Alternate Form	Model Order
H_1	$\bar{\mu}_{i\cdot} = \bar{\mu}_{i'\cdot}$	H_1	C, RC, R or RC, C, R
H_2	$\sum_j n_{ij}\mu_{ij}/n_{i\cdot} = \sum_j n_{i'j}\mu_{i'j}/n_{i'\cdot}$	$H_1 \mid H_7$	C, R, RC or C, R
H_3	$\sum_j n_{ij}\mu_{ij} = \sum_{i'} \sum_j n_{ij}n_{i'j}\mu_{i'j}/n_{\cdot j}$	$H_1 \mid H_7$	R, C, RC or R, C or R
H_4	$\bar{\mu}_{\cdot j} = \bar{\mu}_{\cdot j'}$	H_4	R, RC, C or RC, R, C
H_5	$\sum_i n_{ij}\mu_{ij}/n_{\cdot j} = \sum_i n_{ij'}\mu_{ij'}/n_{\cdot j'}$	$H_4 \mid H_7$	R, C, RC or R, C
H_6	$\sum_i n_{ij}\mu_{ij} = \sum_{j'} \sum_i n_{ij}n_{ij'}\mu_{ij'}/n_{i\cdot}$	$H_4 \mid H_1, H_7$	C, R, RC or C, R or C
H_7	$\mu_{ij} - \mu_{ij'} - \mu_{i'j} + \mu_{i'j'} = 0$	H_7	C, R, RC or R, C, RC

NOTE: Cell means are denoted by μ_{ij} and cell counts by n_{ij}. The average over a subscript is denoted by $\bar{\mu}_{i\cdot}$ and the sum over a subscript by $n_{i\cdot}$. All statements hold for all i, i', j and j' with $i \neq i'$ and $j \neq j'$. The vertical slash should be read "subject to." Thus H_2 is H_1 subject to H_7 being true.

TABLE 4.2

THE [T | t] MATRIX FOR A TWO-WAY LAYOUT WITH FACTOR PARTITIONS

MEAN	OBESERVER		SUBJECT		O x S				
4.69	0.04	-0.11	-0.07	0.00	-0.03	0.05	-0.02	-0.05	-277.37
	1.09	-0.07	-0.12	0.23	0.01	-0.04	-0.01	-0.04	-20.58
		1.93	-0.05	-0.12	-0.01	-0.02	-0.07	0.06	21.19
			1.12	0.02	0.03	0.04	-0.02	-0.04	124.98
				1.85	0.02	-0.02	-0.03	-0.06	-68.40
					0.25	-0.02	-0.01	-0.02	18.74
						0.44	-0.02	-0.04	2.14
							0.47	0.04	2.03
								0.74	18.71

Using (4.9), its expected value is

$$E(SS_{OS}) = 4\sigma^2 + \underline{\beta}_3' T_{33}' T_{33} \underline{\beta}_3$$

which would test the hypothesis

$$H_0 : \underline{\beta}_3 = 0$$

i.e., no interaction. When we consider the expected value of

$$SS_S = \underline{t}_2' \underline{t}_2 = 20298.45$$

we have

$$E(SS_S) = 2\sigma^2 + \underline{\beta}_2' T_{22}' T_{22} \underline{\beta}_2 + 2\underline{\beta}_2' T_{22}' T_{23} \underline{\beta}_3 + \underline{\beta}_3' T_{33}' T_{33} \underline{\beta}_3$$

which involves subject terms ($\underline{\beta}_2$) and interaction terms ($\underline{\beta}_3$) and their cross product as well. We summarize the information in [T | $\underline{t}$] in the sequential ANOVA table given as Table 4.3 where the "Expected Mean Square Coefficients" are given by (4.10) with an appropriate scaling. The hypotheses tested by these sums of squares correspond to H_2, H_6 and H_7 of Table 4.1.

If we wished to fit a submodel containing only the observer effects for example, we merely drop subjects and interaction from the model we would then estimate the model parameters as

$$\hat{\underline{\beta}}_{(1)} = T_{(1)}^{-1} \underline{t}_{(1)}$$

$$= \begin{bmatrix} 4.69 & 0.04 & -0.11 \\ & 1.09 & 0.07 \\ & & 1.93 \end{bmatrix}^{-1} \begin{bmatrix} -277.37 \\ -20.58 \\ 21.19 \end{bmatrix}$$

$$= \begin{bmatrix} -58.75 \\ 10.96 \\ -18.11 \end{bmatrix}$$

If these coefficients were used to obtain estimated cell and marginal means following Bryce, Carter and Scott [1980] the means would be estimated under the constraint of no subject main effect or subject by observer interaction. If hypothesis other than those tested by the sequential sums of squares are desired, they can be obtained by reordering the columns of the T matrix in Table 4.2 and reapplying the Q-operator in a manner analogous to that demonstrated in Section 2. We obtain a new T matrix from which a sequential ANOVA analogous to Table 4.3 may be obtained. The last two rows of the new [T | t] matrix would be

$$\begin{array}{cc|c} \cdot & \cdot & \cdot \\ \cdot & \cdot & \cdot \\ \cdot & \cdot & \cdot \\ 1.07 & -0.07 & 2.53 \\ 0 & 1.90 & 21.50 \end{array}$$

TABLE 4.3

SEQUENTIAL ANALYSIS OF VARIANCE TABLE

SOURCE	DF	SUM OF SQUARES	EMS COEFFICIENTS O	S	OS
MEAN	1	76936.41			
OBSER(O)	2	872.23	7.318	0.122	0.051
SUBJ(S)	2	20298.45	0	7.196	0.047
OS	4	709.90	0	0	2.375
ERROR	13	1204.00			

Thus the sum of squares for testing the equality of the marginal means for observers (H_4) is

$$SS_O = 2.53^2 + 21.50^2 = 468.78$$

The same sum of squares could be obtained by inverting the original T matrix and applying (4.11), i.e.,

$$Q'\left[\begin{array}{cc|c} 0 & 0 & -277.37 \\ & & \\ 0.915 & 0 & -20.58 \\ 0.035 & 0.518 & 21.18 \\ & & \\ 0.104 & 0.024 & 124.98 \\ -0.112 & 0.033 & -68.40 \\ & & \\ -0.020 & 0.012 & 18.74 \\ 0.070 & 0.024 & 2.14 \\ 0.029 & 0.076 & 2.03 \\ 0.041 & -0.041 & 18.71 \end{array}\right]$$

$$= \left[\begin{array}{cc|c} 1.07 & -0.07 & 2.53 \\ & 1.90 & 21.50 \end{array}\right]$$

which yields the same sum of squares as before. Of course, this technique would be most efficient if it was desired to test hypotheses analogous to H_1 or H_4 for all terms in a model. In this case, only one matrix inversion would be required.

4.2 The Mixed Model

The complexity of the model increases when we consider a mixed model, but the computations are essentially the same. We use the model (1.9) as defined. However, instead of $\underline{\beta}$ being a set of fixed unknown constants as in the fixed model, we now assume that it consists of both fixed constants and random variables, i.e.,

$$E(\underline{\beta}) = \begin{bmatrix} \underline{\phi} \\ \underline{0} \end{bmatrix} \quad \text{and} \quad V(\underline{\beta}) = \begin{bmatrix} 0 & 0 \\ 0 & V \end{bmatrix} . \qquad (4.12)$$

Now such a model formulation would generally require the use of some type of maximum likelihood technique to obtain exact solutions. However, what we present here will be an extension of the usual analysis of variance estimators. While these will yield sums of squares which are neither chi-square distributed nor independent, research (Bryce and Allen 1972; Cummings and Gaylor 1974) has indicated that the approximations are fairly good unless the design imbalance is extreme.

Whether the goal of an analysis is estimation or hypothesis testing, some consideration must be given to obtaining expected mean squares for each model fit. In a land mark paper on this subject, H. O. Hartley [1967] said, "However, with regard to [unbalanced designs] ... new and laborious 'algebraic heroics' are required every time a new unbalanced body of data is encountered." He went on to develop the synthesis method which reduced the former "algebraic heriocs" to a simple computational method. Reduced to its essence, the systhesis method simply says whatever algorithm you used to obtain sums of squares from the response variable, use the same algorithm on each of the columns of X to obtain expected mean square coefficients. From (1.10) we see that Q operating on $\underline{y}$ yields $\underline{t}$ from which we can obtain an analysis of variance by summing squares of subsets. It follows, therefore, that Q operating on each of the columns of X will yield the coefficients for the expected values of the sums of squares. Thus, the Q-operator is just a special case of Hartley's synthesis method.

Bryce and Allen [1972] have developed these results from the general point of view of a linear model defined as (1.9) with $\underline{\beta}$ defined as (4.12) and the Q-operations (4.3). In the general case, the expectation of the ith sum of squares is given by

$$E(\underline{t}_i'\underline{t}_i) = d_i\sigma^2 + \sum_{\substack{\text{Random}\\ \text{Factors}\\ j\geq i}} \sigma^2 \operatorname{tr}(T_{ij}'T_{ij}) + \sum_{\substack{\text{Fixed}\\ \text{Factors}\\ j,k\geq i}}\sum \underline{\beta}_j' T_{ij}' T_{ik} \underline{\beta}_k \qquad (4.13)$$

While in the general case this is not to informative, the important point is that all of the information as to the expected value of the sums of squares is contained in T. Furthermore, the reason for the definition of the expected mean square matrix (4.10) becomes obvious since the elements of that matrix now become the expected mean squares coefficients of the variance components associated with the random factors in the model. This property of the triangularization was first recognized by Lucas [1950] and later published by Gaylor, Lucas and Anderson [1970].

All of the foregoing discussion was under the assumption that the model defined by (1.9) was the usual overparameterized less than full rank model. However, in practice it is not necessary to carry the redundant columns of the overparameterized model in computer storage. Bryce, Carter and Scott [1979] have shown that a full rank reparameterization of the random model can be used in conjunction with the Q-operator to obtain a complete ANOVA. The expected mean squares (4.13) are displayed as in (4.10). However, because of the reparameterization some simple adjustments are required and the actual expected mean square coefficients become a linear transformation of (4.10). The primary advantage to the reparameterization approach is that it makes sums of squares typified by (4.11) easily available in random models. As pointed out by Speed [1979] and Bowen [1976] estimates of variance components obtained by using such sums of squares have smaller variances than estimators obtained from other sets of sums of squares.

5.0 COMPUTATIONS FOR LINEAR COMBINATIONS OF PARAMETERS

In this section, we discuss computations for inference on a linear combination of the elements of $\underline{\beta}$ represented by $\theta = \underline{\ell}'\underline{\beta}$. For the moment, we will assume X of (1.9) is of full rank, and thus it is guaranteed that θ is estimable. The basic method is to form the matrix

$$\left[\begin{array}{c|c} T & \underline{t} \\ \hline \underline{\ell}' & \underline{0} \end{array}\right]$$

and then to do row operations to make the $\underline{\ell}'$ position null. The number resulting in the lower right position is the negative of the estimate. The sum of squares of the multipliers is the coefficient of σ^2 in the variance of the estimator. This method is called the elimination procedure and is illustrated in Table 5.1. A comparison of the elimination method with a more conventional method is given by Allen [1977].

The theory behind the results presented in Table 5.1 follow from $\underline{t} \sim N(T\underline{\beta}, I\sigma^2)$ and the rules for the mean and variance of a linear combination of uncorrelated random variables.

In practice, we must decide if X is of full rank. If $[\ T \mid \underline{t}\ ]$ was found using exact arithmetic, then one or more zeros on the diagonal of T would mean X is less than full rank. However, when computations are done in floating-point arithmetic, a nonzero result may be obtained for an expression that is algebraically zero. If a stable algorithm is used to calculate T, then the absolute errors of the computed values from their actual values are small. However, when an element is small in absolute value, a small absolute error may be a moderate or large relative error. This is a potential problem with regard to the diagonal elements of T since a wrong decision with regard to what should or should not be zero also results in a wrong specification of the rank of X. We must also decide what should or should not be zero when testing for estimability. We present some rationale for making these decisions.

TABLE 5.1

The Elimination Procedure for Computing Estimator and Variance

$$\left[\begin{array}{c|c} T & \underline{t} \\ \hline \underline{\ell}' & 0 \end{array}\right] = \left[\begin{array}{ccc|c} 1 & 1 & 2 & 4 \\ 0 & 2 & 1 & 4 \\ 0 & 0 & 3 & 2 \\ \hline 1 & 2 & 2 & 0 \end{array}\right] \begin{array}{c} \text{row} \\ 1 \\ 2 \\ 3 \\ 4 \end{array}$$

$$\begin{array}{l} \text{row 4-(1)row 1} \\ \text{row 5-(1/2)row 2} \\ \text{row 6-(-1/6)row 3} \end{array} \left[\begin{array}{ccc|c} 0 & 1 & 0 & -4 \\ 0 & 0 & -0.5 & -4.5 \\ 0 & 0 & 0 & -4.167 \end{array}\right] \begin{array}{c} 5 \\ 6 \\ 7 \end{array}$$

Estimate of θ = -(-4.167) = 4.167

$$\text{Variance of estimator} = [1^2 + (\tfrac{1}{2})^2 + (\tfrac{-1}{6})^2]\sigma^2$$

$$= 1.2778\sigma^2$$

When T has been computed, we set t_{ii} to zero if

$$t_{ii}^2 \,/\, (\sum_{j=1}^{i} t_{ji}^2) < \varepsilon, \; i = 2, \ldots, p, \qquad (5.1)$$

where ε is a small positive number (its specific value will be discussed presently). The expression on the left hand side of (5.1) is the computed value of the square sine of the i-th column of X and its projection onto the space spanned by the first i-1 columns. A small value indicates that the i-th column is close to a linear combination of the first i-1 columns. If a t_{ii} is redefined to be zero, then an additional orthogonal transformation may be required to make other elements of the i-th row of T null.

If $\underline{\ell}'$ is in the row space of X, then $\underline{\ell}'\underline{\beta}$ is estimable. If T were computed using exact arithmetic, then the row space of T is the same as the row space of X. Thus $\underline{\ell}'\underline{\beta}$ is estimable if $\underline{\ell}$ is orthogonal to the orthocomplement of the row space of T. For every i such that $t_{ii} = 0$, let $\underline{u}_i$ be a vector such that its first i-1 elements are the coefficients for expressing the i-th column as a linear combination of the first i-1 columns, its i-th element is -1, and the remaining elements are zero. The collection of all such vectors is a basis of the orthocomplement to the row space of T.

We realize T is not computed exactly, but we consider θ estimable if

$$(\underline{u}_i'\underline{\ell})^2/((\underline{\ell}'\underline{\ell})(\underline{u}_i'\underline{u}_i)) < \delta \qquad (5.2)$$

for all i such that $t_{ii} = 0$, and δ is a small positive number. This criterion is a different expression for the same criterion proposed by Allen [1974]. When doing the elimination procedure described earlier, $\underline{u}_i'\underline{\ell}$ is the number to be made zero at the ith step and is, therefore, automatically calculated. The factors in the denominator are also easy to calculate.

We now examine the performance of these rules and make suggestions for values of ε and δ. For clarity of presentation we use a small example. The principles also apply to the general case. Suppose that

$$[\,T \mid \underline{t}\,] = \left[\begin{array}{ccc|c} 1 & 1 & 2 & t_{14} \\ 0 & 2 & 1 & t_{24} \\ 0 & 0 & t_{33} & t_{34} \end{array}\right]$$

and that all elements, with the possible exception of t_{33}, are of sufficient magnitude to be calculated with small relative error. We assume t_{33} is not exactly zero for otherwise we would have only the problem of estimability. If (3.1) is not satisfied (i.e., $t_{33}^2/(5 + t_{33}^2) \geq \varepsilon$), we do not alter T. If (5.1) is satisfied, we have

$$[\,T \mid \underline{t}\,] = \left[\begin{array}{ccc|c} 1 & 1 & 2 & t_{14} \\ 0 & 2 & 1 & t_{24} \\ 0 & 0 & 0 & 0 \end{array}\right]$$

and the residual sum of squares is increased by t_{34}^2. We are led to one of three actions:

1. If (5.1) is not satisfied, we estimate θ by

$$\hat{\theta} = \ell_1 t_{14} + \frac{\ell_2 - \ell_1}{2} t_{24} + \frac{-3\ell_1 - \ell_2 + 2\ell_3}{2t_{33}} t_{34}$$

2. If (5.1) and (5.2) are both satisfied, we estimate θ by

$$\hat{\theta}_1 = \ell_1 t_{14} + \frac{\ell_2 - \ell_1}{2} t_{24}$$

3. If (5.1) is satisfied and (5.2) is not, we declare θ not estimable.

In order to evaluate the consequences of these actions, we need the mean square errors of $\hat{\theta}$ and $\hat{\theta}_1$. Let a_{33} denote the true algebraic value of the quantity that has computed value t_{33}. The mean square error of $\hat{\theta}$ is

$$\mathrm{MSE}(\hat{\theta}) = \left[\ell_1^2 + \left(\frac{\ell_2 - \ell_1}{2}\right)^2 + \left(\frac{-3\ell_1 - \ell_2 + 2\ell_3}{2t_{33}}\right)^2\right]\sigma^2$$

$$+ \left(\frac{-3\ell_1 - \ell_2 + 2\ell_3}{2}\right)^2 \left(\frac{a_{33} - t_{33}}{t_{33}}\right)^2 \beta_3^2$$

The mean square error of $\hat{\theta}_1$ is

$$\mathrm{MSE}(\hat{\theta}_1) = \left[\ell_1^2 + \left(\frac{\ell_2 - \ell_1}{2}\right)^2\right]\sigma^2$$

$$+ \left(\frac{-3\ell_1 - \ell_2 + 2\ell_3}{2}\right)^2 \beta_3^2 .$$

When the computed value of $-3\ell_1 - \ell_2 + 2\ell_3$ is zero, the estimators are the same, and (3.2) is satisfied. We exclude this case from further consideration. The estimator $\hat{\theta}_1$ is better than $\hat{\theta}$ (i.e., $\mathrm{MSE}(\hat{\theta}_1) < \mathrm{MSE}(\hat{\theta})$) if and only if

$$\frac{5 + t_{33}^2}{t_{33}^2} > \gamma(2-\gamma)\frac{5 + t_{33}^2}{\sigma^2} = \gamma(2-\gamma)\frac{\beta_3^{*2}}{\sigma^2} \qquad (5.3)$$

where $\gamma = a_{33}/t_{33}$ and $\beta_3^* = \sqrt{(5 + t_{33}^2)}\,\beta_3$ is the standardized coefficient of the third column of X.

There are three possible true situations:

1. $a_{33} \neq 0$.

2. $a_{33} = 0$ and θ estimable.

3. $a_{33} = 0$ and θ not estimable.

Table 5.2 is an evaluation of each action in each situation. The ordered pair (i,j) denotes the use of the ith action in the jth situation. Consequences of (1,1), (2,2), and (3,3) are obvious. When $a_{33} = 0$, so does γ, and (5.3) is always satisfied. Thus (2,2) is at least as favorable as (1,2), and (2,3) is at least as favorable as (1,3). A very high mean square error results from (1,3). The stated consequence of (2,3) is simply an alternate expression for (5.2). In order to take action 2, (5.1) must be satisfied and also $(5 + t_{33}^2)/t_{33}^2 > 1/\varepsilon$. Thus the stated consequence of (2,1) is implied by (5.3). If (5.1) is satisfied and (5.2) is not, then

$$\left(\frac{3\ell_1 + \ell_2 - 2\ell_3}{2t_{33}}\right)^2 > \frac{7(\ell_1^2 + \ell_2^2 + \ell_3^2)}{2(5 + t_{33}^2)}\,\frac{\delta}{\varepsilon} . \qquad (5.4)$$

Since $MSE(\hat{\theta})$ and $MSE(\hat{\theta}_1)$ are increasing functions of the left hand side of (5.4), they are large when δ/ε is large, and thus our statement with regard to (3,1). We want to avoid (3,2).

With these consequences in mind, we consider the choice of specific values for ε and δ. It is clear that we should make ε as large as possible while maintaining

$$\frac{1}{\varepsilon} > \gamma(2-\gamma) \left[\frac{\beta_3'}{\sigma}\right]^2 .$$

Suppose we are willing to assume

$$\left[\frac{\beta_3'}{\sigma}\right]^2 < 1000,$$

then we would take

$$\varepsilon = \frac{.000001}{\gamma(2-\gamma)} .$$

If a stable algorithm such as one based on Householder or Given's transformations is used, if the equivalent of ten or more decimal digits is used in the arithmetic, and if there is no suspicion that the data is highly ill-conditioned, then we would assume $\gamma \doteq 1$ and use $\varepsilon = .000001$. If one or more of these conditions is not satisfied, then we may not be willing to assume $\gamma = 1$ and use $\varepsilon = .0001$. The consequence of (3,2) is serious so we would make δ as large as possible, subject to keeping the consequence of (2,3) in reasonable bounds. We use $\delta = 100\varepsilon$.

Obviously, subjective judgements entered into these choices. Every program for general use should have an option for the user to specify his personal values of ε and δ. If a bad choice of ε leads to (1,2) or (1,3), a large variance should signal the user to try again using a larger ε. Our reasoning behind these suggestions was based more on statistical consideration than numerical considerations.

We would like to conclude this section with a simple message. The usual text book formulas are:

Criterion for estimability	$\underline{\ell} = (X'X)(X'X)^-\underline{\ell}$
Estimate	$\underline{\ell}'(X'X)^-X'\underline{y}$
Variance	$\underline{\ell}'(X'X)^-\underline{\ell}\,\sigma^2$.

Calculation of $(X'X)^-$ may be an ill-conditioned problem while at the same time calculation of the estimates and variances may be a well-conditioned problem. The moral is that using text book formulas literally in computation may introduce an ill-conditioned intermediate step in a well-conditioned problem. The methods presented here are preferred.

TABLE 5.2

EVALUATION OF EACH ACTION IN EACH SITUATION

Action	Situation		
	$a_{33} \neq 0$	$a_{33} = 0$, θ estimable	$a_{33} = 0$, θ not estimable
1 (use $\hat{\theta}$)	The numerically correct action.	This action is never statistically better than (2,2).	This action is never statistically better than (2,3). This action has serious consequence.
2 (use $\hat{\theta}_1$)	This action is statistically better than (1,1) if $\frac{1}{\varepsilon} > \gamma(2-\gamma)\left\lvert\frac{\beta_3'}{\sigma}\right\rvert^2$	The numerically correct action.	We are attempting to estimate $\underline{\ell}'\underline{\beta}$ where $\underline{\ell}$ is a slight perturbation from $\underline{\ell}_1$ and $\underline{\ell}_1'\beta$ is estimable (i.e., $\underline{\ell} = \underline{\ell}_1 + \underline{\ell}_2$) $\underline{\ell}_1'\beta$ is estimable, and $\underline{\ell}_2{}^2/\underline{\ell}^2 < \delta$).
3 (declare not estimable)	If we take $\delta >> \varepsilon$, then this action can occur only when MSE($\hat{\theta}$) and MSE($\hat{\theta}_1$) are large which is tantatmount to not estimable.	This action has a serious consequence.	The numerically correct action.

6.0 REFERENCES

ALLEN, D.M. (1974). "Estimability, Estimates, and Variances for Linear Models" Biometrics Unit Report, BU-529-M, Cornell University, Ithaca, New York.

ALLEN, D.M. (1977). "Computational Methods for Estimation of Linear Combinations of Parameters." Proceedings of the Statistical Computing Section of the American Statistical Association, August, 1977.

BANACHIEWICZ, T. (1937). "Calcul des de'terminants par la me'thode des Eracouiens" Acad. Polon. Sci. (Bull. Int.) A, pp. 109-120.

BELSLEY, D.A., E. Kuh and R.E. Welsh Regression Diagnostics, New York: John Wiley and Sons

BENOIT (From Commander Cholesky) (1924). "Note sur une me'thode de re'solution des e'quations normales . . ." Bulletin Geodesique #2:5-77.

BOYER, C.B. (1968). A History of Mathematics. New York, John Wiley & Sons, Inc.

BRYCE, G.R. and D.M. Allen (1972). "The distribution of some statistics arising from the factorization of the normal equations of mixed models." Department of Statistics Technical Report No. 35, University of Kentucky, Lexington, Kentucky.

BRYCE, G.R. (1979). "A simplified computational procedure for Yates' method of weighted squares of means." BYU Statistics Report No. SD-011-R, Brigham Young University.

BRYCE, G.R., M.W. Carter and D.T. Scott (1979). "Reparameterization of the Cell Means Model and The Calculation of Expected Mean Squares." Proceedings of the Statistical Computations Section of the American Statistical Association, Aug. 1979, (BYU Tech. Report No. SD-012-R.)

BRYCE, G.R., D.T. Scott, and M.W. Carter (1980). "Estimation and Hypotheses Testing in Linear Models -- A Reparameterization Approach to the Cell Means Model." Communications in Statistics -- Theor. Meth. A9(2):131-150.

BRYCE, G.R., M.W. Carter and D.T. Scott (1980). "Recovery of Estimability in Fixed Models with Missing Cells." Invited paper presented at the ASA meetings August, 1980 in Houston, Texas. (BYU Tech. Report No. SD-022-R.)

CHAKRAVARTI, I.M., R.G. Laha and J. Roy (1967). Handbook of Methods of Applied Statistics, Vol. I. New York, John Wiley and Sons.

CHAMBERS, J.M. (1977). Computational methods for Data Analysis. New York, John Wiley and Sons.

CUMMINGS, W.G. and D.W. Gaylor (1974). "Variance Component Testing in Unbalanced Nested Designs." Journal of the American Statistical Associations 69:765-771.

DOOLITTLE, M.H. (1878). "Method employed in the solution of normal equations and the adjustment of a triangulation." U.S. Coast and Geodetic Survey Report (1878). pp. 115-120.

DWYER, P.S. (1941). "The Doolittle Technique." Annals of Math. Stat. 12:449-458.

DWYER, P.S. (1944). "A Matrix Presentation of Least Squares and Correlation Theory with Matrix Justification of Improved Methods of Solution." Annals of Math. Stat. 15:82-89.

DWYER, P.S. (1945). "The Square Root Method and Its Use in Correlation and Regression." Journal of the American Statistical Association. 40:493-503.

DWYER, P.S. (1951). Linear Computations. New York, John Wiley and Sons, Inc.

FRANCIS, I. (1973). "A Comparison of Several Analysis of Variance Programs." Journal of the American Statistical Association. 68:860-871.

GAUS, C.F. (1873). "Supplementom Theoriae Combinationis Obervationum Erroribus Minimis Obnoxiae" Werke, 15 (Gaus lived 1775-1855)

GAYLOR, D.W., H.L. Lucas and R.L. Anderson (1970). "Calculation of Expected Mean Squares by the Abbreviated Doolittle and Square Root Methods." Biometrics, 26:641-655.

GENTLEMAN, W.M. (1972). "Least Squares Computations by Givens Transformations Without Square Roots." University of Waterloo Report (SRR-2062, Waterloo, Ontario, Canada.

GIVENS, W. (1954). "Numerical computation of the characteristic values of a real symmetric matrix." Report ORNL-1954, Oak Ridge Associated Universities, Oak Ridge, Tennessee.

GRAYBILL, F.A. (1969). Introduction to Matrices with Applications in Statistics. Belmont, California: Wadsworth Publishing Company, Inc.

GRAYBILL, F.A. (1976). Theory and Application of the Linear Model, North Scituate, Mass.: Duxbury Press.

HARTLEY, H.O. (1967) "Expectations, Variances and Covariances of Anova Mean Squares by 'Synthesis'" Biometrics, 23:105-114.

HEMMERLE, W.J. (1974). "Nonorthogonal Analysis of Variance Using Interative Improvement and Balanced Residuals." Journal of the American Statistical Association. 69:772-779.

HEMMERLE, W.J. (1976). "Iterative Nonorthogonal Analysis of Variance of Covariance." Journal of the American Statistical Association 71:195-199.

HOCKING, R.R. and F.M. Speed (1975). "A Full Rank Anaysis of Some Linear Model Problems." Journal of the American Statistical Association. 70:706-712.

HOUSEHOLDER, A.S. (1958). Unitary triangularization of a non-symmetric matrix, Journal of A.C.M., 5:339-342.

KENNEDY, W.J. and J.E. Gentle (1980). Statistical Computing, New York: Marcel Dekker, Inc.

KUTNER, M.H. (1974). Hypotheses Testing in Linear Models. The American Statistician, 28:98-100.

LAWSON, C.L. and R.J. Hanson (1974). "Solving Least Squares Problems" New Jersey: Prentice-Hall Inc.

LUCAS, H.L. (1950) "A method of estimating components of variance in disproportionate numbers" Ann. Math. Statist. 21,304.

ROHDE, C.A. and J.R. Harvey (1965). "Unified Least Squares Analyses, Jour. Am. Stat. Assoc., 60:523-527.

SCHUR, J. (1917). "Uber Potenzreihen due cm Innern des Einheitskreises beschrankt sind" Crelle's Journal f. reine u. ange. Math 147:205-232.

SEARLE, S.R. (1971). Linear Models. New York: John Wiley & Sons, Inc.

SEARLE, S. R., G.A. Milliken and F.M. Speed (1979). "Expected marginal means in the linear model." Biometrics Unit Report No. BU-672-M, Cornell University, Ithaca, New York.

SEBER, G.A.F. (1977). Linear Regression Analysis, New York, John Wiley & Sons, Inc.

SPEED, F.M., R.R. Hocking, and O.P. Hackney (1978). "Methods of Analysis of Linear Models with Unbalanced Data." J. Amer. Stat. Assoc., 73:105-112.

STEWART, G.W. (1973). Introduction to Matrix Computations. New York: Academic Press.

Workshop 11

Research Data Base Management

Organizer: Ronald W. Helms, University of North Carolina at Chapel Hill

Chair: Kent Kuiper, Boeing Computer Services

Invited Presentations:

Simple Query Language Requirements for Research Data Management, Gary D. Anderson, McMaster University

Data Editing on Large Data Sets, James J. Thomas, Pacific Northwest Laboratory, and Robert A. Burnett, Pacific Northwest Laboratory, and Janice R. Lewis, Pacific Northwest Laboratory

SIMPLE QUERY LANGUAGE REQUIREMENTS FOR RESEARCH DATA MANAGEMENT

Gary D. Anderson, McMaster University, Hamilton

Increasingly, researchers are managing large and complex data sets with existing packaged software systems designed for use in the research environment. For example, varying degrees of data management capabilities are found in BMDP, P-STAT, SAS, SIR, and SPSS. These systems provide data input, statistical analysis, displays, tables, reports, etc. primarily through the use of build-in procedures, or a high-level retrieval language. Also, with the exception of P-STAT and SIR, these systems were primarily designed for batch usage.

Under the assumption that researchers will continue to have increased interactive terminal access to their computing systems, this paper discusses common requirements that researchers have for query-like access to their data sets to handle the many ad hoc data modification, data listing, and simple tabling or reporting requirements they face. A set of commonly needed capabilities for this purpose are suggested and a query language syntax for providing them is discussed.

Finally, possible approaches to linking these capabilities into existing packaged systems is presented.

Key Words
Data Management, Query language, Retrieval, Display, Reporting.

I. INTRODUCTION

In the process of managing large research studies, we are faced with many clerically-oriented ad hoc tasks which, although they may be considered rather simple in nature, require significant time and personnel resources over the duration of the study. Among these chores are such things as: generating sorted lists of various subsets of the study population; creating sets of address labels and form letters for follow-up contacts; retrieving information on specific study subjects; making routine modifications or additions to the data on a small set of study subjects. In this paper, we will suggest a query language system as a means of addressing these tasks.

Present trends in computer systems indicate that reliable and responsive interactive terminal access is becoming the accepted rule. This being the case, the researcher should certainly be able to expect good interactive access to his data for ad hoc retrievals. However, it is generally the researcher's clerical staff, not the researcher himself, or his technical staff who are responsible for the tasks listed above. If these routine retrievals are to be handled by personnel untrained in computers, software systems must be available which can be learned and reliably used by individuals lacking previous computing background. The systems in common use today for managing research data do not lend themselves to ad hoc use by those unprepared to become reasonably knowledgeable about research data management and analysis.

In the sections which follow, we will begin by reviewing an example set of data typical of multi-center studies and discussing, briefly, alternatives available for meeting the primary data management needs associated with such a data set. We will then outline the type of interactive environment and associated capabilities required to address the ad hoc retrieval needs of this data. Next, a list of commands, felt to be required in a researcher-oriented query language will be presented. This will be followed by several examples of how the proposed commands would be used to solve routine problems on the example data set. Finally, we will comment on the need for research data management system developers to consider the addition of a query language as part of their package.

II. THE DATA SET

Each of the data records created in a typical research study can most easily be conceived of as belonging to one of a series of rectangular or flat data sets. We will call these data sets record types. Each record type consists of multiple observations on a set of variables. A typical study generally involves multiple record types.

To illustrate, consider a set of record types resulting from a hypothetical multi-center study, based in schools throughout the U.S.A., of the effect of regular intake of Vitamin C on the general health of school children. In this study, let us assume that selected children are being given sustained doses of Vitamin C, while matchid controls are receiving a placebo treatment.

The record types collected in this study are as follows:

1. A STATE record is collected with data about the state educational situation and state government (one record per state).

2. A DISTRICT record is collected for each district in the state having schools in the study.

3. A SCHOOL record provides information on each school enrolled in the study (one record per school).

4. As students are enrolled in the study, a STUDENT record is created which describes the student (one record per student).

5. Finally, a FOLLOWUP record is created each time a student is examined regarding his treatment compliance and health (about once every three months).

Using a common convention for describing record types, their contents and interrelationships, we can represent them conceptually as follows:

STATE(STATEID ,EXPEND,NELEM,NHIGH,GOVPARTY,
PCTDEM)

DISTRICT(STATEID,DISTRICT,NDELEM,NDHIGH,
SALARY,NSTAFF,TEACHSAL,TSRATIO,HEALTHPR)

SCHOOL(STATEID,DISTRICT,SCHOOL,ENROLL,TYPE,
PYRS,PSEX,NTEACH,SCHLHLTH,NURSE,STRATA)

STUDENT(STUDID,NAME,STREET,CITY,TREATMNT,GRADE,
SEX,AGE,FAMSIZE,SOCIO,MATCHID,STATE,
ENTDATE,DISTRICT,SCHOOL)

FOLLOWUP(STUDID,FOLLOWDT,COLDS,NURSVST,
SAMPLE,COMPLNCE)

POINTER(STATEID,DISTRICT,SCHOOL,STUDID)

The underscored variable names are those which both make the identity of an individual record occurrence of each type unique within that record type and further establish the relationships between the various record types. We can thus see that the STATE, DISTRICT and SCHOOL record types form a natural hierarchy (i.e. the STATE hierarchy) as do the STUDENT and FOLLOWUP record types (i.e. the STUDENT hierarchy). Thus, one way to organize this data for management by a key oriented data base management system (e.g. SIR[9]) is as two hierarchical case structures (i.e. STATE and STUDENT) connected by network pointers. The network link from the STUDENT cases to the STATE cases is accomplished by the non-key data values for STATE, DISTRICT and SCHOOL within the STUDENT record type. In order to establish a link from the STATE case to the STUDENT cases belonging to each state, the POINTER record type is added within the STATE hierarchy (or STATE case).

If this data is managed by a relationally oriented data management system (e.g. SAS[7] or P-STAT[5]), the POINTER record type would be unnecessary. Instead, the STUDENT records are sorted into STATE, DISTRICT, and SCHOOL order when linkage between the SCHOOL and STUDENT record type is desired.

A detailed discussion of the relative advantages and disadvantages of the hierarchical, network, or relational approaches to managing data of this type is beyond the scope of this paper. For a detailed discussion of these models, the reader is referred to MARTIN[4] and DATE[3]. It is sufficient to assume that whatever data management approach or data base management is being used,it will have the ability to create new record types from the source data record types mentioned above in whatever manner is required. For example, suppose we wish, for each student, the STUDID,NAME,AGE and SEX from the STUDENT record type and the average rate of compliance (based on the variable COMPLNCE) from the FOLLOWUP record type. We will assume the data management system we are using has procedures or a retrieval language able to produce a new record type containing this data.

Consequently, we will limit ourselves to consideration of a query language which acts upon a single record type at one time and which leaves the concurrent processing of multiple record types to the host system being employed for the major data management aspects of the study. Thus, the query language we are proposing will be able to access any one of the original source record types described above or any new record type created by the host data management (or database) system. By limiting access to a single record type, we are able to keep the query system sufficiently simple, straightforward, and limited that it can easily be learned and used by non-computer personnel without danger of their getting into major trouble with it.

In what follows, we will use the term data base when referring to the collected set of records from the study belonging to the original source record types. The term "data base" refers to the complete set of data organized in a clean form, fully documented (e.g. by a schema), fully protected by an adequate set of security and integrity mechanisms and stored in a form which allows it to be conveniently managed by whatever system we may be using for that purpose.

III. REQUIRED SOFTWARE ENVIRONMENT

Before presenting the proposed query language structure, let us consider several aspects of the general computer environment that this author feels are of the utmost importance in achieving a successful research query system.

First, a query system should allow the user whatever flexibility is needed to access and display records. It must do this, however, in a way that honors whatever security is in effect and also in a way which cannot compromise the integrity of the data base. Thus it is important that the ad hoc user work one step removed from the data base. This is best accomplished by providing him with a work area into which records retrieved from the data base are copied. He is then free to display, sort, modify, and add to these record copies without further immediate reference directly to the data base itself. Returning these records to the data base (i.e. to accomplish an update) can then be reserved for a point at the end of a session, after which all desired modifications have been made in the work area. The data update then becomes an occasional volume operation rather than a direct interactive modification process as far as the data base is concerned. Such an approach minimizes contentions for multiple user access directly on the data base. Also, since changes and additions can be carefully checked and verified visually while they are still in the workspace of the query system, this approach reduces the likelihood of making unwanted changes to the data base.

Thus, the environment suggested for a research query system is one in which the user is working on a copy of the records, and not on the data base records directly.

Secondly, the user must be presented with a command environment in which each command given is executed immediately or enters the system into a mode in which it prompts the user for the next step. For query type access, it is unacceptable to have to type in a several line procedure or program which is then run in its entirety. The immediate command execution environment is necessary to provide the great flexibility needed in a query system and to provide a sufficiently dynamic, interesting, and satisfying environment for the user.

IV. COMMANDS IN A SUGGESTED QUERY LANGUAGE

Assuming the existance of a command environment as discussed in the previous section, the following set of commands are proposed for the query language:

(A) A series of commands which set global conditions either at the beginning of a query session or at any time the user wishes to change the global environment. All but the SOURCE command have default values:

SET sets short or long prompts, the prompt character to be used, etc. (default long)

SOURCE specifies which record type is to be referenced.

CASES restricts, by key, the cases to be searched in the SOURCE record type (default all).

SELECT specifies the list of variables to be active in the SOURCE record type for the LIST, DISPLAY, ADD, and UPDATE commands (default all).

HEADING indicates the heading to be used with the DISPLAY command. HEADING* implies use of the variable names as column headings.

OUT directs output from either the LIST or DISPLAY commands to a specified file or the line printer device (default terminal).

SETUP a command which initiates an interactive session to request values for each of the global commands.

SHOW lists the present setting of the global commands.

CLEAR clears present settings of named global parameters, returning them to the default values.

HELP explains all commands or specified commands.

(B) Commands for retrieving from the source record type, and for LISTing, SORTing and externally SAVEing records in the work space.

FIND searches the SOURCE r.t. for requested records and moves a copy of these records to the work space.

LIST provides a tabular listing of variables SELECTed.

DISPLAY provides formatted listings and simple reports.

SORT sorts the records in the work area in order, on any set of variables included in the command.

SAVE saves a copy of the work area.

(C) Commands for ADDing, UPDATEing, and DELETEing records in the work file, and then RETURNing them to the data file:

ADD prompts for all necessary key variables plus SELECTed variables for a set of records to be placed into the work area from the terminal. Any variable not included in the SELECT statement (and not a key), is set to missing (undefined) in the created records.

UPDATE prompts the user to make changes in values of SELECTed variables in specified records presently in the work area.

FORCE this command forces a variable to a specified value on all records in the work area.

DELETE tags selected records for deletion from the data base when the RETURN command is executed. It also removes them from further LISTing, DISPLAYing and UPDATEing.

RETURN applies the set of records in the work area to the data base as a volume data entry operation.

V. EXAMPLES OF USE OF THE PROPOSED RETRIEVAL LANGUAGE

In this section, we consider a series of example problems solved by use of the proposed query system on the STUDENT record type.

(1) Set up an initial global environment: (user-typed statements are underlined)

```
>SETUP
  SOURCE          : STUDENT
  CASES(ALL)      : __
  SELECT(ALL)     : STUDID,NAME,AGE,SEX
  HEAD(none)      : *
  OUT(terminal)   : __
>
```

(2) Move all grade two students from the State of Idaho into the work area:

```
>FIND GRADE IS 2 AND STATE IS 'ID'
  [13 records moved to the work area]
```

(3) Display the first two records in the work area after sorting by NAME

```
>SORT BY NAME
  [work area sorted by NAME]

>DISPLAY 2

 STUDID    NAME              AGE   SEX
 SU16028   Cassidy, Nick     7     M
 SU16497   Johnson, Jane     7     F
>
```

(4) Data has just become available for the students matched to these second graders . Update the student records, presently in the work area, with the identifier of their matched student.

```
>SELECT STUDID,MATCHID
>UPDATE
  STUDID :   SU16028 ?
  MATCHID:   MISSING ?6115

  STUDID :   SU16497 ?
  MATCHID:   MISSING ?3228
     .
     .
     .
>
```

(5) Although normal data entry would likely be handled through a forms entry, key punching, or some other standard process, enter the data on these new matching students into the present workspace to illustrate how an ad hoc data entry could be handled through the proposed query system.

```
>SELECT ALL
>ADD
  STUDID  : SU05325
  NAME    : HANSEN, JIM
  STREET  : 132 SOUTH ST
  CITY    : __
  TREATMNT: 2
     .
     .
     .
>FORCE CITY TO "Coeur d'Alene"
```

(6) Now that desired modifications and additions have been made to the work area, submit the contents of the work area for entry into the data base. Entry of this data set as a one- time batch of data allows control of entry by a write access password, the opportunity to specify where the entry report is to be printed, the assignment of a journalling file, and the naming of a file where copies of the records found to have errors will be placed. Thus, even though it is an ad hoc update, it can still be made to conform to established standards which the data base administrator has set up for controlling modifications to the data base and for reporting these modifications in an acceptable manner.

```
>RETURN ALL
  Update password     : XXXXXXXX
  Update listing      : LP
  Journal file        : VCJRNL
  Error record file   : ER800128
   [38 records moved to data entry]
>
```

(7) Save a copy of the present work area contents for future reference or access by other procedures. Next, move all of the student records from school number 732 in district number 58 of the State of Montana into the work area overwriting the present contents.

```
>SAVE IDGRADE2
   [38 records saved in IDGRADE2]

>FIND STATE EQ 'MT' AND DISTRICT =58 &
>>   AND SCHOOL IS 732
   [12 records moved to the work area]

>SELECT STUDID,NAME,CITY,MATCHID
>DISPLAY ALL

  STUDID   NAME            CITY        MATCHID
  SU07404  Johnson, June   Livingston  27448
  SU11276  Rand, Susan     Livingston  24191
  SU12595  Casidy, Jane    Livingston  26551
     .
     .
     .
>SELECT NAME,STUDID
>SORT BY NAME
   [work area records sorted by NAME]

>LIST NAME='@Susan@'
  NAME   : Chadwick, Susan
  STUDID : SU12902

  NAME   : Rand, Susan
  STUDID : SU11276

  NAME   : Rose, Susan
  STUDID : SU12834
```

The above examples illustrate the ease with which records from a single record type can be accessed, modified, viewed, sorted, and tabulated into simple reports. Obviously many utilities can be added to such a system to act further upon the records once collected into the work area. For example, the CML [1] system in operation at McMaster, from which many of the ideas in the proposed query language have come, provides a built-in utility for interactively specifying a format for address labels which are then printed from the work area records. It also provides a utility for generating form letters making use of the work area data records. For further examples of query systems, the reader is referred to the references DATATRIEVE [2], SEQUEL[8] and QUEL[6].

VI. INTERFACING THE QUERY LANGUAGE

The query language system we have described is extremely limited as a stand-alone system. It provides no ability to access multiple record types concurrently. As such, it addresses only a very limited aspect of data management. It provides only the most rudimentary tabular reports. It does not address the general issues of large-volume data input and updating, but only provides easy access for ad hoc changes. Yet, the ease of use and the friendly nature of such a system will make it acceptable to non-computer personnel.

Consequently, the recommendation of this author is that a query front-end system, such as that described in this paper, be considered for addition to whatever capability is presently provided in existing research data management systems. The ideas embodied in this proposed query system could easily be integrated into existing systems such as P-STAT[5], SAS[7], and SIR[9]. The problem today is not that query systems of the type described in this paper are not already in existance, or that any new and earth-shaking ideas have been put forward here; the problem is rather that existing query systems do not interface well with the packages we use with our research data, and that the developers of research systems have not given, thus far, much attention to the area of simple query languages.

VII. REFERENCES

[1] CML (1980) CML-Computer Managed List, Version II Users Manual, Ahmed, Khursh, et al, Computation Services Unit, McMaster University, 1200 Main Street West, Hamilton, Ontario. L8N 3Z5, 1980.

[2] DATATRIEVE (1980) DATATRIEVE-II Version 2.0 Users Guide Digital Equipment Corporation, Maynard, Ma. 01754, 1980.

[3] DATE (1977) An Introduction to Database Systems, Second Edition, Date C.J., Addison Wesley Pub. Co. Inc., 1977.

[4] MARTIN (1977) Computer Data Base Organization, Martin, James, Prentice Hall, Englewood Cliffs N.J. 07632, 1977.

[5] P-STAT (1978) P-STAT 78 - An Interactive Computing System for File Maintenance, Crosstabulation and Statistical Analysis, Buhler, Roald and Shirrell, P-STAT Inc., P.O.Box 285, Princeton, N.J. 08540, 1979.

[6] QUEL (1975) INGRES - A Relational Data Base Management System, Held, G.D., et al, Proceedings of AFIPS 1975, NCC Vol 44, AFIPS Press, Montvale N.J., pp 409-416.

[7] SAS (1979) SAS Users Guide 1979 Edition, SAS Institute, Gary, N.C., 1979.

[8] SEQUEL (1976) SEQUEL 2: A Unified Approach to Data Definition, Manipulation and Control, Chamberlin, D.D. et al, IBM Journal of Research and Development, Vol. 20, #6, Nov. 1976.

[9] SIR (1980) SIR - Scientific Information retrieval Users Manual, Version 2, Robinson, Barry N. et al, SIR Inc., P.O. Box 1404, Evanston, IL. 60204, March, 1980.

Data Editing on Large Data Sets *

James J. Thomas
Robert A. Burnett
Janice R. Lewis

Pacific Northwest Laboratory

Richland, Washington

Abstract

The process of analyzing large data sets often includes an early exploratory stage to first, develop a basic understanding of the data and its interrelationships and second, to prepare and cleanup the data for hypothesis formulation and testing. This preliminary phase of the data analysis process usually requires facilities found in research data management systems, text editors, graphics packages, and statistics packages. Also this process usually requires the analyst to write special programs to cleanup and prepare the data for analysis. This paper describes a technique now implemented as a single computational tool, a data editor, which combines a cross section of facilities from the above systems with emphasis on research data base manipulation and subsetting techniques. The data editor provides an interactive environment to explore and manipulate data sets with particular attention to the implications of large data sets. It utilizes a relational data model and a self describing binary data format which allows data transportability to other data analysis packages. Some impacts of editing large data sets will be discussed. A technique for manipulating portions or subsets of large data sets without physical replication is introduced. Also an experimental command structure and operating environment are presented.

Keywords: Exploratory Data Analysis, Relational Data Base, Data Editing, Statistics, Data Analysis, Subsetting

I. Data Editing in the Data Analysis Process

The data analysis process can be represented by an iterative sequence of gaining understanding of the data, data cleanup and preparation, hypothesis formulation, hypothesis testing, and summerizing results. Much of this initial process of gaining the basic understanding through hypothesis formulation is called exploratory data analysis[1]. For the purposes of this paper the following simple model of the data analysis process will be utilized(figure 1).

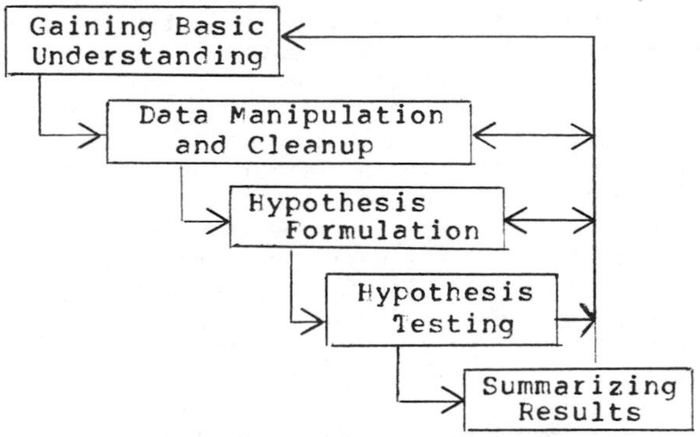

The First step in the model is to gain a basic understanding of the data. The data to be analyzed must come with some minimum syntactic attributes. Most often there is also a minimum semantic description of the raw data. This information along with the initial exploratory phases of data analysis lead to the basic understanding. The next step in the model is to cleanup and prepare the data for analysis. This often involves converting to common

* Work Supported by U.S. Department of Energy
Under Contract DE-AC-06-76RLO 1830

engineering units, handling missing values, categorization, transformations or other data manipulation operations required to develop a hypothesis. After developing a hypothesis the analyst may be required to again transform the data. Then the hypothesis is formulated and tested for validity. This provides the data analyst with a better understanding of the data, leading to the resultant analysis or subsequent iterations. The data editing operations are usually found in the first phases; gaining the basic understanding through hypothesis formulation.

Our experience with data analysis of medium to large data sets indicates that gaining this basic understanding and data preparation phase constitutes a time-consuming and often cumbersome process. This process typically involves the utilization of uniquely written computer programs which only delay and distract the data analyst. Therefor an interactive tool was designed to include base level capabilities from a collection of computer science disciplines to form an interactive data editor specifically tailored to edit large data sets.

The data editor is one of several techniques being developed by a joint group of computer scientists and statisticians at PNL with direction to develop methodologies to Analyze Large Data Sets(ALDS)[2-5]. Further references to the data editor will be by its name, ALE, ALDS Data Editor.

II. Data Editing Requirements

The data editing requirements will be divided into those that were initial requirements and those requirements that a direct result of editing Large Data Sets. The latter will be discussed in Section III.

After observing the data analysts on the ALDS team it was obvious that a highly interactive environment must exist. The process of gaining the basic understanding and data cleanup of the data is often exploratory in nature and therefore a regimented batch process would not fit this phase of data analysis. The majority of the operations in the initial phases of the analysis of large data sets were found not to be exotic. Therefore only the base level capabilities in each functional area were initially required to provide the minimum data editing requirements. Some of the required techniques can be commonly found in relational data management systems, statistical sampling packages, interactive graphics systems, and text editors. A few of the base level capabilities required in an interactive data editing environment provide the data analyst with the capibilities to:

* logically model the data to correspond closely to the data analyst conceptual framework (usually a case by variable model),
* logically define and manipulate groups of data by case, variable, or value,
* modify data on a case basis or on a more global basis,
* modify variables as functions of other variables, constants, and arithmetic functions,
* create and delete variables and cases,
* ramdomly select groups of cases,
* specify general expressions with schemes to tailor them to specific data sets,
* view data graphically.

The data editor can now be defined as a computational tool for the exploration, manipulation, modification, and preparation of data prior to and during the hypothesis formulation phases of data analysis.

III. Observations and Impact of Large Data Sets.

When editing large data sets several observations and subsequent data editing requirements were formulated. These are as follows:

* The initial data manipulation process is awkward. The largeness of the data has an impact on how the data analyst proceeds. Such questions as, "will it fit?", " If not how can I partition the data set meanfully?", are common.
* The data's semantic descriptive information is vital in order to grasp the essential structure of large data sets. Minimum descriptive statistics should be provided as standard descriptive information e.g. range, mean, mode, distributions, missing value indicators, etc.
* Replication of portions of the data set are a significant problem. When subsetting the data into homogeneous clusters one often creates a copy of part if not all of the cases within the cluster. In a multi-million item

data set severe physical limitations will exist.

* Large data sets are often collected over time and under diverse conditions, making missing data the norm rather than the exeption.

* The human interface between the data analyst and the computional tools must be obvious to the data analyst. The constant distraction on an unfriendly interface becomes dominate when working an large data sets. Also different data analysts work in different interaction sytles. One prefers an elegant command syntax while another literally fumbles at the keyboard and prefers a graphic interaction style.

* The data analyst must have the ability to take small samples of the data, perform an operation (i.e. transformation) on that data , and verify the results, all in a temporary environment. Then after verification perform the operation on the entire data set.

* The review of the selection criteria, transformations, etc. is required prior to execution when working with large data sets. An interactive environment also provides a faster technique for making mistakes. The review process should be the default with optionally turned off. Our experience is that it has never been turned off.

* A context sensitive interrupt is required to provide the data analyst the capibility to examine a process in progress and determine its status, intermediate results, etc. Also optionally terminating the process is required. This becomes particularly important when performing global operations on large data sets.

* Automatic command and data modifcation history is essential due to the number of operations possible and flexibility offered in an interactive environment. The history logging should become a part of the file description.

IV. The ALDS Data Editor(ADE)

ADE was developed utilizing a self-describing binary data format[6,7]. This format allows for an easy relational conceptional model of the data. The model is that of a flat file of rows and columns. From the data analyst point of view there are cases and variables respectively. The following paragraphs illustrate a few of the most commonly used facilities in the data editor to manipulate data in the early phases of the analysis of large data sets.

The first and probably the most used facility is the specification of a complex relational clause to apply on subsequent operations. This clause must be a well formed expression i.e.

```
DATE > 1975 & SAMPLE < 200
(VAR1 > VAR2) ! (VAR3 <= 25.5)
```

The following example will list the first 10 cases that satisfy the condition where VAR1 < VAR2:

```
CONDITION VAR1 < VAR2
LIST 10
```

The transform facility allows the data analyst to modify an existing value with a simple assignment or a complex function. This also can be performed on a single item or on the entire data set. A sinple example to change the variable named RADAR1 to 42.5 would be:

```
CHANGE RADAR1 = 42.5
```

A more complex operation to tranform all values in TOTAL1 to be a function of several other variables would be:

```
XFORM TOTAL1 = VAR1 * (1-ABS(X))/VAR4
REPLACE ALL
```

Note the above XFORM statement specifies the transformation while the REPLACE statement performs the operation. The above facilities can now be combined to perform varaible transformations under specified conditions:

```
CONDITION VAR4 > 20.5
XFORM     VAR3 = VAR4 + VAR3
REPLACE ALL
```

The above example replaces all values in VAR3 by the sum of VAR4 and VAR3 where VAR4 greater than 20.5. The ability to create and delete variables and cases is provided. Note that on creation of a variable the data analyst will be requested to provide semantic information about that variable. Early experience indicates that only a few additional variables are created during data analysis. These new variables are usually functions of other variables transformed to facilitate hypothesis testing. The delete operation is seldom used.

When analyzing large data sets it is mandatory to have the capability to define a complex operation and then perform that operation on the entire data set. These types of operations are called global as compared to local operations on individual cases and variables. Both global and local operations are supported. The analyst may perform a single operation or a group of operations by executing a group of commands in a command file. This combined with a synonym capability allows the data analyst to specify a formula in a generalized expression and then specify a synonym to map the expression into the particular variable set for the current file.

A single missing value code is supported in all three data types, real, integer, and character. The data analysts have stated the need for at least five different missing value codes. A missing value is denoted by the '#' symbol. A search for all cases that contain a missing value in variable VAR1 would be:

```
CONDITION VAR1 = #
SELECT ALL
```

Handling large data sets usually requires an exploratory process of cutting the data set i.e. forming subsets. Forming physical subsets or replications of the original data set when handling large data sets becomes a limiting factor in the data analysis process. This limitation may be either a function of external storage space(disk) or access time. These data sets are often temporary and are used to verify a hypothesis. Therefor a technique is introduced to create a virtual subset(V-SET). A V-SET appears from the analyst's perspective to be a separate logical file which all commands can reference. The following command sequence illustrates how a V-SET is formed by first providing a relational condition and then forming the V-SET definition via the TABLE or SELECT command.

```
CONDITION FIRSTVAR < 20 & B > 15
TABLE
```

The V-SET is composed of a table of indexes referring back to the orginal file and associated descriptive information. The analyst can interagate the table directly although most of the references are to the selected cases defined by the V-SET. The V-SET maintains its own sequencing as well as the original sequence from the master file. Either may be used for referencing purposes. The formation of a V-SET can be controlled by individual or group case selection, case selection based on a relational condition, random selection, or any combination of the above. From the data analyst's point of view this provides a straightforward technique to perform stratified sampling. The following command sequence selects 50 cases randomly that satisfy the condition where THIRDVAR = 20.

```
CONDITION THIRDVAR = 20
RAND 50
TABLE
```

The data analyst may perform stepwise refinement analysis by a sequence of TABLE operations on resultant tables. In effect a retable operation. Assuming the above sequence has been executed forming a V-SET, then the following sequence will randomly select 5 cases from the 50 chosen before and form a new V-SET.

```
RAND 5
TABLE
```

When the analyst wants to create a separate physical file the SAVE command can be used. This saves the file as defined by the current V-SET. Note the variables can be selected and/or reordered if necessary. When passing data on to other analysis systems this becomes an important capability. See Appendix A for a brief description of the current available commands.

V. Human Factors and the Editing Environment

The editing environment is designed to be highly interactive. A menu driven syntax or a command syntax could be implemented as the user interface. The current implementation provides a free form command interface with response occurring after a command is entered.

There are two editing modes; MASTER and SUBFILE. MASTER mode provides editing on the master file with access to all the cases in the file. SUBFILE mode provides access to only those cases that are defined by the V-SET definition. On entry the user is asked for a master file name. Then the user may create V-SETS and switch back and forth between editing the master file or the V-SET. Also on entry the user is asked if a backup is desired. This can become a problem when working with large data sets.

Early in the implementation of the data editor we realized that the editing mode and the current case number were not obvious and allowed for confusion. Therefore a Variable Prompt was developed to provide the editor status at all times.

The prompt took the form of:

ADE: $\begin{Bmatrix} M \\ S \end{Bmatrix}$ n>>

where M = master mode
S = Subset mode
n = current case number

ADE:M3>> illustrates editing in master mode on case 3.
ADE:S242>> illustrates editing in subset mode on case 242.

Editing large data sets often involves careful setup of complex operations. It is desirable if not mandatory that before starting a complex operation, the data analyst be provided with a review and verify option for the specified operation. This can be turned off at the users request. An example review response would be:

```
ADE:M1>> SELECT ALL
  CONDITION: A>B & C>20
  TRANSFORMATION: NONE
  VARIABLES SELECTED: NONE
  CASES SELECTED: NONE
  Do you want to continue y/n?
```

Also a dynamic interrupt facility is required to provide the data analyst with the capabilities to suspend operations in progress, determine the current status, and then optionally continue. This is especially important when performing complex operations on large data sets. The following sequence illustrates an interrupt during a REPLACE operation with 85000 cases in the file.

```
ADE:M1>> REPLACE ALL
  Replace set for 85000 cases
  CONDITION A>B & C>20
  TRANSFORMATION: VAR1 = A+B
  VARIABLES SELECTED: NONE
  CASES SELECTED: NONE
  Do you want to continue y/n? Y
 -- CTRL/C During select at case 44381
  Do you want to continue y/n? Y
```

Another implication of dealing with large data sets is that the analyst cannot view all the data on a video screen at one time. One should utilize the terminal as a viewport into the data base. With this concept in mind both vertical and horizontal scrolling is provided in the data editor.

Experience with large data sets also indicates that typing all the variable names needed for an operation can be tedious and error prone. A simple notion of grouped variables helps this problem. The command VARSEL SAM34:SAMLAST will select all variables between SAM34 and SAMLAST inclusively in subsequent operations. The command CASSEL 10:50 specifies that subsequent operations will be performed on cases 10 through 50 inclusive.

A facility that became vital when editing large data sets was an automatic logging of all commands. This provides for a trail to improve data validity as well as a history for file integrity. This also provides the system developers with a tool for logging and verifying program correctness. The log files were also found useful for initial introduction and training purposes.

VI. ADE Command Syntax

The command syntax is designed with the philosophy that it is important for a command to be obvious rather than simple. The commands as implemented in ADE usually reflect the logical phrases in a complex operation.

The commands are divided into three classes:

Specification commands
Information commands
Operation commands

Specification commands provide the ability to specifiy relational conditions, selection of cases and variables, random sampling criteria, etc. These specifications then are applied to a variety of information and operation commands providing multi-dimensional viewing and subsetting on the data set.

Information commands provide the analyst with selected file viewing and printing. Also Information commands provide access to all the specification status i.e. CONDITIONS, SYNONYMS,etc. Listing implies output to the interactive device while printing implies output to the line printer.

Operational commands include the creation of new data variables, subsetting, locating, deleting, replacement, command file execution, synonym definition, and other file manipulation commands. A one sheet command syntax brief is given in Appendix A. Detailed discussion of each command is not given in this paper.

If desired one can combine commands together to form more complex single commands:

```
CON A<25 & B>50 AND CASSEL 10:50 SELECT ALL
        LIST 20 CONDITION A<25
```

VII. Conclusion

The data editor concept does provide a unique approach to solve the initial data preparation phases of data analysis on large data sets. Its capabilities are what the authors define as the minimum functional needs to start exploratory data analysis and effective data manipulation. The concept of a V-SET is conceptually well understood by the data analysts with little instruction. It seems natural in the highly interactive mode of data analysis. The command structure is still under evaluation at PNL. The automatic verification of command specifications and automatic logging of all commands seemed natural and of great benifit both during implementation and analyst use.

Many of the techniques discusssed in specific reference to large data sets may also apply to small and medium size data sets. However, when dealing with large data volumes, these techniques become a necessary part of data manipulation. The techniques described are by no means encompassing of techniques to handle large data sets but hopefully will offer food for thought.

Acknowledgements

The authors wish to thank the statistisions at PNL, Wes Nicholson, Dan Carr, and Dave Hall for their guiding comments. Also the authors wish to thank Harvard Holmes of Lawerence Berkeley Laboratories and George Rogers of George Washington University for their editing comments.

References

[1] Tukey J.W., Exploratory Data analysis, Addison-Wesley Co., Reading Mass., 1977

[2] Thomas J.J. et al, Analysis of Large Data Sets on a Minicomputer, Proceedings of the Computer Science and Statistics: 12th Annual Symposium on the Interface 1979, Univ. of Waterloo, pp. 442-447

[3] Nicholson W.L., Analysis of Large Data Sets, Proceedings of the 1979 DOE Statistical Symposium, Oak Ridge, Tn., September 1980.

[4] Carr D.B., The Many Facets of Large, Proceedings of the 1979 DOE Statistical Symposium, Oak Ridge, Tn., Septmeber 1980.

[5] Burnett R.A., The ALDS Project - Computer Science Research Areas, Proceedings of the 1979 DOE Statistical Symposiom, Oak Ridge, Tn., September 1980.

[6] Thomas J.J. et al, Distributed Data Analysis in a Mobile Real Time and Minicomputer Network Environment, Proceedings Trends and Applications, 1979, IEEE Cat. No. 79CH2402-7C.

[7] Burnett R.A., A Self-Describing Data File Structure for Large Data Sets, Computer Science and Statistics: 13th Symposium on the Interface, Carnegie-Mellon University, Pittsburgh, Penn., 1981.

Appendix A: ADE Command Brief

ALDS Data Editor (ADE)
Command Brief

SPECIFICATION COMMANDS:

```
SEQPTR n               set sequence pointer to n.
CONDITION  r           set relational condition r.
CASSEL (case-list)     set sequential case selection
VARSEL (var.-list)     set variable selection
XFORM  arith.          set transform for variable
RAND   {n,.fr}         select n samples / probability sample function
VERIFY {ON,OFF}        set verify switch
MASTER                 set mode to MASTER
SUBFILE                set mode to SUBFILE (tabled subset)
```

```
   where: n is an integer;
          r is a Boolean expression which may include the
             operators <,>,<=,>=,=,^= (not equal), ! (OR),
             & (AND);
          case-list is a list of case numbers, e.g.  1, 5, 20:40:2 ;
          var.-list is a list of variable names, e.g. V1,V2,V5:V10 ;
          arith. is an arithmetic expression which may include
             +,-,*,/,EXP,LN,LOG,SQRT,SIN,COS,TANH,ATAN
```

INFORMATION COMMANDS:

```
* LIST {n,ALL}         list n (or all) cases on the terminal
* PRINT                print subfile on the line printer
  SHOW  { TABLE(cc)  VERIFY  CONDITION  CASSEL  VARSEL  XFORM
          SEQPTR   MODE   DESCRIPTION   SYNONYM  }
                         show status of command specification.
```

OPERATION COMMANDS:

```
(cc) * TABLE (SELECT)       form a table of cases defining a subset
(cc) * FIND  n              find nth case which meets specifications
(cc) * REPLACE {n,ALL}      replace variables in n (or all) cases
(cc) * DLCAS  n             delete n cases
     * DLVAR                delete variables
       ERASE                erase case table selections
       RESET                reset all specifications
     * SAVE 'filename'      save subset in case table to 'filename'
       RELAB oldlab,newlab  change label "oldlab" to "newlab"
       NWVAR                add a new variable to data set
       NWCAS                add a new case to data set
       CHANGE  arith.       change single variable
(cc)   EXECUTE 'filename'   execute a command file
       SYNONYM equiv,name   define synonym "equiv" for "name"
       EDIT                 close existing file and restart editor
       END                  end the editing session
```

* commands operate only on variables and cases qualified by SPECIFICATION commands.

cc -- command is CTRL/C interruptible

Workshop 12

Graphical Methods and their Software

Organizer: Beat Kleiner, Bell Laboratories

Chair: Beat Kleiner, Bell Laboratories

Invited Presentations:

Census Bureau Statistical Graphics, Lawrence H. Cox, U.S. Bureau of the Census

Mosaics for Contingency Tables, J.A. Hartigan, Yale University, and Beat Kleiner, Bell Laboratories

The Use of Kinematic Displays to Represent High Dimensional Data, David Donoho, Harvard University, and Peter J. Huber, Harvard University, and Hans-Mathis Thoma, Harvard University

CENSUS BUREAU STATISTICAL GRAPHICS

Lawrence H. Cox, U.S. Bureau of the Census

ABSTRACT

For some time, the U. S. Bureau of the Census has published statistical data in graphical as well as tabular formats. These graphical displays include barcharts, piecharts, line graphs, time series plots and univariate and bivariate statistical maps. Such publication graphics are provided both monochromatically and in color. Recently, the Census Bureau has initiated a research program to investigate the application of computer graphics to statistical data analysis. Examples of such analytical graphics are regression and time series plots, scatterplots used in outlier analysis, line graphs depicting rate of change overlayed on barcharts depicting level or value of one or more variables, and color statistical maps. This paper describes the computer graphics hardware and software capabilities of the Census Bureau, our experience in computerized statistical graphics, and research directions for employing computer graphics as an analytical tool in statistical data analysis.

Keywords and phrases: computer graphics, statistical data analysis, graphical presentation of data, U.S. Census Bureau.

Author's footnote: Much of the material presented in this paper was also presented in Barabba, Cox and DesJardins (1980).

INTRODUCTION

Graphics is today at the center of a revolution in information technology and communication. This revolution has been brought about by the disparity between society's recently acquired ability to produce and combine almost limitless volumes of raw data and the problem of developing technologies to keep pace with the resulting ever-increasing need for effective new techniques to communicate and synthesize the information these data embody.

As the nation's primary data provider, the Bureau of the Census is keenly interested and involved in the use of graphical techniques to display and analyze statistical data. Statistical graphics are graphics used in this manner. As an area of study, statistical graphics is rapidly maturing. It poses the question: how can we best combine graphical displays, numeric data and explanatory or interpretive text to most fully and accurately inform policy planners and other data users and analysts?

This paper briefly describes the Census Bureau's current and planned future efforts in statistical graphics. Most of the topics included here are discussed in greater breadth and technical depth elsewhere. Therefore, in lieu of providing technical details, we indicate available references whenever possible.

The Bureau uses graphical displays in two primary program activities, data presentation and statistical data analysis. After describing the brick and mortar of the Bureau's graphics production capabilities, we separately discuss these two areas of statistical graphics. We conclude with a look forward to our future plans and goals.

CENSUS BUREAU COMPUTER GRAPHICS HARDWARE AND SOFTWARE

The Census Bureau is now making graphic arts quality maps and charts in a fraction of the time previously required. Bar charts, trend charts, pie charts and maps may now be produced in finished form within minutes. Non-technical personnel such as clerical staff can sit at a computer terminal, give English-like instructions, and immediately obtain graphic output. Once the desired graphical and artistic form is achieved, production of a publication-quality finished product can be directed to one of several graphics output devices. Making this possible is an English-like computer command language developed in-house through which the user invokes the DISSPLA computer graphics system. In this language, the user specifies to the computer the kind of chart desired and provides appropriate numeric data and labelling, scaling, titling and footnoting specifications. A computer program converts these commands into parameters used by DISSPLA to produce the desired chart on the specified graphics display device.

Bureau graphics displays are currently available through devices such as Tektronix and Chromatics computer terminal display screens and a Xynetics flatbed plotter. Production-quality graphic output is available through a COMP-80 computer-output-to-microfilm device. These graphic devices can produce color or black and white images on a computer terminal screen or on film or bond paper. A small number of microcomputer based graphics workstations for research and data analysis work are projected for acquisition and use this year.

The Bureau is utilizing and exploring new uses of computer graphics in its statistical publications at an increasing rate. In the past, the assistance of a computer programmer was usually required to produce computer graphics. The advent of user-oriented graphics software such as DISSPLA and the graphics workstation concept soon will make more graphics more directly available to a wider class of users.

Bureau graphics software operates in the following manner. Parameter-driven computer programs are capable of producing the types of graphics most commonly used at the Bureau, including barcharts, time series graphs, piecharts, maps and slides. This computer software provides the user with a keyword input mode through which (s)he may specify the appearance of specific features of the chart. The computer programs take default actions on those features not specified. If desired, the user may invoke an interactive input mode in which the computer program asks the user specific questions about the chart and leads him or her through the chart description process. The keyword method, being more concise, is the method used by experienced statisticians, clerks and staff who produce charts more or less constantly. The interactive method had been primarily used in orienting new graphics users to the system and its capabilities. However, as interactive graphics becomes more readily available to our in-house user community, the interactive mode will surely become the method of choice for the legions of subject matter analysts at the Bureau who need to analyze data and prepare charts for publication. The Bureau has also developed sophisticated mapping software. Special purpose statistical graphics software, such as GR-Z and SABL, used in seasonal adjustment of time series, are also available at the Bureau for the expert statistician.

The Census Bureau is in the process of delineating its total computer graphics requirements for the 1980's. These requirements will then be translated into functional specifications for an integrated interactive Bureau computer graphics system. This system, called CBIG (Census Bureau Integrated Graphics) is being designed to serve the Bureau's total graphical needs for statistical data presentation and analysis as well as other graphical applications such as computer-assisted forms design. CBIG is planned to be fully operational during 1983.

GRAPHICS FOR THE PRESENTATION OF STATISTICAL DATA

In their article in The American Statistician, Beniger and Robyn (1978) trace the use of graphics to convey quantitative information as far back as the first known map, a map of Northern Mesopotamia dating to about 3800 B.C. Activities involving collection and presentation of "statistical" data are also quite old. However, systematic attention was not given to the use of graphics for presenting statistical data until 1786 when the Englishman William Playfair published his Commercial and Political Atlas.

In his Atlas, Playfair introduced graphic displays which remain in widespread use today, most notably the barchart. In subsequent publications, Playfair invented the piechart and the circle graph. These and other graphic displays soon became established standards for data presentation.

In the United States, limited use was made of graphics in Census reports prior to 1870; and the few graphics employed up to that time were essentially confined to the field of vital statistics (Funkhauser 1937). However, beginning with the Ninth and Tenth Censuses, conducted under the supervision of Francis Walker, tradition and innovation in the use of graphics for presenting statistical data were established in the United States.

The first statistical Atlas of the United States, the Statistical Atlas of the Ninth Census, was published in 1874. In this atlas, several innovative forms of graphical data presentation were introduced. Among these were bilateral histograms known as "age pyramids" and companion bilateral frequency polygons depicting age-sex comparisons for several variables including marital status and death by various forms of disease.

Subdivided squares were used in the 1874 Atlas in a most innovative fashion to depict ethnicity and migration data. Each state was represented by a square of area proportional to the population of the state. This square was divided by vertical lines into three rectangles of area proportional to the numbers of foreign, native non-white and native white components of the population. The latter two rectangles were also subdivided horizontally to represent proportions of individuals born within the state and those born in other states. In addition, next to each state square was a proportional rectangle of equal height representing the number of persons born in the state who had become resident of other states.

The Atlas also made considerable use of color, pie charts and concentric circle diagrams. While the graphics in the Atlas were not altogether successful in accurately conveying the statistical information to the reader, the Atlas was well-received by the public and marked a milestone world-wide in the use of graphics for presenting statistical data.

Although no atlas was prepared by the Census Bureau for the Tenth Census, the various census reports were replete with graphic illustrations. While the diagrams illustrating the Tenth Census were not of as high technical quality as those of the Ninth Census, according to Funkhauser (1937) the cartographic illustrations were superior.

The Statistical Atlases of the Eleventh and Twelfth Censuses were prepared under the direction of Henry Gannett, the chief geographer of the U. S. Geological Survey. These atlases epitomize the use of graphics in statistical reports and publications during the 19th and early 20th centuries. Their most distinctive feature is their use of many statistical maps which colorfully and ingeniously illustrate a variety of statistical facts and relationships.

In the atlases of Gannett, color and shaded maps displayed counts and proportions of population distributions for variables including ethnicity, race, sex, urban/rural, religiousity, and death by specific causes; economic data such as value of personal and real property and taxation data; and agriculture statistics. A particularly interesting map displayed the location of the center of population during each census year since 1790, at which time the population center of the nation was some ten miles east of Baltimore, from which it began an almost due westward march to Columbus, Ohio during the following century. This form of map display remains today a part of our Decennial Census reports. Age-sex pyramids were used by Gannett to represent proportions of variables such as the slave and native and foreign white components of the population. Barcharts were also prominently featured, illustrating variables such as conjugal condition (marital status).

Although a complete set of standards for use of graphics in the presentation of statistical data does not exist even today, conventions for the use of specific graphic media including barcharts, piecharts, and circle, line and time series graphs can be traced back through the 19th century to the work of Playfair. The Census Bureau continues to utilize these media in its census and survey publications in a more or less conventional manner. In meeting its responsibility to present statistical data to as many users and in the most broadly useful fashion as it can, the Bureau attempts to achieve uniformity and consistency in the statistical graphics it publishes. This policy allows our data users to easily assimilate and combine these data across publication and time boundaries.

Even though the graphic media we employ have remained relatively constant over time, the variety and quality of content and level of graphic communication achieved through these media in Bureau publications continues to mature. In our non-recurrent work, we have experimented with new and expanded graphic formats. This is illustrated by the federal statistical publication STATUS magazine to which the Bureau contributed substantially.

STATUS was an outgrowth of special graphic charts and displays prepared on a regular basis by the federal statistical system to keep the president and vice-president abreast of major economic and social trends. In its final form during the mid 1970's, STATUS was a monthly chartbook of social and economic trends which utilized a variety of graphic media in color. These media include barcharts, piecharts, line, circle and time series graphs and statistical maps. Each map or chart was accompanied by a brief description of the socioeconomic trend it portrayed. In addition to its information value to policy planners and other data users, STATUS provided the Bureau with an opportunity to experiment with innovative techniques for graphic communication of statistical information.

A survey of STATUS readers was conducted to measure its popularity and utility to data users. STATUS received high marks from each user group in all important areas, with the notable exception of the "map of the month" feature. The map of the month was a centerfold choropleth statistical map of the United States produced by the Bureau which displayed frequency class intervals by county for a bivariate distribution of interest. For example, the map of the month for August 1976 displayed for each county the death rate of males due to cartiovascular disease by the percentage of households with more than one person per room. Many users surveyed reported that the bivariate statistical maps were difficult to read and that they did not informatively depict correlations between the single variables. Local area planners polled were more positive in their attitudes toward these maps and their usefulness than were their counterparts in the federal statistical community. This we interpreted as an expression of greater need for this form of data display (particularly on a county basis) on the part of state and local officials. The feedback obtained from this survey was most valuable to the Bureau, reminding us of our own early learning and familiarization experiences with bivariate choropleth maps. Having overcome these cognitive obstacles ourselves, we realized that the future of this form of graphic communication depended upon our ability to educate users in its application and interpretation. To aid users in interpreting the map of the month, as adjuncts to the bivariate map we provided the two corresponding univariate maps on the centerfold, as well as a brief description of how the map may be interpreted. To improve our understanding of the reader's perceptions, we also initiated investigations into the cognitive aspects of color combinations.

From the time of Henry Gannett until the present day, the Bureau has been a leader in innovating and improving graphic display through statistical maps. Our Urban Atlas series consists of statistical maps containing detailed demographic data at the census tract level for over 60 Standard Metropolitan Statistical Areas (SMSAs). Our so-called "satellite map" which depicts the population distribution of the nation by brightness intensity of white light on a blue background has gained singular popularity. Bureau research into new uses of color in statistical mapping during the early 1970's led to our bivariate choropleth maps which portray interactions between two variables by representing class intervals in terms of color differences and intensities defined on a two-dimensional color grid. A recent cooperative effort involving the Bureau and NASA produced DIDS (which originally stood for Domestic Information Display System), an interactive computer system for producing choropleth statistical maps from a variety of socioeconomic databases.

A fact which is sometimes overlooked in discussions and plans for the use of graphics in data presentation is that graphics can almost always attract and draw the reader's attention to the data more quickly and successfully than can tables or text. Because attention is prerequiste to communication and learning, statistical graphics can and should do more than simply convey information and form impressions upon the user—they should be used to create and encourage interest in the data which otherwise might not develop. In so doing, statistical graphics can significantly help familiarize the user with the data and orient and educate him or her to its breadth.

The Census Bureau is keenly interested in developing

new effective techniques of graphic display of statistical data, as well as revitalizing useful old techniques and improving those in current application. Because our continuing responsibilities to the broad user community limits our ability to experiment with new graphic data displays in any substantial way in our periodic publications, we seek to use special publications such as STATUS to maximum advantage to help introduce graphic innovations into the public domain.

The Bureau's tradition of producing census statistical atlases was broken after the Statistical Atlas of the Fourteenth Census was published almost sixty years ago. The period from 1920 to the mid 1960's marks a waning interest in statistical graphics. No less pronounced, however, is the strong resurgence of interest in this area which began approximately fifteen years ago. Consistent with this return to sustained interest in graphics within the statistical community, the Bureau's plans to produce the Census Historical Chartbook in 1983.

The Chartbook will revive graphic traditions and techniques established by Walker and Gannett. It will feature data from the 1880 and 1980 Decennial Censuses and, using modern techniques of computer graphics and statistics, will display these data side-by-side for comparison and analysis. The reader will be able to visually experience and appreciate nineteenth century graphic arts forms as they portray the growth of the nation during the past 100 years, and will have the opportunity to compare the artistic quality of modern computer statistical graphics with the human graphic arts of the nineteenth century. The Chartbook will also include modern statistical analysis of the data so that it will have an historical significance statistically as well as graphically.

The Chartbook, an historical statistical graphics project, should not be confused with the 1980 National Atlas to be produced by the U. S Geological Survey. The Bureau of the Census will prepare approximately 100 statistical maps for the 1980 National Atlas.

GRAPHICS IN STATISTICAL DATA ANALYSIS

For some time, policy planners have been using statistical charts, particularly those portraying change, time series graphs, regression lines with confidence bands, scatterplots and, recently, techniques such as box-and-whisker and quantile-quantile plots to interpret statistical data, identify trends and forecast key variables. This area of statistical data analysis is aimed at interpreting statistical data and discovering its meanings and implications.

The Bureau of the Census is the principal gatherer and provider of statistical information for the nation. The Bureau's primary responsibility to the data user community is to provide useful statistical data which are as timely, complete and accurate as possible. For various reasons, federal statistical policy requires the Bureau to maintain impartiality in its role as data gatherer and disseminator. Consequently, responsiblility for data interpretation and policy analysis resides elsewhere in the federal statistical system, Congress and the Executive Branch of the federal government, with the result that the Bureau performs little or no statistical data analysis of an interpretive, forecasting or policy planning nature on its data (although it routinely provides measures such as standard errors to facilitate the efforts of others in these areas).

However, even in its efforts to maintain impartiality, the Bureau must ensure that it does not divorce itself from the needs and perceptions of its data user community. Therefore, while the Bureau does not statistically analyze its data in an interpretive fashion, such as would a policy planner, it routinely performs descriptive analyses on its data prior to their release. Although these analyses are usually textual adjuncts to published tabular or graphical displays and are not themselves graphical, a brief discussion of them is relevant to the present topic.

There are two principal reasons why the Bureau performs descriptive data analysis. First, the Bureau must remain in close touch with the needs and perceptions of the data user community. The more familiar we are with user requirements for our data and the range and limitations of the analytic techniques users apply to these data, the better able we are to produce data which are as complete and usable as possible for these purposes. Second, because our users vary widely in their statistical sophistication, it is often incumbent upon the Bureau to educate the user to the meaning and limitations of the data. To do so, it is often useful to briefly describe the data and its implications. For example, each chart and map in the publication STATUS was accompanied by a description of the trend it depicted.

At the risk of drawing fine semantic distinctions, we characterize descriptive data analysis as analysis which points out the implications of the data set alone, without reference to auxiliary information, whereas interpretive data analysis is performed by considering the data set in its proper context, be that context socioeconomic, political, ecological or otherwise. The Bureau has traditionally made descriptive analyses of its data available to users whenever this information was deemed appropriate and helpful. When these data are displayed graphically, even more care must be taken in their description.

There is a third major area of statistical data analysis, statistical data analysis for the purpose of verifying and enhancing data quality and for developing and investigating the properties of new and improved statistical techniques. The Bureau of the Census is heavily involved in this area of statistical data analysis and is beginning to use statistical graphics in these efforts. The methodological settings for these applications are quite diverse, including time series and regression analysis, distribution fitting, cluster analysis, data editing and estimation. In the remainder of this section, we describe the Bureau's graphical activities in graphical statistical data analysis. In many cases, the projects discussed are in their initial stages. Plans to increase our use of graphics in these and other areas are discussed in the next section.

The focal point of these graphical statistical data

analysis activities is the newly formed Statistical Graphics Research Staff. This staff is responsible to develop or otherwise acquire statistical graphics techniques and software which promise to improve Bureau data analysis operations and to introduce these new analytical tools to the Bureau data analyst community.

Two areas of mathematical statistics in which graphics play an important role at the Bureau are time series and regression analysis. Traditionally, graphics have played a major role at every stage of time series analysis. These applications are well described in Cleveland and Dunn (1979) and we do not elaborate upon them in the present discussion. Under the ASA/NSF/Census research fellowship program, new work in time series analysis has recently begun at the Bureau, particularly in the area of seasonal adjustment. Computer statistical graphics, particularly the SABL and GR-Z computer software packages developed by Bell Laboratories, are a major component of Bureau time series research. An independent research program aimed at developing an integrated statistical graphics capability for the X-11 time series software package which would be usable by non-programmer time series analysts is currently underway at the Bureau.

Graphical techniques are also finding their place in regression analysis at the Bureau for purposes such as choosing transformations of key variables and fitting distributions to residuals data. In our discussion of the Methods Development Survey later in this section, we cite a recent useful example of the role graphics has played in regression analysis at the Bureau.

Another current and potentially major use of graphics for statistical data analysis at the Census Bureau is in our statistical editing, auditing and data quality assurance activities. Similar to their use in time series and regression analysis, graphical displays of numeric data can often quickly inform data analysts if they are on the wrong methodological track and can draw attention immediately to aberrant values such as statistical outliers, as well as to meaningful clusters of data values. Moreover, graphics often can communicate this information much more quickly and directly than can voluminous statistical summaries.

Statistical editing of individual data involves verifying that data values derived from respondents or other sources such as administrative records meet certain criteria of reasonableness. For example, it is unreasonable (although not impossible) for a female to be 14 years of age and have six children, or for the gross receipts of a firm divided by its number of fulltime employees to be less than $100. At the Bureau, such trivial as well as considerably more complex edit failures of individual data records are detected by analytical computer programs. These edit programs currently operate in batch mode without the aid of graphical displays. Graphics comes into play in the editing process when the resulting statistical summaries and underlying statistical operations are audited to assure their quality. We now illustrate this process in detail.

In sample surveys such as economic surveys, sample data obtained from units within specific industry groups or lines of business are weighted-up (inflated) and aggregated to produce estimates of variables such as total sales or total value of shipments in specific industry groups and geographic areas. One check that is often performed on the accuracy of these current estimates is to graphically display them and compare them with each other and with graphical displays of other data such as prior-year estimates or actual census values. The modes of graphical display typically employed for this purpose are scatterplots or simple or bilateral barcharts, often overlayed by a line chart depicting rate of change.

Examination of these displays may identify current estimates which are spurious because of their unusual values or their value differences with other data. The identification of such questionable estimates triggers further investigation. Typically, sample observatons which contribute to questionable estimates would be examined. Although these sample observations have already passed edit tests of reasonableness, the necessarily broad nature of these tests may have failed to detect an error which has since been magnified when the sample data are inflated-up to produce the survey estimates. If necessary, follow-up with the respondent will be performed, resulting either in verification of the accuracy of the reported observation or a correction.

Even if all reported sample observations are correct, one of these values may be uncharacteristic of the business activity of this firm and its industry, such as if this single firm had undergone a prolonged strike during the past year. Bureau data analysts might then determine that a more accurate estimate of business activity in this industry could be achieved by adjusting its weighted contribution to the estimate. As a component of our statistical standards and methodology program, the statistical analysis triggered by examination of the graphical displays may indicate problems with the sample design or the weighting scheme employed in the estimation process. This would lead to further investigation resulting in improved procedures or modified weighting parameters for future surveys.

The extent to which these analyses are performed in a particular survey of course depends upon resource availability, priorities and schedule constraints. Because statistical graphics have already established an important role in maintaining and improving the quality of our statistical products, we plan to increase their use in our statistical edit and audit activities.

Statistical choropleth and dot maps are also useful in the edit and correction processing of geographically defined data such as county data. Our users have for some time been using statistical maps for interpretive analysis of statistical data. A recent Bureau project made use of computer-generated statistical maps to analyze population projections made at the county level. Choropleth map displays of these projections highlighted regional patterns of growth and quickly drew attention to unusual or erroneous values. These maps simplified and enhanced the process of analyzing data for over 3,000 counties and their important combinations such as states and SMSAs.

From the preceding discussion, we see that graphical

techniques offer enormous potential for improving statistical editing and auditing operations and their quality. Most large-scale editing processes at the Census Bureau are performed by sophisticated computer programs. These programs can process large masses of data efficiently, edit individual data uniformly, and perform a greated varity of complex anlayses than could human experts in the time available. However, somehow the human ability to view individual cases in context becomes blurred and the depth and range of expert knowledge is not completely captured in the design and implementation of "black-box" computer programs. By increasing our use of graphical and interactive techniques in data editing and analysis, we plan to recover and apply this valuable insight and knowledge. The preceding discussion of correcting survey estimates illustrates this point nicely. A second example arises in the area of map editing.

The ability to accurately geocode addresses is fundamental to operations for producing reliable statistics, particularly urban small area statistics. In the early 1970's the Bureau introduced a system called DIME (Dual Independent Map Encoding) for encoding street maps in a computer. The success of DIME is measured by the over 250 cities which have since created and used the corresponding GBF/DIME files, and the many applications to which these files have been put. The DIME system is based upon mathematical principles which facilitate the correction, use and updating of automated street maps and their corresponding computer files. DIME is dicussed at length and in technical depth in: Bureau Technical Paper #48: Toplogical Principles in Cartography (1979).

In terms of analytical applications of graphics, the DIME concept was extended to an interactive computer system called ARITHMICON. ARITHMICON combines mathematical algorithms derived from linear graph theory and two-dimensional topology with computer graphics routines for drawing street maps on a graphics terminal screen. This system allows cartographers with limited computer and mathematical skills to edit and analyze computer maps in an interactive mode. ARITHMICON also has many potential applications such as for producing small area special tabulations, districting and vehicle routing problems, sample selection and a variety of other functions useful to urban and regional planners. This integration of DIME and its underlying mathematical theory with computer graphics exemplifies what can be achieved through the use of analytic and graphic techniques in an interactive man-machine environment.

To test alternative methodologies and concepts for use in our Current Population Survey (CPS), the Bureau devised a survey research vehicle known as the Methods Development Survey (MDS) (an outgrowth of the Methods Test Panel). Seeking to investigate non-sampling phenomena such as rotation group bias (Bailar 1975), the effects of telephone versus personal interviewing, and the choice of respondent rule used in the CPS, the MDS was designed according to the following three experimental treatments: the effect of continued interviewing by the same interviewer versus different interviewers, mode of interview (personal vs. telephone), and type of respondent rule. These treatments result in twelve treatment combinations which must be investigated. For example, one such treatment combination is "same interviewer-personal interview-designated respondent." The interested reader is referred to Cowan, Roman, Wolter, and Woltman (1979) for a more complete description of the MDS.

Graphical techniques for data analysis contributed signigicantly to the MDS at an early stage. Originally, researchers conjectured that the MDS data could be represented as a linear model and, accordingly, a corresponding linear regression model was developed. However, regression plots and other graphic analytical tools soon demonstrated that the proposed linear model was inappropriate, thereby conserving valuable time and resources by leading researchers away from the wrong investigative direction and back into the more promising direction of survey research analysis. While the conclusion to abandon the linear model certainly could have been reached by standard, non-graphical techniques, and indeed was subsequently confirmed by standard statistical tests, the use of graphics in the MDS led researchers to the proper conclusion in a quick, informative and convincing manner. In addition, the process of producing and analyzing the graphical displays helped to sharpen and refine the focus of subsequent analyses.

Following its early success in the MDS, graphics came to be relied upon to a considerable degree in other aspects of this survey. By analyzing simple bar and line charts of the MDS data, data analysts soon realized that statistically significant differences in estimates of key variables such as the unemployment rate could not be detected within single treatments such as between personal and telephone interviews. However, through graphic and analytic means statistically significant differences were detected between treatment combinations. Statistical graphics continue to be used to analyze MDS data.

As a final and topical example of the Bureau's analytical use of graphics, we mention our use of graphics in the operations analysis and control of the 1980 Decennial Census. On a flow basis, statistical maps were produced displaying the return rate of mail-back Census questionnaires. These graphical displays kept Bureau management continually abreast of the progress of this all-important activity during its brief lifetime, thereby enabling them to pinpoint problems and quickly direct resources to their solution. The results of these graphical analyses were both interesting and gratifying: non-response overall was low and no established regional patterns of non-response were detected. This information helped verify that our operational plan was well-conceived and uniformly executed.

FUTURE DIRECTIONS

The preceding sections have dealt with the Bureau's past and present use of graphical techniques in conducting its statistical programs. The contributions of individuals such as Walker, Gannett and modern day Bureau statisticians, computer scientists and cartographers to the advancement of this art reflects the Bureau's commitment to continued progress in our use of graphics to display, communicate and analyze statistical information. This goal is necessary to the Bureau in meeting its responsibility to improve the information quality and content of its data products. Moreover, it is essential for the Bureau to achieve this goal if the statistical community is to meet the information requirements imposed by the growing complexity of the nation's social and economic life and the continually increasing volumes of available data which describe it.

While the Bureau is pleased with its accomplishments in statistical graphics, there remains much more to be done in this important area. The Bureau's goal is to place statistical graphics on an equal footing with its other important statistical program components such as sample design and the study of non-response. Much of this work will be methodological in nature. However, the interplay between this undertaking and the goals, orientation and programs of the Bureau must be given equal attention.

Graphics at the Census Bureau is the shared responsibility of four organizational units which operate in four program areas: statistical and computer graphics methods and software for statistical research and data analysis; the development and support of computer graphics systems and software packages for general Bureau use; automated cartography and map production and cartographic research; and the development and release of statistical graphics in published form. These groups perform individual and coordinated efforts to meet current user graphical needs and identify and address new graphical applications and their requirements. The CBIG effort previously described is currently the most intensive of these coordinated activities.

Through the CBIG Requirements Study and their integrated computer graphics capabilities that CBIG will provide, the Census Bureau will undertake the task of integrating statistical graphics within its overall program structure. Because the Census Bureau is the nation's principal source of statistical data, the resulting improvements this will bring about in Bureau data quality and completeness should have profound effects upon the statistical community. Because of the Bureau's important role in the statistical community, it often finds itself both leading some and following others in separate facets of important areas of growth and change, such as has been the case in survey sampling from the 1950's up to the present day. Communications between participants and concerned observers are vital in such efforts. Accordingly, the Bureau has made the following commitment to several professional groups, including the American Statistical Association and the American Congress on Surveying and Mapping, which it also makes to the INTERFACE community today: through its Statistical Graphics Research Staff, the Census Bureau offers to serve as a "clearinghouse" and point of contact within the professional community for exchanging information and experience in the area of statistical graphics. The Bureau's goal is assuming this role is to develop and maintain lines of professional communication in the emerging area of statistical graphics which, it believes, will help spread and establish conventions for the use of statistical graphics techniques.

As the nation's principal data provider, the Bureau appreciates the positive role it can play in fostering and improving what our previous Director Mr. Barabba once termed the "graphic literacy" of the statistical and data user community. The Bureau of the Census is committed to work with others in discovering and applying the potentialities of this virtually untapped statistical resource.

REFERENCES

Bailar, Barbara (1975), "The Effects of Rotation Group Bias on Estimates from Panel Surveys," Journal of the American Statistical Association, 70, 23-30.

Barabba, Vincent, Cox, Lawrence and DesJardins, David (1980). "Graphics at the Bureau of the Census," American Statistical Association, Proceeding of the Business and Economic Statistics Section (in press).

Beniger, James and Robyn, Dorothy (1978) "Quantitative Graphics in Statistics: A Brief History,"The American Statistician, 32, 1, 1-11.

Cleveland, William and Dunn, Douglas (1979), "Some Remarks on Time Series Graphics, "Time Series Analysis, Surveys and Recent Developments, Institute of Mathematical Statistics (in press).

Cowan, Charles, Roman, Anthony, Wolter, Kirk, and Woltman, Henry (1979), "A Test of Data Collection Methodologies: The Methods Test," American Statistical Association, Proceedings of the Social Statistics Section.

Funkhauser, H.G. (1937),"Historical Development of the Graphical Representation of Statistical Data," Orisis, 1, St. Catherine Press, Brugge, Belgium.

Playfair, William (1786), Commercial and Political Atlas; 3rd edition, Stockdale, London, 1801.

U. S. Bureau of the Census (1979), "Technical Paper #48: Topological Principles in Cartography," U.S. Department of Commerce, Washington, D.C.

MOSAICS FOR CONTINGENCY TABLES

J. A. Hartigan

Yale University
New Haven, Connecticut 06520

B. Kleiner

Bell Laboratories
Murray Hill, New Jersey 07974

ABSTRACT

A contingency table specifies the joint distribution of a number of discrete variables. The numbers in a contingency table are represented by rectangles of areas proportional to the numbers, with shape and position chosen to expose deviations from independence models. The collection of rectangles for the contingency table is called a *mosaic*. Mosaics of various types are given for contingency tables of two and more variables.

Keywords: Contingency Tables, Graphical Methods, Independence, Discrete Variables.

TWO-DIMENSIONAL CONTINGENCY TABLES

Table 1 shows the number of immigrants into the United States in 1961-70 from Great Britain and Ireland classified with respect to two discrete variables: country of origin and period of entry. A table such as Table 1 is known as *a contingency table*, and this 2×2 example is its simplest form.

Independence in Table 1 requires that the ratio of immigrants from the two countries is the same in the two time periods. In this case the immigration frequencies would be proportional to the areas of rectangles obtained by dividing a unit square by a vertical line according to the relative immigrations from the two countries over the whole time period, and by a horizontal line according to the relative immigrations in the two time periods summed over countries (Figure 1).

	GREAT BRITAIN	IRELAND	YEAR TOTAL
1961-1965	107	26	133
1966-1970	95	9	104
COUNTRY TOTAL	202	35	237

Table 1: IMMIGRATION INTO THE UNITED STATES (THOUSANDS)

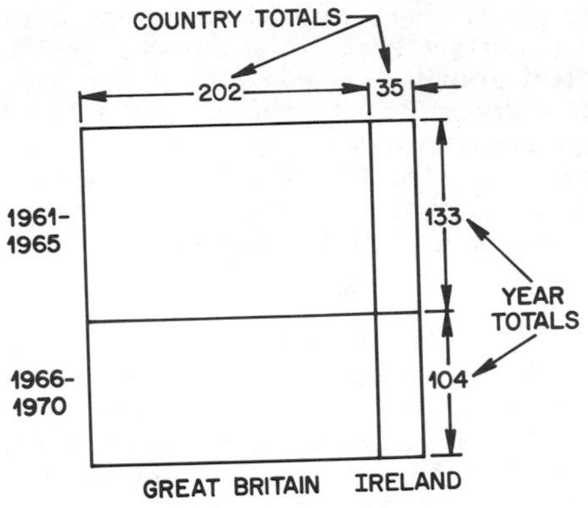

Fig. 1 Expected numbers under independence

The rectangles in Figure 2 are drawn so that their *areas* are proportional to the actual frequency. The bases of the rectangles are the same as for the independence model, here proportional to the overall country frequencies. Under independence, heights for different countries in the same time period would be the same, and so variations in the heights show departures from independence. For each row a horizontal dashed line is drawn at the height all rectangles in that row would have under independence. Therefore Figure 2 shows for each cell the actual frequency (represented by the area of the rectangle ABCD in the right hand upper corner), the frequency expected under independence (rectangle ABFE) and the residual (rectangle CDEF). See also Bertin (1967, p. 229).

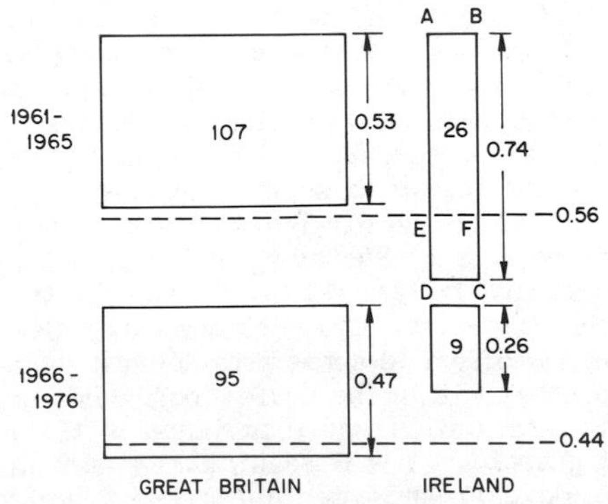

Fig. 2 Column proportion mosaic, separated. Base constant for each column, height proportional to immigration from a given country in each time period.

Figure 3 is a mosaic of Minnesota High School graduates in 1938 classified by 7 occupational levels of their father and by 3 high school ranks (Hoyt et al., 1959). The bases of the rectangles are chosen proportional to the marginal frequencies of the high school ranks.

Comparisons are made between high school ranks within each occupational level. The dotted line corresponds to the height of the rectangle which would occur if the two variables were independent. Derivations from the dotted line thus indicate lack of independence. The picture shows that students for occupation levels 1 and 2 tend to have higher school rank.

In Fig. 4, the heights of the rectangles are chosen proportional to the marginal frequencies of father's position and so comparisons should be made within each high school rank as father's position varies. A general correlation between father's level and high school rank is evident.

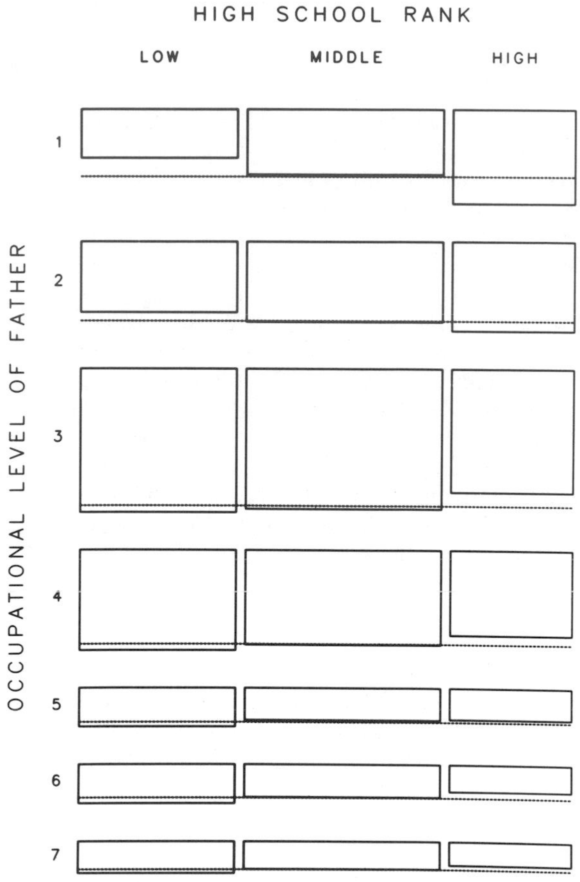

Fig. 3 Minnesota high school graduates classified by 7 levels of occupation at level of father and 3 high school ranks. Bases constant within high school ranks.

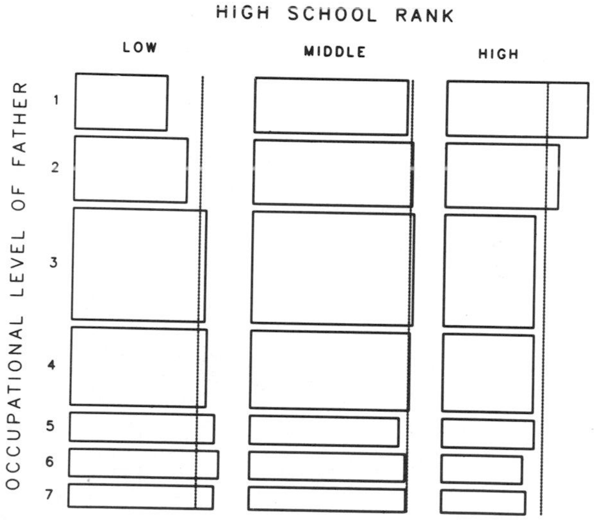

Fig. 4 Minnesota high school graduates classified by 7 levels of father's occupation and 3 high school ranks. Bases constant within occupational level.

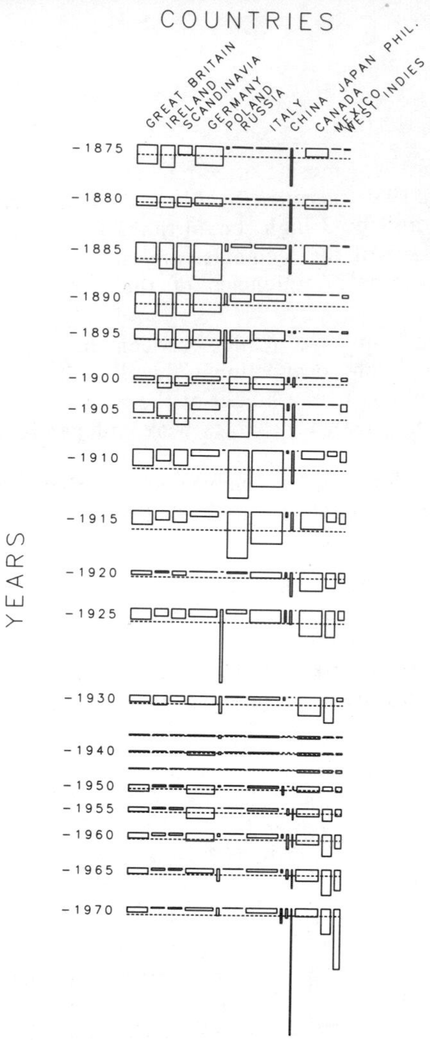

Fig. 5 Immigration into the United States 1871-1970 classified by 20 5-year periods and 13 countries of origin. Bases constant within countries.

Figures 5 and 6 show mosaics for immigration into the United States from 13 countries in 20 5-year periods from 1871 to 1970. (Historical Statistics of the United States, 1975). Territorial transfers have to some extent impaired the comparability of these data within countries. Northern Ireland, for instance, was included with Ireland (Eire) prior to 1925; Poland was included with Austria, Hungary, Germany and Russia during 1899-1919 but reported as a separate country before and afterwards. Different things show up in the two pictures. For example, in Figure 5 the low immigration in 1931-1945 is apparent, while Figure 6 shows the low immigration from Asian countries — in general Figure 5 with fixed country bases gives easy comparisons between and within time periods, and Figure 6 with fixed time bases gives easy comparisons between and within countries. In Figure 6, note the large immigration from Western Europe in 1871-1890, from Russia and Italy in 1891-1915, and from the Western Hemisphere in 1945-1970.

Fig. 6 Immigration into the United States 1871-1970 classified by 20 5-year periods and 13 countries of origin. Bases constant within 5-year periods.

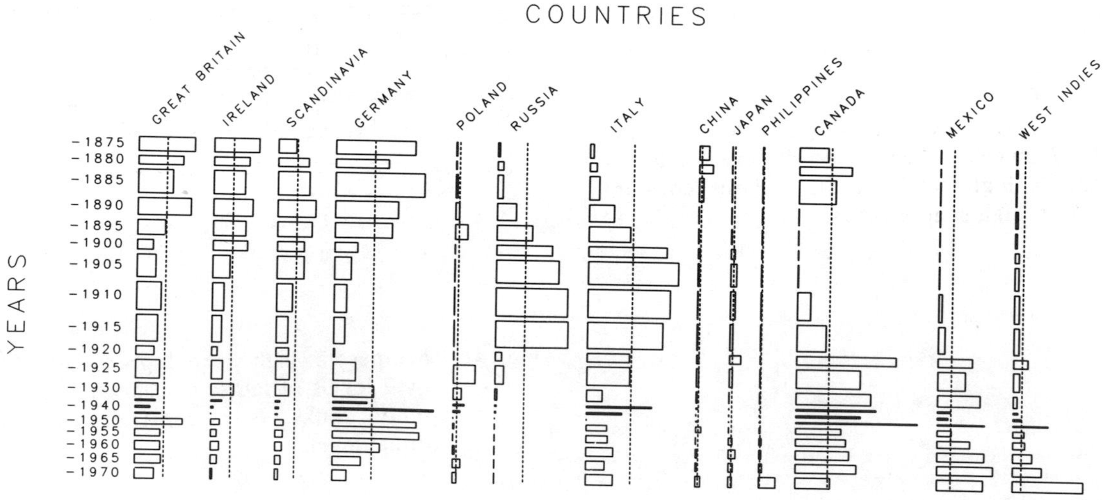

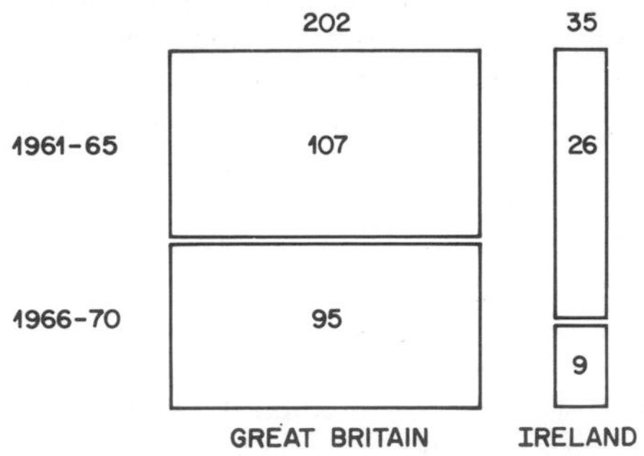

Fig. 7 Column proportion mosaic, separated.

Figure 7 divides the square by a vertical line into two rectangles of area proportional to the country totals, and then divides each of these rectangles by a horizontal line into two rectangles of area proportional to the year count for that country. The final four rectangles are then set slightly apart. A similar mosaic is drawn for the immigration data in Figure 8. Figure 8 may be compared with Figure 6; it is easier to make comparisons between and within years, harder to make comparisons between and within countries, because the countries are no longer aligned (except for the first and last one).

Each dashed line tracks a particular country over time, to make comparisons across years easier. Coding the countries by color is a better way of tracking them across time.

The necessity for setting the rectangles slightly apart is seen in the many zero entries represented by thin lines, which would be invisible without the separation of rectangles.

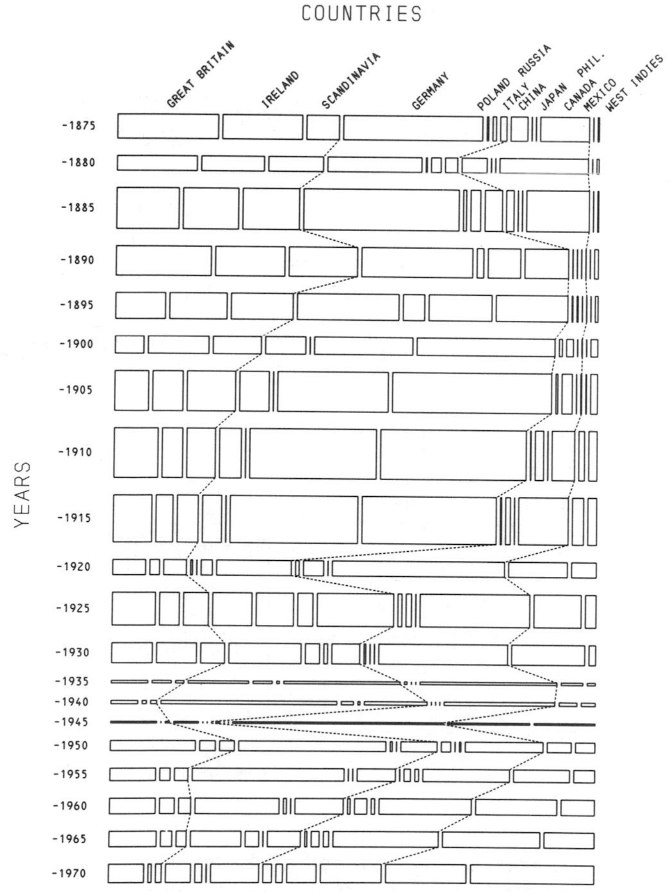

Fig. 8 Immigration into the United States 1871-1970 classified by 20 5-year periods and 13 countries. Each period is subdivided by country proportions.

MULTI-DIMENSIONAL CONTINGENCY TABLES

Figure 9 is a mosaic of 13,968 Minnesota High School graduates, classified by 2 sexes, 7 occupational levels of father, 3 high school ranks at time of graduation in June 1938 (low, middle, high), and 4 post high school statuses in April 1939 (enrolled in college, enrolled in non-collegiate school, employed full time, other) (Hoyt et al., 1959). The mosaic is constructed by dividing a unit square horizontally by sex, then vertically within each of the sex divisions by high school rank, then horizontally within each of the sex by rank divisions by father's occupation, and finally vertically, within each sex by rank by occupation division, by post high school status. The shaded rectangle, for example, corresponds to the 256 males who were in the highest third of their class, whose fathers were in occupation level 1, and who were at college. The final rectangles are separated to make small counts more visible; there are larger separations corresponding to the earlier subdivisions. There are more females than males in the respondents, and the females are mostly in the upper two ranks. Overall there are fewer than expected low ranked students. Relatively more females with fathers in the 3rd occupational level responded. The higher occupational levels have more children in the higher ranks. The high ranked children at the first and second occupational levels tend to be in college more frequently; the occupational levels are incorrectly ordered for predicting college attendance — the correct order, consistent through most combinations of sex and rank, is 1,2,4,5,7,3,6. The females tend to go less frequently to college, more frequently to school other than college, or to work. Different orders of dissecting the square may suggest other dependencies in the table.

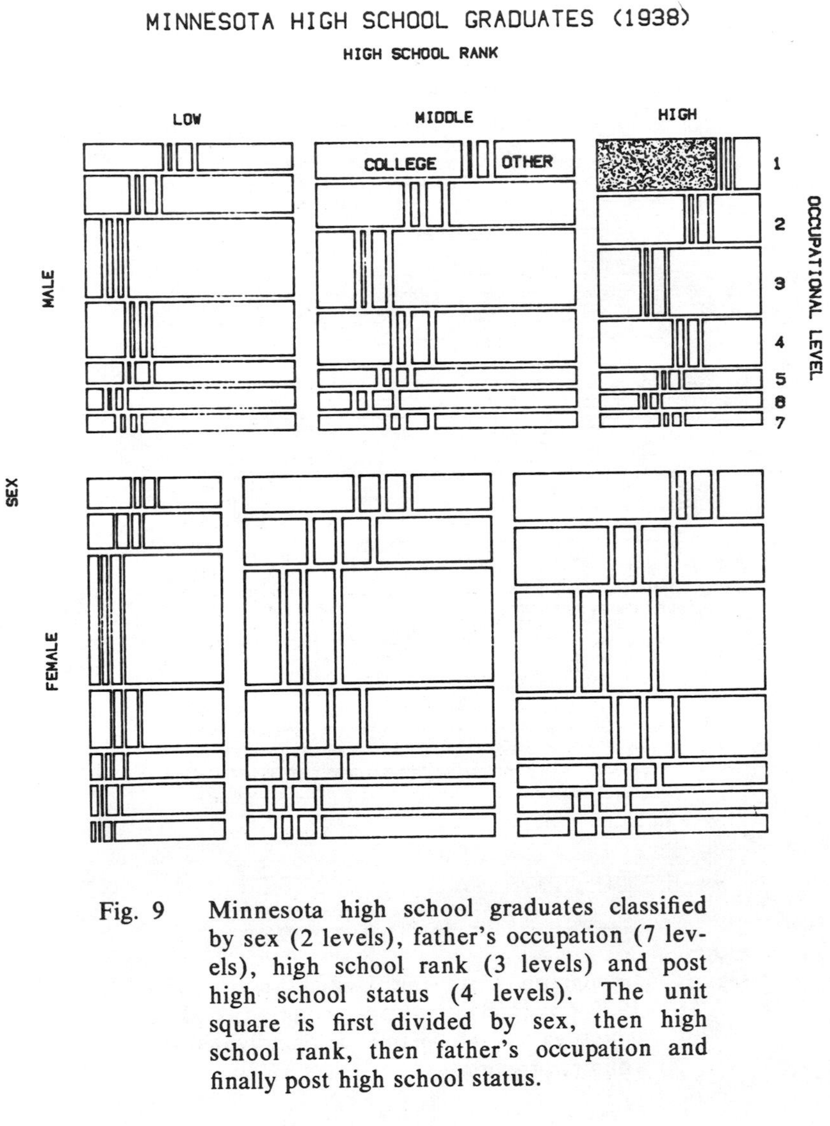

Fig. 9 Minnesota high school graduates classified by sex (2 levels), father's occupation (7 levels), high school rank (3 levels) and post high school status (4 levels). The unit square is first divided by sex, then high school rank, then father's occupation and finally post high school status.

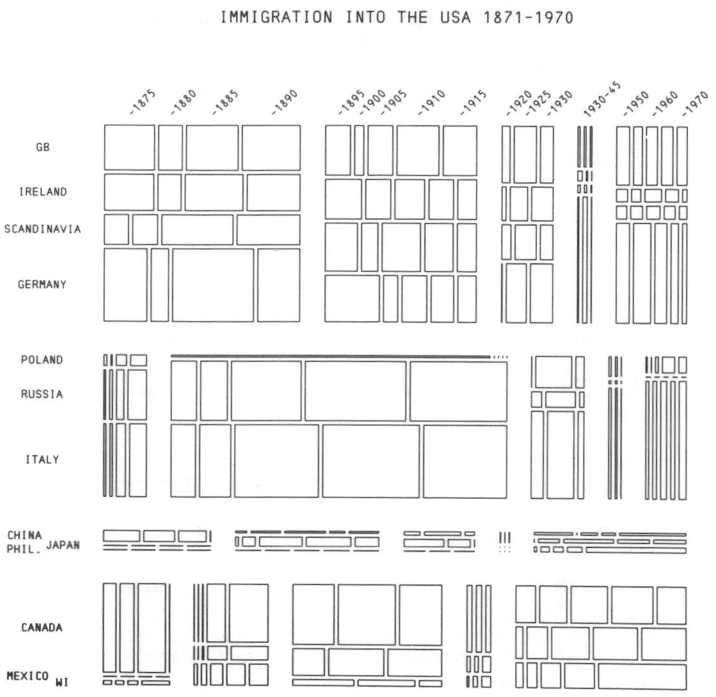

Fig. 10

Immigration into the United States 1871-1970. The unit square is first dissected by region, then era, then country, finally 5-year period.

Figure 10 dissects the square of the immigration data first vertically by region (Western Europe, Eastern Europe and Italy, Asia, Western Hemisphere); then each region division is divided horizontally by era (1871-1890, 1891-1915, 1916-1930, 1931-1945, 1946-1970); each region-era division is divided vertically by country, and each country-era division is divided horizontally by 5-year periods. All rectangles are separated to make the zero entries clear. Separations are larger for the earlier divisions.

Western Europe is the main source of immigrants over the last 100 years; Western Europe was especially dominant in 1871-1890. The majority of immigration from Eastern Europe and Italy occurred between 1891 and 1915. In order to discourage immigration from Eastern Europe, immigration quotas were set in 1924 for each European nation at 2% of the persons with "national origins" from that country residing in the United States in 1890 (*before* the large wave of Eastern European immigrants). As a result, immigration from Eastern Europe dropped dramatically after 1925; the Western hemisphere, which was not included in the quota system, has been the major source of immigration since the passage of the act. In general, most of the legal immigration into the United States in the last 100 years occurred before 1915; the period between 1931 and 1945 had an especially low number of immigrants (partly due to the quota system, the depression and World War II).

Looking at individual countries, we see that recently Ireland and Scandinavia have decreased number of immigrants compared to Great Britain and Germany. The immigration from Russia and Poland has almost disappeared. Note that the zero Polish immigration in 1890 to 1915 is due to the administrative absorption of the Congress Kingdom of Poland into Russia in the 1880's. Overall the Asian immigration is small. The relatively large Chinese immigration before 1885 was stopped by the Chinese exclusion act of 1882; the relatively large Japanese immigration in 1900-1925 was almost completely eliminated by the quota system. In recent times there has been a large increase in immigration from the Philippines. In the Western hemisphere, Canada's share of the immigration is declining, compared to Mexico and the West Indies; also note the zeroes for Canada in 1886-1900 when the limit of settlement in Canada first reached Manitoba and Saskatchewan, opening up large areas for farming.

References

Bertin J. (1967), *Semiologie Graphique*. Gauthiers-Villars, Paris.

Historical Statistics of the United States: Colonial Times to 1970 (1975). Bureau of the Census . Part 1, pp. 97-109 (Series C89-119).

Hoyt C. J., Krishnaiah P. R. and Torrance E. P. (1959), Analysis of Complex Contingency Data, *Experimental Education*, **27**, pp. 187-194.

THE USE OF KINEMATIC DISPLAYS TO REPRESENT HIGH DIMENSIONAL DATA

David Donoho, Peter J. Huber, Hans-Mathis Thoma

Harvard University, Cambridge MA

ABSTRACT

Traditional data presentations deriving from pencil-and-paper techniques are inherently 2-dimensional, while the human visual system effortlessly deals with several more dimensions in an integrated fashion. In particular, moving pictures do impart a strong subjective 3-d effect. We describe a data manipulation and display system designed to tap this human ability for the purposes of statistical data analysis and discuss the main issues, design problems and solutions chosen. An experimental version of our system has been operating on a VAX-computer at Harvard since the fall of 1980. (A videotape illustrating some possible applications was shown at the conference.)

KEYWORDS: Multidimensional data analysis, computer graphics, kinematic displays, PRIM-systems, interactive statistics, exploratory data analysis.

1. Introduction

Statisticians still are very far from exploiting the full capabilities of the human visual system. Typically, statistical graphs are black-and-white line drawings. In distinction to cartography, few of them make use of color in an essential way. We are not aware of any published presentation of three dimensional data by stereo pairs in the statistical literature (a common practice in the crystallographic literature, for example). Astronomers are explicitly using the time dimension in their data analysis: Changes between photographs of the same region of the sky are made visible as a flicker, highlighting variable stars and moving objects. Admittedly, there is a problem with time-dependent graphs: moving pictures can be published as a film strip, a video tape, or maybe even as a hologram, but not in a traditional journal.

Graphs can be used both to explore data and to present results. We are almost exclusively concerned with the former aspect. There the goal is not to create eye-catching art, nor charts for reading off numbers like from a table, but graphics which exploit the built-in pattern discovery capabilities of the human visual system to the largest possible extent. In particular, graphics should facilitate seeing (in a most literal sense!) whether the data contains any unsuspected peculiar feature which might be worth investigating.

Partly because motion yields one of the strongest 3-d cues, partly because of technical constraints, we shall concentrate on the use of monochromatic kinematic displays, which produce a 3-d effect by posting slightly rotated projections in rapid succession on a video screen. But a brief discussion of more general issues would seem to be in order. Compare also [5], [6].

2. The human visual system and statistical graphics.

On the lowest, "hardwired" level, our visual system is able to distinguish at least 7 dimensions:

- 3 for space (horizontal, vertical, depth)
- 1 for time (changes)
- 3 for color (hue, saturation, brightness)

On a slightly higher level, several more variables can be encoded into size, shape, orientation and texture of standardized objects. On a third level, we have structured objects like "glyphs", "castles" and "Chernoff faces". The higher the level, the harder it becomes to digest a scene: one no longer grasps it as a whole, but must synthesize it from its parts.

Let us now concentrate on the seven primary dimensions. They clearly are not all on the same footing. Color information is difficult to perceive in small objects (e.g. in the points of a scatter plot), and there is no natural order among perceived hues. Time is fleeting (our short term memory fades away in a few seconds). Moreover, our visual system confounds several of these dimensions: our depth perception (through accommodation and binocular vision) is so poor that depth cues are borrowed from at least three other dimensions: time (rigid body motion), saturation and brightness. In addition, size and shape (perspective) are confoundable with depth. With scatterplots, binocular vision seems to work rather poorly, and also in other respects they present more perceptual problems than, say, connected polyhedra. This has to do with the fact that we see edges much better than points (there are special "edge detectors" in the visual cortex). Not surprisingly, well-chosen lines, added to a scatterplot, can dramatically enhance our spatial perception -- and may force a suggestive but unwarranted interpretation onto the viewer. Among the different depth cues, rigid body motion (rotation or a rocking motion) in our experience is the strongest and most reliable, but it must be assisted by at least one other cue (brightness, binocular vision, or perspective) in order to avoid annoying perceptional flip-flops between the two possible senses of rotation.

Because of this confounding and because of the poor resolution in some dimensions, it may not be feasible to encode 7 distinct variables into these 7 dimensions. In fact, most people will have difficulties perceiving simultaneously more than the 3 space dimensions, plus 1 to 3 variables which have been discretized into a small number of levels by color, size or shape. Some people may be able to train themselves to grasp a higher number of dimensions while gazing, say, at a cloud of colored objects which "gyre and gimble in the wabe", as John Tukey has put it; but with Alice we suspect that such a data display might be <u>rather</u> hard to understand.

<u>3. Physical display devices</u>.

There are two categories of computer graphics devices (video screen as well as hardcopy):

- line drawing
- raster scan

The former draw lines and points (like a pen on paper), the latter operate like an ordinary TV screen and can produce half-tone pictures.

For the purposes of exploratory data analysis, line drawing systems still seem to be better suited. A majority of the graphs to be produced will be line drawings anyway (scatter plots, curves). Line drawing systems typically offer a high resolution (about 4096x4096 addressable points), crisper pictures, and characters that are readable even in very small fonts. There is relatively little information per frame to be generated and saved, and one can produce scatterplots and wire-cage pictures smoothly turning in real time at an affordable cost.

Raster scan devices must store and display information for every single point or "pixel" on the screen, so there are many fewer distinguishable points (usually not more than 512x512) with a concomitant much lower resolution. In particular, they produce only coarse lines, showing awkward staircase effects, unless one widens these lines even further and modulates the intensity pattern in a sophisticated fashion. Raster-scan devices clearly are great for showing half-tone pictures of solid objects and of surfaces, but then it is indispensable to eliminate hidden surfaces, and there are only a few flight simulators (in the mega-$ range) which can do so in real time.

The hardware to which we have access consists of an Evans & Sutherland Picture System 2, hooked up to a VAX-11/780. This is a line drawing system which contains a hardware matrix multiplier. It is able to show smoothly rotating 3-d objects consisting of several thousand points or lines. Untransformed coordinated and the appropriate transformation matrices are shipped from the VAX to the Matrix Arithmetic Processor of the Picture System, where they are transformed and deposited in the PS-memory, from where the actual picture then is continually refreshed on the screen.

There are color screens available now, but ours is black-and-white only. Color would be beautiful for subsetting (highlighting and masking, see below). For published color scatter plots produced by a PS2, see [2].

<u>4. The PRIM systems</u>.

The first serious attempt to tap the human ability to deal with 3-dimensional pointclouds by using kinematic display techniques was made by Tukey, Friedman and Fisherkeller in the early 1970's [4]. They created the PRIM9 system at the Stanford Linear Accelerator Center. The letters stand for Projection, Rotation, Isolation and Masking, and 9 is the maximum number of dimensions the system could handle.

The meaning of projection and rotation is clear. Isolation allows to separate off subsets (e.g. outliers) for closer scrutiny, and masking allows to blank out points when they are hit by a coordinate hyperplane which moves at a constant speed through the sample space (this allows to transpose into the time domain a dimension not shown on the screen).

The system showed 2-d projections of up to 9-d pointclouds, and a reasonably good 3-d effect (though not smooth motion) was achieved by posting incrementally rotated projections. The system was a tour de force in programming; all transformations were done by software in the host computer, and the transformed coordinates were then shipped over to the minicomputer (Varian 620) driving the IDIIOM display. While the user was viewing the screen, an IBM 360-91 (later a 370-168) was kept fully busy with turning the cloud.

PRIM9 showed the feasibility of the idea. But it was evident that in future implementations the transformations would have to be done in a cheaper way by specialized hardware. In retrospect, we must admire the clever choice of options and the excellent human engineering of PRIM9.

In the summer of 1978, W. Stuetzle and H. M. Thoma created a quick, stripped down version PRIMS for a quite different computer system at the Swiss Federal Institute of Technology in Zurich (DEC-10 plus a PDP-11/34 driving an Evans & Sutherland Picture System 2), utilizing the built-in hardware rotation features of the PS2.

In the summer of 1979, Friedman, Stuetzle and Thoma designed a successor to PRIM9 at SLAC, named PRIM79, which is currently being built. Again, a large IBM machine will be used as the host, but the real-time transformations shall be done in a Motorola 68000 microcomputer, which also drives the raster-scan screen.

In the fall of 1979, we began to construct our system at Harvard, using a computer installation consisting of a VAX-11/780 with a PS2; a first version of the system became operative in the late summer of 1980.

5. The design of the Harvard system: ISP and PRIMH.

PRIM9 had one other serious deficiency (besides its gobbling up of mainframe CP time): it was not coupled with a general purpose interactive data manipulation system. Such a system is essential for several reasons. First, the dynamical range of the eye is only about 30:1, that is, we have difficulties comparing two features when one is 30 times larger than the other. On top of this, we have the logarithmic perception law, which causes us to give too much importance to the sparsely populated fringes of a data cloud. As a consequence, raw data rarely is in a form suitable for direct viewing, but it must be carefully massaged first (i.e. transformed by trial and error). Often, a simple logarithmic or square root transformation of some variables is all that is needed. Examples of more complicated transformations are "sharpening" by excision (drop the most isolated points) or by displacement (replace each point by a weighted average of its k nearest neighbors). One can incorporate the more important of these transformations into the system (this was done in PRIM9), but one cannot possibly anticipate everything. After seeing the data, one invariably wants to try some unanticipated trick -- for example, fit a hyperplane to some subset of the data by principal components and then look at residuals. In other words, it is important that one can freely improvise.

Example of general purpose interactive data manipulation systems are languages like APL, S, ISP or TROLL. We chose ISP, not only for the obvious reason that one of us (Donoho) had developed it as an undergraduate at Princeton, but also because it was the only reasonably portable, compact and expandable language.

5.1. The Interactive Statistical Package (ISP).

In intent and practice, ISP is sort of a glorified pocket calculator operating on arrays. The notation is Fortran-like. There is a growing list of basic statistical operations (regression, time series, exploratory data analysis, etc.). The arrays are kept in a workspace in memory; the workspace can be saved as a file and retrieved. ISP has macro capability, and it is open ended: the user can create his own commands by writing macros calling Fortran routines in a straightforward way. A few examples of how ISP might be used follow.

(1) Subtract row-wise medians from a matrix x and call the resulting matrix x1:

```
reduce/op=med/axis=2   x > rowmed
let  x1 = x-rowmed
```

(2) Create a time series consisting of 400 independent Gaussian random variables, perform a spectrum analysis on it, tapering 10% on each side, and plot the logarithm of the smoothed periodogram:

```
let  x = gauss(array(400))
fft/taper=0.1  x > xre xim
runave/span=7  (xre**2 + xim**2) > spec
plot  (log(spec))
```

(3) Regress a response vector y on an nxp matrix x and plot the residuals against the fitted values:

```
regress  x y > resids
scat  (y-resids) resids
```

Our current version of ISP is entirely written in Fortran and runs on a VAX. The code is largely machine independent, with the following known exceptions:

* The machine must be large enough to accommodate a workspace of about 40000 floating point numbers in real or virtual memory if one is to work with realistic data sets.
* String handling is byte-oriented.
* Error recovery is machine dependent.

5.2. The Harvard PRIM-system (PRIMH).

We already commented on the excellent overall design of the original PRIM9. We followed it closely; the main deviations have to do with our different hardware, in particular with the special 3-d transformation capabilities of the Picture System 2. There is no restriction to 9 dimensions as in PRIM9, but at present, transformations involving more that 3 dimensions simultaneously must be done in ISP. We can identify points by labels, and points can be connected by lines. This allows for instance to indicate a fourth dimension like time through the sequential ordering of the observations, or to show a minimal spanning tree, etc., and it enhances the overall spatial perception of the data. At present, the only means for user interaction are through a tablet, but dials and a joystick should become available soon.

The next version of PRIMH will include a sophisticated dynamic subset scheme, which encompasses in particular the isolation and masking operations of PRIM9.

PRIMH is programmed in Fortran and is highly modular; it is machine dependent on the special hardware features and on the Evans & Sutherland software of the Picture System.

5.3. Interfacing ISP and PRIMH.

Our present software interface between ISP and PRIMH is overly simple but foolproof: the user prepares the data on ISP, saves the workspace, quits, starts up PRIMH, and loads the saved workspace. This helped with debugging by keeping the systems isolated. We are currently adding a command to ISP which triggers the above operations by a single command word, and a menu item to PRIMH which causes a return to ISP.

Interactions with ISP and PRIMH are kept separate both conceptually and physically. The user converses with ISP through a keyboard and an alphanumeric display, with PRIMH through a tablet (and soon through dials) and the Picture System screen.

6. General experiences.

The proposed core graphics standard (as described in [1]) certainly is a helpful source of inspiration, but we found that it is conceived on somewhat too low a level and is not directly applicable to specialized high power graphics hardware.

The main difficulty we encountered in designing a PRIM-like system was that we never were able to plan very far in advance. After playing with a preliminary version for a few hours, one invariably encounters an inadequacy, or a desirable innovation one had not thought of before. Usually, these are in the human engineering domain, and it is not always easy to accommodate the changes within the constraints set by previous design decisions. Of course, any single such change could be added by a kludge, but this amounts to delayed suicide. Even if one takes great care with designing the system modularly, one sooner or later will run into a dead-end street necessitating a major redesign.

The most recent impasse we met was when we realized that we wanted multiple pictures on the screen in sufficient generality to deal at least with (i) stereo pairs, (ii) the option to freeze a particular view in a corner of the screen and to continue working with other views of the same data set (our short term memory fades quickly and needs assistance), (iii) the possibility to manipulate one pointcloud with regard to another, and rotate the two together (e.g. in order to visually identify two approximately homologous subsets of the two clouds). In order to do that neatly, one needs a modifiable data block which contains the entire information necessary to recreate the picture with the currently valid parameters, and we had neglected to provide for that.

We found it quite essential that the user can interact with a kinematic display program without having to take his eyes from the data points he is concentrating upon (in this respect, dials, joysticks and toggle switches have a great advantage over their mere simulation by menu items and a pen). It is very hard to provide for flexible and powerful user interactions under such constraints; in the more distant future, voice interaction might be a way out.

It is awkward to produce hardcopy from a kinematic display. Plots or still photographs are not good enough: when you freeze the picture, depth perception is lost. In particular, this creates a communication problem: you cannot publish insights gained while viewing a kinematic display, at least not in a traditional journal! Filming the screen may give the qualitatively best results, but is relatively awkward because of synchronization problems (beats between the frame frequencies of the display and of the camera), and because there is no immediate control of success. Videotaping is more convenient, but for some cameras it seems to be impossible to adjust the automatic brightness control so that the points of a scatter plot are bright spots on a dark background. The biggest nuisance however is the poor resolution of the ordinary raster-scan TV-screens. The letters of our PS-screen menu would have to be magnified about 3 times (with a corresponding reduction of the number of menu items) if they are to remain readable on the replay of the videotape, and there is no way to show labels of individual points in a large scatter plot.

Our experiences with actual data sets are not very extensive yet and can be summarized by the following, somewhat anecdotal remarks.

(1) Looking at a mass of data becomes a pleasing experience. You are much less tired after playing two hours with data on a PRIM-system, than, say, after wading two hours through a stack of printed SPSS output, and you may afterwards have a much clearer perception of what is really going on in your data set. There is a cumulative beneficial effect of interactivity, graphics and of continuity (one does not have to re-orient oneself in every new projection if there is a smooth transition from one to the next).

(2) The most immediate and assured gain is with problems relating to the ordinary 3-space we live in. An example is the problem of connecting bundles in the stem of a palm tree, as shown in the videotape. In this case, 3-d visual control furnishes absolutely essential feedback for the design of algorithms which should automatically connect bundles.

(3) We found that often a careful look at 1-d and 2-d projections spanned by coordinate axes and pairs of axes reveals the whole story, but only after a considerable effort, involving a mental reconstruction of the 3-d scene from the lower dimensional projections. On the other hand, a PRIM-system will immediately show what is going on, without any effort (the reconstruction is done in our head, too, but at a much lower and much better automated level). A beautiful example has been published by Reaven and Miller [3].

(4) A simple but important use of PRIM-systems is the discovery of multivariate outliers (those which slip through one-dimensional routine checks -- for instance test scores of cheaters, who may tend to avoid extreme values in any single variable, but who do not quite match the overall pattern).

(5) Most statisticians have a good intuition about ordinary straight line regression, that is about fitting a straight line to twodimensional scatter plots. It was somewhat of a shock to us to see, thanks to PRIMH, to what ugly mess of points we had been fitting multiparameter regressions, without having checked for peculiar features in the carrier or in the response.

7. References

[1] Status Report of the ACM Graphics Standards Planning Committee. Computer Graphics 13 (1979) #3.

[2] R. Langridge et al., Real-Time Color Graphics in Studies of Molecular Interactions. Science 211 (1981) 661-666.

[3] G. M. Reaven and R. G. Miller, An Attempt to Define the Nature of Chemical Diabetes Using a Multidimensional Analysis, Diabetologia 16 (1979) 17-24.

[4] J. W. Tukey, J. H. Friedman and M. A. Fisherkeller, PRIM-9, An Interactive Multidimensional Data Display and Analysis System. Proc. 4th International Congress for Stereology, Sept. 4-9, 1975, Gaithersburg, Maryland.

[5] P. A. Tukey and J. W. Tukey, Methods for Direct and Indirect Graphic Display for Data Sets in 3 and More Dimensions. Proc. of the Sheffield Conference, ed. by V. Barnett, Wiley (to be published).

[6] H. Wainer and D. Thissen, Graphical Data Analysis. Ann. Rev. Psychol. 32 (1981) 191-241.

Research supported by NSF Grant MCS-79-08685 and ONR Contract N00014-79-C-0512.

Contributed Papers

The following are written versions of contributed papers that were presented as poster sessions at the 13th Interface Symposium.

ORDER STATISTICS AND AN EXPERIMENT IN SOFTWARE DESIGN

R. S. Wenocur, Drexel University

Abstract. Not long ago, most introductory undergraduate statistics courses were taught without the use of computers. But now, it is common practice to introduce computing as part of such courses. Conversely, with current trends toward teaching techniques of structured programming and requiring ample documentation, it is useful to introduce elementary statistics into an undergraduate's first programming course. As an example, we examine a programming assignment, given to the members of a class studying FORTRAN, to develop methods of analyzing order statistics of random samples, exceedances thereof, and related waiting times. The students, who were required to work in teams and to perform alphanumeric manipulations, were thereby compelled to develop well structured programs and to provide usable documentation.

Key words and phrases. Order statistics, computer science classroom, FORTRAN, teamwork, documentation, structured programming, workshop environment, software design.

Thirty-five students waited lethargically for the arrival of their professor on the first day of the new term. Engineering and business students, potential mathematicians, aspiring social scientists, and those whose specialty would be computer science, gathered in the overheated classroom, anticipating a ten-week term of soporific lectures, tedious reading assignments, standard examinations, and two or three unrelated, artificial programming projects to be completed by individuals, with collaboration among classmates discouraged, if not forbidden. This course in FORTRAN was, for most, a requirement for graduation; consequently, the class was filled to capacity with a group of undergraduates whose academic majors spanned a broad spectrum of disciplines. How could meaningful assignments be given, without custom designing programming problems for individual students? Leaning on elbows and gazing at a hazy sky through large, unopened windows, these young men and women expected, therefore, simply to be subjected to pedantry.

Within a few minutes, she arrived. Photocopies of a classroom memo were quickly distributed; as an introductory gesture, she reviewed aloud the content thereof. An initial paragraph announced her name, office hours, recommended texts, and other vital information. But seventy eyes opened widely as they stared at the second paragraph; incredulous freshmen, sophomores, and upperclasspersons read a description of the course. There would be no examinations, no lectures, and only one programming assignment, which would be developed gradually throughout the term. Under no circumstances would a student be compelled to keep his books closed or his notes hidden. Class time would be used as a workshop period, and she would act not as lecturer, but as software manager. Teamwork would be not only allowed, but encouraged.

The instructor explained her reasons for operating this course in an unconventional manner. When would an employer forbid a computer analyst to use his manual? Does a standard examination reflect any aspect of a realistic working environment? Is it not true that modern software engineering is becoming too complex for development of programs by only one or two people? Would not a workshop environment be more suitable for our heterogeneous group? How could lectures be planned for a class the members of which have dissimilar interests and various levels of programming proficiency? Moreover, are lectures as effective as forcing the programmer to make his own discoveries and errors, and, thereafter, explaining how to code properly? Whenever a programmer encounters specific difficulties and questions, he is thereby motivated to listen and learn, and discussions with his instructor lose any somniferous character.

Thirty-five students were jubilant. What an easy course this would be! No exams! One assignment! They could work together with friends! Despite successive modifications, the project could not be very difficult, since this was not an honors class. (Goodman (1981) discusses an innovative treatment of an honors group.) Anyone who would do anything at all could surely get a high grade.

Two days later, when the class met again, Part I of the programming project was assigned: twenty-four data records were supplied, some elementary computations were to be performed, then, mean and standard deviation for a resultant one-dimensional array were to be computed and printed. The tasks were easier than anyone might suspect. But a warning was issued: structure the program, use subprocedures, keep the number of input records variable, and streamline the operation of your team, since modifications would be required later. Rather than assigning several unrelated programming projects, the instructor attempted to simulate a genuine working environment in which the original program might be modified progressively, as more sophisticated versions were produced. Developing one respectable piece of software with n parts is a better training exercise than writing n tiny programs; students must think creatively, work to the best of their abilities, and learn to interact effectively with other programmers.

When the first set of modifications was introduced, one week later, the need for teamwork and well structured programming became apparent. Documentation was important within the coöperative environment, and users' guides were required. A mini-statistical-package was to be developed: using flexible subroutines, sample correlation coefficients were to be computed and printed, trends analyzed, and confidence intervals determined, based upon tables of the t-distribution. Alphanumeric manupulations became necessary, and the program was complex enough to challenge the capabilities of any single programmer; coöperation became vital. Class time was spent answering questions about the elementary statistical concepts, discussing design of subprocedures, explaining core-to-core I/O, suggesting ways to allocate tasks among team members, and examining other methods needed to complete the job.

Finally, one month before the end of the term, additional tasks were assigned. Order statistics of a random sample, exceedances thereof, and related waiting times (in the manner of Gumbel and von Schelling (1950), Gumbel (1958), Sarkadi (1957), Morgenstern (1972), Galambos (1978), and Wenocur(1981)) were to be analyzed. Actual data would be processed, probability tables produced, and pseudo-random numbers generated. Efficient sorting and searching routines became necessary, and class time was devoted to discussing relative merits of alternative techniques. The study of order statistics led to combinatorial formulae, so that efficient iterative and recursive algorithms were explored.

Now,complaints were rampant. Why should statistics be taught in a computer science course? Why don't we take examinations and write simple little programs as everyone else does? How did all the other members of my team get easier jobs? The unorthodox procedures drove the students crazy. Queues the length of which rivaled unemployment lines formed outside the instructor's office, and a course that was officially scheduled to meet three hours per week began to occupy twenty. Operating this course became overwhelming for even a dedicated professor. But, certainly, such conflicts contributed to the real-life nature of this educational experiment.

Despite engendering protests and requiring much time and effort on the part of students and faculty, introducing statistics into a computer science classroom has many advantages for those studying elementary algorithm design. Statistics is a topic of almost universal interest, and can, thereby, serve to unify a heterogeneous class. Statistical analyses encourage modularization through the development of flexible subprocedures that imitate components of a statistical package, lead to the development of attractive output, and require ample documentation, including users' guides. Numerous techniques of data processing can be introduced in a very natural manner: design of adaptable SUBROUTINES and FUNCTION subprograms, use of COMMON storage, counting methods, sorting, searching, alphanumeric manipulations, handling large data sets, use of arrays, design of iterative and recursive algorithms, numerical integration, random number generation, linear regression, and many other techniques are required to write a program that involves statistical analyses of the types mentioned in earlier paragraphs. But, an equally important effect is that, without actually taking a formal course in probability and statistics, the students gradually develop a feeling for making statistical decisions, interpreting output, recognizing unreasonable results, and investigating alternative statistical methodology. Obviously, when order statistics are studied, sorting and searching become important, and combinatorial problems arise that require careful algorithm design. But, as a consequence of their analysis of ranking, many students develop an interest in statistics beyond the familiar sample mean, standard deviation, and correlation coefficient.

When the term was over and grades reported, the students had discovered that the course had actually been neither easy nor impossibly difficult. They informed their professor that they had learned more by being active rather than passive in the classroom , by making mistakes, and, thereafter, under her guidance, working out solutions. Because of the opportunity to work in teams, most felt prepared to operate effectively as members of a professional software development group. Many had confidence in their programming style, and some, filled with enthusiasm, actually requested additional assignments, simply for their own personal enrichment as future computer professionals. Those extra hours of counselling and preparation, she now felt, were well worth the effort. Of course, had she had the services of an assistant or two, she would have been less exhausted.

The time has come to update the format of computer science courses. Statistics could be introduced in order to syncretize a heterogeneous class or to introduce many techniques of data processing in a natural way. Faculty members or graduate students (see Goodman (1981)) should be available to work closely with teams of undergraduates in a more realistic environment. Although students rarely emerge entirely nescient from a course taught in a conventional manner, meaningful assignments and an interactive approach could mark the difference between someone's entering the job market naïvely or with some degree of sophistication.

References.

Galambos, J. (1978). The Asymptotic Theory of Extreme Order Statistics. John Wiley and Sons, New York.

Goodman, S. E. (1981). An experiment in software engineering. ACM SIGSOFT, Software Engineering Notes, Vol. 6, No. 1, p. 15.

Gumbel, E. J. (1958). Statistics of Extremes. Columbia University Press, New York.

Gumbel, E. J. and von Schelling, H.(1950). The distribution of the number of exceedances. Ann. Math. Stat., Vol. 21, 247-262.

Morgenstern, D. (1972). Überschreitungswahrscheinlichkeiten, das Polyasche Urnenmodel und ein Wartezeitproblem bei Urnenziehungen. Math.-Phys. Semest., Göttingen, 19 (2), 213-215.

Sarkadi, K. (1957). On the distribution of the number of exceedances. Ann. Math. Stat. Vol. 28, 1021-1023.

Wenocur, R. S. (1981). Waiting times and return periods related to order statistics -- an application of urn models. Stat. Distrib. in Scientific Work. To appear.

FURTHER APPROXIMATION TO THE DISTRIBUTIONS OF SOME TRANSFORMATIONS TO THE SAMPLE CORRELATION COEFFICIENT

N.N. Mikhail, Beverly A. Prescott and L.P. Lester, Liberty Baptist College

ABSTRACT

In this article the first eleven moments of r, are derived. They are used to examine the distributions of some of the familiar transformations of r under normal assumptions. Tables provided compare μ_2, skewness (β_1), and kurtosis (β_2) with μ_2^*, β_1^* and β_2^* studied by Subrahmaniam and Gajjar (1980). These results provide further evidence of the usefulness of this work.

Key Words: k-statistics, correlation and Transformation

1. INTRODUCTION

It is assumed that there are two correlated variables x and y. A random sample of n pairs of observations (x_i,y_i) is available, whence the sample correlation is commonly written

$$r = \sum_{i=1}^{n} (x_i-\bar{x})(y_i-\bar{y})/\{\sum_{i=1}^{n} (x_i-\bar{x})^2 \sum_{i=1}^{n} (y_i-\bar{y})^2\}^{\frac{1}{2}} \quad (1.1)$$

The properties of r in the case of the bivariate normal distribution have been studied extensively using the first four moments of the well known results of Hotelling (1953). Here in section 2 of the present paper we have derived the first eleven moments of r under non-normal and normal parent populations. Several authors have suggested transformations that render the distribution of the transformed variable close to normal or Student-t distribution. Four of the well known transformations are of interest to us in the present paper.

(i) $Z = \frac{1}{2} \ln \frac{1+r}{1-r}$ (Fisher, 1921)

(ii) $Z = (r-\rho) \sqrt{n-2}/\{(1-r^2)(1-\rho^2)\}^{\frac{1}{2}}$ (Samiuddin, 1970)

(iii) $Z = (r-\rho)/(1-r\rho)$ (Pillai, 1946)

(iv) $Z = \sin^{-1} r$ (Harley, 1956),

where ρ is the population correlation parameter.

The purpose of this paper is to derive the first four moments for these transformations to $O(n^{-2})$. The study will be limited to the normal case.

2. MOMENTS OF r

Expression (1.1) in terms of bivariate k-statistics and populations cummulants is

$$r = \frac{k_{11}}{(k_{20}k_{02})^{\frac{1}{2}}} = \rho\left(1 + \frac{k_{11}-\kappa_{11}}{\kappa_{11}}\right)\left\{\left(1 + \frac{k_{20}-\kappa_{20}}{\kappa_{20}}\right) \times \left(1 + \frac{k_{02}-\kappa_{02}}{\kappa_{02}}\right)\right\}^{\frac{1}{2}} \quad (2.1)$$

Cook (1951a) obtained the first four moments of r under non-normal and normal assumptions. Her technique is used to derive the first eleven moments for non-normal and normal parent populations, of which the first four moments of Cook's (1951a) results are corrected and checked with Hotelling's (1953) results up to $O(n^{-2})$. In this section are recorded the eleven moments of r for the normal case up to $O(n^{-2})$ of which the first four moments are those of Hotelling (1953) up to $O(n^{-3})$.

$$\mu_1'(r) = \rho - \frac{\rho}{2n}(1-\rho^2) + \frac{\rho}{8n^2}(1-9\rho^2) + \frac{\rho}{16n^3} \times (1-\rho^2)(1+42\rho^2-75\rho^4) + O(n^{-4}) \quad (2.2)$$

$$\mu_2'(r) = \rho^2 + (1-\rho^2)\{\frac{1}{n}(1-2\rho^2) + \frac{2\rho^2}{n^2}(3-4\rho^2) - \frac{4\rho^2}{n^3}(3-14\rho^2+12\rho^4)\} + O(n^{-4}) \quad (2.3)$$

$$\mu_3'(r) = \rho^3 + 3\rho(1-\rho^2)\{\frac{1}{2n}(2-3\rho^2) - \frac{1}{8n^2} \times (20-87\rho^2+75\rho^4) + \frac{1}{n^3}(82-889\rho^2+2000\rho^4-1225\rho^6)\} + O(n^{-4}) \quad (2.4)$$

$$\mu_4'(r) = \rho^4 + (1-\rho^2)\{\frac{2}{n}\rho^2(3-4\rho^2) + \frac{1}{n^2}(3-39\rho^2+104\rho^4-72\rho^6) - \frac{1}{n^3}(6-198\rho^2+968\rho^4-1536\rho^6+768\rho^8)\} + O(n^{-4}) \quad (2.5)$$

$$\mu_5'(r) = \rho^5 + \frac{1}{2n}(20\rho^3-45\rho^5+24\rho^7) + \frac{1}{8n^2}(120\rho-1080\rho^3+2985\rho^5-3250\rho^7+1225\rho^9) + O(n^{-3}) \quad (2.6)$$

$$\mu_6'(r) = \rho^6 + \frac{1}{n}(15\rho^4-33\rho^6+18\rho^8) + \frac{1}{n^2}(45\rho^2-330\rho^4+807\rho^6-810\rho^8+288\rho^{10}) + O(n^{-3}) \quad (2.7)$$

$$\mu_7'(r) = \rho^7 + \frac{1}{2n}(42\rho^5-91\rho^7+49\rho^9) + \frac{1}{8n^2}(840\rho^3-5460\rho^5+12313\rho^7-11662\rho^9+3969\rho^{11}) + O(n^{-3}) \quad (2.8)$$

$$\mu_8'(r) = \rho^8 + \frac{1}{n}(28\rho^6-60\rho^8+32\rho^{10}) + \frac{1}{n^2}(420\rho^4-1232\rho^6+2112\rho^8-2432\rho^{10}+800\rho^{12}) + O(n^{-3}) \quad (2.9)$$

$$\mu_9'(r) = \rho^9 + \frac{1}{2n}(72\rho^7-153\rho^9+81\rho^{11}) + \frac{1}{8n^2}(3024\rho^5$$

$$-17136\rho^7+34959\rho^9-30618\rho^{11}+9801\rho^{13}) + 0(n^{-3}) \qquad (2.10)$$

$$\mu_{10}'(r) = \rho^{10} + \frac{1}{n}(45\rho^8-95\rho^{10}+50\rho^{12}) + \frac{1}{n^2}(630\rho^6-$$

$$3420\rho^8-7740\rho^{10}-5750\rho^{12}+1800\rho^{14}) + 0(n^{-3}) \qquad (2.11)$$

$$\mu_{11}'(r) = \rho^{11} + \frac{1}{2n}(110\rho^{11}-231\rho^{11}+121\rho^{13}) + \frac{1}{8n^2}\, x$$

$$(7920\rho^7-41580\rho^9+79761\rho^{11}-66550\rho^{13}+20449\rho^{15}) +$$

$$0(n^{-3}) \qquad (2.12)$$

The eleven moments of r under a non-normal parent population have been derived and will be used for further study of this problem in another article. In the computation of the moments of r, we noticed that their values considerably decrease with increment of order and sample size n. For this reason, we are using these moments of r to attain asymptotic convergence for the distributions of four transformations.

3. SPECIAL TRANSFORMATIONS

In this section special forms for Z are considered. The first four moments in these special cases are derived in the very general nature. Since moments of r higher than the first four moments of Hotelling (1953) are not yet in the literature (Gajjar and Subrahmaniam, 1978, p. 475 and Subrahmaniam and Gajjar, 1980), our study will be useful in examining the distributions of these transformations and their behavior under normal assumptions.

3.1 Fisher's $\tanh^{-1}r$

The transformation $Z = \frac{1}{2}\ln\frac{1+r}{1-r}$ is suggested by Fisher (1921). Using the moments of r in section 2, the first four moments of this transformation are:

$$\mu_1'(z) = \mu_1'(r) + \frac{1}{3}\mu_3'(r) + \frac{1}{5}\mu_5'(r) + \frac{1}{7}\mu_7'(r) +$$

$$\frac{1}{9}\mu_9'(r) + \frac{1}{11}\mu_{11}'(r)$$

$\mu_2'(z)$, $\mu_3'(z)$, and $\mu_4'(z)$ are derived similarly.

Subrahmaniam and Gajjar (1980) give the three central moments $\mu_2(z)$, $\mu_3(z)$, $\mu_4(z)$ to $0(n^{-3})$ of $\tanh^{-1}r$:

$$\mu_2(z) = \frac{1}{n} + \frac{(4-\rho^2)}{2n^2} + \frac{(22-6\rho^2-3\rho^4)}{6n^3}$$

$$\mu_3(z) = \frac{\rho^3}{n^3}, \quad \mu_4(z) = \frac{3}{n^2} + \frac{(14-3\rho^2)}{n^3} \qquad (3.1.2)$$

as given in Hotelling's paper.

3.2 Samiuddin's Transformations

The transformation $Z = \left(\frac{n-2}{1-\rho^2}\right)^{\frac{1}{2}}(r-\rho)\left[1-r^2\right]^{-\frac{1}{2}}$ is suggested by Samiuddin (1970). The first four moments in terms of r up to r^{11}, are:

$$\mu_1'(z) = \left(\frac{n-2}{1-\rho^2}\right)^{\frac{1}{2}}\Big[\mu_1'(r) + \tfrac{1}{2}\mu_3'(r) + \frac{3}{8}\mu_5'(r) +$$

$$\frac{5}{16}\mu_7'(r) + \frac{35}{128}\mu_9'(r) + \frac{63}{256}\mu_{11}'(r) - \rho\Big(1 + \frac{1}{2}\mu_2'$$

$$(r) + \frac{3}{8}\mu_4'(r) + \frac{5}{16}\mu_6'(r) + \frac{35}{128}\mu_8' + \frac{63}{256}\mu_{10}'(10)\Big)\Big] \qquad (3.2.1)$$

$\mu_2'(z)$, $\mu_3'(z)$, $\mu_4'(z)$ are derived in the same manner.

Subrahmaniam and Gajjar (1980) give the central moment $\mu_2(z)$, $\mu_3(z)$ and $\mu_4(z)$ in which they are using Hotelling's results. These are:

$$\mu_2(z) = (n-2)\left\{\frac{1}{n} + \frac{6-\rho^2}{2n^2} + \frac{36-7\rho^2-2\rho^4}{4n^3} +\right.$$

$$\left.0(n^{-4})\right\} \qquad (3.2.5)$$

$$\mu_3(z) = (n-2)^{3/2}\left\{\frac{3\rho+2\rho^2}{2n^3} + 0(n^{-4})\right\} \qquad (3.1.6)$$

$$\mu_4(z) = (n-2)^2\left\{\frac{3}{n^2} + \frac{(24-3\rho^2)}{n^3} + 0(n^{-4})\right\} \qquad (3.2.7)$$

3.3 Nair's Transformation

The transformation $Z = (r-\rho)(1-r\rho)^{-1}$ was suggested by Pillai (1946). The first four moments in terms of r, up to r^{11}, are:

$$\mu_1'(z) = \mu_1'(r) + \rho\,\mu_2'(r) + \rho^2\mu_3'(r) + \rho^3\mu_4'(r) +$$

$$\rho^4\mu_5'(r) + \rho^5\mu_6'(r) + \rho^6\mu_7'(r) + \rho^7\mu_8'(r) +$$

$$\rho^8\mu_9'(r) + \rho^9\mu_{10}'(r) + \rho^{10}\mu_{11}'(r) - \rho\Big\{1 + \rho\,\mu_1'(r)$$

$$+ \rho^2\mu_2'(r) + \ldots + \rho^{11}\mu_{11}'(r)\Big\} \qquad (3.3.1)$$

$\mu_2'(z)$, $\mu_3'(z)$, and $\mu_4'(z)$ are derived in the same way.

Subrahmaniam and Gajjar (1980) have given the three central moments $\mu_2(z)$, $\mu_3(z)$, $\mu_4(z)$ of this transformation in which Hotelling's results are used. These moments up to $0(n^{-3})$ are:

$$\mu_2(z) = \frac{1}{n} - \frac{\rho^2}{2n^2} + \frac{\rho^2(1-\rho^2)}{2n^3}$$

$$\mu_3(z) = -\frac{\rho(3-\rho^2)}{n^3}, \quad \mu_4(z) = \frac{3}{n^2} - \frac{3(2+\rho^2)}{n^3}$$

3.4 Arcsine Transformation

Harley (1956) studied the properties of this transformation. The first four moments of $Z = \sin^{-1} r$, are:

$$\mu_1'(z) = \mu_1'(r) + \frac{1}{6}\mu_3'(r) + \frac{3}{40}\mu_5'(r) + \frac{5}{112}\mu_7'(r) + \frac{35}{1152}\mu_9'(r) + \frac{63}{2816}\mu_{11}'(r). \quad (3.4.1)$$

$\mu_2'(z)$, $\mu_3'(z)$, and $\mu_4'(z)$ are derived similarly.

In Subrahmaniam and Gajjar's (1980) article, the three central moments $\mu_2(z)$, $\mu_3(z)$ and $\mu_4(z)$ of $z = \sin^{-1} r$, in which Hotelling's results are used, are

$$\mu_2(z) = \frac{(1-\rho^2)}{n} + \frac{(1-\rho^4)}{n^2} + \frac{2(1-\rho^2)(1+\rho^2+4\rho^4)}{3n^3} + 0(n^{-4}) \qquad \mu_3(z) = -3\rho(1-\rho^2)^{3/2}\left\{\frac{1}{n^2} + \frac{1+4\rho^2}{n^3}\right\} + 0(n^{-4}) \qquad \mu_4(z) = 3(1-\rho^2)^2 \cdot \left\{\frac{1}{n^2} + \frac{2(2+13\rho^2)}{3n^3}\right\} + 0(n^{-4})$$

The moments are calculated for which μ_2, β_1 and β_2 are computed for different values of n and ρ.

Here we use $\beta_1 = \mu_3^2/\mu_2^3$ and $\beta_2 = \mu_4/\mu_2^2$. The values of our computation for μ_2, β_1 and β_2 are depicted in tables 1.1, 1.2, 1.3, and 1.4.

4. COMPARISON OF RESULTS

The hypothesis which, in practice, one usually desires to test is that $\rho = 0.0$. For the case $\rho = 0.0$ the values of μ_2, β_1 and β_2 in tables 1.1, 1.2., 1.3., and 1.4. show that the distributions for all these transformations of r are approaching normality very rapidly for large values of n, and this convergence to normality is quite good using Hotelling's results.

Our first set of values of μ_2, β_1 and β_2 for all transformations are obviously better than those of Subrahmaniam and Gajjar's (1980) results, when we use the first eleven moments of r for all values of n and $\rho \le 0.35$.

In all four transformations, whether we are using our approximations or that of Subrahmaniam and Gajjar, the departure from normality becomes clear for large values of n and ρ.

Using the first four moments of Hotelling does not show the important feature of the higher values of β_1 and β_2 for $\rho \le 0.35$ in Tables 1.1, 1.2, 1.3, and 1.4.

5. CONCLUSION

The study of these four transformations of r using the first eleven moments up to $0(n^{-2})$ shows that the asymptotic formulae for these transformations converge to the normal distribution when $\rho = 0.0$ for any sample size $n \ge 20$. This feature of convergence to normality does not need higher orders of moments than those derived by Hotelling (1953).

This study of the distributions of these transformations for values of ρ different from zero shows that it is necessary to use moments of higher than order 4. In tables 1.1 – 1.4, the first set of values using our approximation is clearly better than the second set of values in which we used only Hotelling's results because it shows the tendency of the departure from normality in tables 1.1, 1.2, 1.3, and 1.4 for all values of n and $\rho \le 0.35$.

The departure from normality is very significant for $|\rho| \ge 0.35$ and $n > 100$. The computations are available upon request.

TABLE I.I*, Fisher's $\tanh^{-1} r$: values of μ_2, β_1, and β_2

n	ρ	0.0	0.05	0.25	0.35
	μ_2	0.0545	0.0545	0.0554	0.0580
		0.0454	0.0455	0.0455	0.0456
20	β_1	0.0000	0.0131	0.0249	0.0763
		0.0000	0.0000	0.0000	0.0000
	β_2	2.2725	2.1896	2.5803	5.8703
		4.4763	4.4752	4.4512	4.4273
	μ_2	0.0207	0.0207	0.0209	0.0212
		0.0192	0.0192	0.0192	0.0193
50	β_1	0.0000	0.0010	0.0006	0.0331
		0.0000	0.0000	0.0000	0.0000
	β_2	2.6709	2.6363	3.5228	6.7490
		3.5481	3.5478	3.5397	3.5316
	μ_2	0.0102	0.0102	0.0102	0.0103
		0.0098	0.0098	0.0098	0.0098
100	β_1	0.0000	0.0001	0.0000	0.0174
		0.0000	0.0000	0.0000	0.0000
	β_2	2.8281	2.8111	3.8857	6.9815
		3.2670	3.2668	3.2630	3.2592

TABLE I.2*, Sammiuddin's Transformation: values of μ_2, β_1 and β_2

n	ρ	0.0	0.05	0.25	0.35
	μ_2	1.022	1.020	1.052	1.181
		1.055	1.055	1.54	1.052
20	β_1	0.000	0.017	8.396	47.190
		0.000	0.000	0.000	0.000
	β_2	2.096	2.019	3.815	11.681
		3.055	3.055	3.057	3.061
	μ_2	1.015	1.015	1.030	1.080
		1.021	1.021	1.020	1.020
50	β_1	0.000	0.014	152.283	1427.247
		0.000	0.000	0.000	0.000
	β_2	2.575	2.540	4.740	13.273
		3.076	3.076	3.077	3.077
	μ_2	1.009	1.009	1.017	1.040
		1.010	1.010	1.010	1.010
100	β_1	0.000	0.066	1320.763	13661.706
		0.000	0.000	0.000	0.000
	β_2	2.774	2.756	4.428	10.115
		3.049	3.050	3.049	3.049

TABLE I.3*, Nair's Transformation: values of μ_2, β_1 and β_2

n	ρ	0.0	0.05	0.25	0.35
	μ_2	0.0500	0.0501	0.0512	0.0502
		0.0000	0.0500	0.0500	0.0499
20	β_1	0.0000	0.0146	0.5904	16.2255
		0.0000	0.0000	0.0000	0.0001
	β_2	2.7000	2.6012	2.2640	10.6125
		2.7000	2.6999	2.6987	2.6794
	μ_2	0.0200	0.0200	0.0200	0.0200
		0.0200	0.0200	0.0200	0.0200
50	β_1	0.0000	0.0010	6.0541	226.3250
		0.0000	0.0000	0.0000	0.0000
	β_2	2.8800	2.8444	2.7111	11.4560
		2.8800	2.8800	2.8798	2.8796
	μ_2	0.0100	0.0100	0.0100	0.0100
		0.0000	0.0100	0.0100	0.0100
100	β_1	0.0000	0.0001	46.0593	1750.5106
		0.0000	0.0000	0.0000	0.0000
	β_2	2.9400	2.9229	2.6234	12.0278
		2.9400	2.9400	2.9399	2.9399

TABLE I.4*, Arcsine Transformation: values of μ_2, β_1 and β_2

n	ρ	0.0	0.05	0.25	0.35
	μ_2	0.0500	0.0521	0.0496	0.0469
		0.0500	0.0524	0.0495	0.0462
20	β_1	0.0000	0.0088	0.0100	0.0925
		0.0000	0.0011	0.0270	0.0537
	β_2	2.4725	2.3832	2.0181	4.5433
		2.8933	2.9855	2.9484	3.0004
	μ_2	0.0200	0.0203	0.0192	0.0181
		0.0200	0.0203	0.0192	0.0179
50	β_1	0.0000	0.0002	0.0017	0.0425
		0.0000	0.0004	0.0111	0.0219
	β_2	2.7725	2.7353	2.8471	4.7833
		2.9588	2.9598	2.9827	3.0053
	μ_2	0.0100	0.0101	0.0095	0.0089
		0.0100	0.0100	0.0095	0.0089
100	β_1	0.0000	0.0000	0.0031	0.0224
		0.0000	0.0002	0.0055	0.0110
	β_2	2.8832	2.8645	3.1499	4.8485
		2.9797	2.9802	2.9919	3.0035

* The first set of values is our approximation and the second is Subrahmanium and Gajjar's approximation.

REFERENCES

Cook, M.B. (1951a). Bivariate k-statistics, Biometrika 38, 179-195

Cook, M.B. (1951). Two applications of bivariate k-statistics. Biometrika 38, 368-376.

Fisher, R.A. (1921). On the "probable error" of a coefficient of correlation deduced from a small sample. Metron 1, 1-32.

Gajjar, A.V. and Subrahmaniam, Kocherlakota (1978) On the sample correlation coefficient in the truncated bivariate normal population. Commun. Statist. - Simula. Computa., B7,(5), 455-477.

Harley, B.I. (1956). Some properties of an angular transformation for the correlation coefficient. Biometrika 43, 219-224.

Hotelling, H. (1953). New light on the correlation coefficient and its transforms(with discussion). J. Roy. Statist. Soc., Ser. B, 15, 193-232.

Pillai, K.C.S. (1946). Confidence interval for the correlation coefficient. Sankhya 7, 415-422.

Samiuddin, M. (1970). On a test for an assigned value of correlation in a bivariate normal distribution. Biometrika 57, 461-464.

Subrahmaniam, Kocherlakota and Gajjar, A.V. (1980) Robustness to nonnormality of some transformations of the sample correlation coefficient. J. Multivariate Analysis., 10, 60-77.

USING LINEAR PROGRAMMING TO FIND APPROXIMATE SOLUTIONS TO THE FIELDS TO IMPUTE PROBLEM FOR INDUSTRY DATA

Patrick G. McKeown and Joanne R. Schaffer
University of Georgia

Abstract

Sande has suggested a mathematical programming formulation of the fields to impute problem (FTIP) for continuous data. This formulation seeks to find a minimum weighted sum of fields that would need to be changed to yield an acceptable record by solving a mixed integer programming problem known as the fixed charge problem. While this formulation can and has been solved to find an optimal solution to the FTIP, this approach can be expensive in terms of solution time. In this paper, we demonstrate the use of a heuristic procedure to find an approximately optimal solution to FTIP. This procedure uses the SWIFT algorithm developed by Walker in conjunction with a judicious choice of dummy variable costs to arrive at an approximate solution based on a linear programming solution. We will show that this solution is optimal in many cases. We will also discuss the use of the special structure of FTIP to arrive at an optimal solution to the LP problem.

Introduction

Let the fields to impute problem (FTIP) be written as (after Sande [3]):

$$\text{Min} \sum_{j=1}^{n} w_j \delta(Y_j) + \sum_{j=1}^{n} w_j \delta(Z_j) \tag{1}$$

$$\text{s.t. } AX^o + AY - AZ \leq b \tag{2}$$

$$X^o + Y - Z \geq 0 \tag{3}$$

$$X, Z \geq 0 \tag{4}$$

$$\text{where } \delta_j(\cdot) = \begin{matrix} 1, \text{ if } (\cdot) > 0 \\ 0, \text{ otherwise} \end{matrix} \text{ and} \tag{5}$$

where A is mxn, and b is mx1. X^o is the respondent's answers, Y and Z are deviational variables that are added to or subtracted from X^o to yield an acceptable record, A and b serve to define the linear consistency conditions used in the editing, and w is the set of weights to be used to place more confidence in one field or another. Since X^o is a vector of constants, then (2) becomes

$$AY - AZ \leq b - AX^o. \tag{6}$$

FTIP seeks to find a minimum weighted sum of fields that must be changed to yield an acceptable record.

The problem defined by (1) and (3)-(6) is known as a fixed charge problem (FCP). The FCP is a mixed integer programming problem which has been solved by various integer programming methods. McKeown [1] has used branch and bound methods to solve the FCP. He has also used this same procedure to solve FTIP problems where the edit conditions (A and b) and the test records (X^o) were supplied by the U.S. Census Bureau for the Annual Survey of Manufacturers (ASM) [2].

In this paper, we describe various ways to find optimal or near-optimal solutions to FTIP without going through the search required by branch and bound. To do this we use a heuristic adjacent extreme point procedure developed by Walker [4] in conjunction with a judicious choice of dummy continuous costs to find a "good" solution to the problem. To speed up the solution to the associated LP problem, that is, the form of FTIP without the binary variables, $\delta(\cdot)$, we take advantage of the special structure of the A matrix and treat the constraints $X^o + Y - Z \geq 0$ as implicit upper and lower bounds.

In section 2, we discuss the special structure of the A matrix for ASM editing conditions, and how this structure may be utilized to efficiently solve the LP problem associated with FTIP. In section 3, we discuss the heuristic procedure used to find approximate solutions to FTIP. Finally, we present computational results which demonstrate the efficiency of the heuristic procedure and the relative speed of solving the LP utilizing the special structure versus using a general approach to solving the LP.

2. Special Structure of FTIP for ASM Edit Conditions

For the Annual Survey of Manufacturers, there are two types of edit conditions, ratio edits and balance edits. Ratio edits require that the ratio of the responses in two fields be between upper and lower bounds. In other words, if X_j and X_k are the responses for ratio jk, then

$$\ell_{jk} \leq X_j/X_k \leq U_{jk} \tag{7}$$

where $X_k > 0$. This may be broken up into two linear inequalities, i.e.,

$$X_j - X_k U_{jk} \leq 0 \tag{8}$$

$$X_j - X_k \ell_{jk} \geq 0 \tag{9}$$

In the ASM edit conditions, there are 20 fields and 23 ratio edits. All fields are involved in at least one ratio edit but all pairs of possible ratios do not have bounds. Also, some ℓ_k values are zero which results in the corresponding ratio collapsing into a redundant nonnegativity condition. In most situations, 44 or 45 linear inequalities are generated by the ratio edit conditions.

The second type of ASM edit condition is the balance edit. This edit requires that the sum of two responses must equal a third response. In other words, if X_j, X_k and X_ℓ are the three responses, then

$$X_j + X_k + X_\ell = 0 \tag{10}$$

is the corresponding equality constraint. There are four such balance edit conditions involving 12 fields. No field is involved in more than one balance constraint.

The 44 or 45 inequality constraints resulting from the ratio edit conditions and the four equality constraints give us a total of 48 or 49 constraints and what is initially 20 variables. After adding the deviational variables, Y and Z and subtracting AX^o from b, there are 40 total variables, two for each field. Each constraint will be of the form of (8), (9), or (10) except that the first 20 variables will have the sign shown in (8), (9), or (10) while the second 20 variables will have the opposite sign. For example, for the ratio jk involving fields j and k, the two constraints become:

$$X_j - \ell_{jk}X_k - X_{j+20} + \ell_{jk}X_{k+20} \geq \bar{b}_{jk} \quad (11)$$

and

$$X_j + U_{jk}X_k - X_{j+20} + U_{jk}X_{k+20} \leq \tilde{b}_{jk} \quad (12)$$

Similarly, the balance constraint involving X_j, X_k, and X_ℓ becomes:

$$X_j + X_k - X_\ell - X_{j+20} - X_{k+20} + X_{\ell+20} = b_{jk\ell} \quad (13)$$

where

$$\bar{b}_{jk} = -X_j^o + \ell_{jk}X_k^o \quad (14)$$

$$\tilde{b}_{jk} = -X_j^o - U_{jk}X_k^o \quad (15)$$

and

$$b_{jk\ell} = X_j^o - X_k^o + X_\ell^o \,. \quad (16)$$

In terms of (1)-(6), X_j^o is the response in field j, X_j corresponds to Y_j, X_{j+20} corresponds to Z_j, and $\bar{b}_{jk}$, $\tilde{b}_{jk}$, and $\hat{b}_{jk\ell}$ correspond to the $b_i - \sum_{j-1}^{\Gamma} a_{ij}X_j^o$ terms in (6) for each type constraint.

The constraints (11) and (12) can be broken up into less-than or greater-than-or-equal constraints. First, if $\bar{b}_{jk} \leq 0$ or $\tilde{b}_{jk} \geq 0$, the ith constraint will be of the less-than type. Similarly, if $\bar{b}_{jk} > 0$ or $\tilde{b}_{jk} < 0$, the jk^{th} constraint will be of the greater-than type. Let $\Lambda = \{jk$ constraint: jk is a less-than-or-equal-to constraint$\}$ and $\Gamma = \{jk$ constraint: jk is a greater-than-or-equal-to constraint$\}$. Then, in terms of edit conditions, Γ is the set of edit conditions which are violated by the respondent's record, X^o. In Λ, no change in X^o is required so the Y_j and Z_j can be set to zero and still satisfy the constraints. To satisfy the Γ constraints, it is necessary to make one or more Y_j or Z_j positive in each constraint. Our objective is to make the minimum weighted number of such changes.

For the balance constraints, if $b_{jk} = 0$, the balance constraint is satisfied by X^o. If, on the other hand, $b_{jk\ell} \neq 0$, the balance constraint is not satisfied by X^o and one or more Y_j or Z_j must be changed to satisfy constraint $jk\ell$. Once again, as with the ratio constraint, it is our objective to make $X^o + Y - Z$ satisfy all ratio and balance edits. This corresponds to finding a feasible solution to the constraint set (11)-(13) in which the sum of weighted changes, $W_j\delta(Y_j)$ and $W_j\delta(Z_j)$, is minimized. Finding the optimal solution to this problem requires an integer programming solution. However, any feasible solution to (11)-(13) will yield a set of changes, Y and Z, which, while not necessarily optimal, may be of interest. If we assign dummy continuous costs (as will be discussed later), finding a feasible solution to (11)-(13) corresponds to solving a linear programming problem. This could be done by adding slack variables to the constraints in Λ, surplus and artificial variables to the constraints in Γ, and artificial variables to the four balance constraints and going through a Phase I - Phase II or Big M optimization procedure.

While such a procedure would definitely work, it fails to take advantage of the special structure of (11)-(13), i.e., the sparsity of the constraints (11) and (12), the non-overlapping nature of the variables in (13) and the preponderence of constraints in Λ as compared to those in Γ. We do this by using a three-step approach. First, we remove the artificial variables in the balance constraints by using a modified dual simplex. Second, we do not use artificial variables for the greater-than-or-equal-to constraints, but rather use the dual simplex method to remove any infeasibilities. Finally, we optimize the resulting feasible solution over the dummy continuous costs. All of this is done in a bounded variable context. This is possible since the constraints (3), $X^o + Y - Z \geq 0$, become

$$Z_j \leq X_j^o \quad (17)$$

if $X_j^o \geq 0$ and

$$Y_j \geq -X_j^o \quad (18)$$

if $X_j^o < 0$. So, we do not carry along the constraints (3) explicitly.

While steps two and three are fairly straightforward in their use of dual and primal simplex, step one is somewhat different in the manner in which the outgoing and incoming variables are chosen in the dual simplex method. First, the outgoing variable is chosen as an artificial variable in a constraint where $b_{jk\ell} > 0$. Second, the incoming variable is chosen in such a way that the current level of infeasibility in the constraints in Γ is not increased. This requires choosing a positive element in the row of tableau corresponding to $b_{jk\ell} > 0$ such that when a pivot is made, no additional right-hand-side values become negative. This involves checking the ratio of the pivot element and $b_{jk\ell}$ against all other non-zero elements in the column of the pivot element to determine the effect of pivoting the column into the basis. Since (11)-(13) is so sparse, this is fairly easy to do and has led to a dramatic decrease in the number of pivots necessary to achieve a feasible solution.

Heuristic Procedure

Since finding a minimal value to (1)-(6) involves solving a difficult mixed integer programming problem, it is useful to find a "good" heuristic solution to the problem. Walker [4] has developed a procedure known as SWIFT to find heuristic solutions to fixed charge problems such as (1)-(6). SWIFT works by first solving the fixed charge problem as a linear programming problem. This is usually done by simply dropping the fixed charges and solving the resulting LP. In our case, we have only fixed charges so we have no real continuous costs to use in the LP solution. Our problem is to choose a set of continuous costs that will yield a useful LP solution. By useful, we mean a solution which may be "close" to the true optimal solution.

After an LP solution has been found, SWIFT computes a revised set of reduced costs ($Z_j - C_j$ values) which include the fixed cost for each non-basic variable. This revised set of reduced costs adds in the fixed cost of the incoming variable and subtracts out the fixed cost of the outgoing variable. They also capture the effect of variables moving from degenerate to non-degenerate and vice versa and the effect of variables switching bounds. If any of the reduced costs indicate that the total cost can be reduced, the appropriate nonbasic variable is brought into the basis and a new set of revised reduced costs are computed. When a situation is reached where no improvement is indicated, SWIFT stops with the current solution as the heuristic solution. In our procedure, SWIFT is preceded by a search that one at a time brings each nonbasic variable into the solution to search for a solution that has a lower total cost than the current lowest cost solution.

However, prior to going to SWIFT, it is necessary to solve the LP version of the fixed charge problem. Since there are no continuous costs, several options are possible. First, the LP could be solved with all continuous costs equal to zero. While this will lead to a feasible solution, it may be quite "far" away from the optimal solution. Another approach would be to linearize the fixed costs by dividing each by its upper bound (if known). However, since upper bounds will be known for at most only half of the variables, this does not appear to be the approach to take. Finally, we could assign "dummy" continuous costs to all variables that somehow reflect the variables which a priori appear to be in error and which must be changed. This latter approach is the one which we have chosen to use.

The choice of dummy continuous costs was made as follows:

i) assign all variables a continuous cost of 1.0;

ii) if a field has no response, then give the corresponding Y_j variable a cost of $\hat{C}_j = 0$;

iii) compare the record being edited to the prior year record. Then if X_j^o = current value in jth field and X_j^* = prior year value in jth field and if

a) $X_j^o/X_j^* > \delta$, set $\hat{C}_j = 0$ for the Z_j variable;

b) $X_j^o/X_j^* < 1/\delta$, set $\tilde{C}_j = 0$ for the Y_j variable;

c) if $X_j^* = 0$ and $X_j^o > \Delta$, set $\hat{C}_j = 0$ for the Z_j variable.

The value of δ and Δ are user determined to fit the data being edited. Currently these values are set so that $\delta = 3$ and $\Delta = 999$. These continuous costs seek to force the LP solution to include changes in those fields which have no response or have current year values which are much different from those for the prior year.

Using these values, we then solve the following linear programming problem.

$$\text{Min} \quad \sum_{j=1}^{n} \tilde{C}_j Y_j + \sum_{j=1}^{n} \hat{C}_j Z_j$$

Subject to (6), (3), and (4).

SWIFT is then called to find the heuristic solution.

Computational Results

Two sets of test problems were used in the computational testing. First, a set of seven sample problems provided by the U.S. Census Bureau were run using the procedure described here. These same problems were run on a previously developed procedure which did not take advantage of the special structure of the problem. Two statistics were compared. These were running times and number of fields imputed. Table 1 shows the results of this testing.

Table 1 shows that the procedure described here was dramatically faster than a similar procedure that did not take advantage of the special structure. Also, the special structure heuristic found the optimal solution in all seven of the test problems while the previous procedure missed the optimal solution in two cases.

The second set of data was made up of 164 test problems which were generated by perturbing one or more fields of "good" record. This perturbation was such that the number of failed fields was known prior to any solution being found. Up to 11 fields were perturbed in this manner. The results of this testing are shown in Table 2. Once again, the heuristic procedure was able to find the optimal solution to all 164 problems in an average of .545 CPU seconds.

TABLE 1

Test Data (All $w_j = 1.0$)

Problem	Solution Time		Fields Imputed		
	LP1	LP2	H1	H2	Optimal
1	1.828	.655	11	8	8
2	2.785	.509	11	11	11
3	2.842	.565	13	13	13
4	3.167	.446	9	9	9
5	3.530	.558	14	14	14
6	.556	.336	5	3	3
7	2.044	.333	3	3	3

(All times are CPU sec. on CDC CYBER 70/74 with OPT = 1.)

LP1, H1: results using LP and SWIFT without use of special structure
LP2, H2: results using LP and SWIFT utilizing special structure

TABLE 2

Perturbed Data* (All $w_j = 1.0$)

Number of Problems	% Heuristic Optimal	Ave LP Time	Max LP Time
163	100%	.545 sec.	1.476

*A "good" record with between 1 and 11 fields perturbed to be in error.

REFERENCES

1. P. McKeown, "A Branch and Bound Algorithm for Solving Fixed Charge Problems," submitted to Naval Research Logistics Quarterly.

2. P. McKeown, "A Mathematical Programming Approach to Editing of Economic Survey Data," submitted to J. of American Statistical Association.

3. G. Sande, "An Algorithm for Fields to Impute Problems of Numerical and Coded Data," Working Paper, Statistics Canada.

4. Warren E. Walker, "A Heuristic Adjacent Extreme Point Algorithm for the Fixed Charge Problem," Management Science, 22, 587-596 (1976).

USING COMPUTER-BINNED DATA FOR DENSITY ESTIMATION

David W. Scott, Rice University

ABSTRACT: With real time microcomputer monitoring systems or with large data bases, data may be recorded as bin counts to satisfy computer memory constraints and to reduce computational burdens. If the data represent a random sample, then a natural question to ask is whether such binned data may successfully be used for density estimation. Here we consider three density procedures: the histogram, parametric models determined by a few moments, and the nonparametric kernel density estimator of Parzen and Rosenblatt. For the histogram, we show that computer-binning causes no problem as long as the binning is sufficiently smaller than the data-based bin width $3.5\sigma\, n^{-1/3}$. Another result is that some binning of data appears to provide marginal improvement in the integrated mean squared error of the corresponding kernel estimate. Some examples are given to illustrate the theoretical and visual effects of using binned data.

Key words: Histogram, kernel density estimator, binned data, optimal integrated mean squared error.

1. Introduction. In real-time microprocessor monitoring systems, data are often recorded with less accuracy than available in the hardware to preserve memory. If the data are viewed as a random sample, then the data may be compactly processed into bin counts or, specifically, collected in the form of a histogram with relatively narrow bins. The bin width represents twice the maximum possible error induced by the recording process. A natural question is how to apply standard density estimation for a random sample collected in such a manner.

We shall assume the true sampling density $f \in C^1$ with mean and variance given by μ and σ^2 respectively. Suppose $x_1,\ldots,x_n$ is a random sample from f. Define an equally spaced mesh $\{t_i, -\infty < i < \infty\}$ such that $t_{i+1} - t_i = \delta\ \forall i$, where δ is the bin width. Let n_i be the number of samples falling in the interval $(t_i - \delta/2, t_i + \delta/2]$. Then the computer-binned data are given by the counts $\{n_i\}$ and $\Sigma n_i = n$.

2. Histogram Estimator. If the density estimator to be used is also a histogram, analysis using the recorded data is virtually the same as with the true unbinned data. One simply aggregates adjacent bins in the recorded data to achieve the optimal smoothing desired. The bin width h^* that minimizes the integrated mean squared error is given by

$$(2.1)\quad h^* = [6/\int_{-\infty}^{\infty} f'(x)^2 dx]^{1/3}\, n^{-1/3}$$

where f is the true sampling density and n is the sample size (Scott 1979). This leads to a data-based choice

$$(2.2)\qquad h = 3.5 s\, n^{-1/3}$$

where s is an estimate of the standard deviation. The choice (2.2) is essentially optimal for Gaussian data. For other sampling densities, a slightly smaller choice than (2.2) may be optimal. In particular, correction factors based on skewness and kurtosis have been given in the above reference.

The danger is that bin width of the optimally smoothed histogram may actually be smaller than the bin width chosen for the microprocessor program. This problem may be eliminated by estimating the optimal bin width using the data-based procedure (2.2) with a preliminary estimate of the variance. The microprocessor's histogram could then be chosen with bin width narrower by a factor of at least five. Care must also be taken with respect to the sample range.

3. Parametric Models. Many parametric density models are determined by the values of a few lower order moments. For example, the estimated mean and variance with binned data are given by

$$(3.1)\qquad \bar{x}_\delta = \frac{1}{n}\sum n_i t_i$$

$$(3.2)\qquad s_\delta^2 = \frac{1}{n-1}\sum n_i (t_i - \bar{x}_\delta)^2 .$$

Using method similar to those found in Scott (1979) it is possible to show that $E\bar{x}_\delta = \mu$ and $Es_\delta^2 = \sigma^2 + \delta^2/12 + O(\delta^4)$, assuming that f has sufficient high order contact in the tails; see Kendall and Stuart (1977). These are just Sheppard's corrections for moments computed using frequency data! Therefore, δ may be chosen sufficiently small to limit parameter error induced by binning or used in Sheppard's formulae.

4. Kernel Estimators. Given a kernel K which is a selected symmetric probability density function and a positive parameter h, the kernel estimate of $f(x)$ is given by

$$(4.1)\qquad \hat{f}(x) = \frac{1}{nh}\sum_{i=1}^{n} K\left(\frac{x_i - x}{h}\right),$$

an estimate that is nonnegative and integrates to one. The estimator (4.1) was originally proposed by Parzen (1962) and Rosenblatt (1956). The kernel estimators are superior to histograms and have been widely studied; see Tapia and Thompson (1978). The choice of the (smoothing) parameter h that minimizes the integrated mean squared error is proportional to $n^{-1/5}$. It is worth noting that (4.1) is not an unbiased estimate and that for the optimal choice of h, the resulting estimator contains four times as much variance as squared bias; for a comparison of methods for choosing h, see Scott and Factor (1981). For binned data, the kernel estimate (4.1) becomes

$$\hat{f}_\delta(x) = \frac{1}{nh}\sum n_i\, K\left(\frac{t_i - x}{h}\right) . \tag{4.2}$$

A careful analysis (Scott 1981) shows that the relative bias of the estimator (4.2) as compared to (4.1) is increased by the factor $\delta^2/12h^2$ while the relative variance is decreased by the factor $\delta f(x)$. In fact, the overall error is often smaller for small positive values of δ as compared to $\delta = 0$; that is, the kernel estimator (4.2) using binned data is superior to the kernel estimator (4.1) using the (true) raw data. Intuitively, the improvement is possible because the variance to bias ratio for $\delta = 0$ is greater than one. In Figure 1, with standard Gaussian data using the value of h optimal for (4.1), we see that only modest improvements in error are possible, but that surprisingly large values of δ seem acceptable. However, if we are most interested in the shape of the true density function, we should compare the second derivatives of (4.1) and (4.2) for the same pseudo-random data sets. The range of values of δ where the second derivative is improved is narrower. As an example, we generated 1,000 N(0,1) points from IMSL (1980) routine GGNPM with seed 12345. Using $h^* = .2661$ and $\delta = 0.2$, we computed (4.1) and (4.2) and their second derivatives. The maximum absolute difference between $\hat{f}$ and $\hat{f}_\delta$ was a very small 0.007 at $x = 0.15$. We remark that the maximum absolute difference between f and $\hat{f}$ was .028 at $x = .60$. However, the maximum absolute difference between $\hat{f}''$ and $\hat{f}''_\delta$ was 0.204 at $x = .25$; see Figures II and III.

We conjecture that using binned data with other nonparametric density estimation techniques might also result in reduced integrated mean squared error. We performed a small Monte Carlo simulation using the Fourier estimator of Grace Wahba (1981), estimating the effect of δ. The preliminary estimates indicated that the integrated mean squared error was essentially constant for a range of δ-values (cf. Figure I). Another estimator for which binned data may be useful is the penalized likelihood density estimator; see Good and Gaskins (1972). In fact, Good presented a paper at the interface meeting discussing a new algorithm for solving the penalized-likelihood criterion using fast-Fourier transforms on binned data. Computational advantages are also realizable with other methods as well.

References:

Good, I. J. and Gaskins, R. A. (1972). Global nonparametric estimation of probability densities. Virginia J. Sci. 23 171-193.

International Mathematical and Statistical Libraries (1980). Houston, Texas.

Kendall, M. G. and Stuart, A. (1977). The Advanced Theory of Statistics, Vol. 1, New York: Macmillan.

Parzen, E. (1962). On estimation of a probability density function and mode. Ann. Math. Statist. 33, 1065-76.

Rosenblatt, M. (1956). Remarks on some nonparametric estimates of a density function. Ann. Math. Statist. 27, 832-7.

Scott, D. W. (1979). On optimal and data-based histograms. Biometrika 66:605-610.

Scott, D.W. (1981). Kernel density estimation with binned data. Working paper, Rice University, Houston, Texas.

Scott, D.W. and Factor, L. E. (1981). Monte Carlo study of three data-based nonparametric probability density estimators. J. American Statistical Assoc. 76:9-15.

Tapia, R. A. and Thompson, J. R. (1978). Nonparametric Probability Density Estimation. Baltimore: Johns Hopkins.

Wahba, G. (1981). Data-based optimal smoothing of orthogonal series density estimates. Ann. Statis. 9:146-156.

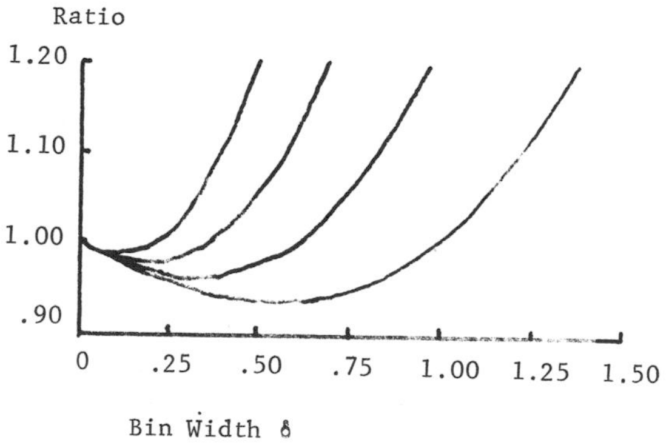

Figure 1. Theoretical integrated mean squared error ratio for binned and unbinned kernel estimates of Gaussian data.

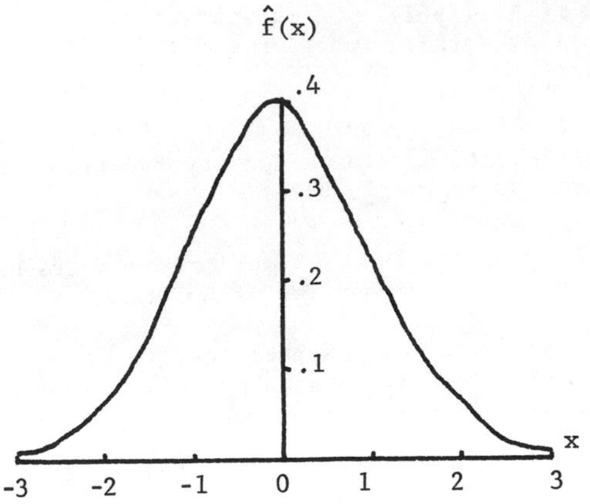

Figure 2. Example of kernel estimate for standard Gaussian data with n = 1000.

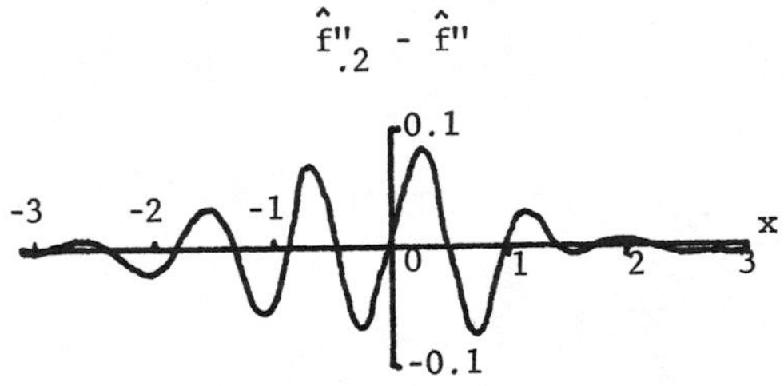

Figure 3. Effect of binning on second derivative of the estimate shown in Figure 2.

ON THE NONCONSISTENCY OF MAXIMUM LIKELIHOOD NONPARAMETRIC DENSITY ESTIMATORS

Eugene F. Schuster and Gavin G. Gregory
The University of Texas at El Paso

Abstract

One criterion proposed in the literature for selecting the smoothing parameter(s) in Rosenblatt-Parzen nonparametric constant kernel estimators of a probability density function is a leave-out-one-at-a-time nonparametric maximum likelihood method. Empirical work with this estimator in the univariate case showed that it worked quite well for short tailed distributions. However, it drastically oversmoothed for long tailed distributions. In this paper it is shown that this nonparametric maximum likelihood method will not select consistent estimates of the density for long tailed distributions such as the double exponential and Cauchy distributions. A remedy which was found for estimating long tailed distributions was to apply the nonparametric maximum likelihood procedure to a variable kernel class of estimators. This paper considers one data set, which is a pseudo-random sample of size 100 from a Cauchy distribution, to illustrate the problem with the leave-out-one-at-a-time nonparametric maximum likelihood method and to illustrate a remedy to this problem via a variable kernel class of estimators.

Key words: nonparametric, density, estimation, kernel estimates, nonconsistency, crossvalidation, maximum likelihood.

1. Introduction. Habbema et al. [3,4] and Duin [2] propose the same maximum likelihood (ML) procedure for choosing the smoothing parameter in the Rosenblatt-Parzen kernel density estimator

$$f_n(x) = \hat{f}(x|\theta;\ x_1,\dots,x_n) = n^{-1}\sum_{i=1}^{n}\theta^{-1}k((x-x_i)/\theta) \tag{1.1}$$

Here $x_1, x_2, \dots$ are independent identically distributed random variables with common density $f(x)$, the kernel k is a fixed density, and θ is the smoothing parameter.

Choosing θ to maximize the nonparametric likelihood

$$\prod_{i=1}^{n}\hat{f}(x_i|\theta;\ x_1,\dots,x_n) \tag{1.2}$$

is useless; (1.2) is unbounded as $\theta \to 0$. To avoid this degeneracy problem, Habbema et al. and Duin replaced $\hat{f}(x_i|\theta;\ x_1,\dots,x_n)$ in (1.2) by the kernel estimator of $f(x_i)$ based on the data with x_i removed. That is, they chose θ to maximize the function

$$L_n(\theta) = \prod_{i=1}^{n}\hat{f}(x_i|\theta;\ x_1,\dots,x_{i-1},x_{i+1},\dots x_n). \tag{1.3}$$

Monte Carlo simulation indicated that estimators (1.1) chosen by this ML method worked well for short tailed distributions. However, they drastically oversmoothed for long tailed distributions. The basic problem for long tailed distributions is that the spacing between extreme order statistics does not converge to zero. As a result, if k has a finite support then the sequence of smoothing parameters $\{\theta_n\}$, chosen, for each n, by the ML method, will not approach zero. This leads, as we show in Section 2, to the nonconsistency of the corresponding kernel estimators for long tailed densities such as the Cauchy and double exponential. Monte Carlo results, reported in Section 3, suggest a remedy for this situation is to use the ML method with so-called variable kernel estimators. Related results are contained in Gregory and Schuster [5].

2. Nonconsistency of the ML Procedure. By nonconsistency we mean that $\sup_x|f_n(x)| \not\to 0$ in probability. This nonconsistency will be demonstrated for a wide class of densities f and kernels k, but we make no attempt to state results for as wide a class as possible. For the sake of argument, attention is placed on the left tail of the distribution.

Let F denote the cdf of the density f and let $h(u) = u/fF^{-1}(u)$, $0 < u < 1$. We assume

$$h \text{ is continuous and } \lim_{u\to 0^+} h(u) = h(0^+) \text{ exists, possibly infinite.} \tag{2.1}$$

We say f has a long left tail if $h(0^+) > 0$. Assume for the present only that k has finite support. Without loss of generality we suppose the support is contained in $[-1,1]$, i.e.

$$k(u) = 0 \text{ if } |u| > 1. \tag{2.2}$$

A basic observation concerning the smoothing parameter $\theta=\theta_n$ which maximizes (1.3) is that for each x_i, $|x_i - x_j| \le \theta_n$ for some x_j with $j \ne i$. In particular

$$x_{2n} - x_{1n} \le \theta_n, \tag{2.3}$$

where $x_{1n}<x_{2n}<\dots<x_{nn}$ are the order statistics of the sample. Let $u_{in} = F(x_{in})$, $i=1,\dots,n$. Then $x_{2n}-x_{1n} = F^{-1}(u_{2n})-F^{-1}(u_{1n}) = h(u_n^*)(u_{2n}-u_{1n})/u_n^*$, where $u_{1n} \le u_n^* \le u_{2n}$. From (2.3) it follows that

$$h(u_n^*)(u_{2n} - u_{1n})/u_{2n} \le \theta_n. \tag{2.4}$$

Using uniformity of $(u_{2n}-u_{1n})/u_{2n}$ and standard arguments (2.4) yields

$$P(\theta_n < b\varepsilon) \le \varepsilon + P(h(u_n^*) < b),\ b,\ \varepsilon > 0. \tag{2.5}$$

Lemma 1. Under (2.1) and (2.2), $h(0^+) > 0$ implies $\theta_n \overset{P}{\not\to} 0$. Furthermore $h(0+) = \infty$ implies $\theta_n \overset{P}{\to} \infty$.

Proof: Choose $0 < b < h(0^+)$ in (2.5).

Lemma 2. If $\theta_n \overset{P}{\to} \infty$ and $\sup_u|k(u)| < \infty$ then $\sup_x f_n(x) \overset{P}{\to} 0$.

Since k is bounded the proof follows from (1.1).

Now Lemmas 1 and 2 combine to give the nonconsistency result for distributions like the Cauchy where $h(0^+) = \infty$. There is no difficulty here; the density estimate flattens out entirely. It is more difficult to establish the nonconsistency for boundary cases where $0 < h(0^+) < \infty$. The double exponential density is one of these and is covered by the following lemma. In addition to (2.2) we will assume that the kernel k is left continuous and of bounded variation on $(-\infty,\infty)$.

Lemma 3. Let θ_n maximize $L_n(\theta)$ of (1.3) for each n. Suppose f is unimodal and $h(0^+) = a$ where $0 < a \leq \infty$. Then $\sup_x |f_n(x) - f(x)| \not\to 0$ in probability.

The proof is in the Appendix.

3. Remedy for Long Tailed Distributions. A remedy we found for estimating long tailed distributions was to extend the kernel class of estimates to two types, the constant kernel type of (1.1) and the variable kernel type (3.1) considered by Breiman et al. [1];

$$f_n(x) = \hat{f}(x|\theta;\ x_1,\ldots,x_n)$$
$$= n^{-1} \sum_{i=1}^{n} (\alpha d_{iw})^{-1} k((x-x_i)/(\alpha d_{iw})) \qquad (3.1)$$

where $\theta_n = (w,\alpha)$, $w \in \{1,\ldots,n\}$, $\alpha > 0$ and d_{iw} is the wth nearest neighbor to x_i in the sample.

We consider the two types of kernel estimates (1.1) and (3.1) as one family and let the maximum of the likelihood in (1.3) choose between them. Notice that for the variable kernel estimator the maximum of (1.3) is found over a two-dimensional space $\{(w,\alpha)\}$. Suppose k has finite support with $k(x) = 0$ for $|x| > c$. Let $\theta_n = (\alpha_n, w_n)$ maximize $L_n(\theta)$ of (1.3) using the variable kernel estimator of (3.1). In this case it is easy to see that the argument, given in the Appendix, that leads to the nonconsistency of the Rosenblatt-Parzen estimates breaks down. Here $k((x_{2n}-x_{1n})/\alpha d_{iw}) > 0$ would say that $\alpha \geq (x_{2n}-x_{1n})/cd_{iw}$. For large w, the right side of this inequality is small even when the spacing $x_{2n}-x_{1n}$ does not converge to zero. Very little is known regarding the theoretical properties of the estimate (3.1). For a more detailed discussion of the empirical performance of the estimator (3.1) with respect to three error measures see [1] and [5].

In Figures 1 and 2 we use the Cauchy sample of 100 given in Table 1 to illustrate the performance of the ML Rosenblatt-Parzen constant kernel estimator as well as that of the ML variable kernel estimate of (3.1). In maximizing (1.3) for f_n of (3.1) we used a range of α of .1(.1)5.5 (i.e. from .1 to 5.5 in steps of .1) and a range of w of 15(5)45. For Figure 1 we used the quartic kernel $k(x) = .9375(1-x^2)^2$ for $|x| \leq 1$ and zero otherwise. This finite support kernel is popular because it is continuously differentiable and readily computable. In Figure 2 we used the same data set but chose a student's t(29) density, which is similar to the standard normal but involves less computing time. We included this second kernel to illustrate that the flattening behavior of the Rosenblatt-Parzen estimator of the density, with smoothing parameter chosen by the ML method, cannot be overcome by taking a kernel with infinite support.

A look at Figures 1 and 2 shows that the constant kernel estimator picked by the ML method is much too flat in both cases. In the combined family the ML method picked the variable kernel estimator with w = 35 for both kernels and with $\alpha = 5.3$ in Figure 1 and with $\alpha = 2.4$ in Figure 2.

Appendix

Proof of Lemma 3. For $a = \infty$, the desired conclusion follows from Lemmas 1 and 2. Suppose then that $a < \infty$. Let (Ω, A, P) be the underlying probability space on which $x_1, x_2, \ldots$ are defined and let F_n be the usual empirical distribution function. Suppose $s(n) = \sup_x |f_n(x) - f(x)| \xrightarrow{P} 0$. Then there exists a subsequence $\{s(n_j)\}_{j=1}^{\infty}$ converging to 0 a.s. We will show that this leads to a contradiction.

We first show that $P(A) > 0$ where $A = \{\lim\sup \theta_{n_j} > 0\}$. In this direction we use Fatou's lemma for nonnegative random variables and the observation that $u(n) = u_{1n}/u_{2n}$ is uniformly distributed over (0,1) to see that

$$E\{\lim\inf u(n_j)\} \leq \lim\inf E\{u(n_j)\} = 1/2.$$

But then

$$E\{\lim\sup (1-u(n_j))\} = 1 - E\{\lim\inf u(n_j)\} \geq 1/2.$$

Hence $P(B) > 0$ where $B = \{\lim\sup (1-u(n_j)) > 0\}$. Using (2.4), we can conclude that

$$P(A) \geq P(B) > 0. \qquad (A.1)$$

Let us write $Ef_n(x) = \int \theta_n^{-1} k((x-u)/\theta_n) dF(u)$. Then upon integration by parts we see that

$$\sup_x |f_n(x) - Ef_n(x)| =$$
$$\sup_x \left| \int_{-\infty}^{\infty} \theta_n^{-1} k((x-u)/\theta_n) d\{F_n(u) - F(u)\} \right|$$
$$= \sup_x \left| \int_{-\infty}^{\infty} \{F_n(u) - F(u)\} dk((x-u)/\theta_n)/\theta_n \right|$$
$$= \sup_x |F_n(x) - F(x)| \cdot v/\theta_n \qquad (A.2)$$

where v is the variation of k on $(-\infty,\infty)$. We will let Ω' be that subset of Ω $(P(\Omega')=1)$ on which $\sup_x |F_n(x) - F(x)| \to 0$, $h(u_{2n}) \to a$, and $s(n_j) \to 0$.

Take $\omega \in A' = A \cap \Omega'$. Then there is a subsequence $\{n_{j_k}\}$ depending on ω with $\{\theta_{n_{j_k}}(\omega)\}$ converging to $\theta(\omega) = \lim\sup \theta_{n_j}(\omega)$. If $\theta(\omega) = \infty$ then $f_{n_{j_k}}(x,\omega) \to 0$ for every x and we can conclude that $s(n_j)(\omega) \not\to 0$. Suppose $\theta(\omega) < \infty$. Let x_0 be the unique mode of f. Then changing variables and using Lebesgue's Dominated Convergence Theorem we see that

Table 1. Cauchy Data

-23.306	0.116
-17.511	0.117
-11.159	0.157
-10.315	0.186
-8.792	0.226
-4.828	0.253
-3.836	0.266
-3.319	0.284
-2.996	0.337
-2.750	0.343
-2.449	0.380
-2.417	0.381
-2.363	0.402
-2.314	0.444
-2.114	0.534
-1.058	0.580
-1.892	0.600
-1.623	0.684
-1.543	0.715
-1.401	0.728
-1.297	0.807
-1.236	0.846
-1.178	0.884
-0.083	0.885
-1.054	0.892
-1.052	0.926
-1.023	1.039
-0.987	1.067
-0.834	1.269
-0.756	1.306
-0.525	1.374
-0.480	1.501
-0.462	1.575
-0.455	1.638
-0.454	1.857
-0.434	1.925
-0.429	2.053
-0.421	2.451
-0.328	2.622
-0.318	3.019
-0.280	3.130
-0.265	3.772
-0.212	4.693
-0.165	4.833
-0.157	5.099
-0.006	6.609
0.051	6.656
0.056	7.594
0.063	12.152
0.084	16.504

Legend for Figures 1 and 2

Δ - ML constant kernel estimator
x - ML combined family estimator
No symbol - Cauchy density

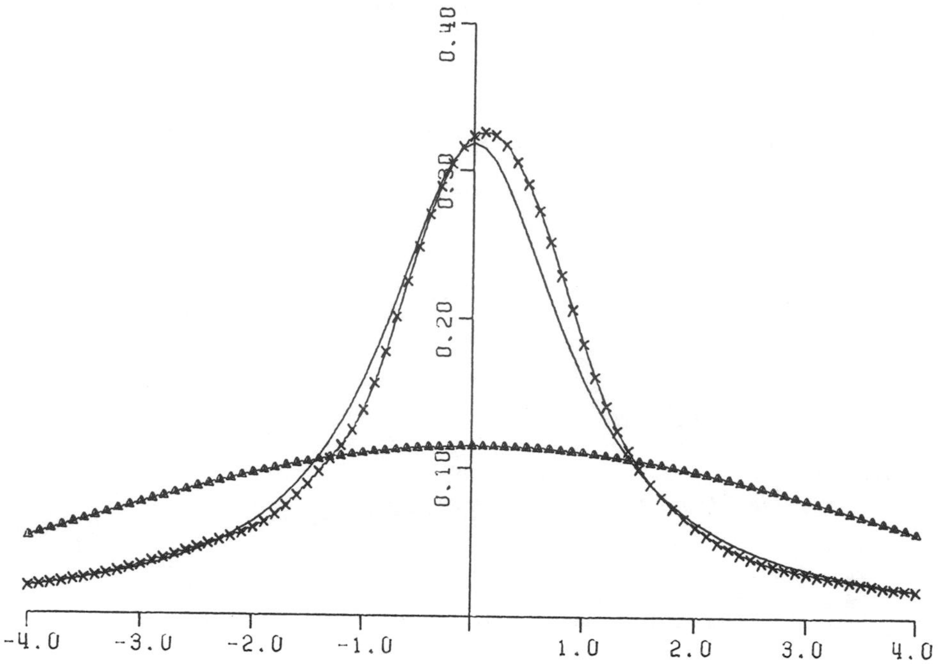

Figure 1. Cauchy Estimation Quartic Kernel

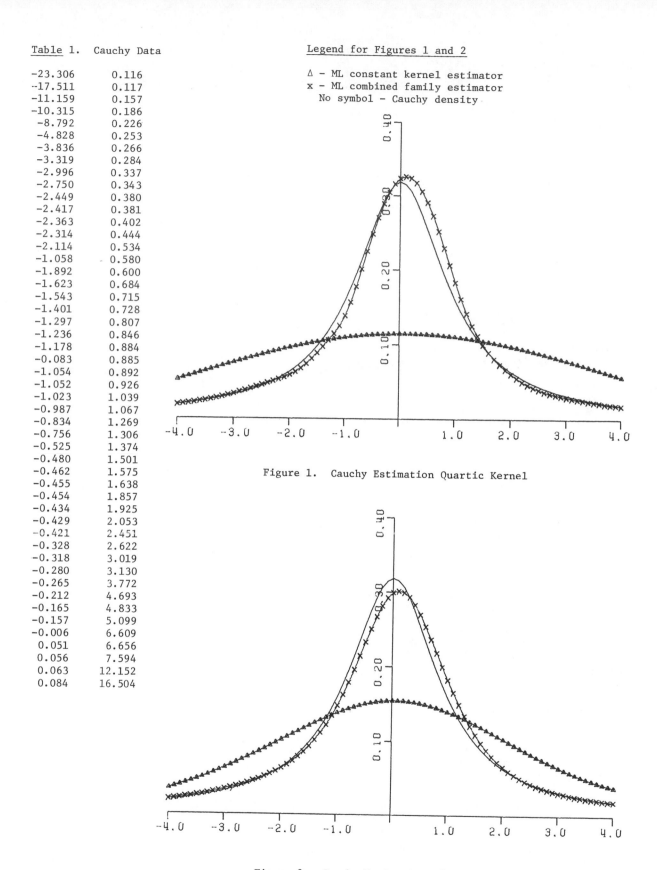

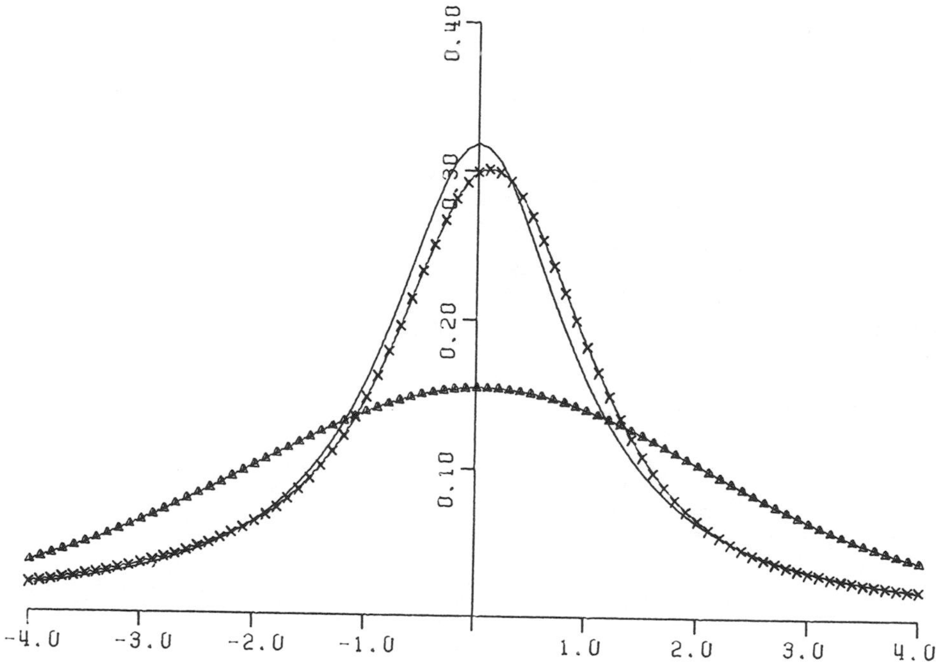

Figure 2. Cauchy Estimation t(29) Kernel

$$Ef_{n_{j_k}}(x_0,\omega) = \int_{-\infty}^{\infty} f(x_0-u\theta_{n_{j_k}}(\omega))k(u)du$$

converges to

$$\int_{-\infty}^{\infty} f(x_0-u\theta(\omega))k(u)du < f(x_0).$$

By the triangle inequality we see

$$|f_n(x_0)-f(x_0)| \geq |Ef_n(x_0)-f(x_0)|-|f_n(x_0)-Ef_n(x_0)|.$$

Using A.1 and A.2 we can conclude that $|f_{n_{j_k}}(x_0,\omega) - f(x_0)| \not\to 0$ and so $s(n_j)(\omega) \not\to 0$.

Thus we have shown that $s(n_j)(\omega) \not\to 0$ for any $\omega \in A'$. Since $P(A') > 0$ this is a contradiction and the proof is complete.

References

[1] Breiman L., Meisel, W., Purcell, E. (1977). Variable kernel estimates of multivariate densities and their calibration. Technometrics. Vol. 19, No. 2, 135-144.

[2] Duin, Robert P. W. (1976). On the choice of smoothing parameters for Parzen estimators of probability density functions. IEEE Transactions on Computers. 1176-1179.

[3] Habbema, J. D. F., Hermans, J. and van den Brock, K. (1974). A stepwise discriminant analysis program using density estimation. Compstat, 1974, Proceedings in Computational Statistics. Wein, Physica Verlag, 101-110.

[4] Habbema, J. D. F., Hermans, J. and van den Brock, K. (1977). Selection of variables in discriminant analysis by F-statistic and error rate. Technometrics. Vol. 19, 487-493.

[5] Gregory, G. G. and Schuster, E. F. (1979). Contributions to non-parametric maximum likelihood methods of density estimation. Proceedings of Computer Science and Statistics: 12th Annual Symposium on the Interface, ed. Jane F. Gentleman, 427-431, University of Waterloo, Waterloo, Ontario, Canada.

COMPUTER PROGRAM FOR KRISHNAIAH'S FINITE INTERSECTION TESTS FOR MULTIPLE COMPARISONS OF MEAN VECTORS

C. M. Cox, C. Fang and R. M. Boudreau
University of Pittsburgh

ABSTRACT

The program FIT performs Krishnaiah's finite intersection test procedure on the mean vectors from k multivariate populations. The test procedure is valid under the following assumptions: a) the k populations are distributed as multivariate normal, b) the covariance matrices of the k populations are equal. We can perform two-sided or one-sided tests. The common covariance matrix, $\Sigma = (\sigma_{ij})$, may be unknown or known. When Σ is unknown, the test statistics are distributed as multivariate F or multivariate t for the two-sided test or the one-sided test respectively. In the case when Σ is known, then the test statistics are distributed as multivariate chi-square or multivariate normal for the two-sided test or the one-sided test respectively. The program FIT computes suitable bounds on the required percentage points of these distributions.

Key words: finite intersection test, multiple comparisons of mean vectors, linear combinations, multivariate F, multivariate t, multivariate chi-square, multivariate normal distributions.

1. STATEMENTS OF THE PROBLEMS

For $i = 1,\ldots,k$, we assume that $\underset{\sim}{X}_i' = (X_{i1},\ldots,X_{ip})$ is distributed as a multivariate normal with mean vector $\underset{\sim}{\mu}_i' = (\mu_{i1},\ldots,\mu_{ip})$ and covariance matrix $\Sigma = (\sigma_{ij})$. The null hypothesis to be tested is

$$H: \underset{\sim}{\mu}_1 = \underset{\sim}{\mu}_2 = \ldots = \underset{\sim}{\mu}_k.$$

Consider the linear combinations

$$\underset{\sim}{\lambda}_g = \sum_{i=1}^{k} c_{gi}\, \underset{\sim}{\mu}_i = \underset{\sim}{c}_g' \mu \quad (g = 1,\ldots,q) \qquad (1.1)$$

where $\underset{\sim}{c}_g' = (c_{g1},\ldots,c_{gk})$ and $\mu' = (\underset{\sim}{\mu}_1,\ldots,\underset{\sim}{\mu}_k)$. The null hypothesis H is decomposed into q sub-hypotheses $H_g (g = 1,\ldots,q)$ such that $H = \bigcap_{g=1}^{q} H_g$, where $H_g: \underset{\sim}{\lambda}_g = \underset{\sim}{0}$. H is accepted if and only if each of H_g $(g = 1,\ldots,q)$ is accepted. When H is rejected, we wish to draw inference on $\underset{\sim}{\lambda}_1,\ldots,\underset{\sim}{\lambda}_q$ which are linear combinations of mean vectors.

Now, let

$$A: \bigcup_{g=1}^{q} A_g, \quad A^*: \bigcup_{g=1}^{q} A_g^*, \quad A^{**}: \bigcup_{g=1}^{q} A_g^{**} \qquad (1.2)$$

where $A_g: \underset{\sim}{\lambda}_g \neq \underset{\sim}{0}$, $A_g^*: \underset{\sim}{\lambda}_g > \underset{\sim}{0}$, $A_g^{**}: \underset{\sim}{\lambda}_g < \underset{\sim}{0}$.

The object of this paper is to develop a computer program to test $H_1,\ldots,H_q$ and H against two-sided and one-sided alternatives. To test H_g's simultaneously, we consider the following model. The conditional distribution of $X_{i,j+1}$ given $X_{i1},\ldots,X_{ij}$ is a univariate normal with variance σ_{j+1}^2 and mean

$$E_c(X_{i,j+1}) = \eta_{i,j+1} + (X_{i1},\ldots,X_{ij})\underset{\sim}{\beta}_j \qquad (1.3)$$

for $j = 1,2,\ldots,p-1$, where

$$\eta_{i,j+1} = \mu_{i,j+1} - \underset{\sim}{\beta}_j'\, \underset{\sim}{\theta}_{ij};\ \underset{\sim}{\theta}_{ij}' = [\mu_{i1},\ldots,\mu_{ij}] \qquad (1.4)$$

and

$$\underset{\sim}{\beta}_j = \Sigma_j^{-1} \begin{bmatrix} \sigma_{1,j+1} \\ \vdots \\ \sigma_{j,j+1} \end{bmatrix},$$

$$\sigma_{j+1}^2 = |\Sigma_{j+1}|/|\Sigma_j| \qquad (1.5)$$

$$= \sigma_{j+1,j+1} - (\sigma_{1,j+1},\ldots,\sigma_{j,j+1})\Sigma_j^{-1} \begin{bmatrix} \sigma_{1,j+1} \\ \vdots \\ \sigma_{j,j+1} \end{bmatrix}$$

In the above equation Σ_j denotes the top $j \times j$ left-hand corner of Σ.

Now, let $H_{gj}: \lambda_{gj} = 0$, where $\lambda_{gj} = \sum_{i=1}^{k} c_{gi}\eta_{ij}$ for $j = 1,2,\ldots,p$ and $\eta_{i1} = \mu_{i1}$. Then $H_g = \bigcap_{j=1}^{p} H_{gj}$. So the problem of testing H_g may be viewed as a problem of testing $H_{g1},\ldots,H_{gp}$ simultaneously. (P. R. Krishnaiah (1979)).

2. FINITE INTERSECTION TEST PROCEDURES

Let $\underset{\sim}{x}_{it}' = (x_{it1},\ldots,x_{itp})$, $(t = 1,\ldots,n_i)$, be a random sample drawn independently from the i-th multivariate normal population, where x_{itj} denotes the t-th observation on the j-th variate of the i-th population $(i = 1,\ldots,k;\ t = 1,\ldots,n_i;\ j = 1,\ldots,p)$. Also let

$$\bar{\underset{\sim}{x}}_{i.}' = (\bar{x}_{i.1}, \bar{x}_{1.2},\ldots,\bar{x}_{i.p}),\quad \bar{x}_{i.j} = \frac{1}{n_i}\sum_{t=1}^{n_i} x_{itj}$$

$$S = \frac{1}{N-k}\sum_{i=1}^{k}\sum_{t=1}^{n_i} (\underset{\sim}{x}_{it} - \bar{\underset{\sim}{x}}_{i.})(\underset{\sim}{x}_{it} - \bar{\underset{\sim}{x}}_{i.})' = (s_{ij}) \qquad (2.1)$$

and $N = \sum_{i=1}^{k} n_i$. To test H_{g1} for $g = 1,\ldots,q$ (i.e. for the first variable) we use the following test statistics:

$$\hat{\eta}_{i1} = \bar{x}_{i.1},\ (i = 1,\ldots,k);$$

$$\hat{\lambda}_{g1} = \sum_{i=1}^{k} c_{gi}\, \hat{\eta}_{i1},$$

$$t_{g1} = \frac{\hat{\lambda}_{g1}}{\{d_{g1}\, s_{11}/(N-k)\}^{\frac{1}{2}}} \tag{2.2}$$

where

$$d_{g1} = \sum_{i=1}^{k} \frac{c_{gi}^2}{n_i},$$

and

$$F_{g1} = t_{g1}^2 \quad (g = 1,\dots,q).$$

If Σ is known $s_{11}/(N-k)$ is replaced by σ_{11}, and

$$N_{g1} = \frac{\hat{\lambda}_{g1}}{\sqrt{d_{g1}\,\sigma_{11}}} \tag{2.3}$$

$$\chi_{g1}^2 = N_{g1}^2 \qquad (g = 1,2,\dots,q)$$

To test $H_{g,j+1}$, for $g = 1,\dots,q$ and $j = 1,\dots,p-1$ we consider the following linear regression model. Let

$$\underset{N\times k}{A} = \begin{bmatrix} 1 & 0 & \cdots & 0 \\ \vdots & \vdots & & \\ 1 & 0 & \cdots & 0 \\ 0 & 1 & 0 \cdots & 0 \\ \vdots & \vdots & & \\ 0 & 1 & 0 \cdots & 0 \\ & & \ddots & \\ 0 & \cdots & 0 & 1 \\ & & & \vdots \\ 0 & \cdots & 0 & 1 \end{bmatrix} \begin{matrix} \}\, n_1 \\ \\ \}\, n_2 \\ \\ \}\, n_k \end{matrix}$$

$$\underset{N\times j}{A_j} = \begin{bmatrix} \underset{\sim}{y}_{11} & \underset{\sim}{y}_{12} & \cdots & \underset{\sim}{y}_{1j} \\ \underset{\sim}{y}_{21} & \underset{\sim}{y}_{22} & \cdots & \underset{\sim}{y}_{2j} \\ \vdots & \vdots & & \vdots \\ \underset{\sim}{y}_{k1} & \underset{\sim}{y}_{k2} & \cdots & \underset{\sim}{y}_{kj} \end{bmatrix} \tag{2.4}$$

where $\underset{\sim}{y}_{i\ell}' = (x_{i1\ell}, x_{i2\ell}, \dots, x_{in_i\ell})$ denotes the n_i observations on the ℓ-th variate of the i-th population for $i = 1,\dots k$, $\ell = 1,\dots,p$. Now let

$$\underset{N\times(k+j)}{Z_j} = [A : A_j], \quad \underset{\sim}{Y}_{j+1} = \begin{bmatrix} \underset{\sim}{y}_{1,j+1} \\ \underset{\sim}{y}_{2,j+1} \\ \vdots \\ \underset{\sim}{y}_{k,j+1} \end{bmatrix}$$

and

$$\underset{\sim}{\theta}_{j+1} = \begin{bmatrix} \eta_{1,j+1} \\ \vdots \\ \eta_{k,j+1} \\ \underset{\sim}{\beta}_j \end{bmatrix} \tag{2.5}$$

Then $E_c\,[\underset{\sim}{Y}_{j+1}] = Z_j\,\underset{\sim}{\theta}_{j+1}$. The least square estimates of σ_{j+1}^2 and $\eta_{1,j+1}, \eta_{2,j+1}, \dots, \eta_{k,j+1}$ and the appropriate test statistics are as follows:

$$s_{j+1}^2 = \underset{\sim}{Y}_{j+1}'\,[I - Z_j(Z_j'Z_j)^{-1}Z_j']\,\underset{\sim}{Y}_{j+1}$$

$$\hat{\underset{\sim}{\theta}}_{j+1} = (Z_j'Z_j)^{-1}Z_j'\underset{\sim}{Y}_{j+1} = \begin{bmatrix} \hat{\eta}_{1,j+1} \\ \hat{\eta}_{2,j+1} \\ \vdots \\ \hat{\eta}_{k,j+1} \\ \hat{\underset{\sim}{\beta}}_j \end{bmatrix}$$

$$\hat{\lambda}_{g,j+1} = \sum_{i=1}^{k} c_{gi}\,\hat{\eta}_{i,j+1} \tag{2.6}$$

$$t_{g,j+1} = \hat{\lambda}_{g,j+1} / \left\{d_{g,j+1} \frac{s_{j+1}^2}{N-k-j}\right\}^{\frac{1}{2}}$$

where

$$d_{g,j+1} = \underset{\sim}{c}_g^{*\prime}\,(Z_j'Z_j)^{-1}\,\underset{\sim}{c}_g^*, \quad \underset{1\times(k+j)}{\underset{\sim}{c}_g^{*\prime}} = [\underset{\sim}{c}_g' : \underbrace{0\dots0}_{j}]$$

and

$$F_{g,j+1} = t_{g,j+1}^2 \quad (j = 1,\dots,p-1;\ g = 1,\dots,q)$$

If Σ is known, $s_{j+1}^2/(N-k-j)$ is replaced with the value

$$\sigma_{j+1}^2 = \frac{|\Sigma_{j+1}|}{|\Sigma_j|}$$

$$= \sigma_{j+1,j+1} - (\sigma_{1,j+1},\dots,\sigma_{j,j+1})\Sigma_j^{-1} \begin{bmatrix} \sigma_{1,j+1} \\ \vdots \\ \sigma_{j,j+1} \end{bmatrix} \tag{2.7}$$

$\hat{\eta}_{ij}$ is replaced by

$$\hat{\eta}_{ij}^* = \bar{x}_{i.j+1} - [\sigma_{1,j+1},\dots,\sigma_{j,j+1}]\,\Sigma_j^{-1} \begin{bmatrix} \bar{x}_{i.1} \\ \vdots \\ \bar{x}_{i.j} \end{bmatrix} \tag{2.8}$$

$$\hat{\lambda}_{g,j+1}^* = \sum_{i=1}^{k} c_{gi}\,\hat{\eta}_{i,j+1}^*$$

and $d_{g,j+1}$ is replaced by

$$d_g = \sum_{i=1}^{k} \frac{c_{gi}^2}{n_i}.$$

and

$$N_{g,j+1} = \frac{\hat{\lambda}_{g,j+1}^*}{\sqrt{d_g\,\sigma_{j+1}^2}} \tag{2.9}$$

$$\chi_{g,j+1}^2 = N_{g,j+1}^2 \qquad (g = 1,2,\dots,q;\ j = 1,\dots,p-1)$$

For the two-sided test against the alternative $A_{gj}: \lambda_{gj} \neq 0$ the hypothesis

$H_{gj}(g = 1,\ldots,q;\ j = 1,\ldots,p)$ is accepted if

$$F_{gj} \le c_{\alpha j}$$

and rejected otherwise, where $c_{\alpha j}$ are chosen such that

$$P[F_{gj} \le c_{\alpha j};\ g = 1,2,\ldots,q;\ j = 1,\ldots,p|H]$$

$$= \prod_{j=1}^{p} P[F_{gj} \le c_{\alpha j};\ g = 1,2,\ldots,q|H] = (1-\alpha) \qquad (2.10)$$

When H is true, the joint distribution of $F_{1j},\ldots,F_{qj}$, $(j = 1,\ldots,p)$ is a central multivariate F distribution with $(1,N-k-j+1)$ degrees of freedom and are independent for different values of j. The simultaneous confidence intervals on $\sum_{i=1}^{k} c_{gi}\eta_{ij}$ for $g = 1,\ldots,q;\ j = 1,\ldots p$ are

$$[\sum_{i=1}^{k} c_{gi}\hat{\eta}_{ij} - \sqrt{\frac{c_{\alpha j} d_{gj} s_j^2}{N-k-j+1}}, \quad \sum_{i=1}^{k} c_{gi}\hat{\eta}_{ij} + \sqrt{\frac{c_{\alpha j} d_{gj} s_j^2}{N-k-j+1}}\ ] \qquad (2.11)$$

For the one-sided test against the alternatives $A^*_{gj}: \lambda_{gj} > 0$, the hypothesis $H_{gj}(g = 1,\ldots,q;\ j = 1,\ldots,p)$ is accepted if $t_{gj} \le c_{\alpha j}$ and rejected otherwise.

where $c_{\alpha j}$ are chosen such that

$$P[t_{gj} \le c_{\alpha j};\ g = 1,2,\ldots,q;\ j = 1,\ldots,p|H]$$

$$= \prod_{j=1}^{p} P[t_{gj} \le c_{\alpha j};\ g = 1,2,\ldots,q|H] = (1-\alpha)$$

The above inequalities are reversed for the alternatives $A^{**}_{gj}: \lambda_{gj} < 0$. (2.12)

When H is true, the joint distribution of $t_{1j},\ldots,t_{qj}$ is a central multivariate t distribution with $N-k-j+1$ degrees of freedom and are independent for different values of j. The simultaneous confidence intervals on $\sum_{i=1}^{k} c_{gi}\eta_{ij}$ for $g = 1,\ldots,q;\ j = 1,\ldots,p$ are given by

$$[\sum_{i=1}^{k} c_{gi}\hat{\eta}_{ij} - c_{\alpha j}\sqrt{\frac{d_{gj} s_j^2}{N-k-j+1}},\ \infty) \text{ for alternatives } A^*_{gj} \qquad (2.13)$$

$$(-\infty,\ \sum_{i=1}^{k} c_{gi}\hat{\eta}_{ij} + c_{\alpha j}\sqrt{\frac{d_{gj} s_j^2}{N-k-j+1}}\ ] \text{for alternatives } A^{**}_{gj}.$$

In the case when Σ is known, the above test statistics F_{gj} and t_{gj} are replaced by χ^2_{gj} and N_{gj} respectively. Under the hypothesis H, the joint distribution of $\chi^2_{1j},\ldots,\chi^2_{qj}$ is a central multivariate chi-square distribution with one degree of freedom, and the joint distribution of $N_{1j},\ldots,N_{qj}$ is a central multivariate normal distribution. The distributions are independent for different values of $j = 1,\ldots,p$. The simultaneous confidence intervals are constructed similarly.

The program gives equal weights to the p variables by choosing $c_{\alpha j}$, $j = 1,\ldots,p$ to satisfy

$$P\{G_{gj} \le c_{\alpha j};\ g = 1,2,\ldots,q|H\} = 1-\alpha^*$$

where $(1-\alpha^*)^p = 1-\alpha$, and G_{gj} is the appropriate statistic.

The preceding simultaneous confidence intervals are for linear combinations on the conditional means. Therefore, in addition the program also provides approximate simultaneous confidence intervals based on the original population means using the method of Mudholkar and Subbaiah (1975).

3. COMPUTATION OF THE CRITICAL VALUES

The exact critical value $c_{\alpha i}$ for the tests are difficult to obtain except for special structures of the correlation matrix of the accompanying multivariate normal distribution (Krishnaiah (1979)). hence bounds on $c_{\alpha j}$ for the appropriate probability integrals must be used.

For the two-sided test with Σ unknown, the critical point $c_{\alpha j}$ is the upper $100\alpha^*$ percentage point of the q-variate multivariate F distribution with $(1,N-k-j+1)$ degrees of freedom and the accompanying correlation matrix $\Omega_j = (\rho^j_{rs})$, $j = 1,2,\ldots,p$. where

$$\rho^1_{rs} = \frac{1}{(d_r\, d_s)^{1/2}} \sum_{i=1}^{k} \frac{c_{ri}c_{si}}{n_i}, \quad d_r = \sum_{i=1}^{k} \frac{c^2_{ri}}{n_i} \qquad (3.1)$$

and

$$\rho^{\ell+1}_{rs} = \frac{\tilde{c}^{*\prime}_r (Z'_\ell Z_\ell)^{-1} \tilde{c}^*_s}{[\tilde{c}^{*\prime}_r (Z'_\ell Z_\ell)^{-1} \tilde{c}^*_r]^{1/2} [\tilde{c}^{*\prime}_s (Z'_\ell Z_\ell)^{-1} \tilde{c}^*_s]^{1/2}}$$

for $\ell = 1,\ldots,p-1$.

For the one-sided test with Σ unknown, the critical point $c_{\alpha j}$ is the upper $100\alpha^*$ percentage point of the q-variate multivariate t distribution with $N-k-j+1$ degrees of freedom, and with the above structure for the accompanying correlation matrix. For the two-sided test with Σ known, $c_{\alpha j}$ is the upper $100\alpha^*$ percentage point of the q-variate multivariate chi-square distribution with 1 degree of freedom and with the accompanying correlation matrix $\Omega = (\rho_{rs})$, where

$$\rho_{rs} = \frac{1}{(d_r d_s)^{1/2}} \sum_{i=1}^{k} \frac{c_{ri}c_{si}}{n_i}. \qquad (3.2)$$

and for the one-sided test with Σ known, $c_{\alpha j}$ is the upper $100\alpha^*$ percentage point of the q-variate multivariate normal distribution with the above accompanying correlation matrix, (Eq. (3.2)).

Alternative bounds on the probability integrals may be computed using the following inequalities. Let $\underset{\sim}{N}' = (N_1,\ldots,N_q)$ be distributed as multivariate normal with mean vector $\underset{\sim}{0}$ and covariance $\sigma^2(\gamma_{ij})$ where (γ_{ij}) is the correlation matrix. Also let $\underset{\sim}{u}' = (u_1,\ldots,u_q)$ be distributed as multivariate normal with mean vector $\underset{\sim}{0}$ and covariance $\sigma^2 I_q$ and s^2/σ^2 is distributed independently of $N_1,\ldots,N_q$ and $u_1,\ldots,u_q$ as chi-square with ν degrees of freedom. Sidak (1967) showed that

$$P[|N_1| \le c_1,\ldots,|N_q| \le c_q] \ge \prod_{i=1}^{q} P[|u_i| \le c_i] \tag{3.3}$$

and

$$P[|N_1|\sqrt{\nu}/s \le c_1,\ldots,|N_q|\sqrt{\nu}/s \le c_q]$$

$$\ge P[|u_1|\sqrt{\nu}/s \le c_1,\ldots,|u_q|\sqrt{\nu}/s \le c_q] \tag{3.4}$$

$$\ge \prod_{i=1}^{q} P[|u_i|\sqrt{\nu}/s \le c_i] \tag{3.5}$$

Inequality (3.3) can be used to obtain upper bound on the critical values $c_{\alpha j}$ which we call the product upper bound of the multivariate chi-square distribution with one degree of freedom. Similarly, inequalities (3.4), (3.5) can be used for the upper bound on the critical value $c_{\alpha j}$ of the multivariate F distribution with $(1,\nu)$ degrees of freedom. We call the upper bound of Eq. (3.4) the Sidak's upper bound and the upper bound of Eq. (3.5) the product upper bound of the multivariate F distribution respectively.

Upper bounds and lower bounds on the critical values may also be computed by using the Poincare's inequality:

$$1-\sum_{i=1}^{q} P(E_i^c) \le P(E_1,\ldots,E_q) \le 1-\sum_{i=1}^{q} P(E_i^c) + \sum_{i<j} P(E_i^c E_j^c) \tag{3.6}$$

where E_i^c denotes the complement of the event E_i. When testing against the alternative hypotheses $A_{gj}:\lambda_{gj} \ne 0$ and $A^*_{gj}:\lambda_{gj} > 0$, we have

$$1-\sum_{i=1}^{q} P[G_i > c_i] \le P[G_i \le c_i;\ i=1,\ldots,q]$$

$$\le 1-\sum_{i=1}^{q} P[G_i > c_i] + \sum_{i<j} P[G_i > c_i,\ G_j > c_j] \tag{3.7}$$

where $\underset{\sim}{G}' = (G_1,\ldots,G_q)$ is the appropriate statistic which is distributed as a multivariate normal, a multivariate chi-square with one degree of freedom, a multivariate t with ν degrees of freedom or a multivariate F with $(1,\nu)$ degrees of freedom. The left-hand side can be used to obtain upper bound on the critical value $c_{\alpha j}$ and the right-hand side can be used to obtain lower bound on the critical value $c_{\alpha j}$. When testing against the alternative hypotheses $A^{**}_{gj}: \lambda_{gj} < 0$, the test statistic is distributed as a multivariate normal or a multivariate t distribution according as Σ is known or unknown respectively, the inequalities are reversed (see Eq. (2.12) and we have

$$1-\sum_{i=1}^{q} P[G_i < c_i] \le P[G_i \ge c_i;\ i=1,\ldots,q]$$

$$\le 1-\sum_{i=1}^{q} P[G_i < c_i] + \sum_{i<j} P[G_i < c_i,\ G_j < c_j) \tag{3.8}$$

The left-hand side of Eq. (3.8) can be used to obtain lower bound for the critical value $c_{\alpha j}$ and the right-hand side can be used to obtain the upper bound on the critical value $c_{\alpha j}$.

DECISION FLOW CHART FOR BOUNDS ON CRITICAL VALUES

PRODUCT: Product Upper Bound
HI: Poincare's Upper Bound
LOW: Poincare's Lower Bound
SIDAK: Sidak's Upper Bound
EXTERNAL: Externally supplied critical values

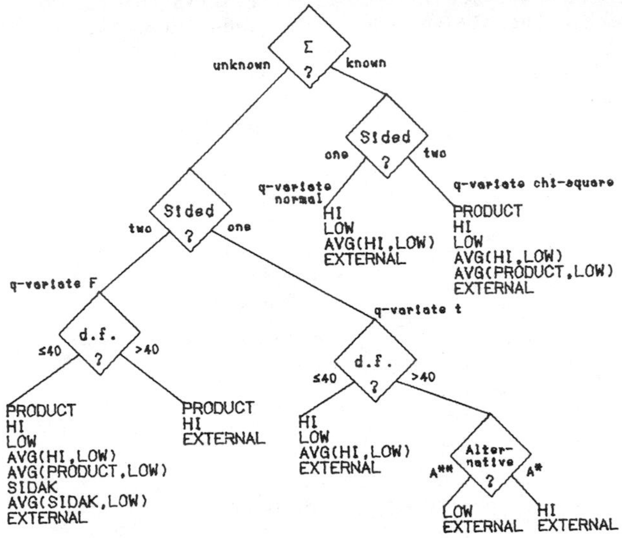

On the DECsystem-10 , the multivariate F and t integrals cannot be computed for degrees of freedom (d.f.) greater than 40 due to overflows.

The Sidak's upper bounds of Eq. (3.4) are sharper than the product upper bounds of Eq. (3.5), however the product upper bounds, while being only moderately good, have the advantage that they use much less computer time. The Poincare's bounds involving the bivariate probability integrals require more computer time than either of the above two upper bounds. They are as sharp bounds as Sidak's bounds (Cox, et al. (1979)).

4. DATA STRUCTURE

FIT is a conversational FORTRAN program. At the time of execution, the following input must be provided in answers to a sequence of questions:

1. k: (integer) the number of populations (maximum 10).
2. POP_i ($i = 1,\ldots,k$): k names, one name for each of k populations, each having five characters and separated by one space. FORMAT (10(A5,1X))
3. n_i ($i = 1,\ldots,k$): (integers) the number of observations on each of k populations (maximum 500 combined).
4. p: (integer) the number of variables (maximum 10).

5. $VAR_i (i = 1,\ldots,p)$: p names, one name for each of p variables, each having five characters and separated by one space.
6. IW: 0 to test all pairwise comparisons (allowable when the number of populations is less than or equal to 7.)
   - -1 to test adjacent means
   - -2 to test against a control using the k-th population as the controlled population
   - or enter the number of linear combinations q (maximum 21), if you wish to specify particular combinations not included in the above options.
7. c_{gi} $(g = 1,\ldots,q;\ i = 1,\ldots,k)$: (real) q rows of k numbers representing the coefficients of the linear combinations (requested by the program only if IW = q).
8. IPRT: 1 to have observations printed
   0 otherwise
9. KNOWS: 1 if Σ is known
   0 if Σ is not known
10. S: (real) p rows of p numbers, the value of Σ (requested by the program only if KNOWS = 1)
11. NSIDED: 0 to test against combination < 0
    1 to test against combination > 0
    2 for two-sided test
12. ALPHA: (real) overall level of significance α.
13. OPTION: An input integer indicator
    1 Calculate product upper bound.
    2 Calculate Poincare's upper bound.
    3 Calculate Poincare's lower bound.
    4 Calculate the average of Poincare's upper and lower bounds.
    5 Calculate the average of product upper bound and Poincare's lower bound
    6 Calculate Sidak's upper bound.
    7 Calculate the average of Sidak's upper bound and Poincare's lower bound.
    8 Calculate the exact critical value for VAR_1 when IW=-2 and $n_1 = \ldots = n_k$. The program then requests another OPTION for the remaining variables.
    9 If the critical values are known and you wish to enter them.

    OPTION 3 to 8 are calculated for the multivariate F and t distributions only when the degrees of freedom ν is less or equal to 40.
14. CALPHA: p values for the critical values (requested by the program only if OPTION=9).

The observations $x_{i\ell j}$ $(i = 1,\ldots,k;\ \ell = 1,\ldots,n_i;\ j = 1,\ldots,p)$ should be stored as follows:

$$\begin{array}{l} x_{111}\ x_{112}\ \cdots\ x_{11p} \\ \vdots \\ x_{1n_1 1}\ x_{1n_1 2} \ldots\ x_{1n_1 p} \\ \vdots \\ x_{k11}\ x_{k12}\ \cdots\ x_{k1p} \\ \vdots \\ x_{kn_k 1}\ x_{kn_k 2} \cdots\ x_{kn_k p} \end{array}$$

5. GENERAL REMARKS

Finite Intersection tests for multiple comparisons of means of normal populations were proposed by Krishnaiah in 1960 in an unpublished report. These results were later published in Krishnaiah (1963) and Krishnaiah (1965a). The multivariate F distributions mentioned in this paper was introduced in Krishnaiah (1965a). Finite Intersection tests for multiple comparisons of mean vectors were proposed in Krishnaiah (1965b).

The authors wish to thank Dr. J. M. Reising for his helpful suggestions in preparation of this paper.

This work was sponsored by the Air Force Flight Dynamics Laboratory, Air Force Systems Command under Grant AFOSR 77-3239. All inquiries regarding this program should be directed to R. M. Boudreau.

REFERENCES

[1] Cox, C. M. et. al. (1979). A study on finite intersection tests for multiple comparisons of means. In Multivariate Analysis-V. North-Holland Publishing Company.

[2] Krishnaiah, P. R. (1963). Simultaneous tests and the efficiency of generalized incomplete block designs. ARL 63-174. Wright-Patterson Air Force Base, Ohio.

[3] Krishnaiah, P. R. (1965a). On the simultaneous ANOVA and MANOVA tests. Ann. Inst. Statist. Math. **17** 34-53.

[4] Krishnaiah, P. R. (1965b). Multiple comparison tests in multiresponse experiments. Sankhya Ser. A **27** 65-72.

[5] Krishnaiah, P. R. (1979). Some developments on simultaneous test procedures. In Developments of Statistics Vol. 2. Academic Press, Inc.

[6] Mudholkar, G. S. and Subbaiah P. (1975). A note on MANOVA multiple comparisons based upon step-down procedures. Sankhya Ser. B **37** 300-307.

[7] Sidak, Z. (1976). Rectangular confidence regions for the means of multivariate normal distributions. JASA **62** 626-633.

APPROXIMATING THE LOG OF THE NORMAL CUMULATIVE

John F. Monahan
North Carolina State University

ABSTRACT

Approximation formulas are obtained for $\ell n\ \Phi$, where Φ is the cumulative distribution function for the normal distribution.

KEY WORDS: Normal cumulative, rational Chebyshev approximation, minimum distance estimation

I. INTRODUCTION

Let the normal cumulative distribtuion function be represented by

$$(1.1)\quad \Phi(x) = \int_{-\infty}^{x} \phi(u)du = (2\pi)^{-\frac{1}{2}} \int_{-\infty}^{x} \exp(-u^2/2)du$$

so that $d\Phi/dx = \phi$. The function of interest here is $\ell n\ \Phi$. The function $\ell n\ \Phi$ is always negative, increases monotonically to zero, and its tail behavior is important:

$$(1.2)\quad \begin{array}{ll} \ell n\ \Phi(x) \approx -\phi(x)/x & \text{for } x >> 0 \\ \ell n\ \Phi(x) \approx -x^2/2 & \text{for } x << 0 . \end{array}$$

The standard library functions ℓn, erf, and erfc can be used to compute $\ell n\ \Phi$:

$$(1.3)\quad \ell n\ \Phi(x) = \begin{cases} \ell n(1+\text{erf}(x/\sqrt{2}))-\ell n\ 2 & \text{for } x \geq 0 \\ \ell n(\text{erfc}(-x/\sqrt{2}))-\ell n\ 2 & \text{for } x \leq 0 . \end{cases}$$

Using (1.3) to compute $\ell n\ \Phi$ suffers in three areas:

1) For $x << 0$, the range is limited to that of erfc. Underflow can occur in erfc although the final result, roughly $-x^2/2$, will be far from overflowing.

2) Most seriously, all accuracy will be lost for large x.

3) It is slow.

To surmount these problems, approximations for $\ell n\ \Phi$ were sought with the following goals:

1) Constant relative accuracy throughout the range of $\ell n\ \Phi$,

2) Faster computation near zero than in the tails,

3) A reasonably accurate and fast approximation that can work when implemented in single precision.

The second goal reflects the application of the approximation that stimulated this work.

In computing minimum distance statistics, the Anderson-Darling distance function [2]

$$d(F_n,F) = -1-n^{-2} \sum_i (2i-1)\{ \ell n\Phi(Z_i)+\ell n\ \Phi(-Z_{n-i+1})\}$$

must be minimized over μ and σ , where $Z_i = (X_{[i]}-\mu)/\sigma$. Here, the Z_i's are often near zero but can take extreme values. The third goal is essentially the challenge of a "stable" approximation, since single precision calculations on the IBM 370 are not very accurate. Furthermore, the implementation on more accurate hardware would improve the accuracy with the same speed.

II. A Priori Computation

To evaluate $\ell n\ \Phi$, two continued fraction expansions are available [1,26.2.14&15]:

$$(2.1)\quad 1 - \Phi(x) = \phi(x)/(xg(x^2)) \quad (x > 0)$$

$$(2.2)\quad \Phi(x) = 1/2 + \phi(x)h(x) \quad (x \geq 0)$$

where $g(w) = \cfrac{w^{-1}}{1 + \cfrac{2/w}{1 + \cfrac{3/w}{1 + \cfrac{4/w}{1 + \cfrac{5/w}{1 + \cdots}}}}}$

and $h(x) = \cfrac{x}{1 - \cfrac{x^2}{3 + \cfrac{2x^2}{5 - \cfrac{3x^2}{7 + \cfrac{4x^2}{9 - \cdots}}}}}$

The first expression (2.1) is best for large values of x and (2.2) for x near zero. A transition point of 2.45 was used for a priori and testing calculations. Recall, of course, $\Phi(-x) = 1 - \Phi(x)$.

III. Approximation

The subroutine IRATCU, the IMSL [6] FORTRAN implementation of the Cody, Fraser and Hart [4] algorithm "Chebyshev", was used to obtain rational function approximations to the following functions:

(3.1) $g(1/z) \approx R_{\ell m}(z)$ for $0 \le z \le 1/4$

(3.2) $\ell n\ \Phi(x) \approx S_{\ell m}(x)$ for $-2 \le x \le 0.6$

(3.3) $\ell n \Phi(x)/\exp(-x^2/2) \approx T_{\ell m}(x)$ for $0.6 \le s \le 2$

Here $R_{\ell m}$, $S_{\ell m}$ and $T_{\ell m}$ are all rational functions of the form $P(x)/Q(x)$, where P and Q are polynomials of degrees ℓ and m, respectively.

The approximating function to $\ell n\ \Phi$, represented here by $\Psi(x)$, is thus given by:

$$\Psi(x) = \begin{cases} -x^2/2-\ell n(r/(-xR_{\ell m}(1/x^2))) & x \le -2 \\ S_{\ell m}(x) & -2 \le x \le 0.6 \\ \exp(-x^2/2)T_{\ell m}(x) & 0.6 \le x \le 2 \\ -U_{\ell m}(r \exp(-x^2/2)/[xR_{\ell m}(1/x^2)]) & 2 \le x \end{cases}$$

where $r = (2\pi)^{-\frac{1}{2}}$ and $U_{\ell m}(z)$ is a Pade approximation of $-\ell n(1-z)$.

Table 1

Relative Accuracy of Approximating Functions

R_{32}	3.3×10^{-7}	
R_{44}	4.8×10^{-10}	
S_{42}	3.2×10^{-7}	
S_{54}	1.7×10^{-10}	
T_{42}	2.3×10^{-7}	
T_{45}	1.4×10^{-10}	
U_{11}	4.3×10^{-7}	(approx)
U_{22}	1.6×10^{-11}	(approx)

Relative accuracy means $\max_x |(\hat{g}(x)-g(x))/g(x)|$.

Other approximations are available from the author.

For a single precision approximation, the rational functions R_{32}, S_{42}, T_{42} and U_{11} were used. The coefficients (21) can be found in the FORTRAN implementations given in the Appendix. When implemented in double precision, the relative accuracy was better than 7×10^{-7}, that is

$$\frac{|\ell n\ \Phi(x) - \Psi(x)|}{-\ell n\ \Phi(x)} \le 7 \times 10^{-7} \quad \text{for all } x.$$

When implemented in single precision on the IBM 370, the relative accuracy falls to 4×10^{-6} for $|x| \le 4$. This deteriorates further for large values of x chiefly because of the error in the single precision computation of exp. The approximating function R_{32} retains most of its accuracy. Note that underflow occurs for $x \ge 18.78$.

For a higher precision approximation, the rational functions R_{44}, S_{54}, T_{45} and U_{22} were used. Again, the coefficients (31) can be found in the FORTRAN implementation in the Appendix. The relative accuracy here was 5×10^{-10}.

The goals of Section I (especially (2)) and the desire to keep the number of constants at a reasonable level influenced the decisions regarding the form of Ψ and the transition points $\pm$ 2 and 0.6. Changing the transition points would involve tradeoffs: reducing the degrees of $R_{\ell m}$ by moving beyond 2 would force $S_{\ell m}$ and $T_{\ell m}$ to a higher degree to maintain the same accuracy level throughout.

The Pade approximants U_{11} and U_{22} were used to replace complicated many-digit constants with simple integers and so reduce the human effort of implementation.

Finally, some work was done for appximating $\ell n\ 2\Phi(x)$ for x near zero but is not reported here.

IV. Acknowledgements

The computations were done on the IBM 370's and the Amdahl 470/V8 at the Triangle Universities Computation Center. The author would like to thank Dennis Boos for instigating this work, E. L. Battiste and David McAllister for helpful suggestions, and the NCSU Computation Center for assistance.

V. References

[1] Abramowitz, Milton and Stegun, Irene, eds. (1965), Handbook of Mathematical Functions,

[2] Boos, D. D. (1981), "Minimum Distance Estimators for Location and Goodness-of-Fit," J. Am. Stat. Assn. (to appear).

[3] Cody, W. J. (1969). "Rational Chebyshev Approximation for the Error Function," Math. Comp. 23, 639-638.

[4] Cody, W. J., Fraser, W., and Hart, J. F. (1968). "Rational Chebyshev Approximation Using Linear Equations," Numer. Math. 12, pp. 242-251.

[5] Hart, J. F., et al. (1968). Computer Approximations, Wiley, New York.

[6] IMSL Library Reference Manual (1969). Inter. Math. and Statis. Libraries, Houston, Texas.

```
      DOUBLE PRECISION FUNCTION ALNPHI(X)
      IMPLICIT REAL*8(A-H,O-Z)
C  COMPUTES THE NATURAL LOGARITHM OF THE NORMAL DF
C  J F MONAHAN    JAN 1981   DEPT. OF STAT., N C S U, RALEIGH, N C 27650
C  DOUBLE PRECISION IMPLEMENTATION OF 'SINGLE PRECISION' ALGORITHM
C  RELATIVE ACCURACY 7E-7 IN DOUBLE PRECISION ON IBM 370
      DATA PR0,PR1,PR2,PR3/.1000000331605568D1, .9071497753602147D1,
     1    .1561614824967968D2,    .2806832609648498D1 /
      DATA QR1,QR2/    .8071680440342330D1,    .9528364246287411D1 /
      DATA PS0,PS1,PS2,PS3,PS4/-.6931469710383208D0,.1044216478038279D1,
     1-.6351060127816162D0,.1877245896446035D0,-.2337240033940884D-1/
      DATA QS1,QS2/  -.3553830981887509D0,    .4795018411050619D-1 /
      DATA PT0,PT1,PT2,PT3,PT4/-.6933961642033521D0,.2207415657453527D0,
     1-.1205628285211437D0,.2474823386424022D-1,-.1994060903697693D-2/
      DATA QT1,QT2/  .8353363773307393D0,   .1675342594125622D0 /
      DATA RSR2PI/0.39894228040143D0/
      ALNPHI=0.
C  UNDERFLOW FOR IBM 370 FOR X > 18.78
      IF(X.GT.18.78) RETURN
      AX=DABS(X)
      IF(AX.LT.2.) GO TO 2
C  DO THE TAILS FIRST
      XX=X*X
      XR32=(((PR3/XX+PR2)/XX+PR1)/AX+PR0*AX)
     1  /((QR2/XX+QR1)/XX+1.D0)
      IF(X.LT.0.) GO TO 1
      OMPHI=DEXP(-X*X/2.)*RSR2PI/XR32
C  OMPHI=1-PHI       NEXT USE PADE APPROX TO GET LN(1-OMPHI)
      ALNPHI=-OMPHI*(6-OMPHI)/(6-4*OMPHI)
      RETURN
    1 ALNPHI=DLOG(RSR2PI/XR32)-X*(0.5*X)
      RETURN
C  NOW DO THE RANGE -2 < X < 0.6
    2 IF(X.GE.0.60) GO TO 3
      ALNPHI=(PS0+X*(PS1+X*(PS2+X*(PS3+X*PS4))))/(1.0+X*(QS1+X*QS2))
      RETURN
C  LAST TAKE CARE OF 0.6 < X < 2
    3 ALNPHI=(PT0+X*(PT1+X*(PT2+X*(PT3+X*PT4))))
     1       *DEXP(-0.5*X*X)/(1.0+X*(QT1+X*QT2))
      RETURN
      END
```

RELATIVE ERROR OF APPROXIMATIONS

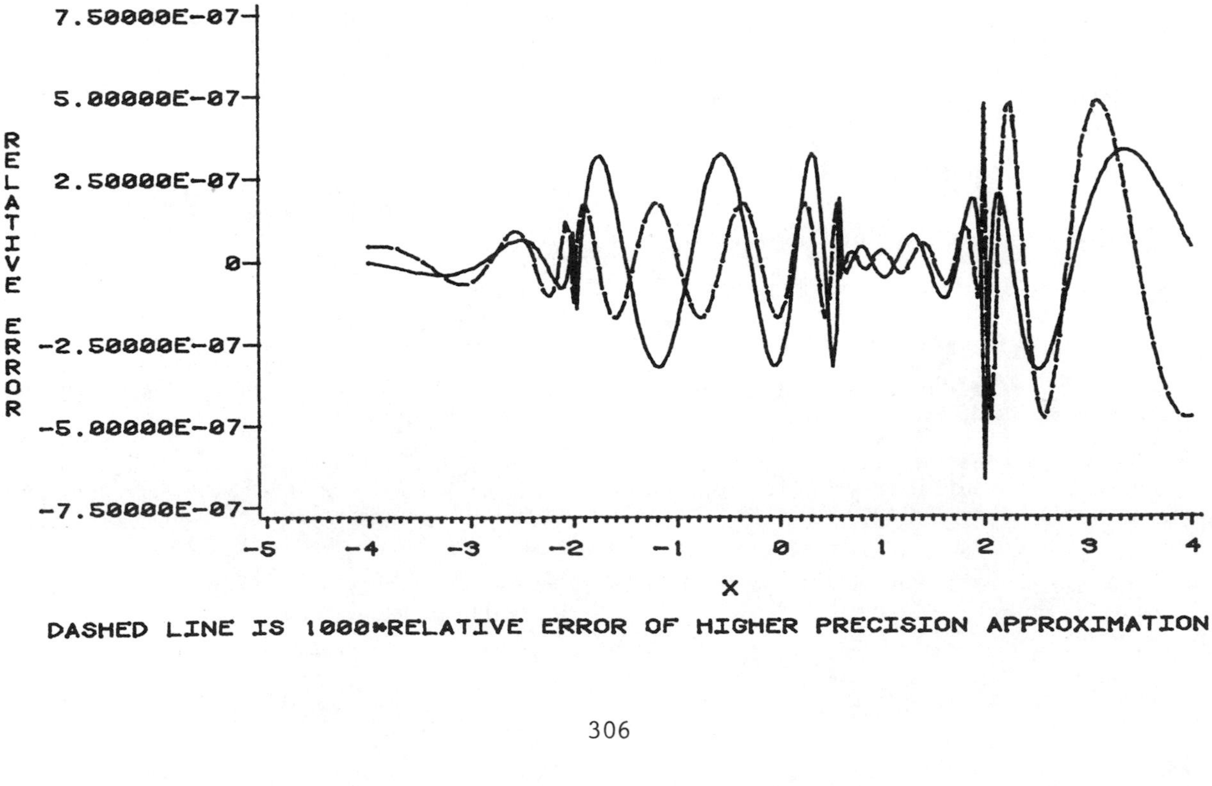

DASHED LINE IS 1000*RELATIVE ERROR OF HIGHER PRECISION APPROXIMATION

```
      DOUBLE PRECISION FUNCTION DLNPHI(X)
C  COMPUTES THE NATURAL LOGARITHM OF THE NORMAL DF
C  J F MONAHAN    JAN 1981   DEPT. OF STAT., N C S U, RALEIGH, N C 27650
C  RELATIVE ACCURACY 5E-10 IN DOUBLE PRECISION ON IBM 370
      REAL*8 X,XX,AX,OMP,XR44,RSR2PI,PR(5),QR(5),PS(6),QS(5),PT(5),QT(6)
      DATA PR/ .1000000000482936D1,  .1985567931985068D2,
     1.1098879956653959D3,.1834429208065582D3,.6506080508395136D2/
      DATA QR/ 1.D0,  .1885568000099099D2,  .9303216221079135D2,
     1              .1181358153137159D3,  .1775722502232874D2/
      DATA PS/  -.6931471804413338D0,  .1116578800666792D1,
     1          -.7599613777275790D0,  .2748377491091320D0,
     2          -.5324292689952796D-1, .4471834925812340D-2/
      DATA QS/ 1.D0,    -.4597786000829981D0,   .1079153770519465D0,
     1                  -.8722606392165005D-2,  .1239415190611190D-4 /
      DATA PT/  -.6931538538914723D0,  .1241706867532580D0,
     1-.1133871231901621D0,.1016145066987096D-1,-.5219185285512414D-2/
      DATA QT/  1.D0,  .9720635314558741D0,  .3228363288804008D0,
     1.5383079749056880D-1,.1999987264080985D-1,.9688449346136640D-2/
      DATA RSR2PI/.3989422804014 3D0/
      DLNPHI=0.
C  UNDERFLOW FOR IBM 370 FOR X > 18.78
      IF(X.GT.18.78) RETURN
      AX=DABS(X)
      IF(AX.LT.2.D0) GO TO 2
C  DO THE TAILS FIRST
      XX=X*X
      XR44=((((PR(5)/XX+PR(4))/XX+PR(3))/XX+PR(2))/AX+PR(1)*AX)
     1 /((((QR(5)/XX+QR(4))/XX+QR(3))/XX+QR(2))/XX+QR(1))
      IF(X.LT.0.) GO TO 1
      OMP=DEXP(-X*X/2.)*RSR2PI/XR44
C  OMP=1-PHI      NEXT USE PADE APPROX TO GET LN(1-OMP)
      DLNPHI=-OMP*(30-OMP*(21-OMP))/(30-OMP*(36-9*OMP))
      RETURN
   1  DLNPHI=DLOG(RSR2PI/XR44)-X*(0.5*X)
      RETURN
   2  IF(X.GE.0.60) GO TO 3
C  NOW DO THE RANGE -2 < X < 0.6
      DLNPHI=(PS(1)+X*(PS(2)+X*(PS(3)+X*(PS(4)+X*(PS(5)+X*PS(6))))))
     1       /(QS(1)+X*(QS(2)+X*(QS(3)+X*(QS(4)+X*QS(5)))))
      RETURN
C  LAST TAKE CARE OF 0.6 < X < 2
   3  DLNPHI=((((PT(5)*X+PT(4))*X+PT(3))*X+PT(2))*X+PT(1))
     1       *DEXP(-X*X/2.D0)
     1       /(((((QT(6)*X+QT(5))*X+QT(4))*X+QT(3))*X+QT(2))*X+QT(1))
      RETURN
      END
```

A kTH NEAREST NEIGHBOUR CLUSTERING PROCEDURE

M. Anthony Wong and Tom Lane, Massachusetts Institute of Technology

Due to the lack of development in the probabilistic and statistical aspects of clustering research, clustering procedures are often regarded as heuristics generating artificial clusters from a given set of sample data. In this paper, a clustering procedure that is useful for drawing statistical inference about the underlying population from a random sample is developed. It is based on the uniformly consistent kth nearest neighbour density estimate, and is applicable to both case-by-variable data matrices and case-by-case dissimilarity matrices. The proposed clustering procedure is shown to be asymptotically consistent for high-density clusters in several dimensions, and its small-sample behavior is illustrated by empirical examples.

Keywords: high-density cluster; kth nearest neighbour density estimation; clustering procedure; set-consistency.

1. INTRODUCTION

1.1 Shortcomings of Clustering Procedures

A recent study by Blashfield and Aldenderfer (1978) shows that numerous clustering methods have been developed in the past two decades. A review of many of these techniques can be found in Everitt (1974) and Hartigan (1975). However, hardly any of the originators of these methods have approached the clustering problem from within a theoretical framework. More often than not, the concept of a real population cluster is vague and is left undefined. Since no statistical evaluation of the sample clusters can be performed under the circumstance, the validity of the clusters obtained by these methods is always questionable. Consequently, the existing clustering procedures are often regarded as heuristics generating artificial clusters from a given set of sample data, and there is a need of clustering procedures that are useful for drawing statistical inference about the underlying population from a sample. In this paper, a clustering procedure based on the kth nearest neighbour density estimate is proposed, and it is shown to be set-consistent for high-density clusters in several dimensions. The set-consistency property of a hierarchical clustering procedure will be defined next.

1.2 A Theoretical Approach to Evaluating Hierarchical Clustering Methods

In order to evaluate the sampling property of a clustering method, it is necessary to have population clusters defined on population probability density functions from which the observations are obtained, and to have some ways of judging how the sample clusters deviate from the population clusters. Let the observations $x_1, x_2, \ldots, x_N$ in p-dimensional space be sampled from a population with density f, taken with respect to Lebesque measure. Using the high-density clustering model given in Hartigan (1975, p. 205), the true population clusters can be defined on f as follows: a <u>high-density cluster</u> at level f_o in the population is defined as a maximal connected set of the form $\{x \mid f(x) \geq f_o\}$. The family T of such clusters forms a tree, in that $A \in T$, $B \in T$ implies either $A \supset B$, $B \supset A$, or $A \cap B = \phi$. A hierarchical clustering procedure, which produces a sample clustering tree T_N on the observations $x_1, \ldots, x_N$, may then be evaluated by examining whether T_N converges to T with probability one when N approaches infinity. A clustering method (or equivalently, T_N) is said to be <u>strongly set-consistent</u> for high-density clusters (or T) if for any A, $B \in T$, $A \cap B = \phi$,

$$P_r\{A_N \cap B_N = \phi \text{ as } N \to \infty\} = 1,$$

where A_N and B_N are respectively the smallest cluster in the sample tree T_N containing all the sample points in A and B. Since $A \subset B$ implies $A_N \subset B_N$, this limit result means that the tree relationship in T_N converges strongly to the tree relationship in T.

Using this definition of consistency, hierarchical clustering methods can be evaluated by examining whether they are strongly set-consistent for high-density clusters. Hartigan (1977a, 1979) has examined the set-consistency of many of the best known hierarchical clustering methods for high-density clusters. It was shown that the complete linkage and average linkage (Sneath and Sokal 1973) methods are not set-consistent, while single linkage (Sneath 1957) is <u>weakly</u> set-consistent in one dimension but not in higher dimensions. Thus most of the relevant evaluative work under the high-density clustering model has been carried out. However, the important problem of developing clustering procedures that are set-consistent for high-density clusters did not receive much attention. In Hartigan and Wong (1979), and Wong (1980), a hybrid clustering method is developed which is weakly set-consistent for high-density clusters in one dimension; and, there exist empirical evidence that similar consistency results hold in several dimensions. However, although the hybrid method has the advantage of being practicable for very large data sets, it is not well-suited for small samples ($n < 100$) and it is only applicable to case-by-variable data matrices. In this paper, a strongly set-consistent clustering procedure is developed which is applicable to both case-by-variable data matrices and case-by-case distance matrices, and its development is outlined next.

1.3 Development of the kth Nearest Neighbour Clustering Procedure

Under the high-density clustering model, density estimates can be used to generate sample clusters, namely the high-density clusters defined on the estimates. And a clustering procedure is expected to be set-consistent for high-density clusters if it is based on a uniformly consistent density estimate. Consider the kth nearest neighbour density estimate: the estimated density at point x is $f_N(x) = k/(N V_k(x))$, where $V_k(x)$ is the volume of the closed sphere centered at x containing k sample points. Such a density estimate is uniformly consistent with probability 1 if f is uniformly continuous and if $k = k(N)$ satisfies $k(N)/N \to 0$ and $k(N)/\log N \to \infty$. (See, for example, Devroye and Wagner 1977.)

Wishart (1969), in an attempt to improve on the single linkage clustering technique, developed a procedure entitled Mode Analysis which is related to the kth nearest neighbour density estimate. However, Wishart's procedure was not designed to obtain the high-density clusters defined on the density estimate, and its computational algorithm is very complicated. In this paper, a clustering algorithm for deriving the tree of sample high-density clusters from the kth nearest neighbour density estimate is developed. A detailed description of this clustering procedure is given in Section 2. In Section 3, it is established that the proposed method is strongly set-consistent for high-density clusters. Empirical examples are given in Section 4 to illustrate the small-sample behavior of kth nearest neighbour clustering.

2. A kth NEAREST NEIGHBOUR CLUSTERING PROCEDURE

The proposed nearest neighbour clustering algorithm consists of two stages:

2.1 The Density Estimation Stage

The kth nearest neighbour density estimation procedure is used in this stage of the clustering procedure because it provides a strongly uniform consistent estimate of the underlying density. Let $x_1,\ldots,x_N$ be independent, identically distributed random vectors with values in R^p, $p \geq 1$, and with a common probability density f. If $V_k(x)$ is the volume of the smallest sphere centered at x and containing at least k of the random vectors $x_1,\ldots,x_N$, then the kth nearest neighbour density estimate of f at x is

$$f_N(x) = k/(NV_k(x))$$

And in Devroye and Wagner (1977), the following strong uniform consistency result of this estimate is shown:

<u>Lemma</u> (Devroye and Wagner, 1977):

If f is uniformly continuous on R^p and if $k = k(N)$ is a sequence of positive integers satisfying:
(a) $k(N)/N \to 0$, and
(b) $k(N)/\log N \to \infty$, as $N \to \infty$,
then

$$\sup_x |f_N(x) - f(x)| \to 0 \text{ with probability 1.}$$

One purpose of the kth nearest neighbour clustering method is to discover the population high-density clusters given a random sample from some underlying distribution F with density f. In this first step of the proposed procedure, a uniformly consistent estimate of f is obtained. The clusters defined on the estimated density f_N can then be used as sample estimates of the population high-density clusters defined on f. These hierarchical sample high-density clusters are constructed in the second stage of the proposed clustering algorithm.

2.2 The Hierarchical Clustering Stage

In this stage, a distance matrix $D(x_i,x_j)$, $1 \leq i, j \leq N$, for the N observations is first computed using the following definitions:

Definition 1: Two observations x_i and x_j are said to be <u>neighbours</u> if

$$d^*(x_i,x_j) \leq d_k(x_i) \text{ or } d_k(x_j),$$

where d^* is the Euclidean metric and $d_k(x_i)$ is the kth nearest neighbour distance to point x_i.

Definition 2: The distance $D(\cdot,\cdot)$ between the observations x_i and x_j is

$$\begin{aligned} D(x_i,x_j) &= (1/2)[1/f_N(x_i) + 1/f_N(x_j)] \\ &= \frac{N}{2k}[V_k(x_i) + V_k(x_j)], \text{ if } x_i \text{ and } x_j \text{ are neighbors;} \\ &= \infty, \text{ otherwise.} \end{aligned}$$

Hence, finite distances are defined only for pairs of observations which are in the same neighbourhood in R^p, and the defined distance between a pair of neighbouring observations is inversely proportional to a pooled density at the point halfway between them. The single linkage clustering technique is then applied to this distance matrix D to obtain the tree of sample high-density clusters. (See Gower and Ross (1969), and Hartigan (1975) for computational single linkage algorithms.)

Single linkage clustering is used in this step of the proposed procedure because it has the following property: at every stage of the clustering, the single linkage clusters are the maximal linked sets if objects x_i and x_j are said to be linked whenever $D(x_i,x_j)$ is no greater than a given distance D_o. Now, since the distance D between two "neighboring" observations is reciprocal to the density estimate f_N at the midpoint between them, every cluster obtained by applying single linkage to D has the property that the density estimates over the objects in this cluster are greater than a certain density level f_o. Moreover, as the distance measure D is defined only for pairs of "neighbouring" observatons, the resultant single linkage clusters correspond to maximal connected sets of the form $\{x|f_N(x) \geq f_o\}$, which are the high-density clusters defined on f_N.

The computational requirements for stage 1 and stage 2 are O(p N log N) and O(Nk) respectively. Hence, unlike the hybrid clustering method (Hartigan and Wong, 1979 and Wong, 1980), this procedure is not practicable for large data sets; but, it is better suited for small samples, and is applicable to both case-by-variable data matrices and case-by-case dissimilarity matrices.

3. STRONG SET-CONSISTENCY OF kTH NEAREST NEIGHBOUR CLUSTERING

The asymptotic consistency of the kth nearest neighbour clustering method for high-density clusters in R^p, $p \geq 1$, is given in the following theorem:

Theorem: Let f denote a positive, uniformly continuous function on R^p such that $\{x|f(x) \geq f_o\}$ is the union of a finite number of compact subsets of R^p for every $f_o > 0$. Let T be the tree of population high-density clusters defined on f. Suppose that A and B are any two disjoint high-density clusters in T with connected interiors. Let $x_1, \ldots, x_N$ be a random sample from f and let T_N be the hierarchical clustering specified by the kth nearest neighbour clustering algorithm. Then, provided that $k = k(N)$ satisfies

(a) $k(N)/N \to 0$, and
(b) $k(N)/\log N \to \infty$,

as $N \to \infty$, there exist $A_N, B_N \in T_N$ with $A_N \supset A \cap \{x_1,\ldots,x_N\}$, $B_N \supset B \cap \{x_1,\ldots,x_N\}$ and $A_N \cap B_N = \phi$, with probability 1.

Proof: Since T_N is the tree of high-density clusters for f_N, this theorem is a direct consequence of the Lemma, which states that

$$\sup_x |f_N(x) - f(x)| \to 0, \text{ with probability 1.} \quad (3.1)$$

By definition, for any two disjoint high-density clusters A and B in T, there exist $\delta > 0$, $\varepsilon > 0$ and $\lambda > 0$, such that

(i) $f(x) \geq \lambda$ for all $x \in A \cup B$, and (3.2)
(ii) each rectilinear path between A and B contains a segment, with length greater than δ, along which the density $f(x) < \lambda - 3\varepsilon$. (3.3)

From (3.1), we have for N large,

$$\sup_x |f_N(x) - f(x)| < \varepsilon \text{ w.p. 1.}$$

Thus, it follows fromm (3.2) and (3.3) that for N large, with probability 1,

(iii) $f_N(x) > \lambda - \varepsilon$ for all $x \in A \cup B$, and
(iv) each rectilinear path between A and B contains a segment, with length greater than δ, along which the density estimate $f_N(x) < \lambda - 2\varepsilon$.

Since A and B are disjoint, it follows from (3.4) and (3.5) that high-density clusters of the form $\{x|f_N(x) \geq \lambda - \varepsilon\}$ separate the observations in A and B. The theorem follows.

4. EMPIRICAL STUDY OF THE SMALL-SAMPLE BEHAVIOR OF THE kth NEAREST NEIGHBOUR CLUSTERING PROCEDURE

To illustrate the small-sample behavior of the kth nearest neighbor clustering procedure, an empirical study was performed in which the procedure is applied to various generated data sets. Results of two experiments, in which bivariate data were used, are reported here.

1. Experiment One: 30 observations were generated so that two spherical clusters of observations are present in this data set. The scatter-plot of this sample set is shown in Figure A, in which the observation numbers are plotted next to the observations. This data set is useful for illustrating the effectiveness of the proposed procedure in identifying spherical clusters. The dendrogram giving the hierarchical clustering obtained by the kth nearest neighbour method (using k = 4) is shown in Figure B. It is clear that, in this experiment, the kth nearest neighbour clustering indicates the presence of two modal regions of clusters.

2. Experiment Two: 58 observations were generated so that two elongated, elliptical clusters of observations are present in this data set. The scatter plot of this sample set is shown in Figure C, in which the observation numbers are plotted next to the observations. This data set is useful for illustrating the effectiveness of the proposed clustering procedure in identifying non-spherical clusters. The dendrogram giving the hierarchical clustering obtained by the kth nearest neighbour method (using k = 4) is shown in Figure D. Two disjoint modal regions, corresponding to the two elliptical clusters of observations shown in Figure C, can be identified in this dendrogram. However, observations 51, 58, and 37 form a minor modal region within one of the two

clusters; and, observations 22, 23, and 2 form a minor modal region in the other cluster.

5. REFERENCES

Blashfield, R.K., and Aldenderfer, M.S. (1978), "The Literature on Cluster Analysis", Multivariate Behavioral Research, 13, 271-295.

Devroye, L.P., and Wagner, T.J. (1977), "The strong uniform consistency of nearest neighbour density estimates", Annals of Statistics, 5, 536-540.

Everitt, B.S. (1974), Cluster Analysis, Halsted Press, New York: John Wiley.

Gower, J.C. and Ross, G.J.S. (1969), "Minimum spanning trees and single linkage cluster analysis", Applied Statistics, 18, 54-64.

Hartigan, J.A. (1975), Clustering Algorithms, New York: John Wiley & Sons.

______________(1977), "Distributional problems in clustering", in Classification and Clustering, ed. J. Van Ryzin, New York: Academic Press.

______________(1979), "Consistency of single linkage for high-density clusters", unpublished manuscript, Department of Statistics, Yale University.

______________, and Wong, M.A. (1979), "Hybrid Clustering", Proceedings of the 12th Interface Symposium on Computer Science and Statistics, ed. Jane Gentleman, U. of Waterloo, Press. pp. 137-143.

Sneath, P.H.A. (1957), "The application of computers to taxonomy", Journal of General Microbiology, 17, 201-226.

______________, and Sokal, R.R. (1973), Numerical Taxonomy, San Francisco: W.H. Freeman.

Wishart, D. (1969), "Mode Analysis" in Numerical Taxonomy, edited by A.J. Cole, New York: Academic Press.

Wong, M.A. (1980), "A hybrid clustering method for identifying high-density clusters", Working Paper #2001-80, Sloan School of Management, Massachusetts Institute of Technology.

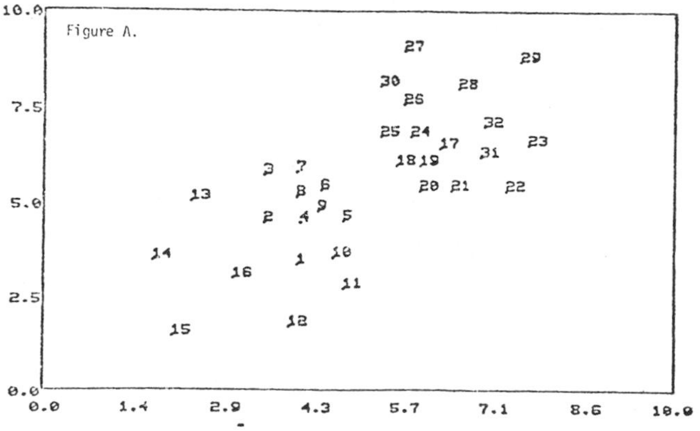

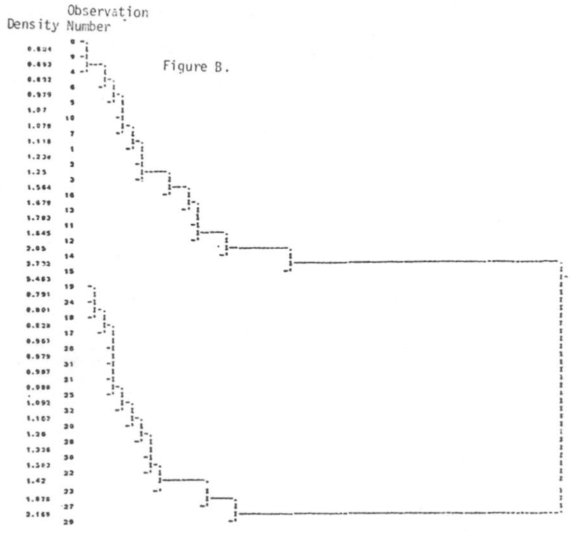

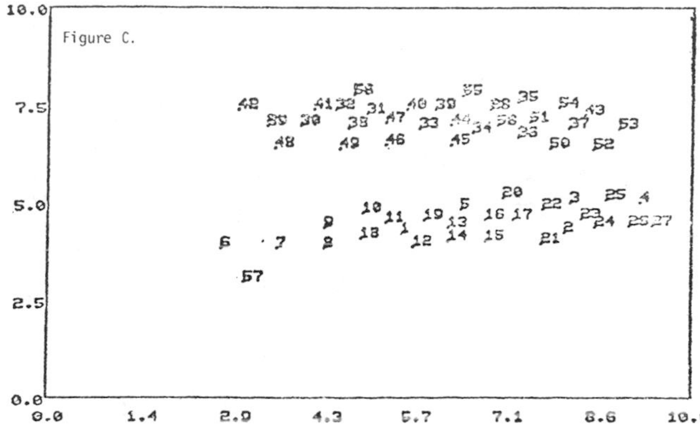

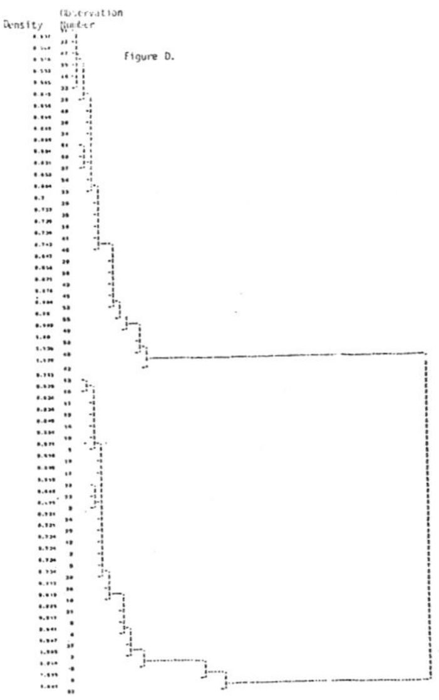

INTERACTIVE STATISTICAL GRAPHICS: BREAKING AWAY

Neil W. Polhemus, Princeton University
Bernard Markowicz, Princeton University

ABSTRACT

The proper role of the computer in data analysis is one of increasing the ability of the analyst to extract information. This requires a high degree of user control within an interactive environment, not only over data manipulation and computations but also over graphical display. An interactive statistical graphics system such as that described in this paper allows the user to break away from constraints imposed by most statistical packages.

KEY WORDS

Graphics, Statistical Analysis, Computer Methods, APL

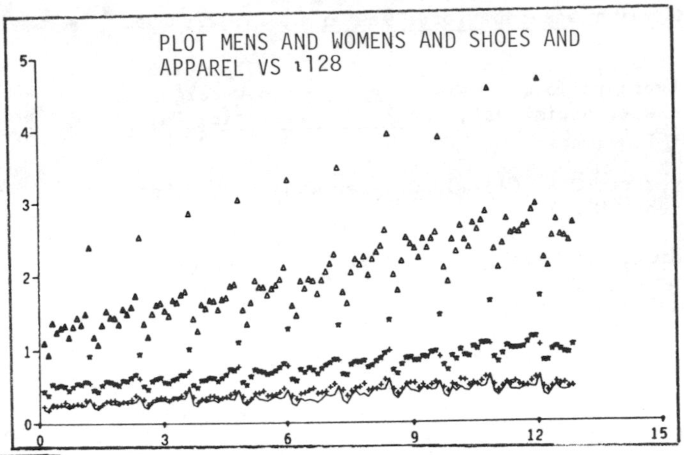

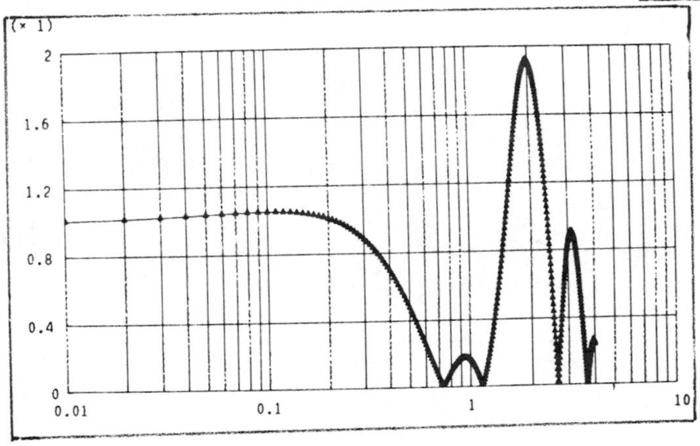

INTRODUCTION

In recent years, many of the most popular statistical packages have added graphical capabilities. While this has expanded significantly the ability to display data in certain standard ways, adding to the interpretation of numerical results, the user is still severely restricted in adapting graphics to specific cases. One of the benefits of interactive computing is the flexibility given to the user to modify the analysis environment and maintain user control rather than package control. This benefit ought to extend into the area of graphical display as well as data manipulation and numerical computations.

Within the Civil Engineering Department at Princeton, an environment exists which encourages breaking away from canned statistical packages. The needs of the users fall into two primary areas: education and research. The educational need clearly requires considerable trial and error, which necessitates an interactive system. The research is primarily of an experimental or analytical nature, which requires considerable user flexibility. In both cases, careful design of a user-machine interface that promotes user efficiency is paramount.

To achieve an adequate interface, three major components were employed:

(i) a powerful language to minimize programming time (APL)
(ii) a basic graphics software system as an output mechanism (usable within APL)
(iii) a large machine for computational speed (IBM 3033 under VM370).

The intent of this paper is to demonstrate the type of user-oriented statistical analysis which is possible under such a user-oriented environment.

DATA INPUT, EDITING, AND DISPLAY:

Data is handled in dynamically defined variables which can be edited, transformed or combined at will. Basic plotting functions can be used to display the data.

```
X←1.2  3.5  2.7  6  5.2  6.3
X[4]←5.9
Y←X*.5
Z←X+Y
```

BASIC STATISTICS:

More commonly used methods are included in functions which can be executed using any data vector. Scaling options are supplied by the user; if unappealing, the graph can be regenerated immediately using different options.

USER CONTROL:

The user's ability to control the plotting window and combine functions can yield displays specific to the data at hand. Here are modified multiple box and whisker plots of aircraft lateral deviations from route centerline as a function of distance along the route.

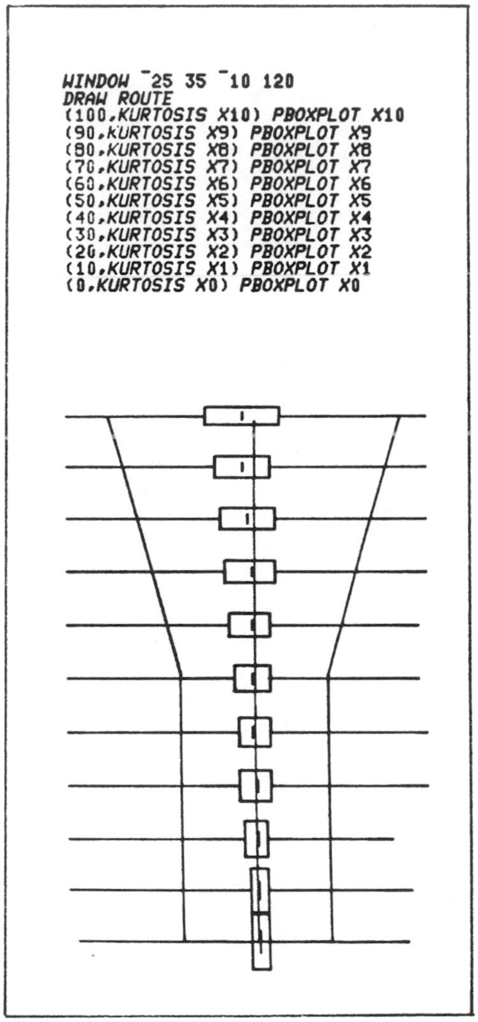

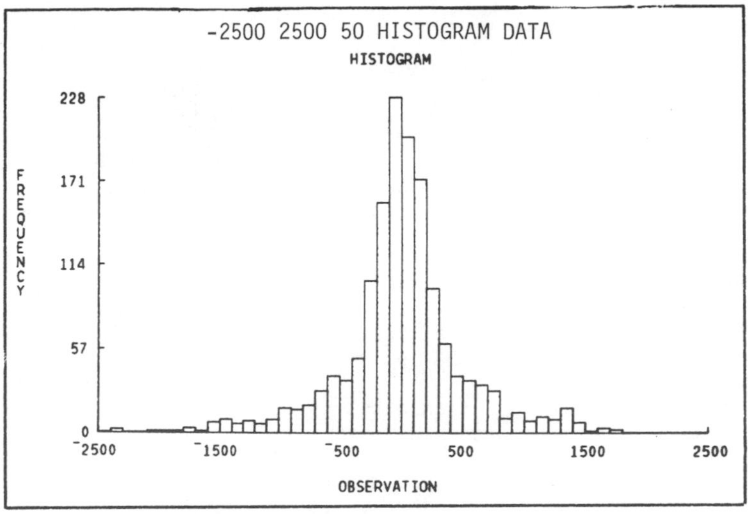

TIME SERIES ANALYSIS

Visual display of autocorrelation functions and periodograms aids in interactive time series modeling. Various operations can be strung together in a single statement using the right-to-left syntax of APL.

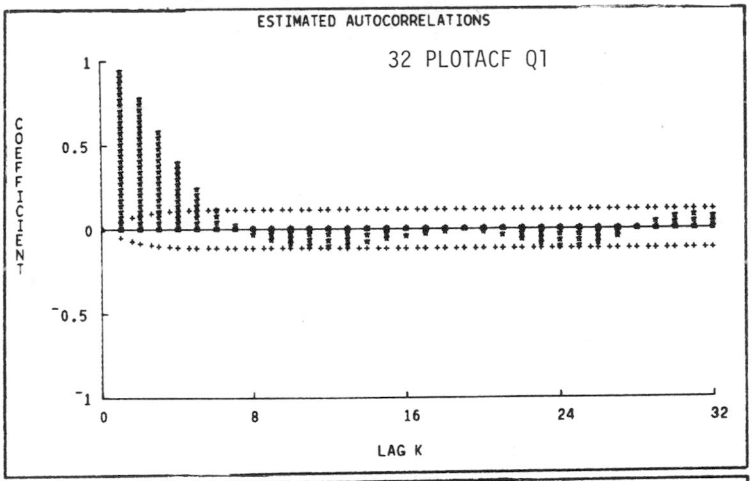

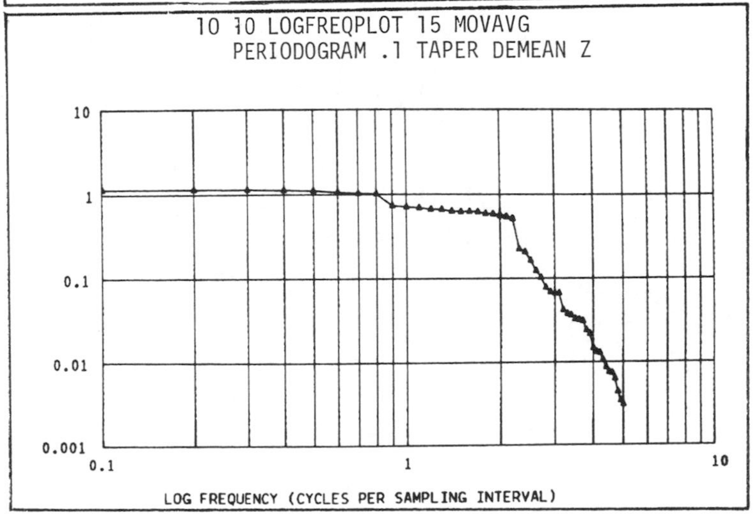

THREE-DIMENSIONAL DISPLAY:

Graphics output devices allow for interactive display in three dimensions. On the right is a display of crosscorrelation matrices for the four time series displayed on the first page. Below is a three dimensional histogram of the aircraft lateral deviations at ten-mile increments along the jet route.

With a 1.2 mega-baud data transmission rate such plots can be produced in a fraction of a second.

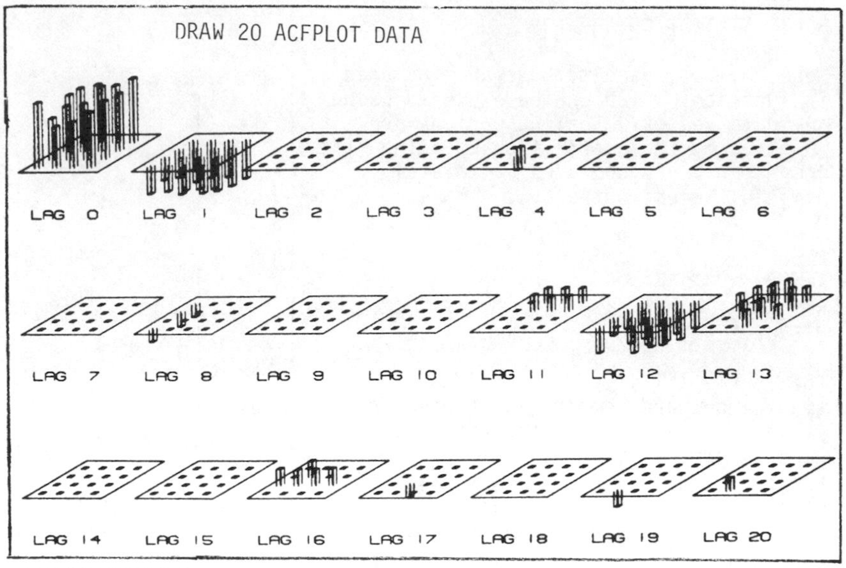

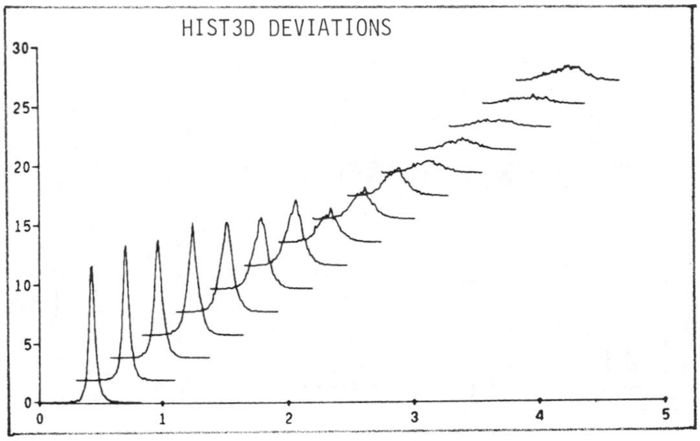

DISPLAY OF LOCATIONAL ATTRIBUTES

Data is displayed on a zonal basis (left below). The map is a simulated aerial view of the region and bars are scaled automatically. The scale indicates minimum, maximum and average. Mixed display is used in the right figure where bars are indexed and referred to the locational summary.

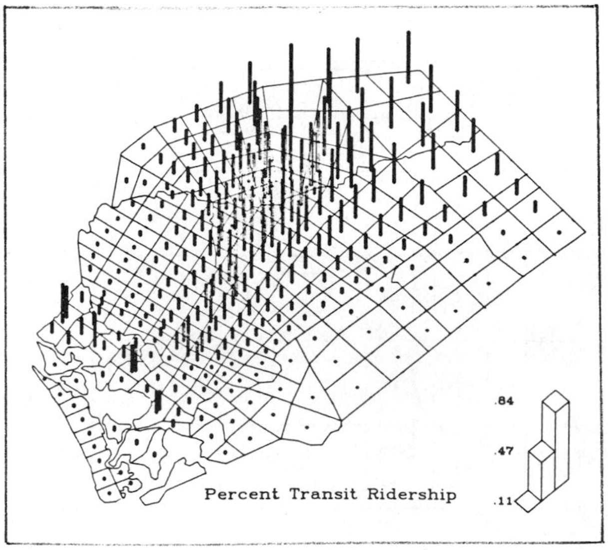

Percent Transit Ridership

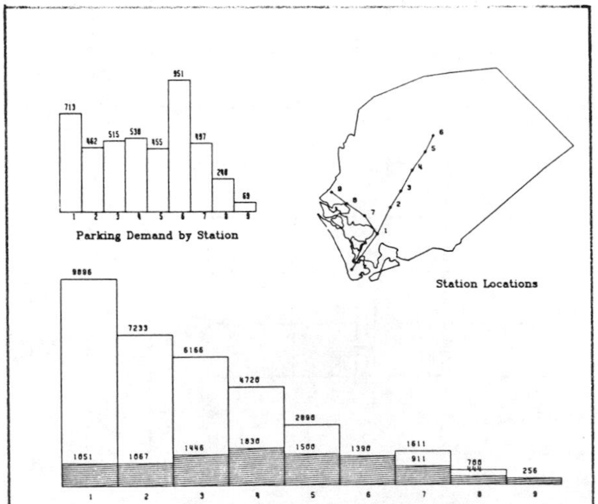

On–Line Ridership & Station Loadings

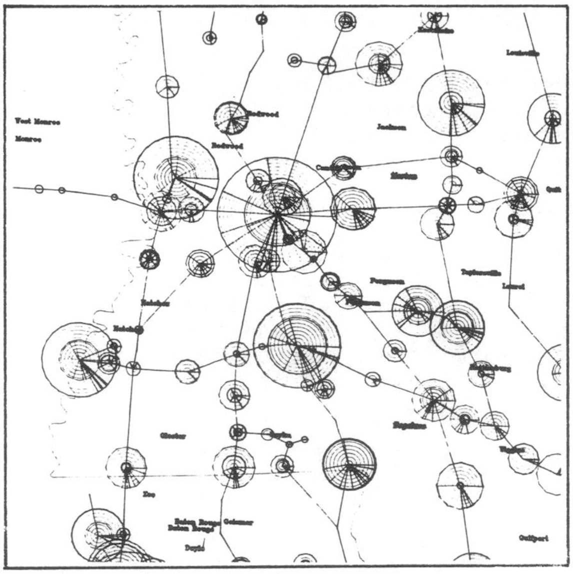

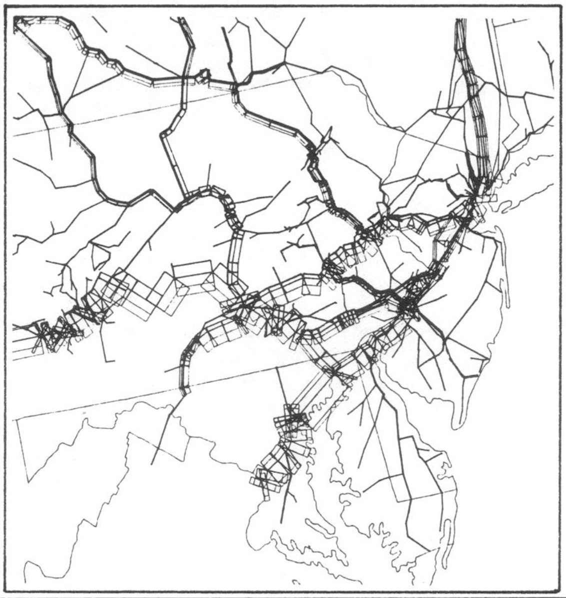

For more complex networks (here above the U.S. rail network) pie charts are used to display node attributes and (right) boxes to display link attributes.

The maps are produced interactively on the Tektronix screen, then, if selected, sent to the Calcomp plotter where colors are added.

"BUSINESS" GRAPHICS;

The need to communicate results to non-technical people calls for more graphically sophisticated displays. Both high information content and quick overall understanding are required.

Shading is used (right) to differentiate between the entities analyzed, but could be used to display additional data characteristics. The chart and shading program, written in APL, is 17 line long and comprises no loop. The use of fancy character fonts adds to the overall quality.

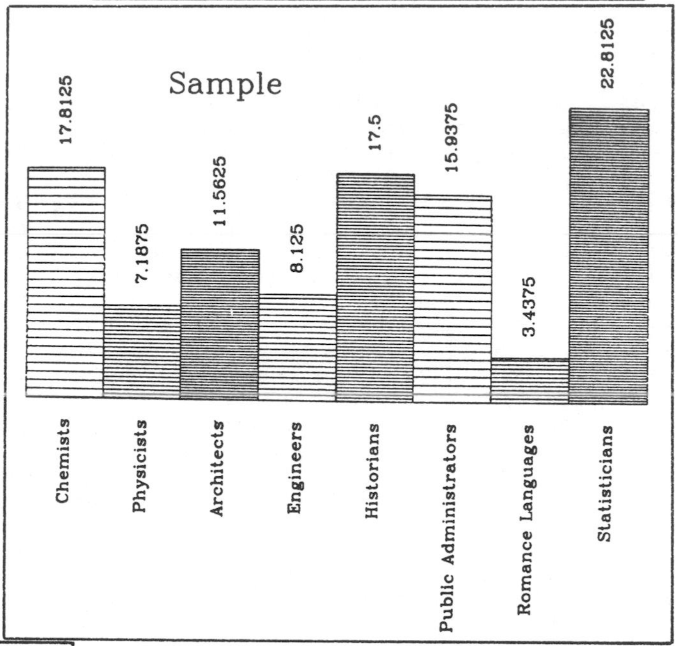

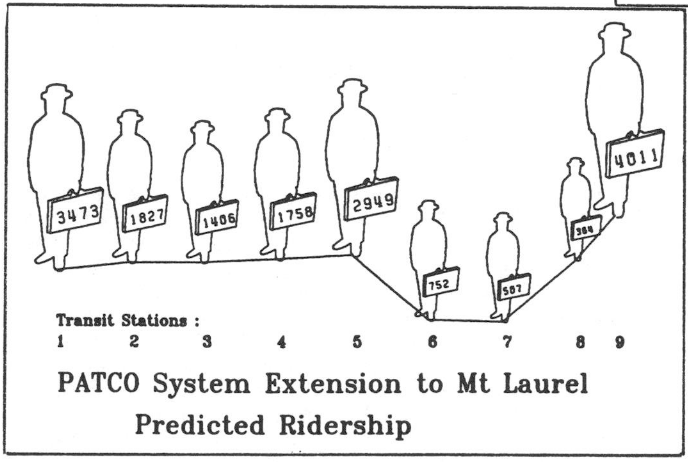

In the last figure, Mr. Commuter is used to display predicted ridership along a transit line. In order to maintain readability a quadratic transformation is used for the character and size.

More information about the techniques illustrated in this paper can be obtained from the authors, through the School of Engineering & Applied Science, Princeton University.

The basic graphics system was developed by the Interactive Computer Graphics Laboratory at Princeton University.

INTERACTIVE GRAPHICAL ANALYSIS FOR MULTIVARIATE DATA

Robert B. Stephenson, University of Waterloo
John C. Beatty, University of Waterloo
Jane F. Gentleman, University of Waterloo

A Fortran program, called CLUSTER, has been implemented which interactively assists the statistician in exploring multivariate data, displaying projected data on a Tektronix 4010, Tektronix 4027 or Hewlett-Packard 2648 graphics terminal. The program is designed to be portable, and isolates device dependent display code so as to simplify the addition of device drivers for other graphics terminals. Both keyboard and cursor input are supported, and emphasis has been placed on a good human interface.

Key Words: Interactive Graphics, Multivariate Data Analysis

1. Introduction

CLUSTER is a program which interactively assists the statistician in exploring multivariate data. It enables the user to select and view a sequence of scatter plots. Each scatter plot is the projection of the data points onto some plane in the data space, and each scatter plot can be viewed separately or as part of a sequence of plots.

The scatter plot is a common approach to showing the data distribution over two dimensions. Gnanadesikan [6] discusses general circumstances under which one may be interested in such dimensionality reduction of multiple response data. The movie produced by Fisherkeller, Friedman, and Tukey [4] and the videotape produced by Donoho, Huber, and Thoma [2] show sequences of scatter plots in which the parallax resulting from motion enables the user to perceive a third dimension.

The well known Andrews high dimensional data plots [1] can be interpreted as plots of projections onto a particular sequence of lines, while the present program permits investigation of projections onto a sequence of planes which may be user defined.

2. Program Description

The program performs two separate tasks: i) the selection of projection planes; and ii) the detailed viewing of selected projections.

The program displays a set of projections. A projection is selected from this set or defined by a user's calculation. The selected projection can then be used as a seed to generate a new set from which to make another selection. This cycle is repeated until the user is satisfied with some projection, for example when separate clusters of data points have become apparent.

In its simplest mode, CLUSTER provides a set of projections from which a selection can be made. This set consists of the principle and interpolated projections. The principle projections are defined by CLUSTER to give a variety of data views. These views remain constant throughout the search. On the other hand, the interpolated projections change with each new selection and are defined to fall between the last two selected projections.

During a search, the user may pause to examine any particular projection in detail. Once an interesting projection has been selected, a variety of manipulations may be performed on the data and on the way in which it is displayed. For example, only a default rectangular "window" in each projection plane is displayed during a search. It is possible that this window may not contain all the projected points on a plane. It may also happen that projected points overcrowd parts of the window. Hence the size and position of the window being displayed may be altered to accommodate either of these possibilities.

2.1. Basic Features

CLUSTER is controlled by issuing commands from one of two command sets: the primary command set and the plot command set. The primary command set is used during the search for and selection of projections. The plot command set is used to view a selected projection. CLUSTER's keyboard interface was taken from the ST package for interactive statistical graphics (Dunn and Gentleman [3]) since these routines provide a flexible and error tolerant input mechanism.

With the primary commands, the program can be instructed to display the principle projections, to display the interpolated projections, to select one of the above projections, to accept a user defined projection, to place the data points in groups, to change the size of the window, to scale any coordinate of the data, and to list important variables used by the program.

Display of the principle projections acts as a preliminary scan of the data space. From this display, the user can obtain a feeling for the data distribution. The scan is displayed in frames of a hundred projections each. Each frame covers the whole screen. A projection can be selected from the scan by its frame number, its horizontal position and its vertical position. For a 6-dimensional data space, for example, CLUSTER defines 992 principle projections and therefore will produce 10 scan frames. Figure 1 is the first such frame produced from the measurements on teeth data in Andrews [1]. There are ten six-dimensional data points, two of which were obtained from human subjects, six of which were obtained from apes, and two of which are of unknown origin. Each data point was obtained by measuring six characteristics of the teeth of a given subject. In Figure 1, the second projection in the bottom row shows the data points for human beings to be in a separate cluster from the data points for apes, with the unknowns in the former cluster. This might lead the data analyst to conclude, as did Andrews, that the unknowns were human.

A display of interpolated projections is obtained by performing an interpolation from the previous to the current projection. The number of interpolation steps and the number of projections displayed in each frame can be varied. Doing so allows the user to view one projection at a time or any number at a time (up to seeing all the interpolated projections on one frame). Figure 2 is an interpolation with 23 projection steps and 25 projections per frame. Interpolation starts with the previous projection in the lower left and ends with the current projection in the upper right. A projection is identified by the number of steps, the number of projections per frame, its frame number, its horizontal position and its vertical position.

The plot commands are the means by which any portion of the projection plane can be displayed. These commands specify operations on the current display to produce the next display. For example, a "zoom" command expands to full screen size a specified box on the screen, a "mooz" contracts to one sixteenth screen size a specified box on the screen, a "scale" expands or contracts the display by any amount, and a "centre" translates the display by centring a specified point on the screen. Also included in the plot commands are operations to list the points near a specified spot, and to display the default window area of the projection plane.

As a general rule, positions and areas on the screen are specified by means of a cross-hair cursor whose position is controlled either by keys or by rotating "thumbwheels", depending on the terminal in use. The same functions can be performed completely from the keyboard, although this is more cumbersome.

CLUSTER has a set of options which can be enabled or disabled on any command in either command set. These options pertain to the listing of information and to the displaying of projections. The list option controls the copying to a file of all lists produced. These copied lists can then be read by other analysis programs. The display options control the symbols plotted for each data point. The data points may be represented by dots, by unique identifying numbers (their numeric position in the data file), or by group numbers.

2.2. Special Features

Several special features have been added to aid in data analysis.

CLUSTER is capable of executing a user supplied subroutine. The data analyst using this feature supplies a subroutine which defines the next projection as a function of the current one. For example, a criterion for selecting the next projection could be a statistic such as E (Friedman and Rafsky [5]), which measures the relative extent to which the projection onto 2 dimensions has distorted the true distances among the points in k dimensions. The value of E could be computed for a selection of planes "near" the current one, and the plane yielding the best preservation of distance would be selected for the next plotted projection.

The trace of an execution of CLUSTER may be written into a text file. Such a file contains all the user's responses and can be used in place of terminal input to control another execution of CLUSTER. This facilitates the preparation of terminal demonstrations and of movies.

CLUSTER can be used with a variety of graphics terminals. Currently the projected data can be displayed on a

Tektronix 4010, Tektronix 4027 or Hewlett-Packard 2648. To simplify the addition of device drivers for other terminals, the device dependent display code has been isolated. Shortly a device driver will be added for a 480 line by 640 pixel by 24 bit color frame buffer equipped with a high speed bit-slice microprocessor. This will enable the use of a vastly larger number of colors than are available on the Tektronix 4027, and will enable more rapid sequencing through an interpolation sequence.

On terminals with color capability, each group of data points can be displayed in a distinct color. This can be extremely useful in searching for clusters. For example, the data points for human beings in Figure 1 could be displayed in one color, data for apes in another, and data for unknowns in a third.

Instead of outputting to an online display terminal, the graphic output can be directed to a text file. The resulting file may be read by a post-processor to produce plots on an offline graphic device. In particular, it is intended that a post-processor be written which will read such a file and drive a Dicomed D48 color film recorder. This will enable the production of high quality color 35mm and 4X5 output, and the production of 16mm movies illustrating the use of CLUSTER to explore multivariate data.

Techniques for graphics package design are discussed in Newman and Sproull [7].

3. Program Use

The user initially designates any known subgroupings of the data by assigning group numbers. For example, in the data discussed above, the user can set the group numbers for human data points to one number, for ape data points to another number, and for unknowns to a third number. This separation is an aid later when projections are displayed, since the group numbers can be used as plotting symbols. On color devices, each group of data points is also plotted in a different color. A plot which appears to have no clustering can in fact show a visual separation when plotted with group numbers and/or with color. Such is the case when the group symbols of the different groups appear in separate areas and are not intermixed.

As data exploration continues, the user can try the scan display of the principle projections. This may provide a feeling for the data distribution. It is possible that from this display more patterns will be observed.

New projections can be obtained from the principle projection scan, from the interpolation between two previous projections, from a user defined function of the current projection data, or directly from the user.

Each new projection can supply more information about the data. Each can be displayed at full screen size, and using the plot commands, it is possible to view any portion of the projection plane. The display options allow the projection to be plotted with dots, with group numbers or with identification numbers. The distribution of each group can be observed, and the identity of any point on the screen can be obtained. Using the list plot command, the identities of a point or a group of points can be displayed. CLUSTER also provides lists of the projection data as a further aid in the analysis of a projection.

4. Work in Progress

This paper is part of a Master's thesis in Computer Science by R.B. Stephenson, supervised by J.C. Beatty and J.F. Gentleman. It is hoped that the software will be developed further and used by statisticians to analyze multivariate data and to investigate new techniques for such analysis.

5. References

[1] Andrews, D.F, (1972). "Plots of High-dimensional Data," Biometrics, 28, pp. 125-136.

[2] Donoho, D., Huber, P.J., and Thoma, M. (1981). "Interactive Graphical Analysis of Multidimensional Data," videotape, Harvard Univ., Cambridge, Mass.

[3] Dunn, Robert M. and Gentleman, J. F. (1976). "The ST Interactive STatistical Plotting Package," Proc. of the 7th Ontario Universities Computing Conference, Waterloo, pp. 306-317.

[4] Fisherkeller, M.A., Friedman, J.H., and Tukey, J.W. (1979). "PRIM-9," 16 mm color film, Bin 88 Productions, Stanford Linear Accelerator Center, Stanford, California.

[5] Friedman, J.H. and Rafsky, L.C. (1981). "Graphics for the Multivariate Two-Sample Problem," J. American Statistical Assoc., to appear in June.

[6] Gnanadesikan, R. (1977). Methods for Statistical Data Analysis of Multivariate Observations, Wiley, p. 6.

[7] Newman, William M. and Sproull, Robert F. (1979). Principles of Interactive Computer Graphics, second edition, McGraw-Hill.

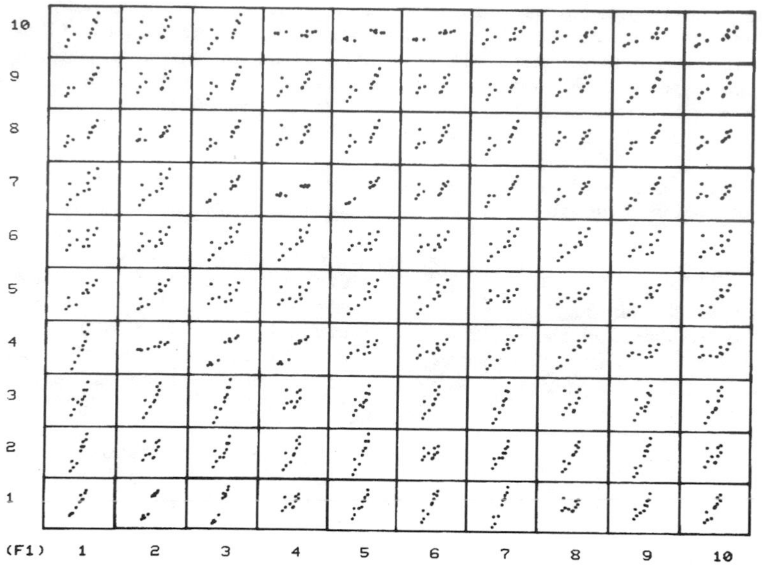

FIGURE 1. A FRAME OF THE PRINCIPLE PROJECTION SCAN.

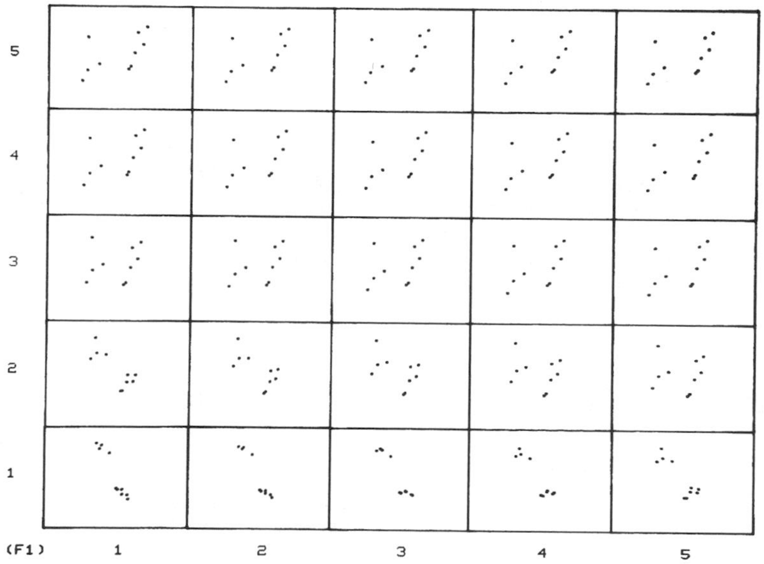

FIGURE 2. A PROJECTION INTERPOLATION.

SLANG, A STATISTICAL LANGUAGE FOR DESCRIPTIVE TIME SERIES ANALYSIS

M. Nicolai and R. Cheng

United Nations Industrial Development Organization

Abstract

SLANG is a language designed to provide easy access to a statistical database for users who have little or no programming experience. This language, which can operate in both interactive and batch mode, allows retrieval and descriptive analyses of time series. For more complex analyses, the retrieval capability of SLANG can also be used as a bridge between a statistical database and those commercially available statistical analysis packages which support only sequential input. Design highlights presented in the paper are: a syntax which allows only a small number of statement types, simple data types and a library of functions which users can extend to increase the power of the language. User experience and efficiency considerations are also discussed. SLANG currently operates under IBM's MVS on a database of international statistics managed by ADABAS.

Key Words and Phrases: retrieval languages, time series analysis, statistical data bases.

1. INTRODUCTION

The Statistical Language SLANG was designed to satisfy the need of UNIDO's Division for Industrial Studies for an efficient, powerful and easy to use package for descriptive time series analysis. The project began in 1978 shortly after a commercially available DBMS (ADABAS) was chosen to manage UNIDO's statistical data. Owing to the general nature of ADABAS, it soon became obvious that the retrieval language provided with it was inadequate for the statistical analyses demanded of the system. At the same time, direct interface of the database with statistical analysis packages, which normally expect sequential input files, became difficult. Today, SLANG is used to perform directly most of the statistical analyses required in UNIDO and provides a powerful interface between the database and statistical packages for more complex tasks.

2. A BRIEF DESCRIPTION OF SLANG

2.1 Variables

SLANG supports two types of variables: "system" variables and "temporary" variables. The former identify time series, or sets of time series, in the statistical database while the latter are defined by the user for holding results of computations.

2.1.1 System Variables

UNIDO's statistical database consists of time series containing yearly data on industrial production, national accounts and external trade for all UN member countries. A SLANG system variable identifies:

(i) a single time series, or

(ii) a set of closely related time series.

The approach of allowing variables to identify either a single time series (i.e. a vector) or a set of time series (i.e. a table) was adopted because several time series in the database are closely related and often viewed, and used, as a single entity (for example, time series on employment in various industrial sectors).

Conversely, for those instances where the user wishes to refer to a single time series in a set, SLANG allows the use of subscripts to select the desired subset. Since system variables can be either vectors or tables, a maximum of two subscripts are allowed for this selection. Because the dimension time is common to all variables, the "year" was chosen as the first subscript; the second subscript is dependent on the nature of the variable.

Examples: (i) EMP

(ii) EMP(,311)

(iii) EMP(1970,311)

In example (i) above, the variable EMP identifies a set of time series on employment in various sectors of industry (ISIC). Example (ii) selects, from this set, the time series for the food processing sector (311), while example (iii) refers to a single observation: employment data for 1970 for sector 311. This last example also illustrates that, by the use of subscripts, SLANG can operate on single observations (i.e. scalars) in addition to time series and sets of time series. Since they really only identify data on the database, system variables cannot be used as output fields in arithmetic expressions or functions. For this purpose SLANG allows the variable type introduced in the next section.

2.1.2 Temporary Variables

SLANG users can define temporary variables to hold the results of computations. Once defined, these variables can be used in further computations, can be displayed or can be written into an output file for processing by other systems. However, unlike system variables, they cease to exist upon completion of the program. Temporary variables can be scalars, vectors or tables, and their dimensions are automatically determined by SLANG on the basis of the context in which they are first used.

2.2 Language Statements

Statements are the basic unit of the SLANG language. Structured according to a very simple syntax, they specify the data selection and analysis required by a user. The number of SLANG statements is very small, a feature which makes the language easy to master and programs simple to write and maintain. Although few, SLANG statements are powerful so that most statistical analyses are well within the scope of the system.

2.2.1 FOR Statement

Most statistical studies done by UNIDO are performed on data for a single country or a group of countries. The FOR statement allows a user to select these countries either by listing their names or by using pre-defined group names. The latter identify countries which share a common characteristic, such as a similar economy or income level. Group names can also be combined to form logical expressions by the use of the logical operators AND and OR. As a final refinement, it is also possible to modify the list of countries selected by a logical expression by using the operators EXCLUDING and PLUS.

Examples: (i) FOR EUROPE AND MARKET-ECONOMY EXCLUDING FRANCE ITALY

(ii) FOR EUROPE OR NORTH-AMERICA PLUS JAPAN

Example (i) selects all countries in Europe which have a market economy with the exception of France and Italy, while example (ii) selects all countries in Europe, North America plus Japan.

2.2.2 SELECT Statement

The SELECT statement allows a user to select specific observations in a time series or specific time series in a set. This is essentially the same function as the one performed by variable subscripts, with the difference that a SELECT statement provides a convenient shorthand notation for this function but, once specified, applies to all variables in a program.

2.2.3 AGGREGATE Statement

No data for groups of countries is stored in the UNIDO database, so that aggregate data must always be derived dynamically from single-countries' data. The AGGREGATE statement is made available in SLANG to perform this function for all the countries selected by a FOR statement. Optionally, the AGGREGATE statement can also convert a system variable to a common currency before aggregating. An aggregation exception report informs users of any missing observations found during the aggregation. Users can, on the basis of this information, decide whether a modification of the program is necessary.

2.2.4 COMPUTE Statement

Two types of COMPUTE statement are supported in SLANG. The first type provides basic computing power and is designed to solve those computational problems which can be reduced to a simple arithmetic expression. The second type is designed to solve more complex tasks by invoking user-written functions. The library of functions available to a user can be easily extended to satisfy new requirements.

2.2.5 DISPLAY Statement

Since most users of SLANG were expected to be unfamiliar with programming techniques, a simple way of displaying the results of statistical analyses was deemed essential. The DISPLAY statement achieves this goal by requiring, in its simplest form, nothing more than a list of the variables to be displayed. System defaults are then used to determine the output format, headers and spacing between variables. More experienced users can override these defaults to tailor the output to their particular requirements.

2.2.6 END Statement

This statement terminates a SLANG program and initiates its interpretation.

2.3 Language Structure

A SLANG program consists of a set of SLANG statements beginning with a FOR statement and ending with an END statement. The FOR statement, which initiates and controls the program, selects the countries whose data is used in the analysis. Once this selection is made, processing begins one country at a time. Because of this sequential operation, computations involving data from two different countries are impossible, unless data for one of the two countries is previously stored in a temporary variable. In order to allow this data storage, a program may consist of several program steps, with each step having its own FOR statement. Results computed in a step are then available for further computation in subsequent steps. An example of a two-step SLANG program is given below:

```
FOR ─┐            ─┐
 .   │  step 1     │
 .   │             │
 .  ─┘             │
FOR ─┐             │   program
 .   │  step 2     │
 .   │             │
 .  ─┘             │
END               ─┘
```

Figure 1

2.3.1 Structure of Program Steps

Statements contained in a program step must follow a precise sequence as indicated below:

```
[FOR
[SELECT     ]      ]
[AGGREGATE  ]      ]  non-executable statements
[COMPUTE    ] ...  ]
[DISPLAY    ]      ]  executable statements
```

Figure 2

In the above figure, brackets indicate optional statements while ellipses indicate statements that can occur more than once per step. At least one DISPLAY statement or one COMPUTE statement must be present in a program step. Execution of a program step can proceed in two ways depending on the presence or absence of an AGGREGATE statement.

Case 1: If an AGGREGATE statement is specified, SLANG first aggregates the requested data for all the countries selected by the FOR statement then executes once only the "executable" statements (Figure 2) present in the step. Naturally, any operation will be on aggregate data.

```
Example: FOR EUROPE
         AGGREGATE
         COMPUTE $GDPCAP1 = GDP$ / POP
         END
```

Upon completion of the above program the variable $GDPCAP1, identified as temporary by the dollar prefix, will contain the GDP per capita in Europe. The system variables GDP$ and POP will contain the total GDP and population respectively for all European countries.

Case 2: If an AGGREGATE statement is not specified, SLANG executes any executable statement once for each country selected by the FOR statement. Any operation, therefore, will be on country data.

```
Example: FOR EUROPE
         COMPUTE $GDPCAP2 = GDP$ / POP
         END
```

Execution of the above program will provide the GDP per capita for each European country.

Combination of steps with AGGREGATE statements and steps without AGGREGATE statements is possible in the same program.

Example:

```
FOR EUROPE                              ]
AGGREGATE                               ]  step 1
COMPUTE $GDPCAP1 = GDP$ / POP           ]
FOR EUROPE                              ]
COMPUTE $GDPCAP2 = GDP$ / POP           ]
COMPUTE $RATIO = $GDPCAP2 / $GDPCAP1    ]  step 2
DISPLAY $GDPCAP2 $RATIO                 ]
END                                     ]
```

In the above example, the temporary variable $RATIO gives, for each European country, the ratio of its GDP per capita to that of Europe as a whole. Two things should be noted in this example. First, the temporary variable $GDPCAP1 computed in step 1 is also available in step 2 of the program. Second, the system variables GDP$ and POP while employed in an apparently similar manner in the two steps, do in fact contain different data: aggregate data in step 1 and country data in step 2. This example also shows that only temporary variables can be used to pass results between program steps.

2.4 Arithmetic Operations

Operands of arithmetic operations can be scalars, vectors or tables. When the two operands have the same dimension, the arithmetic operation takes place between all corresponding terms. In the case of dissimilar operands, such as scalars and vectors or tables, the operation takes place between the scalar and every term of the other operand. The following table summarizes the possible combinations of operands and the results produced:

Operand 1 / Operand 2	Scalar	Vector	Table
Scalar	Scalar	Vector	Table
Vector	Vector	Vector	Not Allowed
Table	Table	Not Allowed	Table

It is possible to overcome the restriction on operations between vectors and tables by first expanding the vector to two dimensions and then performing the operation. A function EXPAND is provided for this purpose.

2.5 Missing Observations

An important feature of any time series analysis package is the treatment of missing observations. Normally in SLANG a term of the result is calculated only if there is a corresponding term for each of the operands. In this case, the operands are referred to as 'strong' operands. For those instances where the above mode of operation would be too restrictive, the user can specify 'weak' operands. For operations on weak operands a term of the result is calculated as soon as there is a valid term for at least one of the operands. The missing term of the other operand is replaced by a "0" (zero) if the operation is an addition and by a "1" (one) if the operation is a multiplication. To indicate that a variable should be treated as weak, the user must enter a question mark after the variable name.

3. DESIGN CONSIDERATIONS

SLANG is an interpreter which operates in two separate phases. The first phase analyzes the complete program and translates it into an internal form. The second phase interprets or executes this internal form of the source program. This approach minimizes the time necessary to decode or analyze each statement in order to execute it.

SLANG is written in PL/I and operates interactively under TSO or in batch under IBM's MVS operating system. The choice of the programming language PL/I, perhaps not one of the most commonly used, was dictated by its ability to support advanced programming techniques such as dynamic allocation of storage.

Many design features were the product of environmental constraints, however, special attention was given to achieve the design goals outlined below.

Modularity. One of the primary design considerations was to produce a system that would be easy to maintain, modify and extend. To achieve this aim, SLANG is composed of modules which follow as strictly as possible major system functions. Additionally, some of these modules, such as the scanner and the parser, are table-driven so that changes in the language require no modification to the source code but only to the contents of these tables. In particular, the modularity of the system requires that each function invoked by a COMPUTE statement be contained in a separately compiled module. Only if reference to the function is made in the program is the module loaded dynamically. This approach permits the addition of new functions without requiring a recompilation of the whole system and keeps program size to a minimum since only those functions necessary in a program are loaded into memory.

Routines interfacing with the data base management system are also contained in a single module to facilitate possible migration to other DBMS in the future.

Efficiency. Special consideration was given to efficient programming on account of the size of the database and the fact that any aggregation would have to be done dynamically. Comparisons between SLANG programs and programs written in the DBMS' own retrieval language have shown that SLANG is faster by factors of two to four. These tests were limited to simple display programs which were the only ones that could be meaningfully compared.

Dynamic Storage Allocation. Storage for variables is allocated dynamically. Such allocation takes place in the program step in which a variable is first named. Variables are also dynamically freed following the step in which they are last named. Other data structures that are dynamically allocated and freed are the structures used to represent the internal form of the source program. While there is a certain cost in allocating storage dynamically, it was felt that such cost is justified by the consequent savings in program size. To give an indication of storage requirements, the minimum storage required for SLANG is about 300 Kbytes and a single variable can require up to 7 Kbytes of storage. Since kilobytes-minute is one of the units used in UNIDO to calculate computer costs, keeping program size to a minimum also provided immediate financial savings.

4. USER EXPERIENCE

User experience with SLANG has shown that even users without previous data processing knowledge can successfully write SLANG programs after only a few training sessions. This fact can, in large part, be attributed to the simplicity of the language but is also probably due to the interactive procedure (IBM's CLIST) which controls the entire SLANG system. This procedure allows a user to create and edit source programs, execute SLANG interactively or in batch, view the results of analyses at a terminal and request the printing of these results on the line printer. Once a user initiates the procedure, he is led "by the hand" through the various functions described above. He is asked, upon completion of one function, whether he wishes to proceed to the next function or go back to a previous one in order to introduce changes or corrections. At no point in this process is the user concerned with the computer's job control language or other system conventions, factors which normally deter inexperienced users.

The introduction of SLANG has drastically reduced the need for small ad-hoc statistical analysis programs thus freeing programming staff for more important tasks. At the same time, it has brought an entirely new category of users into direct contact with the data. This has resulted in increased efficiency and greater general appreciation of the data, its potential and limitations.

ON THE PARAMETER ESTIMATION IN QUEUEING THEORY

Jacob E. Samaan, Rehab Group, Falls Church, Va.
Derrick S. Tracy, University of Windsor

ABSTRACT

Two estimators are introduced for estimating the number m of servers for the multiserver queueing system M/M/m with infinite waiting room. One estimator is the maximum likelihood estimator and the other is more accurate for estimating large values of m. Simulation is used to simulate the system and compare the two estimators numerically.

KEYWORDS

Multiserver queueing system, parameter estimation, number of servers, finite differences, simulation algorithm.

INTRODUCTION

Most papers in queueing theory have been concerned with problems considering various distributions for arrival and service times, various number of service channels and a variety of queue disciplines [1]. Little literature is available about estimation of continuous parameters in queueing theory, viz., the arrival rate, the service rate or the mean waiting time. In this paper, the problem of estimating discrete parameters which take on integral values only is investigated.

Consider the system M/M/m with infinite waiting room. The arrival rate λ and service rate μ are assumed to be known. Estimates for the number m of servers are required.

The system is said to be in state i if i customers are in the system. Let the arrival and departure time of each customer be observed over a time period of length T. Let a_i denote the number of transitions from the state i to the state i + 1, b_i denote the total number of transitions from the state i to the state i - 1, and T_i the total time spent in the state i during the time interval (0,T).

MAXIMUM LIKELIHOOD ESTIMATOR FOR m

The likelihood function of the system is, [3],

$$L_T(m) = A \ln \lambda + B \ln \mu - \lambda T + C - \mu S$$

where $A = \sum_{i=0}^{\infty} a_i$, $B = \sum_{i=1}^{\infty} b_i$,

$$C = \sum_{i=1}^{m} b_i \ln i + \sum_{i=m+1}^{\infty} b_i \ln m ,$$

$$S = \sum_{i=1}^{m} iT_i + \sum_{i=m+1}^{\infty} mT_i .$$

Since the unknown parameter m is discrete and takes on integral values only, the theory of Billingsley [2] is not applicable here. The technique of finite differences is used to find the value of $\hat{m}$ which maximizes the likelihood function $L_T(m)$. Since the first three terms of the likelihood function are independent of m, it suffices to consider the function $F(m) = C - \mu S$. The maximum of this function occurs at $m = \hat{m}$ which satisfies $\Delta F(\hat{m}) < 0 < \Delta F(\hat{m}-1)$, where $\Delta F(m)$ is the first difference of $F(m)$. Hence, the estimator $\hat{m}$ is the value of m satisfying:

$$\sum_{i=m+1}^{\infty} (b_i \ln \frac{m+1}{m} - \mu T_i) < 0 < \sum_{i=m}^{\infty} (b_i \ln \frac{m}{m-1} - \mu T_i) .$$

In other words, $\hat{m}$ is the smallest value of m satisfying

$$\sum_{i=m+1}^{\infty} (Kb_i - \mu T_i) < 0$$

where $K = \ln \frac{m+1}{m}$.

UNIQUENESS OF $\hat{m}$

Each of the random variables $b_i/\lambda T p^*_{i-1}$, T_i/Tp^*_i and $a_i/\lambda T p^*_i$ converge stochastically to one, where p^*_i, i=1,2,..., is the stationary probability that the system is in state i.

Hence for large values of T, we have the following approximation for the sequences:

$$b_i \cong \lambda T p^*_{i-1}, \quad T_i \cong Tp^*_i , \quad a_i \cong \lambda T p^*_i .$$

Using these approximations the uniqueness of $\hat{m}$ can be established by considering the function

$$F(K) = \sum_{i=1}^{K} (b_i \ln i - i\mu T_i) + \sum_{i=K+1}^{\infty} (b_i \ln K - K\mu T_i)$$

and showing that it has a unique maximum, which occurs at K = m, the true value of the number of servers.

To prove that F(K) attains its maximum at K = m, it can be shown that

$\Delta F(K) > 0$ for all $K < m$, and
$\Delta F(m) < 0$.

Similary, to prove that this maximum is unique, it is shown that $\Delta F(K) < 0$ for all $K > m$.

SIMULATION AND NUMERICAL EXAMPLES

Since the estimator $\hat{m}$ does not have an explicit formula to enable us to investigate its statistical properties analytically, we simulate the system and calculate the estimated value $\hat{m}$ to illustrate the properties of this estimator.

An algorithm is constructed to simulate the Markovian system M/M/m and generate the arrays a_i's, b_i's and T_i's. Hundred samples are generated using this algorithm and a FORTRAN program is written to calculate the estimator $\hat{m}$

from each sample. This procedure is repeated for different values of λ, μ, m and the observation time T. The program counts the number $\hat{N}$ of times in which $\hat{m}$ has the true value m, and the number $\hat{N}_1$ of times in which m has a value in the range (m-1,m+1). Then

$$\hat{p} = P(\hat{m}=m) = \frac{\hat{N}}{100},$$

$$\hat{p}_1 = P(m-1 \le \hat{m} \le m+1) = \frac{\hat{N}_1}{100}.$$

Table 1 shows the values of $\hat{p}$ and $\hat{p}_1$ for $\lambda = 1$, $\mu = 0.5$, m = 4 and different values of the observation time T.

T	$\hat{p}$	$\hat{p}_1$
50	0.32	0.78
100	0.43	0.70
250	0.56	0.84
500	0.78	0.98
1000	0.95	1.0
10000	1.0	1.0

Table 1

Clearly, the accuracy of $\hat{m}$ increases by increasing T, and m has the true value m for all the samples when T = 10000. This shows the consistency of m (numerically).

The values of $\hat{p}$ and $\hat{p}_1$ for m = 4 and T = 1000 and different values of λ and μ are given in Table 2.

λ	μ	ρ	$\hat{p}$	$\hat{p}_1$
1.5	0.6	0.62	1.0	1.0
1.5	0.8	0.47	0.93	1.0
1.0	0.6	0.42	0.85	0.97
1.5	1.0	0.38	0.87	0.97
1.0	0.8	0.31	0.69	0.87
1.0	1.0	0.25	0.61	0.87
1.0	1.5	0.17	0.44	0.85

Table 2

From this table we observe that the accuracy of this estimator is very high when ρ is large (say $0.5 \le \rho < 1$) and even for ρ as small as 1/6 the estimator gives the true value of the number of servers 44 times out of the hundred estimates.

Finally Table 3 shows the effect of increasing m while ρ is fixed. The table gives the values of $\hat{p}$ and $\hat{p}_1$ for T = 1000, $\rho = 0.5$ and different values of λ, μ and m.

λ	μ	m	$\hat{p}$	$\hat{p}_1$
1.0	1.0	2	1.0	1.0
1.0	0.67	3	0.98	1.0
1.0	0.50	4	0.96	0.99
1.0	0.40	5	0.81	0.97
1.0	0.33	6	0.72	0.95
1.0	0.22	9	0.43	0.82
1.5	0.25	12	0.22	0.67
1.5	0.20	15	0.05	0.24

Table 3

We notice that $\hat{m}$ is a good estimate for m when we expect m to be small, but it is very poor when m is large (m > 9). This shows the necessity of introducing another method for estimating m in such cases.

ESTIMATING THE LARGE m

In this section we propose another method for estimating the number of servers for the Markovian system M/M/m when we expect m to be large. This method is based on some properties of the sequence T_i, i = 1,2,... . Let n_i be the number of times the system visits the state i, i = 0,1,.... Then

$$n_0 = 1 + b_1,$$

$$n_i = a_{i-1} + b_{i+1}, \quad i = 1,2,\ldots .$$

The time T_i spent in the state i upto total observation time T is the sum of n_i random variables, each exponentially distributed with parameter $\lambda + K\mu$ where K = min (i,m). Then by Renyi's theorem the random variables

$$X_i = \begin{cases} \dfrac{T_i - \dfrac{n_i}{\lambda + i\mu}}{\dfrac{\sqrt{n_i}}{\lambda + i\mu}} = \dfrac{(\lambda+i\mu)T_i - n_i}{\sqrt{n_i}}, & i = 1,2,\ldots,m \\[2ex] \dfrac{(\lambda+m\mu)T_i - n_i}{\sqrt{n_i}}, & i \ge m . \end{cases}$$

are asymptotically N(0,1) as $T \to \infty$.

Now we consider the sequence Z_i of andom variables defined by

$$Z_i = \frac{(\lambda+i\mu)T_i - n_i}{\sqrt{n_i}}, \quad i = 0,1,2,\ldots,$$

and propose the largest value of i for which Z_i is N(0,1) as an estimator for m and denote it by $\bar{m}$. In other words, if L is the maximum number of customers in the system during the time T of observation, then to find $\bar{m}$ we calculate the values of Z_i, i = 0,1,...,L, from the sample and test each Z_i for normality. The largest value of i for which Z_i is N(0,1) is the required value $\bar{m}$.

But each sample produces only one value Z_i for each $i = 1,2,\ldots,L$, which is not enough to test normality. For this method to work, we need r samples ($r \geq 100$). For the jth sample we calculate the sequence $Z_1^{(j)}, Z_2^{(j)}, \ldots, Z_L^{(j)}$, $j = 1,2,\ldots,r$. Then we test the normality of each sample $Z_i^{(1)}, Z_i^{(2)}, \ldots, Z_i^{(r)}$ for each $i = 0,1,2,\ldots,L$, using the Kolmogorov-Smirnov test for one sample. The largest value of i for which this sample is $N(0,1)$ is the estimate $\bar{m}$.

The accuracy of this method is also affected by increasing m , but it is still much better than the maximum likelihood method as Table 4 shows. This method can also be used for estimating small values of m but the number of observations and computation time required are much larger than those required for the maximum likelihood method. So it is not recommended for estimating small values of m .

For illustration, 120 samples are generated with $T = 100$, $\mu = 0.04$, $\rho = 0.5$ and different values of λ and m . The results are summarized in Table 4 below.

λ	m	$\bar{m}$
0.6	3	3
1.8	9	10
3.0	15	14

Table 4

The accuracy of this method also increases by increasing T or ρ . For $\lambda = 0.9$, $\mu = 0.2$ and $m = 9$, i.e., $\rho = 0.5$ but $T = 1000$, the method gives the true value of m , that is $\bar{m} = 9$, Also for $\lambda = 1.89$, $\mu = 0.3$, $m = 9$ and $T = 100$, i.e., $\rho = 0.7$, the method gives $\bar{m} = 9$.

The complete computer output for the case of $\lambda = 1.8$, $\mu = 0.4$, $m = 9$, $T = 100$, $r = 120$, $\rho = 0.5$, is presented in Table 5. In the second column we present the values S_i calculated from the sample for the statistics

$$D_i^{(r)} = \sqrt{r} \sup_{-\infty<X<\infty} |F_i^{(r)}(X) - F(X)| ,$$

where r is the sample size, $F_i^{(r)}(X)$ is the empirical distribution function and $F(X)$ is the theoretical distribution function which is $N(0,1)$. The probability P_i that the statistic $D_i^{(r)}$ has value greater than S_i , if the hypothesis is true, is given in the third column where

$$P\{D_i^{(r)} \leq Z\} = \begin{cases} 0 & \text{if} \quad Z < 0 \\ \sum_{K=-\infty}^{\infty} (-1)^K e^{2K^2Z^2} & \text{if} \quad Z > 0 \end{cases}$$

i	S_i	P_i
1	0.438008	0.990777
2	0.775606	0.584291
3	0.593508	0.872761
4	0.967145	0.306895
5	0.710593	0.693537
6	0.542598	0.930056
7	1.018330	0.250867
8	0.894109	0.400916
9	0.657465	0.780360
10	0.857195	0.454455
11	1.761112	0.004047
12	3.358957	0.000000
13	4.554768	0.000000
14	5.020791	0.000000
15	5.203375	0.000000
16	5.385943	0.000000
17	5.385949	0.000000

Table 5

ALGORITHM

The algorithm used to simulate the Markovian system M/M/m and to generate the arrays T_i's, a_i's, and b_i's, as also the number LL representing the maximum number of customers in the system during the time of observation is presented in Figure 1. Logical operators are shown with circles. If the condition tested by a given logical operator is satisfied, the arrow denoting the direction of control is qualified by the Index 1, and in the opposite case by Index 0.

Table 6 shows a sample generated by this algorithm for $\lambda = 1$, $\mu = 0.5$, $m = 4$ and $T = 1000$. The theoretical stationary probabilities p_i^* , $i = 1,2,\ldots,LL$, are presented in the last column.

LL = 14

i	a_i	b_i	T_i	p_i^*
0	143	0	152.088547	0.130435
1	268	142	264.988281	0.260870
2	254	268	274.045898	0.260870
3	146	254	160.487350	0.173913
4	56	146	68.597824	0.086957
5	30	56	27.831100	0.043478
6	26	30	17.428741	0.021739
7	17	26	13.686196	0.010870
8	12	17	8.510647	0.005435
9	9	12	6.259319	0.002717
10	1	9	3.370247	0.001359
11	2	1	2.100105	0.000679
12	1	2	0.267342	0.000340
13	2	1	0.472652	0.000170
14	0	2	0.057339	0.000085

Table 6

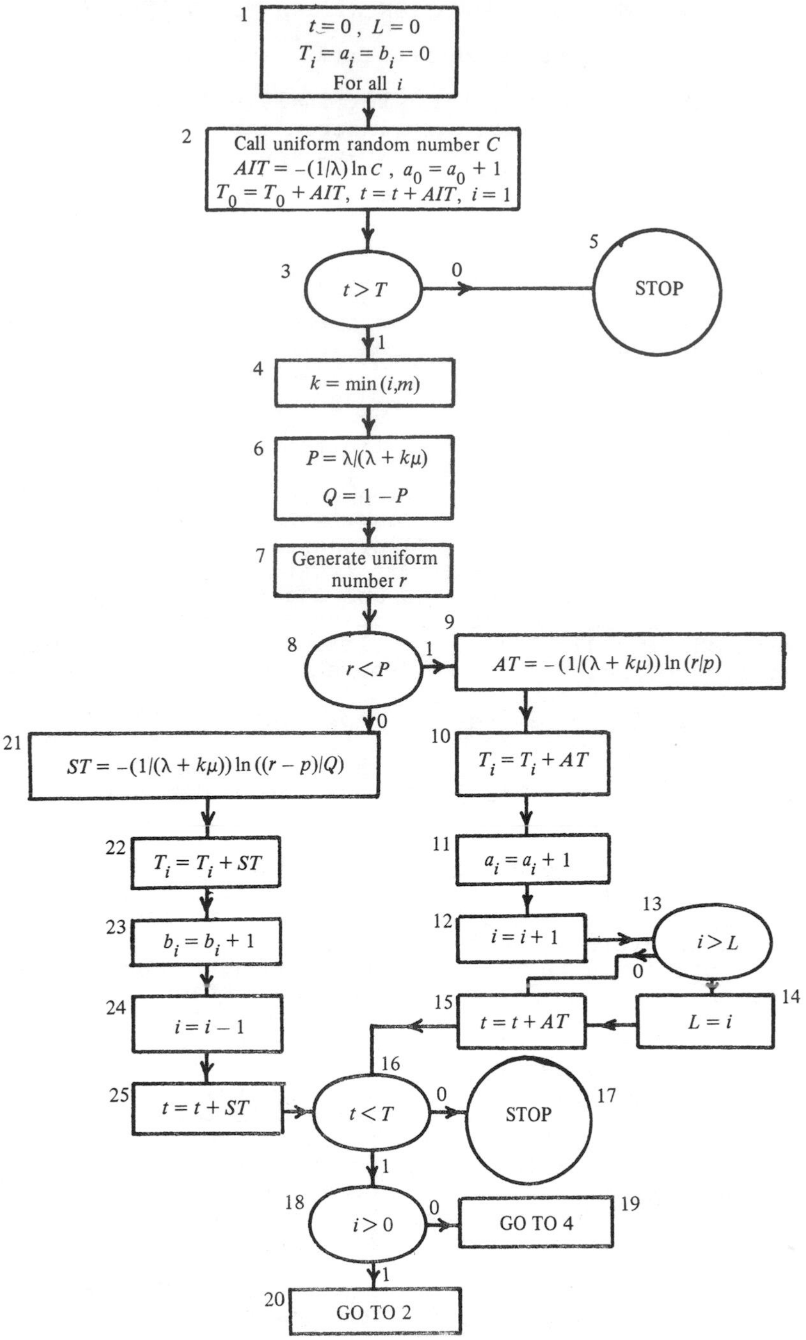

Figure 1

REFERENCES

1. Bhat, U.N. (1969) "Sixty Years of Queueing Theory". Manag. Sci. 15, B280-B292.

2. Billingsley, P. (1961) "A Statistical Inference for Markov Processes". Univ. of Chicago Press.

3. Huybrechts, S. (1965) "Inference Statistique Dans Les Proce's De Markov-Applications Dans Les Problemes De Files D'Attente". Queueing Theory Recent Developments and Applications, A Conference under the Aegis of NATO Science Committee. 127-142.

STATISTICAL COMPUTATION WITH A MICROCOMPUTER

J. Burdeane Orris, Butler University

ABSTRACT: The development of microcomputer systems in recent years has given many individuals the capability for statistical computation that would have been previously impossible. The proper role of microcomputers and their advantages and limitations is discussed. The MICROSTAT system is described as an example of the capability of a microcomputer statistical package. Sample printouts illustrate the degree of computational accuracy that can be achieved with an 8-bit microprocessor.

KEY WORDS: microcomputer, statistical computation, MICROSTAT, computational accuracy, Longley data

In discussing the role of personal computers on statistical practice, Thisted (1981) reached the conclusion that microcomputers are useful for new methods of exploratory data analysis and display. However, he also concluded that microcomputers were not very effective for traditional types of analysis involving 'number crunching'. While agreeing with his major thesis regarding new applications, I must disagree with his latter conclusion regarding the computational capabilities of microcomputers. Not only is there a need for a traditional statistics package for microcomputers, with appropriate selection of hardware and software, microcomputers can do a very commendable job with statisitcal computation.

What is a microcomputer? There is almost a continuous scale of computers varying in size and sophistication from those that cost only a few hundred dollars to ones that cost several hundred thousand dollars. There is not much point in arguing whether a given system is a 'micro', 'mini', or 'macro'. For purposes of this paper we will assume that a system that costs around $4000 or less (without printer) and is based on an 8-bit microprocessor would certainly be considered a microcomputer (or 'personal computer') by anyone's standards.

Microcomputers afford a great deal of performance for their cost and they have provided many people with computational capabilities that would have been impossible a few years ago, and even for those who have access to larger computers, microcomputers provide convenience and easy access. Relative to batch operation on a large computer, the interactive nature of microcomputers provides immediate 'turnaround'. Compared to a terminal, microcomputers provide independence from a central computer and lower cost.

Microcomputers obviously have limitations; however, these limitations are generally quantitative rather than qualitative. In other words, they may not be able to handle as much data or operate as fast, but they can provide results that are as numerically correct as large computers in most instances.

Microcomputers are constrained by limitations of operating memory (RAM) and off-line storage. The RAM of microcomputers is limited to 64K, with many users having only 32K or less, and a good proportion of this is used by systems software (operating sytem, interpreter) and the user's program. Off-line storage is usually provided by 5.25 or 8 inch" flexible disks. Although these disks have limitations in the amount of data they can store, they usually do not represent a limiting constraint. If a data file is large enough to fill a disk, the processing time and disk access time would probably be considered more restrictive than the storage limitation.

The operating speed of a microcomputer is another limitation determined by several factors. One factor inherent in any system is the clock speed. The fastest microcomputers operate at 4 MHz, but many popular systems operate at 1 to 2 MHz. Also the 8-bit microprocessors used in most microcomputers do floating point arithmetic via software rather than hardware and thus are slower. But the biggest speed constraint for most systems is the fact that they use BASIC as their primary programming language. Extended BASIC can be quite a powerful and flexible language, but since it is an interpreter rather than a compiler it is inherently slower. Although other languages are gaining in popularity, BASIC is still the most universal language for microcomputers.

The size and speed limitations of microcomputers do have certain solutions. RAM can be increased by bank selecting, disk capacity and access time can be improved by hard disks, computation speed can be improved by hardware floating point arithmetic processor boards, and programs can be made more efficient by writing in assembler or a compiled language. However, all of these 'fixes' add considerably to the cost and complexity of a microcomputer system.

Note, however, that computational accuracy was not mentioned as one of the limitations of microcomputers. Computational accuracy depends on system software and the user's program. With proper system software and careful programming, computational accuracy can be quite good as illustrated by the MICROSTAT system.

MICROSTAT

MICROSTAT (Orris, 1979; Burns, 1981) is a general purpose statistical package designed specifically for microcomputers. It was written to utilize the advantages of microcomputers while keeping in mind the limitations of such systems. In order for such a system to be useful certain 'policy' decisions had to be made before any programming was even considered. Most importantly we had to ask, "Who would use the system?" and "What programs would be most appropriate?" Our conclusion was that most users who were sophisticated enough to need advanced techniques (especially multivariate procedures) would probably have access to larger computers. Also, storage capacity and speed limitations would make certain types of analyses with large data sets intolerably slow. However, it was felt that there was a need for a microcomputer statistics package that would do standard analyses on moderate size data sets.

Typical users were assumed to be research labs, students, small colleges and businesses, and departments in larger universities without large computer budgets. There was also a desire to make the system sufficiently flexible and computationally accurate enough to appeal to statistics professionals who might find it convenient for certain types of analyses even though they had access to large systems.

To make the system functional, it had to be file-oriented. Any statistics program that requires direct data entry is very limited. By storing data in disk files it can be edited, transformed, ranked, sorted, and augmented. Thus it was decided to make MICROSTAT "semi-batch". I.e., all communication with the system and the data entry is interactive, but the analysis programs read data from previously created files which, in addition to data, contains a header indicating size, title, and variable names.

In order to appeal to the largest number of users, the system had to be written for the 'lowest common denominator' with respect to hardware and software and had to avoid using special features of particular CRT's and printers. We assumed standard hardware to consist of 32K RAM and dual 5.25" disk drives. BASIC was chosen as the language since it is most universal, and, indeed, two years ago it was just about the only language widely available on microcomputers. A parameter file allowed the flexibility of customizing the programs for different hardware configurations by storing such information as screen size, screen clear codes, form feed codes, and etc.. The parameter file is also used to maintain the name of the most recently accessed file as different programs are chained into RAM.

Figures 1 and 2 show examples of program output which illustrate the computational capabilities of microcomputers, or, to be more precise, the computational capabilities of MICROSTAT with NorthStar BASIC (8 digit precision, BCD).

The hypergeometric distribution (Figure 1) was selected as an example because it requires many computations involving very large and very small numbers, and the fact that the individual probabilities converge to 0 as the cumulative probabilities converge to 1 gives confirmation of the correctness of the computation. (Note that the output stops when the probabilities become less than .000005.) The first probability is calculated directly and subsequent probabilities are calculated recursively. The algorithm first checks for an 'easy' solution by cancelling factorials but if necessary the factorials are calculated.

Obviously the factorials involved in the example in Figure 1 are quite substantial, far beyond the standard multiplication method. MICROSTAT uses three different methods for calculating factorials. Up to 49 the standard method is used. From 50 to 300 factorials are calculated by accumulating logarithms which is simple enough, but required a bit of extra work to maintain accuracy. In summing the logs, as the integer part of the sum (representing the magnitude of the factorial) became larger, the number of digits in the fractional part became smaller thus reducing accuracy. To avoid this, after each log was calculated, its integer and fractional components were summed in separate variables. This procedure of splitting the logarithms is called the "Lincoln logs" algorithm.

Above 300, Stirling's approximation is used to calculate the log of the factorial (300 was arbitrarily selected due to speed considerations). Again there was the problem of the resulting variable being dominated by up to 6 digits representing the characteristic of the logarithm without much accuracy with respect to the mantissa. This resulted from the necessity of multiplying a very large number by a very small number. The problem was solved with some 'do it yourself' extended precision multiplication accomplished by converting the numbers to strings and 'multiplying' the strings with the product placed in another string with more than eight characters. Incidentally, the total computational time for the example in Figure 1 was 22 seconds.

Figure 2 shows the output of the multiple regression program using the Longley test data (Longley, 1966; Beaton, 1976). Accuracy is maintained by centering the data and using an algorithm specified by Forsythe (1979). The computational time for the example in Figure 2 was 38 seconds.

CONCLUSION

It must be emphasized that the above examples are for particular programs using a particular version of BASIC. Not all microcomputers would perform as well and perhaps some would perform better. A large majority of users of large computers and large statistical packages had no choice in the selection of the system, but with personal microcomputers users must become aware of the capabilities of different systems in order to make the optimal choice of hardware and software for their individual needs.

REFERENCES

Beaton, Albert E., Rubin, Donald M., and Barone, John L., The Acceptability of Regression Solutions: Another Look at Computational Accuracy. Journal of the American Statistical Association, March 1976, Volume 71, Number 353, pp 158-168.

Burns, Wm., MICROSTAT: Interactive Stats Package, IW Reports/Software, Infoworld, March 16, 1981, Volume 3, Number 5, p. 20ff.

Forsythe, Allen, Elements of Statistical Computation. Byte, January 1979, Volume 4, Number 1, p. 182-184.

Longley, James W., An Appraisal of Least Squares Programs for the Electronic Computer from the Point of View of the User, Journal of the American Statistical Association, September 1966, Volume 62, Number 319, pp. 819-841.

Orris, J. Burdeane, MICROSTAT: An Interactive Statistical Package for Microcomputers, The American Statistician, August 1980, Volume 34, Number 3, pp. 188-189.

Thisted, Ronald A., The Effect of Personal Computers of Statistical Practice. Paper presented at Computer Science and Statistics: Thirteenth Annual Symposium on the Interface. Carnegie-Mellon University, Pittsburgh, PA, March, 1981.

Figure 1. Sample output from MICROSTAT: hypergeometric distribution.

HYPERGEOMETRIC DISTRIBUTION

THE POPULATION OF SIZE 144126 OBJECTS CONTAINS 7293 POSSIBLE OCCURENCES.

THE SAMPLE SIZE IS 87

X	P(X)	CUMULATIVE PROBABILITY
0	.01090	.01090
1	.05057	.06147
2	.11596	.17744
3	.17518	.35262
4	.19611	.54873
5	.17352	.72225
6	.12638	.84863
7	.07793	.92655
8	.04152	.96807
9	.01941	.98749
10	.00807	.99555
11	.00301	.99856
12	.00101	.99957
13	.00031	.99988
14	.00009	.99997
15	.00002	.99999
16	.00001	1.00000
17	.00000	1.00000
18	.00000	1.00000

E(X) = 4.40234
STD. DEV. = 2.04379
VARIANCE = 4.17708

Figure 2. Sample output from MICROSTAT: multiple regression with Longley data.

```
--------------------- REGRESSION ANALYSIS ---------------------
HEADER DATA FOR: LONGLEY,2    LABEL: JASA, V.62, P.819-841
NUMBER OF CASES: 16    NUMBER OF VARIABLES: 7    SIZE: 4 BLOCKS

---------------------------------------------------------------
                LONGLEY DATA (UNPERTURBED)


INDEX          NAME            MEAN          STD.DEV.
  1           --X1-         101.681          10.792
  2           --X2-     387,698.440      99,394.936
  3           --X3-       3,193.313         934.464
  4           --X4-       2,606.688         695.920
  5           --X5-     117,424.000       6,956.102
  6           --X6-       1,954.500           4.761
DEP. VAR.:    --Y--      65,317.000       3,511.968

---------------------------------------------------------------
DEPENDENT VARIABLE: --Y--

VAR.    REGRESSION COEFFICIENT   STD. ERROR   T(DF=  9)     BETA
--X1-             15.0479          84.9122        .177      .0462
--X2-               -.0358           .0335      -1.069    -1.0136
--X3-              -2.0202           .4884      -4.136     -.5375
--X4-              -1.0332           .2143      -4.822     -.2047
--X5-               -.0511           .2261       -.226     -.1013
--X6-            1829.1009         455.4651      4.016     2.4796
CONSTANT  -3482157.3000

STD. ERROR OF EST. =   304.853
         R SQUARED =     .995
        MULTIPLE R =     .998

                 ANALYSIS OF VARIANCE TABLE

  SOURCE     SUM OF SQUARES   D.F.  MEAN SQUARE     F RATIO
REGRESSION   184172400.000     6   30695400.000    330.287
RESIDUAL        836420.000     9      92935.556
TOTAL        185008820.000    15

                                                   STANDARDIZED RESIDUALS
       OBSERVED   CALCULATED   RESIDUAL -2.0                   0                  2.0
   1  60323.000  60055.666   267.334  |                   |        *         |
   2  61122.000  61215.996   -93.996  |                *  |                  |
   3  60171.000  60124.709    46.291  |                   |*                 |
   4  61187.000  61597.131  -410.131  |     *             |                  |
   5  63221.000  62911.305   309.695  |                   |         *        |
   6  63639.000  63888.314  -249.314  |          *        |                  |
   7  64989.000  65153.030  -164.030  |             *     |                  |
   8  63761.000  63774.172   -13.172  |                   *                  |
   9  66019.000  66004.719    14.281  |                   *                  |
  10  67857.000  67401.602   455.398  |                   |              *   |
  11  68169.000  68186.247   -17.247  |                  *|                  |
  12  66513.000  66552.038   -39.038  |                  *|                  |
  13  68655.000  68810.561  -155.561  |             *     |                  |
  14  69564.000  69649.667   -85.667  |               *   |                  |
  15  69331.000  68989.069   341.931  |                   |          *       |
  16  70551.000  70757.774  -206.774  |           *       |                  |

DURBIN-WATSON TEST =   2.5595
```

ON THE EXACT DISTRIBUTION OF GEARY'S U-STATISTIC AND ITS APPLICATION TO LEAST SQUARES REGRESSION

by

N.N. Mikhail and L.P. Lester

Liberty Baptist College

ABSTRACT

In this paper the first four moments of U-Statistic are derived, to which a two-moment graduation shows that neither a beta distribution nor a scalar multiple chi-square distribution is a good fit to the actual distribution of U-Statistic.

On the other hand, a two moment graduation using the exact four moments of U^2-Statistic as a scalar multiple chi-square distribution is a very good approximation to the actual distribution of U^2-Statistic.

A comparison of our results with Gastwirth and Selwyn's (1980, 139) results are given.

From the computations of β_1 and β_2 for the fitted scalar multiple chi-square and for the actual distribution, we can recommend the fitted scalar multiple chi-square distribution for any statistical tests in practical situations.

Key Words: Proportions, moments, chi-square distribution, sample, population.

1. INTRODUCTION

Geary (1970) proposed a nonparametric test similar to the run test (Wald and Wolfowitz, 1940). He found that a simple count of sign changes is nearly as efficient as the familiar Durbin-Watson test (1971) of autoregression of residuals.

Gastwirth and Selwyn (1980, p. 138, 139) studied Geary's U-Statistic for the simplest regression model.

The mathematical form for the exact distribution based on the number of runs of Geary's U-Statistic when the error random variables are normal derived by Gastwirth and Selwyn (1980, p. 139) is complicated for practical purposes. In their study they have also shown that the U-Statistic is asymptotically normally distributed.

Mikhail and Lester (1981) derived the first four moments of U-Statistic which follows a hypergeometric distribution from which they have found that the exact distribution of U-Statistic has moderate values of skewness and kurtosis. They also pointed out that the approximation of U-Statistic given by Geary (Gastwirth and Selwyn, 1980, p. 138) is equivalent to the distribution of the sum of deviations from the grand median in the case of K Independent samples, and have found that both distributions are asymptotically normally distributed.

In this article, the first four moments of U^2-Statistic under non-normal assumptions are derived using the results of Mikhail and others (1976, 1979, 1981). The scalar multiple Chi-square distribution is found to be a very good approximatior of the actual U^2-Statistic, while the beta distribution is a much poorer approximation.

2. THEORY

Here we will study Geary's Statistic for the simplest regression model $y = \mu + e_i$ with each residual is $r_i = y_i - \bar{y}$

If:

$$z_i = \begin{cases} 1 \text{ if } r_i \text{ and } r_{i+1} \text{ are of different sign} \\ 0 \text{ if } r_i \text{ and } r_{i+1} \text{ are of the same sign,} \end{cases}$$

$$\text{for } i = 1, 2, \ldots, n-1 = n^* . \qquad (2.1)$$

The Geary's sign change statistic is defined by:

$$U = \sum_{i=1}^{n^*} z_i \qquad (2.2)$$

Geary's approximation is given by

$$P(U = k) = \binom{n^*}{k}/N \quad , k = 1,2,\ldots, n^* , \qquad (2.3)$$

where

$$N = 2^{n^*} - 1.$$

Using Mikhail and others (1976, 1979, 1981) we can easily see that

$$U = \sum_{i=1}^{n^*} z_i = (1) \quad \text{and} \quad E_N(U) = \frac{n^*}{N}\sum_{i=1}^{N} Z_i =$$

$$\frac{n^*}{N}(1)_N \qquad (2.4)$$

Where (1) and $(1)_N$, are power sums for a given sample of size n^*, and a finite population of size N respectively.

The **first** four moments of U-Statistic follow a hypergeometric distribution and are given in Mikhail and Lester (1981).

The exact distribution of U-Statistic obviously follows:

$$P(U = x) = \binom{NP}{x}\binom{NQ}{n^*-x}/\binom{N}{n^*} \qquad (2.5)$$

$, x = 1,2,\ldots,n^*$, which is a hypergeometric distribution with a population proportion $P = 1 - Q$ is unknown. (Using the first four moments of U-Statistic a two-moment graduation of a scalar multiple chi-square distribution is found to be the best approximation to the actual distribution when $P = 1/2$. This approximation is shown in Tables 1 and 2, to be very close or nearly identical to the hypergeometric distribution with $P = 1/2$, and also to Geary's approximation.)

The values of β_1 (skewness) $= \mu_3/\mu_2^{3/2}$, and β_2 (kurtosis) $= \mu_4/\mu_2^2$, are depicted in Table 1.

Since $U^2 = (1)^2$, is the square of a sample power sum the first four moments of U^2-Statistic are

$$m_1'(U^2) = (e_1 - e_2)A + e_2 A_2$$

$$m_2'(U^2) = (e_1 - 7 e_2 + 12e_3 - 6e_4)A + 4(e_2 - 3e_3 + 2e_4)A^2 + 3(e_2 - 2e_3 + e_4)A^2 + 6(e_3 - e_4) A^3 + e_4A^4$$

$$m_3'(U^2) = (e_1 - 31e_2 + 180e_3 - 390e_4 + 360e_5 - 120e_6)A + 6(e_2 - 15e_3 + 50e_4 - 60e_5 + 24e_6)A^2 + 15(e_2 - 8e_3 + 19e_4 - 18e_5 + 6e_6)A^2 + 10(e_2 - 6e_3 + 13e_4 - 12e_5 + 4e_6)A^2 + 15(e_3 - 7e_4 + 12e_5 - 6e_6)A^3 + 60(e_3 - 4e_4 + 5e_5 - 2e_6)A^3 + 15(e_3 - 3e_4 + 3e_5 - e_6)A^3 + 20(e_4 - 3e_5 + 2e_6)\cdot A^4 + 45(e_4 - 2e_4 + e_6)A^4 + 15(e_5 - e_6)\cdot A^5 + e_6A^6$$

$$m_4'(U^2) = (e_1 - 127e_2 + 1932e_3 - 10206e_4 + 25200e_5 - 31920e_6 - 20160e_7 - 5040e_8)A + 8(e_2 - 63e_3 + 602e_4 - 2100e_5 + 3360e_6 - 2520e_7 + 720e_8)A^2 + 28(e_2 - 32e_3 + 211e_4 - 570e_5 + 750e_6 - 480e_7 + 120e_8)A^2 + 56\cdot(e_2 - 18e_3 + 97e_4 - 240e_5 + 304e_6 - 192e_7 + 48e_8)A^2 + 35(e_2 - 14e_3 + 73e_4 - 180e_5 + 228e_6 - 144e_7 + 36e_8)A^2 + 280\cdot(e_3 - 7e_4 + 19e_5 - 25e_6 + 16e_7 - 4e_8)A^3 + 280(e_3 - 10e_4 + 35e - 44e_6 + 42e_7 - 12e_8)A^3 + 210(e_3 - 9e_4 + 27e_5 - 37e_6 + 24e_7 - 6e_8)A^3 + 168(e_3 - 16e_4 + 65e_5 - 110e_6 + 84e_7 - 24e_8)A^3 + 28(e_3 - 31e_4 + 180e_5 - 390e_6 + 360e_7 - 120e_8)A^3 + 56(e_4 - 15e_5 + 50e_6 - 60e_7 + 24e_8)A^4 + 420(e_4 - 8e_5 + 19e_6 - 18e_7 + 6e_8)A^4 + 840(e_4 - 6e_5 + 13e_6 - 12e_7 + 4e_8)A^4 + 840(e_4 - 5e_5 + 9e_6 - 7e_7 + 2e_8)A^4 + 105(e_4 - 4e_5 + 6e_6 - 4e_7 + e_8)A^4 + 420(e_5 - 3e_6 + 3e_7 - e_8)A^5 + 70(e_5 - 7e_6 + 12e_7 - 6e_8)A^5 + 210(e_6 - 2e_7 + e_8)A^6 + 56(e_6 - 3e_7 + 2e_8)A^6 + 28(e_7 - e_8)A^7 + e_8A^8.$$

Where $(1)_N = A$ and $e_s = \frac{n^{*(s)}}{N^{(s)}}$; $n^{(s)} = n(n-1)\ldots(n-s+1)$. The values of β_1, and β_2 for U^2-Statistic are given in Table 3.

3. THE DISTRIBUTION OF U^2-STATISTIC

The actual study of the first four moments of U-statistic shows that the actual distribution can follow very closely a scalar multiple chi-square distribution with $P = Q = 1/2$, using a two-moment graduation. This result confirms Geary's approximation of $P = 1/2$ in section 2. On the other hand, we have found that for different values of $P = 1/2$, the actual distribution of U-Statistic with moderate values for skewness and kurtosis does not follow a beta Model or a scalar multiple chi-square distribution.

It seemed worthwhile to examine the distribution of U^2-Statistic. We have found that the actual distribution of U^2-Statistic is very accurately fitted by a scalar multiple chi-square distribution using a two-moment graduation. The actual values of β_1 and β_2 for U^2-Statistic are given as the first value in Table 3. These values of β_1 and β_2 can be compared with their respective second and third values in Table 3, for a scalar multiple chi-square and beta distributions, respectively.

Furthermore, an approximate value for $A = N/(2 + n/5)$ is found to give a very good approximation of a scalar multiple chi-square distribution for U^2-Statistic. The values of β_1 and β_2 in the case $A = N/(2 + n/5)$ are depicted in Table 3.

3.1 Chi-Square Model

Given $U^2 = (1)^2$, for a sample of size n^* drawn without replacement from a finite population of size N, the scalar multiple of chi-square distribution with parameters ν and C is given by: $U^2 \sim C\,\chi^2_\nu$ or $U^2/C \sim \chi^2_\nu$

Where ν stands for the degree of freedom (d.f.) The p.d.f. for this model is

$$f(\chi^2) = \frac{1}{2^{\nu/2}\,\Gamma(\frac{\nu}{2})} (\chi^2)^{\frac{\nu}{2}-1} e^{-\chi^2/2} \quad ; \; 0<\chi^2<\infty$$

With U^2/C as a chi-square variable.

The first two moments of U^2 are:

$$E(U^2) = m_1 = C\nu$$
$$\mathrm{Var}(U^2) = m_2' - m_1'^2 = 2C^2\nu$$

Hence:

$$\hat{\nu} = 2m_1'^2 / (m_2' - m_1'^2)$$
$$\hat{C} = (m_2' - m_1'^2)/2m_1'$$

The values of β_1 and β_2 for chi-square model are depicted in Table 1 and 3.

4. COMPARISON OF RESULTS

In this paper the theoretical results in section 2 for U^2-Statistic are used to graduate a distribution model of which a scalar multiple chi-square distribution is found to be the best, especially when $A = N/(2+ n/5)$. This value of A is used as an empirically estimated value for the population proportion $P = A/N$. The values for β_1 and β_2 in this case are depicted in Table 3. These values for β_1 and β_2 in Table 3, for the multiple scalar chi-square distribution are very close to those for the actual distribution.

In Table 1, we are comparing the values of β_1 and β_2 in the case of $P = 1/2$ of the actual distribution, the chi-square model and Geary's approximation. The comparisons show that Geary's approximation as well as chi-square model are very close to the actual distribution of U-Statistic and this closeness becomes better as the sample size becomes larger. An important remark we would like to record is that the actual distribution of U-Statistic does not follow a chi-square or beta model when $P \neq 1/2$.

Finally Table 2 shows that $P = 1/2$ gives the best approximation whether in the case of the hypergeometric distribution, which is the exact distribution of U-Statistic, or to the binomial distribution, which is Geary's approximation. Both probability distributions for $P = 1/2$ are very close to one another, and both of them are different from the probability distribution for an exact distribution for U-Statistic based on the number of runs (Gastwirth and Selwyn, 1980, p. 139). From Table 1, we have seen that, for $P = 1/2$, the hypergeometric and binomial distribution are very close to the actual distribution. This result shows that the exact distribution for U-Statistic based on the number of runs given by Gastwirth and Selwyn (1980) needs further study.

5. CONCLUSION

We can conclude that the approximation of a chi-square model to U^2-Statistic is very good as is the hypergeometric distribution with $P = 1/2$, or Geary's binomial approximation for all practical situations of any statistical test to residuals of least squares regression.

Table 1.* For U-Statistic: The values of β_1 and β_2 for different values of N, n and $p = 1/2$

N	n	β_1	β_2
15	5	0.0000 0.8864	2.705 4.179
1.638×10^4	15	0.0000 0.5343	2.857 3.428
1.678×10^7	25	0.0000 0.4082	2.917 3.250
1.718×10^{10}	35	0.0000 0.3430	2.941 3.176
1.759×10^{13}	45	0.0000 0.3015	2.955 3.136
1.801×10^{16}	55	0.0000 0.2722	2.963 3.111
1.845×10^{19}	65	0.0000 0.2500	2.969 3.094
1.889×10^{22}	75	0.0000 0.2325	2.973 3.081

*The first value is for the actual distribution and the second is for chi-square distribution.

Table 2: Comparison of probability distributions for U-Statistic for $n = 10$, $N = 511$, and $P = 1/2$

x \ P	G - S* Exact	Approximation of Hypergeometric Distribution	Geary's Approximation
1	.010	0.017	0.018
2	.039	0.070	0.070
3	.142	0.164	0.164
4	.214	0.248	0.248
5	.281	0.248	0.248
6	.187	0.164	0.164
7	.099	0.070	0.070
8	.025	0.017	0.018
9	.003	0.002	0.002

* G - S. Exact; is an exact distribution for U-Statistic based on the number of runs, given by Gastwirth and Selwyn (1980).

Table 3.* For U^2-Statistic:
The values of β_1 and β_2 for different values of N, n and $P = 1/(2 + n/5)$.

N	n	P	β_1	β_2
15	5	1/3	1.64	6.22
			2.05	9.32
			-7.10	22.20
1.64×10^{4}	15	1/5	1.96	5.77
			1.94	8.63
			-97.00	551.00
1.678×10^{7}	25	1/7	1.91	4.70
			1.85	8.14
			47.00	-330.00
1.718×10^{10}	35	1/9	1.87	4.38
			1.81	7.92
			28.70	220.00
1.759×10^{13}	45	1/11	1.85	4.25
			1.79	7.80
			23.50	190.00

*The first value is for the actual distribution: the second value is for the chi-square distribution, and the third for the beta model.

REFERENCES

Durbin, J., and Watson, G.S. (1971), "Testing for Serial Correlation in Least Squares Regression. III," Biometrika, 58, 1-19.

Geary, R.C. (1970), "Relative Efficiency of Count of Sign Changes for Assessing Residual Autoregression in Least Squares Regression," Biometrika, 57, 123-127.

Gastwirth, J.L. and Selwyn, M.R. (1980), "Robustness Properties of Two Tests for Serial Correlation." J.A.S.A., 75, 138-141.

Mikhail, N.N., and Ali, M.M. (1976), "Sampling from Fimite Population and its Application to k-Statistics," Journal of Statistical Research, University of Dacca, Bangladesh, 10, 77-84.

Mikhail, N.N., Lester, L.P. and Weaver, T.L. (1979), "Sampling Distribution of Proportions and Percentages with Applications for Finite and Infinite Populations," Submitted to Communications in Statistics Journal.

Mikhail, N.N. and Lester, L.P. (1981), "The Exact Distribution of the K Independent Sample Median with Applications," Submitted for publication.

Wald, A., and Wolfowitz, J. (1940), "On a Test Whether Two Samples Are From the Same Population," Annals of Mathematical Statistics, 11, 147-162.

EXPOSURE TO THE RISK OF AN ACCIDENT: THE CANADIAN DEPARTMENT OF TRANSPORT NATIONAL DRIVING SURVEY AND DATA ANALYSIS SYSTEM, 1978-79

Delbert E. Stewart, Transport Canada, Ottawa, Canada

ABSTRACT - This paper is the first of a series reporting on the methodology and results of a comprehensive twelve-month, nationwide survey conducted in Canada during 1978-1979. There were approximately 22,700 households sampled using a 7-day driver trip diary recording instrument. The surveyed information consists of 3 dependent variables and 59 main independent variables classified into 5 different record types. A data analysis system was designed to provide for maximum flexibility through the implementation of three sub-systems. Part I discusses objectives, design and methodological features for both the survey and data analysis system. Subsequent parts will focus on further system enhancements, i.e. linkage with the Canadian traffic accident data base and implementation of a detailed linear modelling system for statistical analyses. This part presents exposure information for various driver, vehicle and trip variables and examines relative risk ratios that are a function of accidents/fatalities and exposure (travel distance or travel time). These "exposure-sensitive" measurements can be used to study the diverse groups of independent variables surveyed, to identify significant variations, and to provide a basis for the implementation of effective traffic safety countermeasures.

KEYWORDS - traffic safety, countermeasures, modelling, risk analyses, driving exposure, survey, transportation, information system, data base, data analysis.

OBJECTIVES - The main objectives of this study are to obtain a nationally representative estimate of the extent of travel by Canadian motorists and a definitive description of the driver-vehicle-environment patterns existing within the transportation system. This information can be linked with the equivalent accident data to compute measurements of accident risk[1] that are "exposure sensitive". These indexes can then be used to analyze accident involvement rates for the diverse groups of driver-vehicle-environment independent variables, to identify significant variations, and to provide a basis for the implementation of effective countermeasures(1).

SURVEY METHODOLOGY - The methodology was designed by Statistics Canada in consultation with Transport Canada and Canadian Facts Limited which performed both the data collection and the data capture procedures(2). The study commenced in May of 1978 and was completed by April of 1979. The target population consisted of all licensed drivers, 16 years of age and older, resident in Canada. The vehicle target population was comprised of all passenger cars, motorcycles with engine size of 50 C.C. or larger, and all other motor vehicles with a Gross Vehicle Weight (GVW) of less than 10,000 lbs. Since seasonal variations in vehicle miles of travel exist; provincial differences were expected; and further variations among urban/rural population density classifications were anticipated; the national sample was drawn over the period of one full year with stratification by month, province, and population density classifications. The stages of sample selection were as follows: first enumeration areas (EA's) were selected, then blocks (clusters) within EA's were chosen and finally households within blocks were sampled. Individuals within households were selected in a random manner from a listing of eligible drivers. To ensure adequate representation of drivers between the ages of 16 and 24 inclusive, an over-sample of this age group was taken. By randomizing the trip recording start-dates of respondents throughout the days of the month it is possible to avoid confounding a non-response bias with day of the week.

SAMPLE SIZE DETERMINATION - The main dependent variable to be estimated was Driver Miles of Travel (DMT). Based on the parameters including design effect, t-statistic, population standard deviation and population mean, sample sizes necessary for relative margins of error equal to 5%, 10% and 15% were computed(2). The results are given in Table 1.

Table 1: Sample Sizes for 5%, 10% and 15% Relative Margins of Error

Relative Margin of Error (%)	Sample Size
5	968
10	242
15	108

Therefore it appeared that an overall sample size of 9,000 driver respondents would produce sufficiently reliable information for most two-way and three-way cross-tabulations and some cells of four -way cross tabulations. Furthermore, the total national annual DMT would have a relative margin of error equal to 1.6% with 95% confidence, given a sample size of 9,000.

MEASUREMENT METHOD - The method of approach and measurement instruments have been documented in detail (3,4). The survey approach was as follows: a letter of introduction was handed to an adult household member; a listing of all household members possessing a valid drivers' license was compiled; the randomly selected driver(s) was contacted and driver/vehicle information recorded on a questionnaire; the Trip Diary was introduced

[1] See references (4, 6, 7, 8, 9, 10)

and explained to the respondent; an appointment was arranged to review and retrieve the Trip Diary after the 7-day recording period had elapsed; a telephone call was made to the respondent after the first three days of trip recording to encourage and motivate the respondent, re-iterate the importance of the information, answer any question, resolve problems, and supply respondents with more Trip Records (if required); and finally, the interviewer visited the respondent to retrieve the completed Trip Diary and record, wherever possible, any missing data gaps. Interviewers were permitted three successive appointments for both the initial contact and trip diary retrieval with a respondent.

PRE-TEST - During the period March 6-24, 1978, a pilot test of the survey measurement methodology and instruments was carried out. A total of 52 blocks including 46 urban and 6 rural were selected in four cities of southern Ontario and all but one were visited. Since 10 households were approached per block, there were a total of 510 households selected, and information was retrieved from 509. Upon completion of the pilot test, difficulties encountered were discussed between interviewers and respondents resulting in a number of significant changes to both the methodology and instruments. The non-response analysis of the pre-test resulted in sample design changes to ensure that the pre-determined sample size of 9,000 driver/respondents completed diaries would be obtained. Therefore the number of EA's was increased by 20% (1,500 to 1,800) and the number of households per block to be selected increased from 10 to 12.

MAIN SURVEY - The survey utilized all procedures and instruments described earlier, with appropriate modifications resulting from the pilot test. Field supervisors attended a comprehensive two-day briefing covering all field work procedures and operations (5). Through telephone interviews by both field supervisors and consultant managers, recorded information for one in every three respondents was verified. To verify the correctness of geographical locations and household listings a large sample of locations were randomly selected and visited by Head Office Staff and Field Supervisors. The results of the checks showed that only 15 locations of the 1,894 selected were omitted from the sample and there were slight delays in diary recording start dates for only 171 of the remaining 1,879 locations in the sample.

PROCESSING - The consultant was responsible for the data processing of the "Household Questionnaire", "Participant Questionnaire" and the "Trip Diary Records" (these three documents constitute the computer records) and delivering a clean computer tape each month to Transport Canada. The editing, coding, keypunching and computer editing procedures have been thoroughly documented (15).

RESPONSE ANALYSES - Since stratification by month, province and population density classifications was implemented, and record types include household, driver, vehicle, diary and trip information (with numerous variable categories for each type of record) many types of response rates can be examined (15).

The actual and expected sample allocations by province - month strata indicate, through the use of the X^2 goodness-of-fit test at the .05 level of significance, that the actual sample size was lower than expected. Significant over-reporting was experienced during May and June while the winter months, December to April inclusive, showed significant under-reporting. On a provincial level, significant over-reporting took place in Saskatchewan, British Columbia and New Brunswick with all other provinces experiencing significant under-reporting.

Figure 1 indicates that the household response rate at the Canadian level was .633; that is, there was at least one driver identified in 63.3% of the 22,716 households sampled. Seasonal variations were not too significant with the lowest rate of 59.1% occurring in December and the highest rate (67.0%) in March. The household response rates by province however revealed variations more significant in nature with the highest recorded in Prince Edward Island (75.3%) and the lowest, 55.1%, took place in Nova Scotia.

Figure 1: Household Response Analysis for Canada May 1978 - April 1979

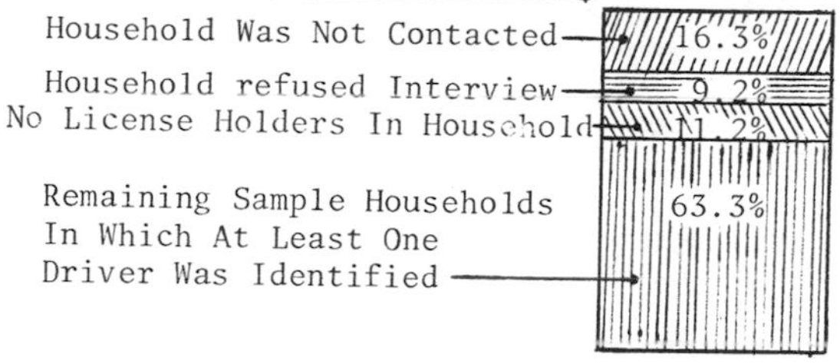

From the 14,379 responding households in which at least one driver was identified, 15,961 potential driver/respondents were selected to participate in the survey (Figure 2). There were 2,573 (16.1%) that did not complete the first interview and another 2,440 (15.3%) selected respondents who refused to accept the diary. Therefore trip diaries were placed with 10,948 driver respondents representing 68.6% of the initial 15,961 selected for the survey. There was a total of 743 drivers who could not be reached to retrieve the trip diary and of the remaining 10,205 total diaries retrieved, 8,773 (86.0%) had diary reviews that were complete. This represents 55.0% of the original potential respondents selected for the survey.

Figure 2: Driver Response Analysis for Canada May 1978 - April 1979

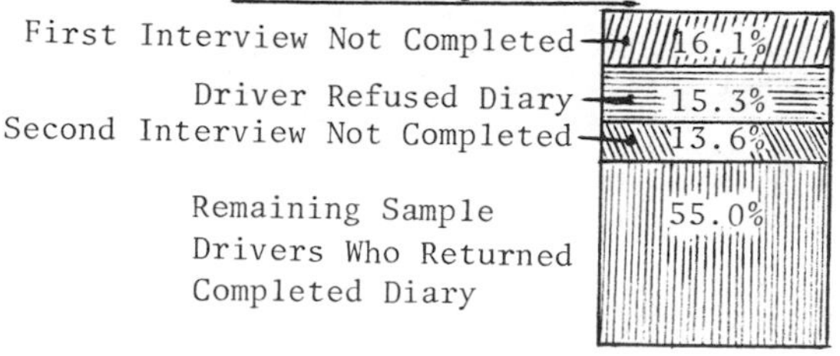

The variation in response rates due to seasonality showed marginal effects and the trip diary refusals were highest during winter months. Since younger drivers have been observed to be involved in a disproportionately larger number of accidents relative to their representation in the driver population, more younger drivers were observed than might be expected through the use of an over-sample on the age group 16-24. The over-all response rates (respondents who returned completed diaries) for the two age groups in the main sample showed negligible variation, 56.2% for the "16-24" age group and 56.0% for the "over 24" age group. The 16-24 over-sample response rate however was considerably lower with only 44.9% of the initial respondents selected returning complete diaries. The main factor contributing to this significantly lower response rate was the inability to make contact, i.e. 31.7% of the 1,573 respondents identified to participate in the 16-24 age group over-sample never completed the first interview.

WEIGHTING PROCEDURES - The weighting methodology implemented in the survey has been documented in detail (2,3,15) and its' verification has been addressed (15). The main aim of the survey was to estimate the total distance driven throughout Canada, limited only by the vehicle class restrictions imposed. Comparing the National Driving Survey weighted percentage distribution of drivers by province with that of Statistics Canada revealed figures of negligible difference.

DATA ANALYSIS SYSTEM - The design, development and implementation of a (sequential) data base system was carried out by Systems Approach Consultants Ltd. (SACL) to provide for further processing of the information (15). Specifically, the main criteria for system design required: (i) a flexible, yet efficient, multi-dimensional tabular reporting capability, (ii) response/non-response data analyses (with province selections), (iii) the ability to evaluate the sampling error for any tabulation request, and (iv) a linkage of SPSS with the data base to permit various levels of data analyses. Complete documentation for the data analysis system has been prepared (11,12,13,14).

SYSTEM DESIGN - The final system design permits production/analysis runs via RJE using either T.S.O. or standard batch processing. Standard batch submission is recommended for tabular reports analyses (e.g. SPSS, coefficient of variation analyses, linear modelling, etc.), while T.S.O. is recommended for production jobs in data manipulation (e.g. editing, massaging, and copying of data, etc.). The data analysis system is presently comprised of 3 sub-systems, namely data manipulation and data base up-date, reporting systems, and SPSS linkage with the data base. Figure 3 gives an overview of the system design features and data flow.

FUTURE SYSTEM ENHANCEMENTS - A linear model system has been developed and tested (16). There are plans to link this model building capability to the system to provide a means of studying significant variations and interactive effects among the independent variables. The model provides a number of capabilities, a few of which include: multi-variate analysis (analysis of variance), graphs, residual analyses, component effect analyses, hypotheses testing, confidence interval estimation.

Methods are being considered to link the National Driving Survey Data Base with the Department of Transport national data file composed of all reportable motor vehicle traffic accidents for the years 1974 to 1979 inclusive. Exposure information merged with equivalent accident information will provide a basis upon which to compute "accident risk ratios" that are exposure sensitive.

Figure 3: National Driving Survey System Design

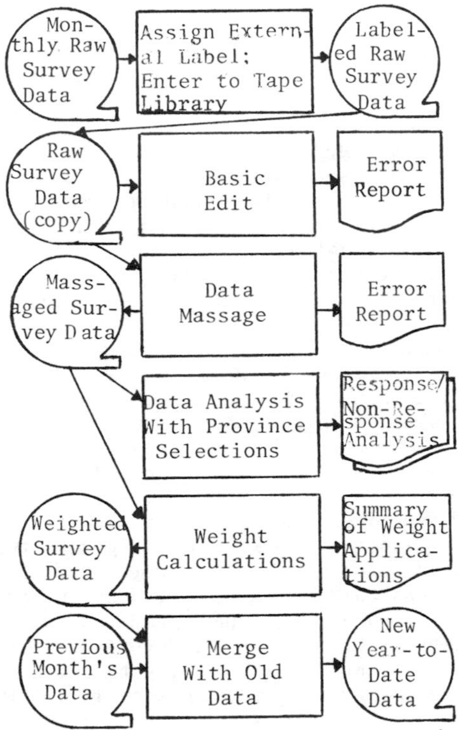

DATA ANALYSES - The analyses of the National Driving Survey Data Base is still in its preliminary stages with special tabulations being produced to perform descriptive analysis. Some of the key variables presently being studied are driving exposure by: sex, age, driving experience, vehicle size, day of week, month, province and trip purpose.

Examination of driving exposure by driving experience shows that 48.4% of the total annual kilometres driven are recorded by license holders with at least 15 years of driving experience.

Almost half, 46.7%, of the total kilometres driven in Canada annually are attributable to the 25-44 driver age group. In fact, 75% of all kilometres driven are due to license holders between the ages of 25 and 64 inclusive, 6.5% belong to the 16-19 age group, 13.1% for the 20-24 age group and 5.3% for the age group 65 and over.

The total annual kilometres driven within Canada in various vehicle sizes showed the following percentage distribution: large - 27.53%, compact - 19.03%, intermediate - 18.50%, light truck - 18.16% subcompact - 15.53%, specialty - 0.68%, motorcycle - 0.31%, not stated - 0.23%.

Driving exposure by day of week indicated a small increasing trend from Sunday to Friday with Sundays accounting for 12.27% of the total annual kilometres driven and 16.26% occurring on Fridays. There are 14.4% of the total annual kilometres driven on Saturdays.

Analyzing driving exposure by trip purpose revealed that 27.6% of the total annual kilometres were attributable to social reasons, 26.5% for commuting purposes, 20.7% for personal/family business, 12.7% for business travel and 12.5% were due to shopping trips.

RELATIVE RISK RATIOS - Exposure information provides a standardized measurement unit with which to compute relative risk ratios for fatalities/injuries/accidents, thereby yielding indexes that are "exposure sensitive".

Driving exposure by sex of driver shows that male drivers out-distance female drivers, with males accounting for 71.4% of the total annual kilometres driven. Statistics Canada (17) have reported that 82% of all motor vehicle accidents in Canada in 1975 involved males. The relative risk ratios are computed as follows:

$$RRR_j = p_{1j}/p_{2j} \quad \text{where,}$$

RRR_j = the relative risk ratio for the jth class,

p_{1j} = the proportion of total accident involvements for the jth class,

p_{2j} = the proportion of driving exposure for the jth class.

The resulting indexes are 1.15 and 0.63 for males and females respectively. From this one would be lead to believe that males are worse drivers than females. Perhaps other independent variables such as age groups, type of vehicles, driver education, driving occupation, roadway conditions, trip purpose, trip times, etc. may shed light on this issue. That is, the combination(s) of factors experienced by males in their driving patterns may or may not indicate a subjection to greater risk?

The National Driving Survey, therefore, contains a wealth of information available for research and analysis. It provides a flexible capacity to produce detailed reports of household, driver, vehicle, diary and trip characteristics. Through the use of this system it is hoped to identify some of the exposure variables that are most important in determining accident risk, and provide a basis for the implementation of effective countermeasures.

REFERENCES

1. Rochon, J., L. Swain, and S. O'Hara: Exposure to the Risk of an Accident: A Review of the Literature, and the Methodology for the Canadian Study, Ottawa, Department of Transport, September, 1978.
2. Swain, L.: National Study on Exposure to the Risk of an Accident: Final Sample Design Specifications, Ottawa, Statistics Canada, May,1978
3. Swain, L.: National Study on Exposure to the Risk of an Accident (National Exposure Study): Estimation Procedures, Ottawa, Statistics Canada, May, 1978.
4. Foldvary, L.A.: "Road Accident Involvement Per Miles Travelled - I", Accident Analysis and Prevention, Vol. 7, pp. 191-205 (1975).
5. Lynn, A. and J. Rochon: Exposure to the Risk of an Accident: A Description of the Data Collection and Processing Procedures, and Analyses of Their Effectiveness, Ottawa, Department of Transport, August, 1979.
6. Hauer, E.: On Some Research Needs in Safety Measures of Effectiveness, University of Toronto, Department of Civil Engineering, December, 1978.
7. Foldvary, L.A., D.W. Potter, and M.H. Cameron: Accident Risk Due to Age and Year of Manufacture of the Car, Australian Road Research Board
8. Lawson, J.J.,: Three Viewpoints on the Collection of Travel Behaviour Data Over Time: II: The Canadian Department of Transport National Driving Survey, Ottawa, Department of Transport, May, 1980.
9. Leitch, K., G. Middleton, and L. Mudge: "Characteristics of Road Travel in Queensland 1973", ARRB Proceedings, Vol. 8, pp. 15-22 (1976).
10. Carroll, P.S., William L. Carlson, and Thomas L. McDole: "Identifying Unique Driving-Exposure Classifications", HIT LAB Reports, 1971.

11.,12.,13.,&14.: 1978-1979 National Driving Survey, Processing System User Guide, Processing System Production Manual, Processing System Program Documentation, Processing System Program Source Code, Road and Motor Vehicle Traffic Safety Branch, Transport Canada, Ottawa (1979).

15. Systems Approach Consultants Ltd., Development and Implementation of a Data System for the National Driving Survey, Prepared for Road and Motor Vehicle Traffic Safety Branch, Transport Canada, Ottawa, March, 1979.
16. Stewart, D.E.: A Linear Model Programming System for Use in Experimental Design Analyses, Ottawa, Department of Transport, April (1975), Revised in February (1981).
17. Statistics Canada, Motor Vehicle Traffic Accidents, 1975, Cat. No. 53-206, Statistics Canada, April (1977).

CONCOR: AN EDIT AND AUTOMATIC CORRECTION PACKAGE

Robert R. Bair, U. S. Bureau of the Census

For over ten years the International Statistical Programs Center (ISPC) of the U.S. Bureau of the Census has been involved in the development and dissemination of generalized computer software products for use by statistical organizations in developing countries. In response to critical needs for improvement in data processing capabilities during the 1970 World Census Program ISPC developed a general cross-tabulation system (CENTS) which has been continually enhanced over the years and is presently installed at over 90 computer centers worldwide.

To meet similar software needs in the developing world for the 1980 World Census Program ISPC recently released a new version of CONCOR (Consistency and Correction System), a generalized tool for editing and automatically correcting census and survey data. CONCOR provides a powerful, high-level command language which allows the computer or subject-matter specialist to easily develop executable edit specifications utilizing techniques like interrecord inspection, hot deck, imputation, and tolerance control.

KEYWORDS: Statistical Computing, Software Packages, Editing, Imputation, Tolerance Control.

1. INTRODUCTION

During the mid-1960's, as the 1960 World Census Program drew to a close, it became increasingly clear that statistical organizations holding the responsibility for collecting, processing, and reporting the results of their censuses were being thwarted in their attempts to produce reports in a timely fashion by a common bottleneck. That bottleneck was data processing. The combination of lack of computing equipment and the shortage of programming expertise in organizations which had computing equipment stalled many efforts in getting out the population counts in a reasonable period of time. In fact, some statistical offices abandoned attempts to produce reports from their census after nearly ten years of effort because of insurmountable processing problems.

In the early 1970's the development of generalized tabulation packages such as CENTS and COCENTS by the Census Bureau, and X-TALLY by the United Nations Statistical Office provided evidence that dramatic increases in statistical computing capability could be realized with the introduction of this type of software tool. CENTS, COCENTS, and X-TALLY enabled organizations with the smallest of computer and programmer resources to proceed a long way in publishing cross-tabulated statistical reports. Nevertheless, it was still evident at the end of the 1970 World Census Program in the mid-1970's that the data processing bottleneck had not been fully resolved; many countries were still unable to publish census results after years of effort, and some even gave up trying.

Unfortunately, what was not available until very recently was software for the validation and correction of data prior to the tabulation process. As early as 1974-75, meetings were held to address the obvious need to provide standard editing software in time for the 1980 round of censuses. Existing packages were examined and each, for different reasons, was deemed unacceptable for use worldwide.

Representatives of various international organizations grouped resources and began to design and implement the software. Computer specialists of the United Nations Statistical Office developed the UNEDIT system to serve the editing needs of organizations with selected small to medium-sized computers.

Through a cooperative effort among the United Nations Latin American Center for Demographic Studies (CELADE), the International Statistical Programs Center (ISPC) at the U.S. Bureau of the Census, and the Office of Population of the U.S. Agency for International Development it was decided to develop a COBOL version of the CONCOR editing system which CELADE had earlier produced in IBM assembler language. This COBOL version of CONCOR is designed to address the complex editing needs of users having medium to large scale equipment.

In January 1980 the latest version of COBOL CONCOR (Version 2) was released by ISPC at a workshop in Washington. The international community of statistical software users represented at that workshop were in agreement that this software would be extremely useful for census and general survey editing in the developing world.

2. HISTORY OF DEVELOPMENT

The CONCOR editing system was originally developed at CELADE (the United Nations Demographic Center for Latin America) to meet the needs of national institutions processing censuses and sample surveys. The original CONCOR language (called Version 0) was completed by CELADE in September 1975. This version was written in assembler language for IBM System 360 and 370 computers. It was used by CELADE for the editing of various surveys and for an editing test on the population census of Haiti. CELADE also used CONCOR for the editing and correction phase of the World Fertility Survey (WFS) for the Dominican Republic in late 1975. The system has since been used for other WFS surveys by CELADE and has been adopted by the World Fertility Survey staff in London as the primary editing system to be used on WFS surveys.

The International Statistical Programs Center (ISPC) of the U.S. Bureau of the Census had for some time been interested in producing an editing package for censuses and surveys that would operate on the many different computer systems located in countries around the world. In mid-1976, ISPC decided that the best model for such an editing system would be the CONCOR package. Based on the experience of CELADE and the World Fertility Survey, Version 1 of CONCOR was the result of the cooperation between CELADE and ISPC in producing an enhanced editing and imputation tool.

Version 1 of CONCOR was written entirely in COBOL and contained some improvements over the assembler language version which were determined by both CELADE and ISPC. The more important of these improvements included an expanded data dictionary, a more readable command language, the allowance of variable names up to forty-five characters in length, and some new editing capabilities.

Version 1 of CONCOR was a joint product of the Computer Services Division of CELADE and the Computer Methods Laboratory of ISPC. The work was performed under a Resources Support Services Agreement with the Office of Population, Bureau for Development Support, U.S. Agency for International Development.

In mid-1979 after several regional workshops had been given by ISPC on the use of CONCOR Version 1, it became increasingly apparent from feedback of workshop participants that Version 1 was not yet an adequate tool for computerized editing of national census data. A review of these comments and evaluations prompted ISPC to commence an in-depth analysis of the design of Version 1, taking into consideration its capabilities, its reliability, and its efficiency. Based on the findings of this study, ISPC developed a proposal to implement major modifications to Version 1, both adding new capabilities and upgrading reliability. The U.S. Bureau of the Census provided financial support for this implementation.

The current release of COBOL CONCOR, Version 2 (December 1979), represents a virtual total reimplementation of the system. Version 2 represents a rewrite of more than 95% of Version 1 comprising 20 programs totaling approximately 48,000 lines of COBOL code.

3. FUNCTION OF PACKAGE

The CONCOR (for 'CONsistency and CORrection')data editing system is designed to be a general purpose software package for the identification and correction of invalid or inconsistent data in various types of censuses and surveys. It is capable of identifying errors in the structure of the responses to an individual question, and consistency between responses to different questions, involving intrarecord and interrecord relationships within the questionnaire. CONCOR can be thought of as a language for preparing files for tabulation and analysis.

Although CONCOR is a relatively new language for data cleaning, the concepts addressed by the language design should be familiar to all those acquainted with the problems of editing census and survey data.

The following editing concepts were incorporated into the design of the CONCOR system:

Structural Checks:

These checks are designed to ascertain whether all the records that should be present for a particular questionnaire or interview are in fact supplied. The language constructs of this version of CONCOR do not address this problem directly, but do provide the facility to make the required structural determinations.

Range Checks:

These checks are designed to find out whether a variable has a value that is outside the limits specified for that variable. These checks are concerned with one variable at a time. This step is often known as 'data validation'.

Consistency Checks:

In these checks two or more variables are compared and the consistency of their values checked. CONCOR provides the tools so that these variables may be in the same record, may be in different records, or may be computed values. A variable may have a correct value when checked on its own, but when compared with other related variables may be found to be in error. An example of this would be a woman 16 years old who reports 10 children born alive. Both variables satisfy the individual range checks for that variable, but taken together it is obvious that a mistake has occurred, either in the reporting of age or the reporting of fertility.

Automatic Corrections:

When censuses or large surveys are being processed, the user will find it essential to have the computer make some or all of the corrections automatically. Routines are included in the system to handle imputations based on a 'hot-deck' approach.

CONCOR Outputs:

The aim of the editing operation is the creation of an 'error-free' output file and a report of the actual errors found in the data and the corrections made to the data. CONCOR can provide all of these as described below:

"Error-Free" Output File:

CONCOR can automatically create an output file containing the data records from the input file with automatic corrections having been applied. This file will be identical in format to the input data file and provides the user with the ability to re-enter CONCOR with the 'edited' file to inspect the corrected data for possible inconsistencies introduced during the correction process (edit recycling). This is an invaluable tool for testing the user's edit specification statements in the CONCOR program before initiating production processing of the actual input data file.

Auxiliary Output File:

CONCOR provides the ability for the user to specify the creation of a special output file which may contain any conbination of input data items, CONCOR internal identifiers, and computed variables. This allows the user to create derevative files containing recoded data or original variables output in a new format. This procedure also permits the creation of 'flat' (non-hierarchical) files.

Statistical Reports:

CONCOR will generate comprehensive statistics about the edit tests performed on data items as well as the errors detected. This release of CONCOR allows the user to specify the level and organization of statistics produced as well as performance of quality tolerance checks against computed error rates.

4. DESCRIPTION OF SYSTEM

To the user the CONCOR system can be properly viewed as four logical subsystems; each subsystem consists of one or more COBOL programs which when executed together perform a major function in the system.

Command Language Analyzer Subsystem:

This subsystem is responsible for the inspection and validation of the user's command language statements which are coded to define the attributes of the data file to be processed (the Dictionary Division) and to define the edit specifications which form the algorithm for data inspection and error correction (the Execution Division).

The user's command statements are read by a parsing routine which separates each statement into recognizable component elements. The classified elements are then read by an analyzing routine which reconstructs the elements into statements and scrutinizes the syntax of each statement. These routines issue detailed messages when user coding errors are uncovered and a more comprehensive explanation of the cause of the error and the probable solution can be found in the CONCOR Diagnostic Message Guide. The cross-reference listing produced by this subsystem lists alphabetically each user-identifier defined in the Dictionary and the source statement line number(s) on which each identifier is referenced in the Execution Division. The parsing and syntactical analysis routines of this subsystem assure users that if their command statements passed through this analysis without producing diagnostic messages, the source COBOL program generated by the following CONCOR subsystem will not produce COBOL compiler errors.

COBOL Source Program Generation Subsystem:

This subsystem is responsible for translating the user's CONCOR command language statements into basic COBOL code. Generally, it can be stated that CONCOR transforms the user's Dictionary Division command statements into the Identification-Division, Environment-Division, Data-Division, and the input/output control sections of the Procedure-Division of the generated source COBOL program. Likewise the user's Execution Division command statements are transformed by CONCOR into the executable logic of the Procedure-Division of the generated source COBOL program. This generative approach to executing user commands has several major advantages over the interpretive approach in that it creates a custom program tailored to the user's needs and, therefore, executes more efficiently on the computer and requires only minimum amount of primary storage in which to reside.

EDITOR Subsystem:

This subsystem is responsible for the reading and editing of the input data file; the accumulation of edit statistics; the creation of an edited output data file and/or a user-specified output data file; and the generation of a listing of execution control statistics.

The link-edited COBOL program (called EDITOR) which is generated out of the previous subsystem is, by itself, the EDITOR subsystem. The execution of this COBOL program is the objective of the CONCOR system; that is, the inspection of data items on the input data file and the automatic correction of bad or missing items encountered. The command statement routines coded by the user in the Execution Division are the basis for the Procedure Division of this program, which controls the execution logic commands against the input data.

The EDITOR program controls the creation of an edited output data file which contains the automatic corrections made to the data. This file is produced in the same format as the input data file. EDITOR also controls the creation of a special output data file the format and content of which is left entirely to the discretion of the user. As the inspection and correction of the data is taking place, EDITOR accumulates user-specified statistics about the quality of the data and the degree and type of corrections applied. This internal system information is passed to the next subsystem, the Report Subsystem, to produce user-specified statistical reports.

EDITOR also automatically maintains control counts of the number of questionnaires and records read, the number of questionnaires and records written to the output files, and the number of execution error conditions that occurred in the run.

REPORT Subsystem:

This subsystem is responsible for the inspection and validation of the user's command language statements which are coded to control the creation of edit statistic reports, the calculation of error rates on the data edited, and performance of tolerance checking on individual work units or user-defined control areas.

The user's command statements are read by a parsing routine which separates each statement into recognizable component elements. The classified elements are then read by an analyzing routine which reconstructs the elements into statements and scrutinizes the syntax of each statement. These routines issue detailed messages when user coding errors are uncovered.

Depending upon the user-specified edit statistics generated in the EDITOR program and the command language statements coded in the Report Division, this subsystem will produce listings reporting statistics about the frequency of edit tests made, the frequency of failed edit tests, and the percentage of cases failing tests. The user controls the organization of the reports produced. They may be summarized by edit commands or user-identifiers, or itemized by each occurrence within a questionnaire or interview.

This subsystem will also calculate the rate of error encountered over all edit tests performed in EDITOR. The user can specify a tolerance level which will be compared against the error rate, and this subsystem will indicate which control area(s) from the input data file had a rate of error greater than the specified tolerance. This subsystem will capture, at the user's request, the identification codes of each control area failing the tolerance check.

5. COMPUTER REQUIREMENTS

COBOL CONCOR has been designed in such a manner as to be easily converted for operation on virtually every computer system which as available:

- a COBOL compiler
- 128K, 6 or 8 bit characters (bytes) of primary storage
- 4 million characters (bytes) of direct access disk storage
- a utility sort program

As of this writing COBOL CONCOR Version 2.2 is operational on the following equipment:

- IBM 370/OS, DOS
- IBM 30XX
- IBM 43XX
- IBM plug compatibles
- HONEYWELL BULL
- NEC 500 (ACOS-4)

Versions of COBOL CONCOR to be available in the near future are:

- UNIVAC 1100
- ICL 1900/2900/2950
- NCR CENTURY
- NCR 8200
- DEC SYS/10
- WANG VS-100

6. ADVANTAGES

COBOL CONCOR has specific features which allow the user to save time in developing an edit program. Also, the package provides for easier maintenance of the user's program. These features include:

- A data dictionary system for one time definition of the data file;
- Edit commands which automatically inspect data items, maintain statistics on errors found, and make corrections where desired;
- Edit diaries which allow the user to choose the level and organization of statistics captured and displayed;
- Data management facilities which control all houskeeping aspects of presenting each questionnaire or interview to the user's program for inspection.

7. PROCESSING SPEED

In benchmarking COBOL CONCOR it was found that the system can process data at a rate of 345 logical records (card images) per cpu second. This test was executed on an IBM System 370/168. The user program contained 571 command statements describing the attributes of the data file and 1377 command statements defining the edits to be done. Each interview on the file contained 82 unique data items.

8. POTENTIAL USE

At present, approximately 64% of developing countries have a computer which can in theory support CONCOR. These installations are constantly being upgraded and the percentage of potential users should rapidly increase over the next few years. As is true of any newly released software product only the test of time will tell how useful, reliable and well accepted the CONCOR system will be. ISPC is looking forward to participating with counterpart statistical organizations around the world in the applied use and evaluation of this software over the next several years.

BGRAPH: A PROGRAM FOR BIPLOT MULTIVARIATE GRAPHICS

Michael C. Tsianco, K. Ruben Gabriel
Charles L. Odoroff and Sandra Plumb
University of Rochester

Abstract. BGRAPH (Tsianco, 1980) is an interactive conversational program to perform biplot multivariate graphics. The program generates two- and three-dimensional biplot displays based on the singular value decomposition (SVD) of a matrix and the resulting rank 2 or 3 approximations. Three dimensional displays may be either orthogonal projections, perspective projections, stereograms or analyglyphs. Other capabilities of the program include subset selection, selective labeling of points, rotation of axes, construction of ellipsoids of concentration, MANOVA biplots and plot storage.

Keywords: Biplot, Singular Value Decomposition, Three Dimensional Graphics, Perspective Projection, Stereograms, Analglyphs, Multivariate Statistical Analysis.

1. Introduction. Biplots (Gabriel, 1971, 1980, 1981) have been found useful in eyeballing data and in diagnosing models to fit data (Bradu and Gabriel, 1978).

To biplot the matrix Y, BGRAPH inputs the matrices P, Λ, Q of the SVD which satisfy $Y = P'\Lambda Q$, $PP' = QQ' = I_y$ (y = rank Y) and Λ being diagonal. A flexible set of commands offers the opportunity to manipulate the form of the graphical output, rotate displays, project, transform, label row and column markers, and select subsets of row and column markers for display.

Display 2.1

Legend: ● $\underline{a}_u$ is u-th row marker; ↗ $\underline{b}_v$ is v-th column marker

$$\underset{Y}{\begin{bmatrix} 2 & 2 & -4 \\ 2 & 1 & -3 \\ 0 & -1\frac{1}{2} & 1\frac{1}{2} \\ -1 & -\frac{1}{2} & 1\frac{1}{2} \end{bmatrix}} = \underset{A}{\begin{bmatrix} 2 & 2 \\ 2 & 1 \\ 0 & -1\frac{1}{2} \\ -1 & -\frac{1}{2} \end{bmatrix}} \underset{B'}{\begin{bmatrix} 1 & 0 & -1 \\ 0 & 1 & -1 \end{bmatrix}}$$

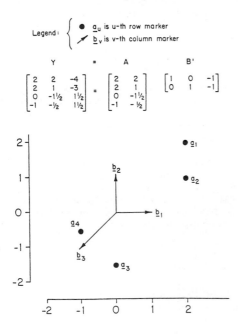

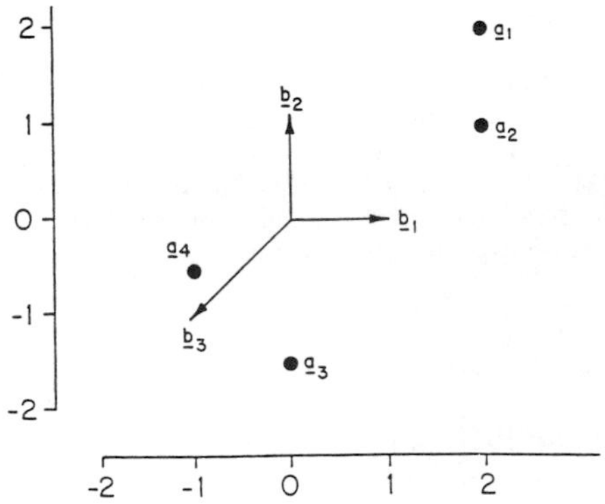

2.1 A Simple Example. Display 2.1 shows a matrix $Y_{(4x3)}$ which can be factorized as $A_{(4x2)}\ B'_{(2x3)}$. The biplot displays the rows of A, and of B. The first row of A, i.e., the vector (2,2) is displayed as the point $\underline{a}_1$; the second row (2,1) is displayed as the point $\underline{a}_2$, and the other two rows as points $\underline{a}_3$ and $\underline{a}_4$. The columns of B' are displayed as arrows $\underline{b}_1, \underline{b}_2$, and $\underline{b}_3$.

2.2 Factorization, Approximation and Display. The matrix Y could be biplotted exactly because it was of rank two. An exact biplot is possible only of matrices of rank one or two, because the biplot itself is planar. For a matrix of higher rank, several steps have to be taken in order to display it by an approximate biplot. The first step approximates the matrix Y by a matrix $Y_{[2]}$ of rank 2. The second step factorizes $Y_{[2]}$ as a product of $A_{(nx2)}$ and $B'_{(2xm)}$. The third step is to use each row of matrix A as a row marker $\underline{a}$ and each column of matrix B' as a column marker $\underline{b}$. These markers are then plotted as an approximate biplot of the original matrix Y.

We next consider each of these three steps. The best known method for lower rank approximation is the Householder and Young least squares method for fitting rank two $Y_{[2]}$ to Y. Factorization of $Y_{[2]}$ is possible; matrices $A_{(nx2)}$ and $B_{(mx2)}$ that satisfy $Y_{[2]} = AB'$ always exist. However, such a factorization is not unique. This has some advantages for the statistician, who may choose a factorization which has desirable data analytic or statistical features. For instance, one particularly attractive factorization is referred to as the GH' factorization of a matrix of deviations from column means. This has orthonormal columns for G and therefore satisfies $Y'Y = HH'$. If the rows of Y represent individuals and the columns represent variables, then Y'Y is n times the estimated variance-covariance matrix; hence the inner products of the rows $\underline{h}$ of H represent the covariances, and the squared lengths of the $\underline{h}$'s represent the variances. The cosines between $\underline{h}$-vectors therefore represent the correlations between the variables. The resulting GH' biplot is useful in many statistical applications.

3. Description of BGRAPH. BGRAPH carries out the graphical display of the biplot of an n x m matrix Y. The graphical display is accomplished via the algorithms described in Newman and Sproull (1973) and Pearce (1977). A summary of BGRAPH commands is given in Display 3.1. A fuller description is available from the authors.

4. How BGRAPH Represents Three-Dimensional Displays

4.1. Orthogonal, Oblique and Perspective Views. Orthogonal, oblique and perspective views all represent a three-dimensional object on a plane by extending imaginary lines of sight from the object to an observation point (the location of the eye of the observer). Each view consists of the

Display 3.1

Summary of BGRAPH Commands

Input and Output:	READ, ENTER, SAVE, NO SAVE, DISPLAY, NO DISPLAY.
Graphical Display of Two-dimensional Bimodel:	BIPLOT, MULTI.
Graphical Display of Three-dimensional Bimodels:	PERSPECTIVE, STEREO.
Orientation of Three-dimensional Bimodels:	ROTATE, FIRST, LAST, SIDE, SIDE3, SIDE6.
Position of the Observer:	OBSERVER, IOD, PLANE, SHIFT.
Special Viewing Commands:	COLOR, INTENSITY, VIEWPORT, WINDOW.
Displaying Subsets of Row or Column Markers:	RPICK, CPICK, ROMIT, COMIT, RALL, CALL, RONLY, CONLY, BOTH.
Labeling Graphical Output:	TITLE, RLABEL, CLABEL, RPOINT, CPOINT, RVECTOR, CVECTOR, AXES, FORMAT.
Manipulating Output:	PROJECT, TRANSFORM, COORDINATES, CIRCLE, ELLIPSE, MEANS.
System Commands:	HELP, DUMP, RESTORE, BRIEF, NO BRIEF, RESET, PRINT.

intersection of a plane and the lines of sight. In orthogonal views, the observation point is at infinity, making the lines of sight parallel. In addition, the lines of sight are perpendicular to the viewing plane. In oblique views, the observation point remains at infinity, but the parallel lines of sight are not perpendicular to the viewing plane. In perspective views, the observation point is no longer at infinity. Thus, the lines of sight are no longer parallel, but converge at the observation point.

4.2 Analyglyphs and Stereograms. Examples of analglyphs can be found in Pearce (1977), and examples of stereograms can be found in Fraser and Kovats (1966) and Rohlf (1971). Both use pairs of perspective views to create three-dimensional images. Each perspective view represents an object as one of the observer's eyes would see it. Each approach provides a pair of views which, with the aid of viewing devices and certain visual and mental processes, can be fused together to form a three-dimensional image.

In analglyphs each view is displayed in a different color (typically red and green). The three-dimensional image is obtained when these are viewed through color filters--so that the filter for the left eye filters out the view for the right eye and vice versa.

In stereograms the two views are displayed side by side and viewed through stereoscopic glasses. Color need play no part in a stereogram, although one might wish to use color for labeling. The optimal distance between the displays of the views depends on characteristics of the glasses being used.

4.3 Further Display Options. In BGRAPH the three-dimensional points to be plotted are scaled and translated to fit inside a unit cube centered at the origin. The cube and its contents can be rotated about any axis through its center (ROTAT). Orthogonal views of the contents of the rotated cube can be obtained (no longer scaled and translated to fit inside the cube). The user can select views from three default directions (SIDE3), from six default directions (SIDE6), or from any desired direction (SIDE).

The user can produce a perspective view or a stereo pair of perspective views of the cube and its contents (PERSP, STEREO). As a depth cue in perspective views and stereo pairs, the size of the characters in marker labels is inversely proportional to their squared distance from the observation points. Furthermore, on a display device and system which permit the varying of intensity, one can create a display in which the intensity of markers and their labels is inversely proportional to their squared distance from the observation point (INTEN).

BGRAPH allows selection of a variety of options which make it possible to produce three-dimensional displays with very little effort. Using the PLANE and SHIFT commands it is possible to create analglyphs with the STEREO command. (The viewing plane should be behind the cube.) The PLANE, SHIFT and IOD commands can be used to tune stereograms to be viewed through different types of stereoscopic glasses. (The viewing plane should be between the cube and the observation point.) One can obtain maximum resolution by electing to draw stereo pairs on separate pages and using a copier to reduce them to a size appropriate for viewing.

Plots produced with BGRAPH can be immediately displayed on a CRT and/or saved as metacode files which may be used later to reproduce the plot on a CRT or by a flat bed plotter. BGRAPH makes it possible to produce color displays on CRT's and systems with color options. Using color in producing a plot on the plotter is a matter of pausing to change pens. It is technically feasible to produce black and white or color slides and movies from metacode files.

Stereograms are easier to tune than are analglyphs. To tune a stereogram so that it can be viewed, simply involves finding adequate settings for the PLANE, IOD and SHIFT commands. Tuning analglyphs requires, in addition, that the color filters used match shades of ink or hues available on film. Furthermore, the use of color may hinder publication. However, in the future we would like to experiment with projecting three-dimensional images on a wall or screen, and we suspect analglyphs may be better suited for this purpose. For this reason, we have made BGRAPH flexible enough to create either stereograms or analyglyphs.

5. Concentration Ellipsoids, Comparison Circles and Projections onto Bimodels. Using BGRAPH's ELLIP command one can define up to 10 groups of row markers, and for each group compute two-dimensional or three-dimensional concentration ellipsoids. These provide a convenient summary of the

location and variability of subsets of markers.

Gabriel (1981) uses comparison circles on the biplot to carry out approximate simultaneous testing of all pairs of samples in MANOVA. The logic behind using such a display to test for a difference between samples is as follows. The Euclidean distance between row markers is an approximation to the numerator of the Hotelling test statistic one would use, while the sum of the radii of the circles centered at these two markers is an approximation to the denominator times the critical value for the test. If these circles do not overlap, it means that the approximate value of the test statistic was greater than its critical value and we would reject the hypothesis of no difference between those samples. BGRAPH has an option PROJECT to project additional data points onto an existing biplot.

6. Implementation of BGRAPH. BGRAPH (Tsianco, 1980) is written in FORTRAN and is implemented on a DEC 10 under the TOPS 10 operating system. Graphic display is carried out through the systems plot package of the National Center for Atmospheric Research (NCAR) graphics package. Graphics display can be on any device which is supported by the NCAR package.

The NCAR subroutines utilized by BGRAPH perform only low level functions such as point and line drawing, so such functions from other graphics packages could be substituted.

The singular value decomposition of the data matrix required by BGRAPH is not performed within the program. A routine to prepare the SVD for input to BGRAPH is available from the authors.

7. An Example. We use an example extensively analyzed in Gabriel (1980), to demonstrate the features of BGRAPH.

Historical data of annual precipitation in Illinois were used to simulate a weather modification situation. Suppose a "cloud seeding" operation had taken place in the years 1955-1960 in the southern Illinois area, and that another such operation had been carried out during 1970-78 in the northwestern part of Illinois. Also, suppose that no cloud seeding was carried out in Illinois at any other time or place. Central Illinois precipitation could, therefore, serve as concomitant observations to indicate "natural" precipitation. (The quotes are used since the data relate to simulated "operations", not to real ones.)

To evaluate the effect of both "operations" is evaluated on 50 years' data, 1928-78, for the following five stations: Dubuque and Moline to represent northwestern Illinois, "seeded" in Period IV--1970-78; St. Louis to represent southern Illinois, "seeded" in Period II--1955-60; Peoria and Springfield to represent central Illinois--never "seeded." These 50 years also provide two "unseeded" periods for comparison, i.e., I - 1929-1954 and III - 1961-1969: Note that these are actual precipitation data except that in the "operational" years each "target" station's precipitation was augmented to simulate effects of "seeding."

Display 7

Period	No. of Years	Areas and Periods of "Operations" and Comparisons		
		Southern Illinois "Target" (St. Louis)	Northeastern Illinois "Target" (Dubuque, Moline)	Central Illinois Control (Peoria, Springfield)
I. 1929-54	26	Unseeded	Unseeded	Unseeded
II. 1955-60	6	"Seeded"	Unseeded	Unseeded
III. 1961-69	9	Unseeded	Unseeded	Unseeded
IV. 1970-78	9	Unseeded	"Seeded"	Unseeded
Total	50			

Display 7.1 is the biplot of the row and column markers using the rank 2 approximation to the data matrix. The row markers are labelled with the year and the column markers with the station name. Display 7.2 repeats the display but adds concentration ellipses (1.5 standard errors) for each period; the row markers are suppressed and the display is reflected about the Y-axis and rotated to orient the station markers to conform to their geographical configuration.

Display 7.3 gives a stereographic view of a 45° rotation of 7.2. The stereogram is accomplished by reducing a full-sized plot made on a Tektronix 4662 on a Xerox 9200 using 64% and 75% reduction. The display should be viewed through stereoscopic glasses (for example, Stereoscope, available from the Hubbard Scientific Company, Northbrook, Illinois).

Display 7.4 is a perspective display of MANOVA biplots. The front view has been rotated -45° and projected on the observer's viewing plane using the SIDE command.

This work was supported in part by ONR Contract N00014-80-C-0387.

Bradu, D. and Gabriel, K.R. (1978). The biplot as a diagnostic tool for models of two-way tables, Technometrics, 20, 47-68.

Gabriel, K.R. (1971). The biplot - graphic display of matrices with application to principal component analysis, Biometrika, 58, 453-467.

Gabriel, K.R. (1980). Biplot, Encyclopedia of Statistical Sciences, New York: Wiley.

Gabriel, K.R. (1981). Biplot display of multivariate matrices for inspection of data and diagnosis, in Barnett, V., Ed., Multivariate Data, New York: Wiley.

Fraser, A.R. and Kovats, M. (1966). Stereoscopic models of multivariate statistical data, Biometrics, 22, 358-67.

Newman, W.M. and Sproull, R.F. (1973). Principles of Interactive Computer Graphics, New York: McGraw-Hill.

Pearce, G.F. (1977). Engineering Graphics and Descriptive Geometry in 3-D, Toronto: Macmillan.
Rohlf, F.J. (1971). Stereograms in numerical taxonomy, Systematic Zoology, 20, 246-55.
Tsianco, M.C. (1980). Use of biplots and 3D-bimodel's in diagnosing models for two-way tables, Ph.D. Thesis, The University of Rochester.

Division of Biostatistics, Box 630, University of Rochester Medical Center, Rochester, New York 14642.

Display 7.1

Commands: RLABEL, CLABEL, CVECTOR, BIPLOT

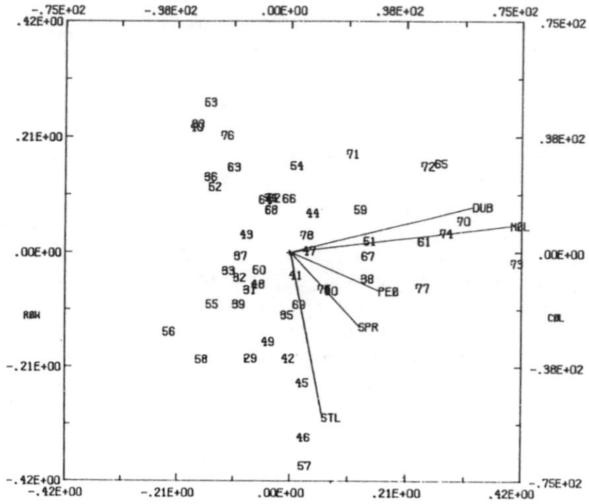

Display 7.2

Biplot with concentration ellipses for each period row markers supressed, X-Y plane.

Commands: BIPLOT, ELLIPSE, ROTATE

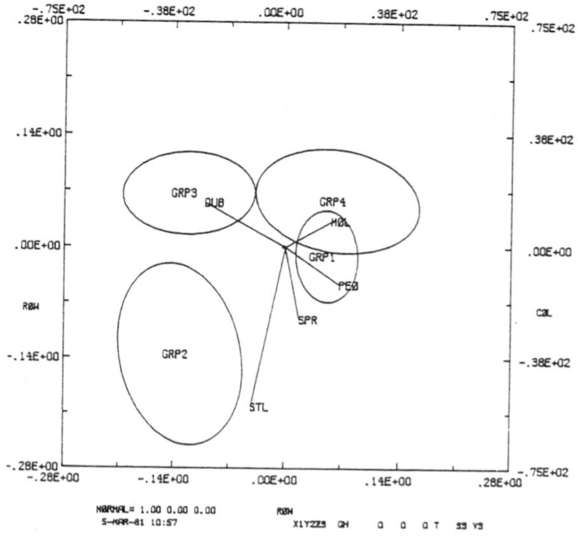

Display 7.3

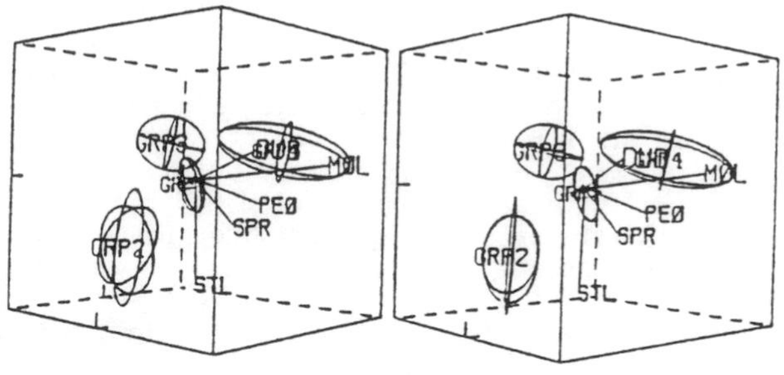

Display 7.4

Perspective plot of MANOVA biplot, rotation -45° about Y-axis, projection onto observation plane.

Commands: PERSP, ROTATE, SIDE

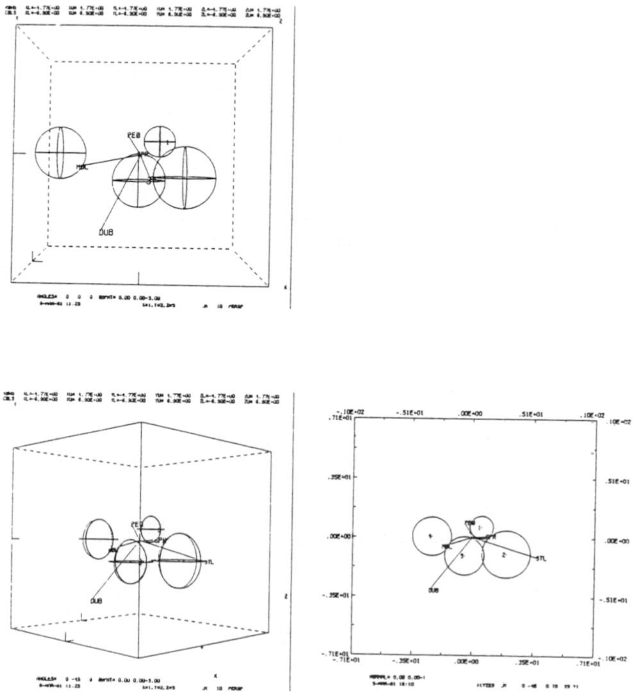

MONCOR--A PROGRAM TO COMPUTE CONCORDANT AND OTHER MONOTONE CORRELATIONS

George Kimeldorf[1], Jerrold H. May, and Allan R. Sampson[2]

University of Texas at Dallas, University of Pittsburgh, and University of Pittsburgh

ABSTRACT

The new interactive FORTRAN program MONCOR is described. MONCOR computes the concordant monotone correlation, discordant monotone correlation, isoconcordant monotone correlation, isodiscordant monotone correlation and their associated monotone variables. Data input can be finite discrete bivariate probability mass functions or ordinal contingency tables, both of which must be given in matrix form. The well-known British Mobility data are used to illustrate the input and output options available in MONCOR.

KEY WORDS: Interactive program, correlation, monotone correlation, concordant monotone correlation, discordant monotone correlation, isoscaling, ordinal contingency tables, nonlinear optimization.

1. General Background

MONCOR is a set of routines to compute monotone correlation measures for two dimensional probability distributions of random variables X and Y given by matrices of the form

$$\begin{bmatrix} p_{11} & p_{12} & \cdots & p_{1J} \\ p_{21} & p_{22} & \cdots & p_{2J} \\ \vdots & \vdots & \ddots & \vdots \\ p_{I1} & p_{I2} & \cdots & p_{IJ} \end{bmatrix}$$

where $p_{ij} = \text{Prob}(X = x_i, Y = y_j)$, $i = 1, \ldots, I$ and $j = 1, \ldots, J$. These probabilities may be theoretical probability distributions or may be estimates derived from ordinal contingency tables.

The monotone correlation between any random variables X and Y (see Kimeldorf and Sampson (1978)) is defined by

$$(1.1) \qquad \rho^*(X,Y) = \sup \rho(f(X), g(Y)),$$

where the supremum is taken over all monotone functions f,g, for which $0 < \text{Var } f(X) < \infty$ and $0 < \text{Var } g(Y) < \infty$. This concept can be refined by measuring separately the strength of the relationship between X and Y in a positive direction and the strength of the relationship in a negative direction, that is, to measure separately the extent of concordancy and of discordancy between X and Y. If in (1.1), f and g are both restricted to be increasing, or equivalently both decreasing, (the terms increasing and decreasing are used nonstrictly), the resulting measure is called the concordant monotone correlation. When f is restricted to be increasing and g decreasing (or equivalently f decreasing and g increasing), we find it convenient to examine $-\sup \rho(f(X), g(Y))$, which in turn can be expressed as $-\sup \rho(f(X), -g(Y))$, where both f and g are increasing. This leads naturally to defining the discordant monotone correlation by $\inf \rho(f(X), g(Y))$, where f and g are both restricted to be increasing.

Sometimes the situation occurs when X and Y should have the same scaling. If in (1.1) we require $f = g$, the resulting measure is called the isoconcordant monotone correlation. Analogous to the DMC definition, the isodiscordant monotone correlation is given by $\inf \rho(f(X), g(Y))$, where $f = g$.

The actual functions that maximize the correlations (assuming they exist) are called monotone variables with their specific interpretation depending on which monotone correlation measure is used in their derivation.

The general approach followed in MONCOR is based upon the algorithm given in Kimeldorf, May and Sampson (1980). MONCOR computes the concordant and discordant monotone correlations, and their isoscale versions, by formulating a linearly constrained nonlinear programming problem whose optimum value is the desired monotone correlation measure. A specially modified form of the nonderivative nonlinear programming algorithm QRMNEW (see May (1979)) is used to solve the nonlinear program.

2. Program Structure

MONCOR consists of two basic components: routines specific to monotone correlation and routines adapted from the QRMNEW program. The primary goal throughout the coding of MONCOR is simplicity of presentation. The components of QRMNEW adapted for MONCOR are written in part in a non-structured style, but with readability enhanced by comment lines.

Since MONCOR was designed to be used in an interactive mode for small problems, dense matrix storage is utilized throughout. Array storage is made available to all routines through labelled common only. All dimensions are explicitly set in each subroutine. MONCOR version 2.3 is dimensioned so as to handle probability matrices of dimension up to 10 x 10. The source listing comments indicate how to redimension the arrays for larger problems.

The original source code for MONCOR consists of 1932 statements written in DEC FORTRAN-10 and is implemented on the University of Pittsburgh's DEC 1099.

Double precision arithmetic is used throughout, and both variables and arrays are set by an IMPLICIT statement at the beginning of each routine.

The main program calls the FORTRAN-10 basic external function RAN, which returns a [0,1] random number. The call to RAN may be modified to accomodate any other random number generator.

3. Program Description

In all sample MONCOR outputs, those items inputted by the user are underlined. On this output, references to material in Section 3 are typed in the left margin.

3.1 Example

To describe the use of MONCOR, the British Mobility data example is employed. (See Kimeldorf, May and Sampson; the original data are in Glass and Hall (1954, p. 183).) The estimated probability matrix $\{p_{ij}\}$ derived from Table 5.1 of Kimeldorf, May and Sampson is given in Table 1.

Table 1

Estimated Probability Matrix
British Mobility Data

		Son's Occupational Status				
		S1	S2	S3	S4	S5
	S1	.014	.013	.002	.005	.002
Father's	S2	.008	.050	.024	.044	.016
Occupational	S3	.003	.022	.031	.064	.027
Status	S4	.004	.043	.053	.205	.128
	S5	.001	.012	.021	.091	.117

In this example I = J = 5.

3.2 Data Input

The program requires input in the following order:

(a) the number of rows of $\{p_{ij}\}$: I,

(b) the number of columns of $\{p_{ij}\}$: J,

(c) the device number that the matrix is to be read in from (entering a 5 allows reading from terminal),

(d) the values for $\{p_{ij}\}$ read in a row at a time (it is assumed $\max(I,J) \leq 20$).

After $\{p_{ij}\}$ is inputted,

(e) the matrix is printed by MONCOR under the heading INPUT MATRIX.

An example of this procedure for the British Mobility data is given in Output 1.

3.3 Output

(a) After the matrix $\{p_{ij}\}$ has been output by MONCOR, the user must specify which one measure of monotone correlation is desired. The user needs only to enter one number (1, 2, 3, or 4) according to the MONCOR prompts.

(b) For each monotone measure calculation, the user has the option of calling for

(i) intermediate point output during the optimization process

(ii) inputting a starting point for the optimization process.

Responses to MONCOR questions concerning (i) or (ii) are YES or NO.

For the remainder of the output description in Subsection 3.3, it is assumed that neither option (i) nor (ii) is requested.

(c) MONCOR then evaluates the correlation at the monotone extreme points (see Section 4, Kimeldorf, May and Sampson). The number of such monotone extreme points is given by MONCOR (if the concordant or discordant monotone correlation is requested, the number of monotone extreme points is (I-1)(J-1); if the isoconcordant or isodiscordant monotone correlation is requested, the number of monotone extreme points is (I-1)).

(d) MONCOR computes the largest correlation among these monotone extreme points, and

(e) prints one monotone extreme point at which this maximum is achieved.

(f) MONCOR then prompts the user to enter a five digit number which is used to warm up the FORTRAN-10 basic external function random number generator RAN. RAN is called that number of times before being used to generate the ten random starting points (see Kimeldorf, May and Sampson (Section 4)).

(g) The estimated monotone correlation measure is given based upon the 10 random starting point calculations as well as the monotone extreme points (Kimeldorf, May and Sampson (Section 4)).

(h) A point (monotone variable) at which the monotone correlation is achieved is printed by MONCOR. MONCOR, without loss of generality, sets X(1) = Y(1) = 0 and X(I) = Y(J) = 1. If an iso-measure is selected, only the X monotone variable is printed. (The Y monotone variable is identical by definition.)

(i) MONCOR then asks if the user wants to use the inputted $\{p_{ij}\}$ matrix to calculate other measures of monotone correlation. If the user responds YES, it begins at 3.3(a) again. If the user responds NO, an opportunity is given to input another $\{p_{ij}\}$, (see 3.2(a)) or to terminate MONCOR.

For illustration, the calculation of the isoconcordant monotone correlation for the British Mobility data is given in Output 2.

3.4 Option: Inputting a Starting Point

Unless otherwise specified, MONCOR generates 10 random starting points from which the optimization procedure starts. However, the user has the option of inputting a monotone starting point by replying YES to the question "DO YOU WISH TO INPUT A STARTING POINT?" The user must then input X(2), ..., X(I-1) and Y(2), ..., Y(J-1). (If an iso-measure is requested, then only X(2), ..., X(I-1) must be input.) MONCOR sets X(1) = Y(1) = 0 and X(I) = Y(J) = 1. If this option is employed, no monotone extreme points are evaluated (see 3.3(c)) and no random starting points are used (see 3.3(g), 3.3(f)).

For the British Mobility data, Output 3 illustrates inputting a pair of starting points for the concordant monotone correlation.

3.5 Option: Requesting Intermediate Point Output

Unless otherwise specified MONCOR does not print out the intermediate steps in the iterations of the nonlinear optimization (see Section 4 of Kimeldorf, May and Sampson). However, the user has the option of requesting intermediate output by replying YES to the question "DO YOU WISH TO SEE INTERMEDIATE POINT OUTPUT?"

If the user has not inputted a starting point, MONCOR will print the intermediate evaluations for each of the 10 random starting points. If the user has inputted a starting point, MONCOR will print out the intermediate evaluations based upon the given starting point.

For each iteration, MONCOR outputs

(a) the iteration number (with iteration number 1 corresponding to the random or inputted starting point),

(b) the number of evaluations, which is the number of times QRMNEW has evaluated the objective function cumulatively up to the given iteration.

(c) the correlation based upon the current iteration,

(d) the gradient which measures the degree of the local steepness at that point (see a description of QRMNEW in May for details),

(e) the coordinates of the current point (X(1), X(I), Y(1), Y(J) for the current point are suppressed, and for iso-measures, Y(2), ..., Y(J-1) are not printed).

(f) When the algorithm stops, MONCOR prints "OPTIMAL SOLUTION FOUND" and provides the optimizing point, the value for the final iteration, and the total cumulative number of function evaluations (see 3.5(b)).

Again for the British Mobility data, the intermediate point output for the isoconcordant monotone correlation is given in Output 4 for random point number 2 (no starting point was input).

3.6 Comment

In order to compute the (Pearson) correlation for a given set of values of X(2), ..., X(I-1) and Y(2), ..., Y(J-1), one can use Options 3.4 and 3.5, so that the first iteration evaluation will provide the desired correlation. For example, to compute Spearman's rank correlation, set X(i) = (i-1)/(I-1) for i = 2, ..., I-1, and Y(j) = (j-1)/(J-1) for j = 2, ..., J-1.

3.7 Program Availability

The MONCOR program is available for distribution. For specific details contact Professor Jerrold May, Graduate School of Business, University of Pittsburgh, Pittsburgh, PA 15260.

FOOTNOTES

[1]The work of this author (George Kimeldorf) was supported by the National Science Foundation under Grant MCS-8002152.

[2]The work of this author (Allan R. Sampson) is sponsored by the Air Force Office of Scientific Research, Air Force Systems Command, under Contract F49620-79-C-0161.

REFERENCES

Glass, D.V. and Hall, J.R., (1954). "Social Mobility in Britain: A study of inter-generation changes in status." In Social Mobility in Britain. Ed. by D.V. Glass. Routledge and Kogan Paul Ltd., London.

Kimeldorf, G., May, J.H., and Sampson, A.R., (1980). "Concordant and discordant monotone correlations and their evaluation by nonlinear optimization." Technical report.

Kimeldorf, G., and Sampson, A.R., (1978). "Monotone dependence." Annals of Statistics 6, 895-903.

May, J.H., (1979). "Solving nonlinear programs without using analytic derivatives." Operations Research 27, 547-484.

Output 1

```
3.2(a) HOW MANY ROWS IN YOUR MATRIX?        >5
3.2(b) HOW MANY COLUMNS IN YOUR MATRIX?       >5
3.2(c) ON WHAT DEVICE IS YOUR MATRIX?  ENTER 5 IF YOU  WISH TO TYPE IT IN NOW.     >5
       ENTER THE MATRIX A ROW AT A TIME, MAX 20 NOS. PER LINE
      >.014 .013 .002 .005 .002
      >.008 .050 .024 .044 .016
3.2(d)>.003 .022 .031 .064 .027
      >.004 .043 .053 .205 .128
      >.001 .012 .021 .091 .117

                      INPUT MATRIX

       .0140 .0130 .0020 .0050 .0020
       .0080 .0500 .0240 .0440 .0160
3.2(e) .0030 .0220 .0310 .0640 .0270
       .0040 .0430 .0530 .2050 .1280
       .0010 .0120 .0210 .0910 .1170

                      END OF INPUT MATRIX
```

Output 2

```
3.3(a) WHICH ONE OF THE FOLLOWING DO YOU WISH TO SEE?
            1=CONCORDANT MONOTONE CORRELATION
            2=DISCORDANT MONOTONE CORRELATION
            3=ISOCONCORDANT MONOTONE CORRELATION
            4=ISODISCORDANT MONOTONE CORRELATION
       >3


       ******************************************** ISOCONCORDANT MONOTONE CORRELATION
3.3(b)(i) DO YOU WISH TO SEE INTERMEDIATE POINT OUTPUT?       >NO
3.3(b)(ii)DO YOU WISH TO INPUT A STARTING POINT?         >NO
3.3(c)    BASED ON       4  MONOTONE EXTREME POINTS, THE
3.3(d)    OPTIMUM OF CORRELATIONS AMONG MONOTONE EXTREME' POINTS IS          0.40656150 AT
3.3(e)     X( 1)=  .00000   X( 2)=1.00000   X( 3)=1.00000   X( 4)=1.00000   X( 5)=1.00000


3.3(f)    ENTER A 5 DIGIT INTEGER        >48147
3.3(g)    ESTIMATED ISOCONCORDANT MONOTONE CORRELATION IS        0.49682067   AT
3.3(h)     X( 1)=  .00000   X( 2)=  .62778   X( 3)=  .84644   X( 4)=  .92409   X( 5)=1.00000
3.3(i)     ANOTHER RUN WITH THE SAME MATRIX?       >YES
```

Output 3

```
       DO YOU WISH TO INPUT A STARTING POINT?       >YES
       X( 2)======>.25
       X( 3)======>.5
3.4    X( 4)======>.75
       Y( 2)======>.25
       Y( 3)======>.5
       Y( 4)======>.75
```

Output 4

```
       -----------------------------------------------------------  RANDOM POINT NUMBER    2
3.5(a)( ITN EVAL    CORR    GRAD
3.5(b)|   1     10  0.4549 2.3D-01
3.5(c)|   2     25  0.4864 1.1D-01
3.5(d)|   3     36  0.4967 1.5D-02
3.5(e)(   4     46  0.4968 4.8D-04

      (XXXXXXXX     OPTIMAL SOLUTION FOUND
3.5(f){  X( 1)=  .00000   X( 2)=  .62778   X( 3)=  .84644   X( 4)=  .92409   X( 5)=1.00000
      (
           CORRELATION COEFFICIENT=   0.496821              56    FUNCTION EVALUATIONS
```

COMPUTER OFFERINGS FOR STATISTICAL GRAPHICS -- AN OVERVIEW

Patricia M. Caporal and Gerald J. Hahn, General Electric Company

CRT display devices, special plotters and other graphics output devices which communicate with small or large host computers provide analysts an opportunity to obtain automatically high quality graphical displays of data and results of statistical evaluations. This paper gives an overview of programs and subroutine libraries for statistical graphics. Different offerings are identified and their technical features compared. The purpose is to provide a working guide to those who wish to use statistical graphics.

Key Words: CRT display devices, plotters, statistical graphics, data analysis, business graphics.

I. INTRODUCTION

Computer graphics has been used extensively as a tool in engineering drafting and design. It has an equally important role to play in data presentation and analysis, where it is related to "business graphics". The applications of statistical graphics are, however, not limited to business problems. These tools can help engineers and technical people, as well as managers and business analysts, to better understand complex sets of data and to present results more clearly. They can, for example, provide fast and clear feedback for on-line quality control, yield comprehensive presentations of the most important facts for management review, and even help make statistics more understandable. Users of automated statistical graphics can obtain an informative display of the latest information in the data base, update plots as new data become available and make desired changes rapidly. Figures 1 to 6 show some recent applications with which we have been involved.

The availability of graphics output devices, such as CRT display terminals and special plotters, have been a major factor in advancing automated statistical graphics. These output devices communicate with small or large host computers or, in some cases, have their own computing capabilities. Their use has resulted in the improved quality of graphics which, along with decreasing hardware cost and increasing capabilities, has been a significant factor in accelerating progress.

The additional link needed to make the extensive use of automated statistical graphics a reality is the development of easy-to-use computer graphics software. In recent years, important strides have been made in the development of needed programs and subroutine libraries. In this report we indicate some recent offerings. Additional detail is provided in a company report (Ref.1).

II. WHAT IS AUTOMATED STATISTICAL GRAPHICS AND HOW IS IT USED?

In the distant past, computers, and desk calculators before that, crunched numbers and produced a variety of statistical information, ranging from

Figure 1. Results of evaluation program for new ultrasonic equipment.

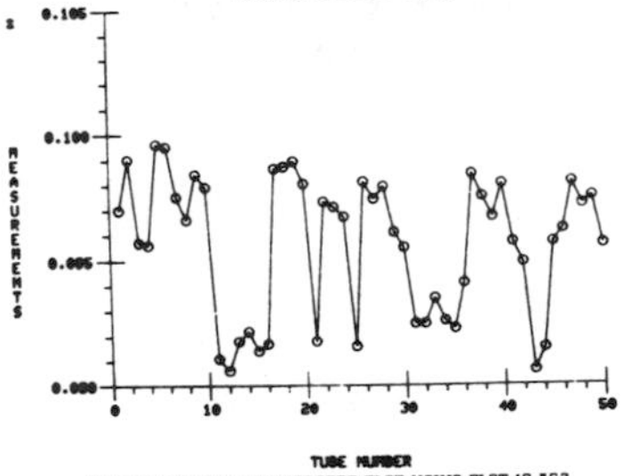

SOURCE: SPECIALLY GENERATED PLOT USING PLOT-10 AG2 AND PLOT-10 TCS

Figure 2. Status of casting development program.

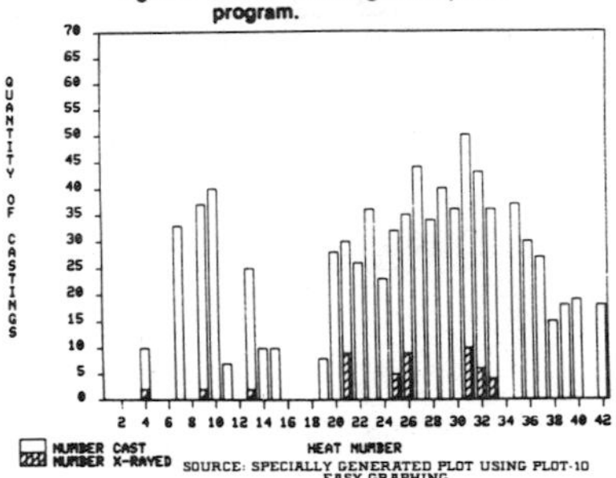

SOURCE: SPECIALLY GENERATED PLOT USING PLOT-10 EASY GRAPHING

Figure 3. Sequential plan decision limits for deciding between two processes and observed results.

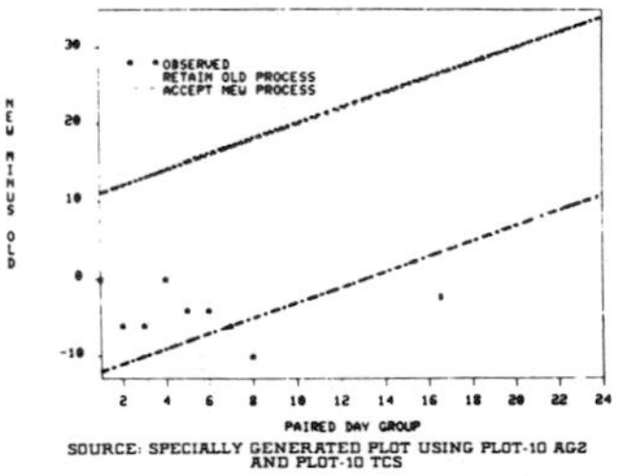

SOURCE: SPECIALLY GENERATED PLOT USING PLOT-10 AG2 AND PLOT-10 TCS

Figure 4. Face plot of multivariate integrated circuit data.

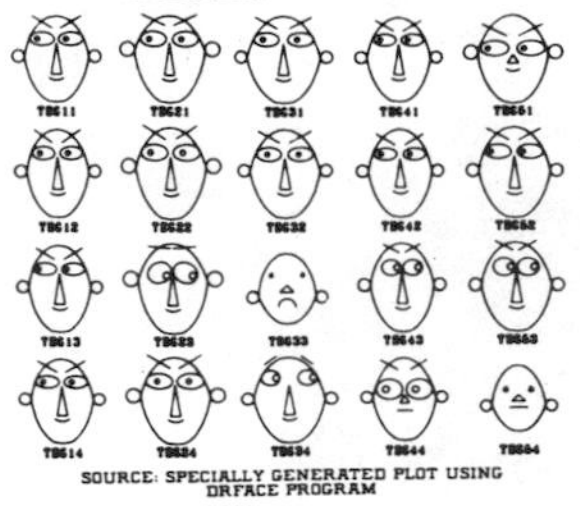

SOURCE: SPECIALLY GENERATED PLOT USING DRFACE PROGRAM

Figure 5. Quality control chart for rotor performance.

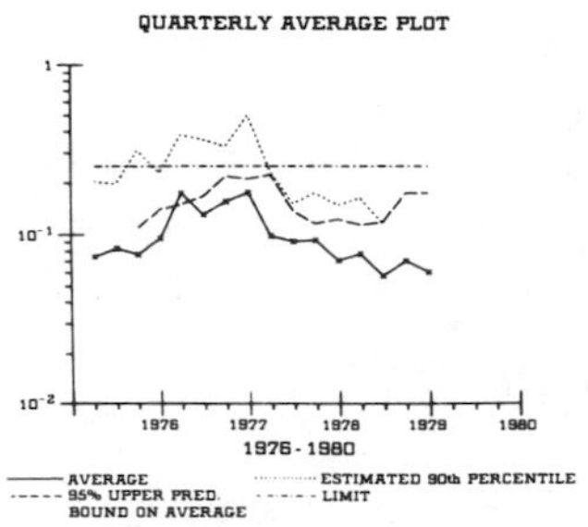

SOURCE: SPECIALLY GENERATED PLOT USING PLOT - 10 AG2 AND PLOT - 10 TCS

Figure 6. Different Weibull distributions for representing product time to failure.

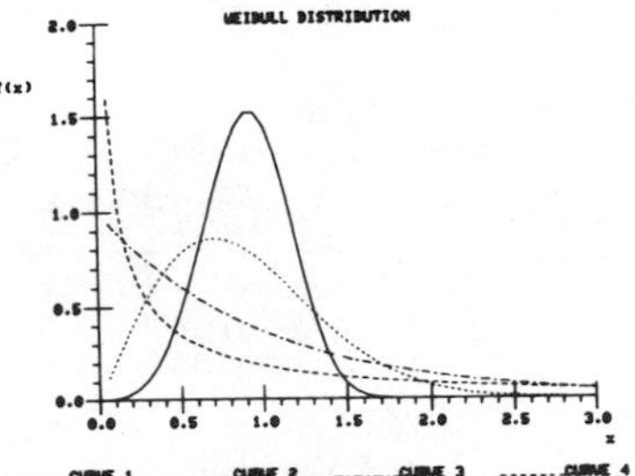

SOURCE: SPECIALLY GENERATED PLOT USING PLOT-10 AG2 AND PLOT-10 TCS

TABLE 1 - SOME SOFTWARE FOR AUTOMATED STATISTICAL GRAPHICS

SOFTWARE AND COST*	SOURCE	HARDWARE REQUIREMENTS AND COMMENTS
DATAPLOT Less than $1000 (Still to be specified)	James J. Filliben Admin. Bldg., A337 Natl. Bur. of Standards Washington, DC 20234	Host computer independent. Requires Tektronix 40XX terminals or Tektronix pen plotters for continuous graphics and various pen plotters and line printers for discrete graphics.
DI-3000 $8000-$12,000	Precision Visuals, Inc. 1518 9th St. Boulder, CO 80202	Host computer independent. Output device independent.
DIGRAF $8000	Univ. Computing Center Univ. of Colorado, B-45 Boulder, CO 80309	Host computer independent. Output device independent.
DISSPLA $25,000-$47,000	Integrated Sofware Syst. 4186 Sorrento Valley Blvd San Diego, CA 92921	Host computer independent. Output device independent. Also available on GEISCO MARK 3000 service; planned on GEISCO MARK III service at a later date.
GCS No Charge	US Army Eng. Waterways Experiment Station PO Box 631 Vicksburg, MS 39180	Host computer independent. Output device independent. There is little or no support available to GCS user by the program distributor.
HEWLETT-PACKARD PLOT/21 $100	Hewlett-Packard 1501 Page Mill Road Palo Alto, CA 94304	Requires the host computer to be an HP 3000 Series II, III or 33 computer, or DEC PDP-11 computer with RT-11 operating system. Requires HP 7221 Pen Plotter as output device. Available on GEISCO MARK III.
PLOT-10 AG2 (Advanced Graphing II) $2000	Tektronix, Inc. PO Box 500 Beavertown, OR 97077	Host computer independent. Requires Tektronix output device. Also available on GEISCO MARK 3000. PLOT-10 AG2 is based on the PLOT-10 TCS subroutine library.
PLOT-10 EASY GRAPHING $1850	Tektronix, Inc. Address given above	Host computer independent Requires Tektronix output device. Available on GEISCO MARK 3000 service.
PLOT-10 IGL (Interactive Graphics Library) $3300-$8500	Tektronix, Inc. Address given above	Host computer independent. Requires Tektronix output device. Available on GEISCO MARK 3000 service.
PLOT-10 IGP (Interactive Graphing Package) $2250	Tektronix, Inc. Address given above.	Host computer independent. Requires Tektronix output device. Also available on GEISCO MARK 3000 service. PLOT-10 IGP is the interactive version of PLOT-10 AG2.
PLOT-10 TCS (Terminal Control System) $1500	Tektronix, Inc. Address given above.	Host computer independent. Requires Tektronix output device. Available on GEISCO MARK 3000 service.
PLOT-50 $475-$3500	Tektronix, Inc. Address given above.	Requires Tektronix 4050 series stand-alone graphics terminal.
PLOT II	GEISCO 401 N. Washington St. Rockville, MD 20850	Requires GEISCO MARK III service. Output device independent.
SAS/GRAPH $2500 first year $1000 annual renewal	SAS Institute, Inc. Box 8000 Cary, N.C. 27511	Requires IBM 360/380 mainframe or compatible machines under OS, OS/VS, and VM/CMS. Requires selected Tektronix, HP or Calcomp graphics output device. Used with SAS ($4500 first year, $1500 annual renewal fee). Also available on GEISCO MARK 3000 service.
SPSS GRAPHIC OPTION $8000 first year $4000 annual renewal	SPSS, Inc. 444 N. Michigan Ave Chicago, IL 60611	Requires IBM /OS host computer. Output device independent. Used in conjunction with SPSS ($6000 for first year and $3000 annual renewal fee).
TELL-A-GRAF $25,000-$40,000	Integrated Software Syst. Address given above	Requires large IBM, DEC, or VAX host computer. Output device independent. Developed as user-oriented version of DISSPLA. Also available on GEISCO MARK 3000 service
TEMPLATE $25,000	Megatek Corp. 3931 Sorrento Vall. Blvd. San Diego, CA 92921	Host computer independent. Output device independent; enhanced version of GCS, compatible with GCS but claimed to contain many more capabilities.
VGM (Virtual Graphics Machine) $5000-$11,000	Bell-Northern Research PO Box 3511, Station C Ottawa, Canada K1Y4H7	Host computer independent. Output device independent; being released in parts; some capabilities in Tables 3 & 4 may not be available yet.
ZETA FUNDAMENTAL PLOTTING SUBROUTINES $250	Nicolet Zeta Corporation 2300 Stanwell Drive Concord, CA 94520	Host computer independent. Requires Zeta output device. There are other Zeta graphics packages available.

*Approximate cost as of Nov. 1980. All costs are one-time costs unless otherwise noted. Price ranges are due to different available options.

simple frequency tables to summary statistics to exotic statistical analyses. To show the results graphically (e.g. to display a fitted regression line with confidence and prediction intervals), the analyst had to plot the computer results by hand or engage a graphic artist.

Somewhat later, various plotting features were introduced into statistical programs, e.g., extensive capabilities to obtain histograms, probability plots, and two-way plots. However, because the outputs from these programs are usually obtained from standard line printers, they are limited to stars, letters, numbers, or other textual characters to produce the plots; they cannot draw lines or circles directly. Thus, the analyst often still had to manually obtain high-quality graphical presentations of the results. More recently, various output devices, such as CRT displays, special plotters, and other equipment that allow points to be joined by lines or curves have been developed. Some of the output devices also permit multicolor screen displays and hard copy plots to be obtained. The quality of these plots has gradually improved. In addition, the CRT output devices and other equipment permit "on-line" assessments to be obtained. A recent booklet (Ref.2) gives a good overview of currently available hardware.

For automated statistical graphics to become practical, appropriate computer graphics software was required. Thus, programs and subroutine libraries for preparing displays such as bar charts, line graphs, pie charts, and contour charts had to be developed. This has been done by various software houses, suppliers of hardware, and others. Further details are provided in the following sections. The software is written, most often, for small or large computers that drive a graphics output device. In most cases, the "host computer" (e.g., Honeywell 600/6000, VAX, etc.) is different from the graphics output device (e.g., Tektronix 4014, CALCOMP plotter, etc.). However, there is also stand-alone equipment that can be programmed directly to produce the desired displays on an associated terminal. The Tektronix 4050 series and the HP 9845 series are examples.

III. TYPES OF SOFTWARE FOR AUTOMATED STATISTICAL GRAPHICS

Table 1 lists the software offerings that we have considered, including where they can be obtained, current cost, and hardware requirements. Offerings available on GE timesharing via the General Information Services Company (GEISCO) are indicated.

Graphics software can be classified in various ways. Three categorizations are:

- o Computer dependent versus computer independent
- o Output device dependent versus output device independent
- o Subroutine libraries versus stand-alone programs

Further comments are provided in Ref. 1.

IV TECHNICAL AND STATISTICAL FEATURES OF THE PROGRAM

Table 2 summarizes technical features and statistical capabilities of the graphics software listed in Table 1. These features are discussed in detail in Ref. 1 and illustrated there by figures using outputs from the various offerings.

Table 1 shows a wide disparity in the costs of the software offerings. Ref. 1 includes a comparison of two popular programs with large price differences (Textronix PLOT-10 EASY GRAPHING versus ISSCO's TELL-A-GRAF).

V GRAPHICS SOFTWARE WITH SPECIAL STATISTICAL ANALYSIS CAPABILITIES

Introduction

Those who need analyze data would find especially useful an offering that allows varied statistical analyses to be conducted and then provides graphical displays of the results. Many graphics off-

Table 2

SUMMARY OF TECHNICAL FEATURES AND STATISTICAL CAPABILITIES OF SOFTWARE

	Color	Automatic 3-D Plotting	Cross-Hatching	Multiple Line Patterns	Multiple Fonts	Varying Line Thicknesses	Blanking Capabilities	Automatic Bar Charts	Automatic Pie Charts	Automatic Cross-Plots	Polynomial Fit	Confidence Limits on Regression Line	Spline Fit	Time Series Capabilities	General Statistical Analyses
DATAPLOT	X	X		X	X			X	X	X	X	X	X	extensive	extensive
DI-3000	X	X	X	X	X	X	X							none	none
DIGRAF	X	X		X	X	X	X			X				none	none
DISSPLA	X	X	X	X	X	X	X	X	X	X	X	X	X	limited	none
GCS	X	X		X	X	X	X	X	X	X	X		X	limited	none
HP PLOT/21	X		X	X	X		X							none	none
PLOT-10 AG2	*		X	X	*		X	X		X				limited	none
PLOT-10 EASY GRAPHING	X		X	X				X	X	X				none	none
PLOT-10 IGL	X	X	X	X	X	X	X				X		X	none	none
PLOT-10 IGP	*		X	X	*		X	X		X				limited	none
PLOT-10 TCS	*			X	*		X							none	none
PLOT-50			X	X	X	X	X	X	X	X	X	X	X	extensive	extensive
PLOT II	X		X	X				X	X	X	in STATII***	in STATII***		in STATII***	in STATII***
SAS/GRAPH	X	X	X	X	X		X	X	X	X	X	X	X	limited	in SAS
SPSS GRAPHICS OPTION	X		X	X	X		X	X	X	X	in SPSS	X	X	none	in SPSS
TELL-A-GRAF	X		X	X	X	X	X	X	X	X	X		X	limited	none
TEMPLATE	X	X	X	X	X	X	X	X	X	X	X		X	limited	none
VGM	X	X	X	X	X	X	X	X	X	X				none	none
ZETA FUNDAMENTAL PLOTTING SUBROUTINES	X			X	X					X			X	none	none

X indicates capabilities available as of November 1980.
* Available only on the Tektronix 4662 Interactive Digital Plotter.

erings do not have extensive statistical analysis capabilities. Those that do can be classified as:

- o Stand-alone graphics equipment
- o Interfaces with, or extensions of, existing statistical packages
- o Specially-developed statistical graphics packages
- o Specialized offerings

Each of these types are discussed briefly below. Some of the offerings (e.g., DATAPLOT, SAS/GRAPH, Chernoff Faces) are currently being reviewed more extensively by us and will be reported upon separately.

Stand-Alone Graphics Equipment

Stand-alone graphics equipment may include a statistical analysis library. A specific example is the PLOT-50 Statistics Program Library for the Tektronix 4054. This includes various relatively elementary statistical analysis features, such as assessment of distribution probabilities, multiple linear regression, polynomial regression, nonlinear regression, t-tests, and up to three-way analysis of variance.

Interface with, or Extensions of, Existing Statistical Packages

The last decade has seen the development of extensive computer packages for statistical data analysis, such as BMDP, MINITAB, PSTAT, SAS, and SPSS. An ideal combination for statistical analysts is to be able to use the powerful statistical capabilities of these packages to analyze the data and then display the results using advanced graphics devices, as described in this report. The following graphics interfaces with existing packages have been developed to date:

- o An interface between the STATII*** package and the PLOT II program on the GEISCO MARK III service.
- o The SAS/GRAPH extension to the SAS package.
- o The SPSS Graphics Option for the SPSS package.
- o An interface between the SAS package and TELL-A-GRAF (This is not included in Table 1. Information is available from AUI Data Graphics, 1701 K St., N.W., Washington, D. C. ,20006).

Specially-Developed Statistical Graphics Packages

The next logical step is the development, from scratch, of statistical graphics packages. Specific current offerings include:

- o DATAPLOT, an extensive statistical package, developed at the National Bureau of Standards, that interfaces with PLOT-10 TCS to produce statistical graphics
- o GR-Z and S, two packages developed at Bell Labs (these are not included in Table 1 - see Refence 3 for further details)

Specialized Offerings

Additional software is available to provide specialized statistical graphics. An example is the program DRFACE to describe multivariate data using faces, an approach originally suggested by Chernoff (Ref. 4). In a fact plot, each observation is shown by a face and the different variables are represented by different facial features. An application of DRFACE to the evaluation of integrated circuit performance data is described in Ref. 5 and illustrated in Figure 4.

VI CONCLUDING REMARKS

The more we became involved in reviewing offerings for automated statistical graphics, the more we became aware of the immensity of the task and the many improvements that are constantly taking place. We have reviewed those offerings about which we have been able to get descriptive information and which we felt to be of greatest interest. We make no claim for completeness and invite readers to let us know about important omissions.

The overall picture, however, is clear. The availability of automated statistical graphics provides the analyst numerous opportunities for presenting data more clearly, more comprehensibly, and more rapidly than had been previously possible. Adapting a well-known saying, one or two well-chosen graphical displays are often worth reams of statistical tabulations. Such displays can now be obtained relatively easily, thanks to modern graphical display devices and the accompanying software.

REFERENCES

1. Caporal, P.M. and Hahn, G.J. An overview of Tools for Automated Statistical Graphics, General Electric Company TIS Report 81CRD024, Schenectady, N.Y. 12345, February 1981.*

2. AUI Data Graphics, Business Graphics Hardware Survey, Fall 1980. AUI Data Graphics, 1701 K Street, N. W., Washington, D.C. 20006.

3. R. A. Becker and I. M. Chambers, "Design and Implementation of the S System for Interactive Data Analysis", Proc. IEEE Compsac 78, 1978, pp. 626-629.

4. H. Chernoff, "The Use of Faces to Represent Points in k-Dimensional Space Graphically," J. Am. Stat. Assoc. 68, pp. 361-68, 1973.

5. G. J. Hahn, C. B. Morgan, and W. E. Lorensen, Graphical Display of Product Performance with Computer-Generated Face Plots, General Electric Company TIS Report 81CRD023, Schenectady, N.Y. 12345, April 1981*

*Available from Technical Information Exchange, Bldg. 81, Room A133, Schenectady, N.Y. 12345.

COMPUTING PERCENTILES OF LARGE DATA SETS

Jo Ann Howell
Los Alamos National Laboratory
Los Alamos, New Mexico 87545

Abstract.

We describe an algorithm for finding percentiles of large data sets (those having 100,000 or more points). This algorithm does not involve sorting the entire data set. Instead, we sample the data and obtain a guess for the percentile. Then, using the guess we extract a subset of the original data through which we search for the true percentile.

A. Sampling for an Estimate

A popular method for computing percentiles in large data sets is to locate percentiles in a random sample from the data, although this method may not produce the true percentile. It is merely an estimate, whereas the method described here locates the Tth largest number in a set of N numbers, where N may be very large. More detail is given in Ref. 4.

We define a large data set as one with enough points that analysis by conventional methods is either very difficult or impossible. The number of points may range from thousands to millions or more. Although the algorithm works for a smaller number of points, when the number is less than about 5000, it is more efficient to sort the entire set. This cutoff depends on the memory size involved.

There are several parameters (Table I) involved in the algorithm that the user may vary to meet the particular needs of the problem. They are described in more detail in the text.

The first step in our algorithm is to predict the Tth largest number or the Pth percentile, where P=100*(N-T)/N. The prediction is made by taking a random sample of size KSAMP from the original N data points. The Pth percentile in this sample is used as an estimate to find the Pth percentile in the original data.

TABLE I
ALGORITHM PARAMETERS

Parameter	Description
N	An integer; the number of input data points.
T	An integer; we compute the Tth largest number from the large data set.
P	P=100*(N-T)/N; P is a percentile.
Y	An array of length KSAMP of data points that have been sampled from the large data set.
KSAMP	An integer; length of the array Y.
PCNTL	A number from 0.00 to 1.00; used to construct a window in the large data set.
PCNTH	A number from 0.00 to 1.00; used to construct a window in the large data set.
INDEXL	An integer index into the array Y.
INDEXH	An integer index into the array Y.
CUTOFL	Y(INDEXL); defines the lower bound for the data to be extracted.
CUTOFH	Y(INDEXH); defines the upper bound for the data to be extracted.
I	An integer; the number of data points larger than CUTOFH.
ICTR	An integer; the number of data points extracted.
KBUF	An integer; the size of the buffer area used to move the large data set in and out of memory (typically 5000 to 10,000).

In our program we call the array containing the sample Y. Assuming that Y is sorted from smallest to largest, we compute the indices INDEXL = max ((N -

T)/N * KSAMP - PCNTL * KSAMP , 1) and INDEXH = min ((N - T)/N * KSAMP+PCNTH * KSAMP , KSAMP).
These indices mark positions (100*PCNTL)% below and (100*PCNTH)% above the estimated percentile.

For example, if we select a sample of 1000 points from a large set of 100,000 data points, and we want to find the 50,000th largest number (the 50th percentile), then

N = 100000 ,
T = 50000 ,
KSAMP = 1000 , and
P = 50 .

If we let PCNTL = PCNTH = 0.05, then INDEXL and INDEXH are respectively 450 and 550. The max and min are included to prevent the indices from pointing outside of the array Y, that is, to exclude nonpositive indices or indices larger than KSAMP.

Next, cutoff values are computed using the indices

CUTOFL = Y(INDEXL)

and

CUTOFH = Y(INDEXH).

These values are used as cutoffs on the original large data set to extract some data that is then searched for the true percentile.

In our example, the number Y(500) is the estimate for the 50th percentile for the large data set. The upper and lower cutoffs are Y(550) and Y(450) which we use to extract data from the large data set.

If we want to be reasonably sure that the Pth percentile for the large data set lies in the interval (CUTOFL,CUTOFH), we must use some care in the selection of PCNTL and PCNTH. These values can be chosen by constructing a distribution-free confidence interval for the Pth percentile in the sample (Ref. 3). If PR is the probability that the percentile P of a set of size N lies in the interval (CUTOFL,CUTOFH), then

$$PR = \sum_{i=INDEXL}^{INDEXH} \binom{N}{i} P^i (1-P)^{N-i}$$

Using the approximation,

$$PR = 1/SQRT(2*PI) \int_a^b e^{-t^2/2} dt$$

where

$$a = (INDEXL-N*P)/SQRT(N*P*(1-P))$$

and

$$b = (INDEXH-1-N*P)/SQRT(N*P*(1-P)),$$

we construct a table of values (Table II) to use for PCNTL and PCNTH for the case KSAMP=1000.

TABLE II
SUGGESTED WINDOW VALUES

KSAMP	P	INDEXL	INDEXH	PR	PCNTL	PCNTH
1000	0.50	468	532	0.95	0.032	0.032
1000	0.75	720	775	0.95	0.030	0.025
1000	0.90	881	919	0.95	0.019	0.019
1000	0.95	936	964	0.95	0.014	0.014
1000	0.98	970	989	0.95	0.010	0.009
1000	0.99	984	998	0.96	0.006	0.008

B. Locating the True Percentile

From the original data, we extract all data points that are greater than or equal to CUTOFL and also less than or equal to CUTOFH. We have considerably fewer data points in this extracted data set than in the original set. If the sample of size KSAMP closely resembles the original large data set in distribution, then the extracted data will probably contain the true percentile. That is, the interval (CUTOFL,CUTOFH) contains the Pth percentile of the large data set. We can then use this smaller set to begin our search for the true percentile.

The number T is adjusted by the number of points in the original data set that were larger than CUTOFH. That is, if I is the number of points larger than CUTOFH, then we let t = T - I and search for the t-th largest number in the extracted data set. For this search we use the Blum algorithm. (See Refs. 1, 2, and 5.)

C. Error Conditions

Several error conditions are signaled in the program by an error message. One of these is a missed true percentile in the extracted data. That is, I > T or I + ICTR < T. At this point, the user can increase KSAMP in the hope of getting a more representative sample or increase PCNTL or PCNTH. It is easy to determine by how much we missed the percentile and in which direction we missed it.

REFERENCES

[1] A. V. Aho, J. E. Hopcroft, and J. D. Ullman, The Design and Analysis of Computer Algorithms, Addison-Wesley, Reading, MA, 1974, pp. 97-99.

[2] M. Blum, R. W. Floyd, V. Pratt, R. L. Rivest, and R. E. Tarjan, Time Bounds for Selection, J. Comput. and System Sci. 7 (1973), pp. 448-461.

[3] H. A. David, Order Statistics, John Wiley, New York, 1970, pp. 13-15.

[4] J. A. Howell, An Algorithm for Computing Percentiles of Large Data Sets, Los Alamos National Laboratory, report LA-UR 81-179.

[5] D. E. Knuth, The Art of Computer Programming, vol. III, Addison-Wesley, Reading, MA, 1973, pp. 216-217.

A SELF-DESCRIBING DATA FILE STRUCTURE FOR LARGE DATA SETS*

Robert A. Burnett, Pacific Northwest Laboratory, Richland, Washington

ABSTRACT

A major goal of the Analysis of Large Data Sets (ALDS) research project at Pacific Northwest Laboratory (PNL) is to provide efficient data organization, storage, and access capabilities for statistical applications involving large amounts of data. As part of the effort to achieve this goal, a self-describing binary (SDB) data file structure has been designed and implemented together with a set of data manipulation functions and supporting SDB data access routines. Logical and physical data descriptors are stored in SDB files preceding the data values. SDB files thus provide a common data representation for interfacing diverse software components. This paper describes the data descriptors and data structures permitted by the file design. Data buffering, file segmentation and a segment overflow handler are also discussed.

Keywords: Data Structures, Data Description, Large Data Bases, Data Management, Statistics, Self-Describing Data Files

Analysis of Large Data Sets

A primary goal of the Analysis of Large Data Sets (ALDS) research project [1,2] at Pacific Northwest Laboratory (PNL) is to develop advanced methodologies and computer software for interactive analysis, management, and display of large volumes of scientific energy-related data. The project, funded by the Applied Mathematical Sciences Research Program of the United States Department of Energy, involves joint interdisciplinary research between statisticians and computer scientists at PNL. The project is guided by an external advisory panel of experts from both disciplines. This paper discusses some of the data management aspects of the ALDS project, with specific emphasis on the design and implementation of a data file type known as Self-Describing Binary (SDB).

Data Management Requirements

A system designed for interactive statistical analysis of large data sets must be supported by data management software which meets a number of functional and performance requirements. A major functional requirement is the ability to store and retrieve data base descriptions. Descriptive information is needed to logically define and characterize the various data elements and their relationships to each other. It is also important to provide a convenient means of naming data structures and collections of data structures. A further requirement is the ability to create and maintain a history of updates and transformations made to the data base as the data analysis process evolves. This capability becomes invaluable in data analyses which are performed in several stages over a long period of time.

In order to avoid time-consuming data conversions from one format to another as different software components (statistics packages, graphical routines, etc.) are utilized, all working data sets must be stored in one common format or representation. This becomes extremely important when working with large data sets. All software components must then include routines to access the data in its common storage representation.

Efficient system performance and high data throughput rates are critical factors in the effectiveness of the interactive data analysis process. Thus the data management software must provide efficient storage and rapid data retrieval. The software must also support data base updates; however, statistical operations generally require fewer and less frequent updates than other types of applications. The software must therefore be optimized for retrieval efficiency at the possible sacrifice of update efficiency.

It became apparent to the ALDS researchers and our advisory panel that existing generalized data base management systems, with all of their built-in overheads, would not meet the above requirements in an effective manner. What is needed is a set of streamlined base-level data manipulation functions which meet the requirements, avoid unnecessary overheads, and can be custom-tailored for statistical operations on large amounts of data.

Self-Describing Binary (SDB) Files

The concept of a self-describing data file was developed to provide a common data representation for ALDS software and to provide the nucleus of a data management system which would meet the requirements of large data set analysis as set forth above. An SDB file logically represents an integrated data set. The file design

*Work supported by the U.S. Department of Energy under Contract DE-AC-06-76RLO 1830.

provides for the storage of file (data set) descriptors, data element descriptors, data element creation/update histories, and data values. Each of these file components is stored in a different logical partition of the file. Pointers and other physical file parameters are also stored in the file to provide the data management software with sufficient information to locate and access the data items. SDB files are binary direct-access files: binary to minimize data storage space, and direct-access to maximize access speed.

A key feature of SDB files is that data values are sequentially stored by column (variable) rather than by row (record). This "transposed" file structure [3,4] was chosen for several reasons. Statistical operations tend to require data values for many or all observations of only a few of the variables at a time. Since all values of a given variable are stored contiguously, only those variables required by the current statistical operation need to be accessed. The transposed file structure will thus require fewer data base accesses than a record-oriented file structure for this type of retrieval. For those situations where logical record-wise retrievals are required, the use of multiple data buffers in the SDB software helps to reduce the number of column-wise physical data accesses.

Another advantage gained by transposed data storage is that data compression techniques can be more easily accommodated. For variables which are restricted to a fixed number of values or categories (e.g., SEX: male,female), bit encoding and radix compression methods can be used to pack multiple data values into a single word of memory. This is not generally possible in record-oriented files where the variables in a record have different cardinalities and/or data types.

A compromise between transposed and record-oriented file structures is achieved by a clustered transposed structure. Several variables which are often used together can be defined to be members of the same cluster. Data values for each variable in the cluster can then be stored row-wise in small sub-records. SDB files will accommodate sub-record clusters, provided that all variables in the cluster have the same data type and cardinality.

Logical File Organization

SDB files can be logically viewed as being composed of multiple variable-length partitions. The logical partitions are completely independent of the physical direct-access records. A logical file partition can span several physical records. Conversely, a physical record may contain portions of several partitions. A single data value (e.g., a 4-byte real number) can even be split across two direct-access records (e.g., one byte at the end of record 29 and the remaining three bytes at the beginning of record 30).

The first partition of every SDB file contains a set of internal file parameters which allow the software to identify and located all other components of the file. Succeeding partitions contain either a vector of data values or a specific type of data description. A data vector partition exists for each variable in the file. Partitions also exist for each of the following types of descriptors:

- Text description of the file (data set).
- File creation/update history: Contains an entry made at file creation time and an additional entry which is added each time the file is updated. Each entry contains the date and time of the update (creation), the name of the person making the update, the software program used to make the update, and text describing the nature of the modification.
- Category value labels (indexed labels). These are concatenated variable-length character strings which are indexed by integer category indices stored in variables of fixed-category data type.
- Logical integrity constraints: Impossible combinations of variable values within a case (record).
- Variable descriptors. These include, for each variable:
  - Label (limited to 16 characters or less)
  - Full text description of the variable (indefinite length)
  - Data type. One of the following: Single precision real, double precision real, integer, fixed length character string, variable length (indexed) character string, fixed categories, linear series (e.g., time series).
  - Number of categories and location of value labels (for fixed-categories data type)
  - Max. and min. allowable values (for numeric data types)
  - Initial and incremental values (for linear series)
  - Missing data codes
  - Summary statistics (mean, min, max, variance)

- Compression indicator
- Data structure description

Data Structures

Although the data values of each variable in an SDB file are stored as a linear vector, this vector by itself or in combination with other variables can represent one of several different types of data structures. The data structure description of a variable defines the type of structure represented by the variable and the relationships of that variable to other variables in the file.

The most common data structure in statistical applications is a simple rectangular case-by-variable (row-by-column) array. Each such array can also be thought of as composed of records which belong to a specific record type. Thus several different rectangular arrays (record types) can exist in the same SDB file. A record type number is stored in the data structure description of each variable belonging to a rectangular array. Variables which form all or part of the data base key for their record type are flagged as such. An example of an SDB representation of a rectangular array is given in Table 1.

Fully crossed data structures (indexed multi-dimensional arrays) are commonly encountered in data analysis. In Table 2, census population figures for a given geographic region are categorized (indexed) by three fixed-category variables: SEX, RACE, and AGE_GROUP. POPULATION is stored as a single variable; its data structure description tells the data management software to interpret the data vector as a three-dimensional array having SEX, RACE, and AGE_GROUP as its ordered indices.

Other types of data structures supported by the SDB file design include one-to-many relations between different record types (hierarchical repeating groups) and many-to-many relations. Pointer vectors, bit maps, repeated value suppression, and other data compression techniques are used to store each data structure as compactly as possible while maximizing retrieval efficiency.

Software Features

To meet the functional and performance requirements of a system for interactive analysis of large data sets, a set of basic data manipulation functions was defined. These functions provide for the creation of SDB files and the addition, deletion, and modification of descriptors and data values within SDB files.

The data manipulation software is implemented at three distinct levels. At the highest level, application interface routines process requests to perform the required data manipulation functions. These requests are defined in terms of the user view of the basic data structures supported by the system and are independent of the details of file design and implementation. The intermediate level routines, however, require detailed specifications of the SDB file implementation. Based on these specifications, they translate logical data requests from the high-level routines into physical data access requests. The lowest level routines implement the data access method. Direct-access reads/writes, data buffering, and file segmentation are performed at this level.

The low level data access software includes a partition overflow handler to accommodate unplanned additions to the partitions of an SDB file. Each partition of a file is composed of one initial segment and zero or more expansion (overflow) segments. The initial segment is allocated a certain amount of space at file creation time based on the initial number of elements (descriptor values or data values) and an anticipated number of future additions as specified by the user-supplied data base definition. Subsequently, if more than the anticipated number of additions to a given partition occur such that the initial segment becomes filled, the overflow handler creates and appends an overflow segment to the end of the file and sets a pointer in the initial segment linking it to the overflow segment. Additional overflow segments can similarly be created as needed. The overflow handler manages all linkages among segments of a logical file partition during updates to and retrievals from the partition such that it appears to the high-level application routines as if each partition is one contiguous indefinitely expandable sub-file within the SDB file.

Data buffering is accomplished via a multiple buffer pool which is managed by a least recently used (LRU) algorithm. Each buffer holds one physical direct-access record. The buffer pool is available uniformly to multiple files; in the current implementation, up to ten SDB files can be open concurrently.

Current Status and Future Directions

A simplified version of the SDB file design and an initial set of data manipulation functions have been implemented on the ALDS computing facility, a Digital

Equipment Corporation VAX-11/780 virtual memory minicomputer. The simplified SDB files are limited to rectangular case-by-variable data structures. These files serve as the common data interface between components of an initial ALDS data analysis system. The initial system includes an expanded and enhanced version of the MINITAB statistical package (including an SDB file interface), a high resolution color graphics capability, and an interactive data editor with powerful subsetting and sampling capabilities [5].

Implementation of the fully expanded SDB file design as described in this paper is just beginning. A data directory capability is also being developed to integrate related data sets stored in multiple SDB files into a common data base and to provide a flexible user view of the data structures and relationships. Plans for the future also include optimization of the data buffering and data access mechanisms, a capability to directly search compressed data, and an interface to a non-procedural query language.

Conclusions

Even in its simplified form, the self-describing file concept has demonstrated its potential for meeting the requirements of a system for interactive analysis of large data sets. Data management software built on this concept and implemented on the virtual memory minicomputers of today and tomorrow promises to provide exciting data management and data analysis capabilities that are affordable by a wide range of users.

References

[1] Thomas, James J., et.al., "Analysis of Large Data Sets on a Minicomputer", Computer Science and Statistics: 12th Symposium on the Interface, University of Waterloo, Waterloo, Ontario, Canada, May 10-11, 1979, pp. 442-446.

[2] Burnett, Robert A., "The Analysis of Large Data Sets Project--Computer Science Research Areas", Proceedings of the 1979 DOE Statistical Symposium, Oak Ridge, TN, September 1980, pp. 205-208.

[3] Turner, M.J., et.al., "A DBMS for Large Statistical Data Bases", Proceedings of the Fifth International Conference on Very Large Data Bases, Rio de Janeiro, Brazil, October 3-5, 1979, pp. 319-327.

[4] Batory, D.S., "On Searching Transposed Files", ACM Transactions on Database Systems, Vol. 4 No. 4, December 1979, pp. 531-544.

[5] Thomas, James J., et.al.,"Data Editing on Large Data Sets", Computer Science and Statistics: 13th Symposium on the Interface, Pittsburgh, PA, 1981.

Table 1. Simple Rectangular (Case-by-Variable) Structure

Variable Label:	STATE_FIPS_CODE	STATE_NAME	POPULATION_1970
Descriptors:	(Integer/Record Type 4/Key)	(Char./Rec.Type 4)	(Integer/Rec.Type 4)
	01	ALABAMA	3444165
	02	ALASKA	302173
	04	ARIZONA	1772482
	⋮	⋮	⋮

Table 2. Fully Crossed (Indexed) Array

Category Variables:

SEX Category	SEX Value Label	RACE Category	RACE Value Label	AGE_GROUP Category	AGE_GROUP Value Label
1	Male	1	White	1	0 to 14
2	Female	2	Black	2	15 to 29
		3	Other	3	30 to 44
				4	45 to 64
				5	65+

Indexed Array Variable:

POPULATION (by SEX, RACE, AGE GROUP)	
1481	(Male, White, 0 to 14)
1552	(Female, White, 0 to 14)
464	(Male, Black, 0 to 14)
498	(Female, Black, 0 to 14)
55	(Male, Other, 0 to 14)
29	(Female, Other, 0 to 14)
1702	(Male, White, 15 to 29)
⋮	⋮

NONLINEAR ESTIMATION USING A MICROCOMPUTER

John C. Nash, Nash Information Services, 188 Dagmar, Vanier, Ontario, Canada

Abstract: The estimation of nonlinear models, possibly involving constraints, can be carried out quite easily using contemporary computers. The estimation process is illustrated using both real-world and artificial problems, the largest problem involving no less than 1250 nonlinear parameters. The formulation of nonlinear estimation problems is presented and various algorithms are suggested for their solution. The particular numerical methods suitable for microcomputer environments are sketched. A discussion of the role of scaling is given. Performance figures are presented for various problems using a North Star Horizon computer and the Radio Shack/Sharp Pocket Computer. It is noted that the microcomputer was able to solve a 41 parameter econometric problem in relatively little time after a service bureau budget had been exhausted in seeking parameter estimates without success.

Keywords: nonlinear estimation, nonlinear least squares, model fitting, function minimization

Problem formulation

Our objective in nonlinear estimation is that, given a set of equations and constraints (the model) which purport to describe a real-world system, and an objective function which measures the success of this model in describing the observed data for the real-world system, find the set of model parameters which maximize the "success" as measured by the objective function.

In stating our problem in this fashion, we ignore the thorny questions of the appropriateness of the model specification or the choice of the objective function. Usually, our objective function will be a loss function which is to be minimized. In this paper, it will be exclusively a least squared error function for the real-world problems. Given that our objective is a loss function, then we can write it as a function of the parameters $\underline{b}$ and the observed data X, that is,

$$S\ (X,\ \underline{b})$$

Nonlinear least squares problems require the definition of the deviations or residuals $\underline{f}$ of the model from some observed value for each observation point, that is,

$f_i(X_i,\ \underline{b})$ is the ith residual so that

$$S\ (X,\ \underline{b}) = \underline{f}^T\ \underline{f}$$

which is a sum of squared terms. Very efficient methods have been developed for the solution of this form so that practitioners of nonlinear estimation frequently try to reduce more general objective functions to this form by transformations.

A number of factors enter into the approach to the solution of nonlinear least squares problems. First, are derivatives of $\underline{f}$ available so that one can form the Jacobian matrix whose (k,j) element is the partial derivative of f_k with respect to b_j? The Jacobian is the central calculation in all the methods of the Gauss-Newton type, such as those associated with the names Hartley and Marquardt. If the Jacobian is not available, can the gradient of S be computed (perhaps numerically), allowing gradient methods for the "general" loss function to be used? Or must we contend with a situation where only function values can be computed because of the complexity or discontinuity of the objective or the model? Similar considerations apply to more general loss functions, with the general, discontinuous (no derivatives) objective function usually considered the most difficult from the point of view of likely success of estimation methods.

As examples of the origin of problems of various types consider the following:

- nonlinear least squares -- fitting a curve to points on a graph in the L_2 norm
- problem transformable to nonlinear least squares -- some maximum likelihood estimations
- general loss functions -- some robust estimators, loss functions defined as an integral over some domain, minimax or L_1 approximation
- discontinuous loss function -- specific choice must be made in model between two operating modes, such as the siting of a product facility.

Including constraints

Constraints often turn a relatively straightforward problem into an almost impossible one. Given the limited memory and computing power of a microcomputer, only a few methods for their inclusion are appropriate. If the constraints are equations, then it is possible in principle to solve for one of the parameters $\underline{b}$ and thus reduce the dimensionality of the problem. Where obvious substitutions present themselves, this is the best technique to use, but the solution of the model equations to find one parameter in terms of the rest may so complicate the system of equations as to render this approach impractical. Projection methods aim to automate the solution process in some approximate way to carry out the "substitution" for us. They can be extended to both equality and inequality constraints, but to date have required large programs unsuitable for small machines. This leaves the penalty and barrier function methods. Continuous and differentiable penalty functions can be added to the loss function to approximate the constraint. Usually a penalty parameter is employed which would, if infinite, force the exact constraint. One then attempts to solve a sequence of problems with increasingly large penalty values in order to approach the desired solution. Unfortunately, this approach may be from the infeasible "side" of the constraint; worse, the large magnitudes of the penalty functions may result in slow convergence of the numerical methods used to minimize the loss function. A discontinuous barrier, created by adding a very large number to the loss function whenever the constraint is violated, is an obvious means for including inequality constraints into our model (it is not advised for equality constraints). However, such barrier functions generally cause havoc for numerical methods requiring derivatives. Even if the computer program used does not specifically call for the user to provide code to calculate derivatives, many methods approximate them internally, and great caution must be exercised in employing barrier functions. Conversely, it should be noted that gradient methods can often be organized to avoid "stepping" into infeasible regions, thereby using a barrier function implicitly. Note that all of the methods discussed use some transformation to alter a constrained problem to give an unconstrained one. So far, mathematical programming -- which would attack the constraints first, then the objective -- has not been developed for micro computer application.

Examples of the types of constraints which can be dealt with successfully are:

b_j positive -- substitute b_j^2 for b_j or make loss function large when parameter negative (barrier)

$b_j\, b_k = b_l\, b_m$ -- solve for one of the four parameters and reduce original problem

By combining the set of constraints into a set of inequations, we can express the overall problem as:

Find the model $\underline{f}\,(X, \underline{b}) = 0$ which satisfies constraints $\underline{c}\,(X, \underline{b}) \geq \underline{0}$ by minimizing S.

<u>Problems considered</u>

Logistic : $x_i = b_1 \,/\, (1 + b_2 \exp(-ib_3))$

Gompertz : $x_i = b_1 \exp(\, b_2 \exp(\, i\, b_3))$

both for observations given by Nash (1979) p. 121.

Houthakker & Taylor Dynamic Linear Expenditure Model: a multi-equation consumer demand model which tries to explain consumer behaviour in terms of a number of elasticities of demand for a series of commodities whose prices and quantities consumed are the known data. The elasticities as well as stocks (habits) are the parameters to be determined. The particular example dealt with had data for 4 commodities (8 variables P and Q) as well as income (total expenditure) over 26 time periods (25 when lags invoked). The model had 4 elasticities for each commodities (16 parameters) and 25 habit parameters for a total of 41. Attempts using (unsuccessful) iterative methods and direct function minimization on an IBM 3033 in Fortran cost over $1000 at a service bureau without yielding satisfactory results.

Large test functions:

$$S(\underline{b}) = \sum_i (b_i - i)^2 + \left(\prod_i (\exp(-(b_i - i)^2 - 1) \right)^2$$

$$T(\underline{b}) = \sum_i Q_i^2 \quad \text{where } Q_i = b_{i+1} - b_i \;;$$

$$Q_N = b_1 - b_N \quad ; \; N = \text{order of problem}$$

$S(\underline{b})$ is actually very easy to solve using conjugate gradients. $T(\underline{b})$ can yield a solution by linear methods, but has multiple solutions which are the eigenvalues of a special matrix.

Rosenbrock's function (banana shaped valley):

$$S = 100\,(b_2 - b_1^2)^2 + (1 - b_1)^2$$

<u>Numerical methods used</u>

The list is incomplete in the sense that others are under consideration and development.

1) sum of squared functions
   a. with Jacobian calculated analytically or numerically, Nash modified Marquardt method (Algorithm 23 of Nash (1979))
   b. with the gradient of the entire sum of squares computed analytically or numerically
      - variable metric algorithm of Fletcher (Algorithm 21 of Nash (1979))
      - conjugate gradients method of Fletcher and Reeves (Algorithm 22 of Nash (1979))
   c. barriers added or no gradients available
      - Hooke and Jeeves method
      - Nelder-Mead polytope method (Algorithms 19 and 20 of Nash (1979))

2) general loss function
   a. with derivatives of S available numerically or analytically
      - variable metric method (Algorithm 21) for modest problems (N.LT.40)
      - conjugate gradients method (Algorithm 22) for large problems
   b. derivatives unavailable
      - Hooke and Jeeves method of storage severely limited (e.g. on a Pocket Computer)
      - Nelder-Mead polytope otherwise (up to N=40 despite "bad press" in the literature.)

<u>Safeguarding the calculations</u>

It has been the author's experience that many users of nonlinear estimation software suffer unnecessary failures due to simple errors or oversights. Many problems have regions where the parameters are clearly invalid -- negative arguments for SQRT or LOG, out of range arguments for EXP -- or are unreasonable in the context of the real problem being solved -- interest rates which are negative or larger than 30%, distances covered by vehicles on a single fuelling. In estimation, it is not only advisable to take account of these considerations, in many cases it is essential. The argument is made by example: Consider the logistic model fit, formulated as

$$f_i = b_1/(1 + b_2 \exp(ib_3)) - x_i$$

The methods proposed above usually fail to find a minimum of the sum of squares function unless a check is made that the argument of the exponential function is less than 50. More stringent checks would include the case $b_3 = 0$ when $b_2 = 1$. It is not necessary to implement the response to violations of such checks as constraints, since most programs can "step back" from undesirable regions of the parameter space if started in a feasible region. Gradient techniques can and should employ checks of this sort.

Scaling

No automatic techniques for scaling have been devised which are resistant to all forms of poor scaling. However, it is generally acknowledged that well-scaled problems are easier and less costly to solve. It is desirable that parameters be roughly equal in magnitude, e.g.

$$f_i = 100\, b_1/(1 + 10\, b_2 \exp(-0.1\, b_3\, i)) - x_i$$

or that the derivatives of S or $\underline{f}$ be roughly equal in magnitude.
The reason such conditions are desirable is that they avoid human errors in entering data (misplaced decimal point), convergence tests are comparable for all parameters, addition of quantities vastly different in magnitude is avoided in a limited precision environment, and geometric methods such as the Nelder-Mead polytope and Hooke and Jeeves avoid highly distorted search regions. The major difficulties in scaling are that the ideal scaling is determined by the unknown solution point and the scale of the problem away from the solution may be very different from this ideal. Nevertheless, Table 1 shows the value of scaling on the two forms of the logistic function given above. Table 2 gives the performance figures on a variety of test problems to indicate what may be accomplished in nonlinear estimation using a microcomputer.

Conclusions and recommendations

The following points may be made as a result of the experience reported here and much which cannot, for reasons of space, be included.

1. Microcomputers are capable of finding the solution to large nonlinear estimation problems.
2. Safeguards on the function evaluation must be included to avoid "unreasonable" regions of the parameter space. Such safeguards both permit solutions to be found and speed the solution process.
3. Safeguards need not be imposed via constraints but can be implemented by simple flags within function minimization methods.
4. Direct search (derivative free) algorithms can be used effectively for quite large problems as shown by the Dynamic Expenditure Model example.
5. Hooke and Jeeves method on the Pocket Computer allows real-world problems to be attacked on a truly portable machine, though the performance is slow.
6. Direct search algorithms allow simple barrier function constraints to be used.
7. BUT (!!) do not try them on gradient methods or implicitly gradient methods.
8. Scaling reduces errors and speeds up the solution process.
9. Marquardt type algorithm is best for modest to medium-sized nonlinear least squares problems.
10. Conjugate gradient methods can be applied to very large problems if gradient can be computed.

Reference

Nash, J.C. 1979, Compact numerical methods for computers, Adam Hilger: Bristol, and Halsted Press: New York (US rights only).

Table 1. Scaling of nonlinear estimation problems.

Method	Starting $\underline{b}$	Function	Value	Time	Fn. value
Nash-Marquardt	1,1,1	unscaled	24349	111 secs.	2.5872735
	1,1,1	scaled	10685	82 secs.	2.5872698
Variable metric	1,1,1	unscaled	24349	144 secs.	2.5872713
	1,1,1	scaled	10685	103 secs.	2.5872675
Nelder Mead polytope	1,1,-1	unscaled	23521	113 secs.	9205 (failed!)
	1,1,1	scaled	10685	187 secs.	2.5872677

Table 2. Performance of microcomputers on nonlinear estimation problems

Computer:
North Star Horizon - 8 digit hardware floating point BASIC

Small problems:

Rosenbrock from (-1.2, 1.0) to solution

19 Jacobian and 25 function evaluations with Nash-Marquardt	22.6 secs.
48 gradient and 183 function evaluations with conjugate gradients (modified)	35.1 secs.
39 gradient and 59 function evaluations with variable metric	23.5 secs.
202 function evaluations (step to create polytope 0.3) with polytope	36.2 secs.

Rosenbrock + b_i^2 i=3,4,..., 10 from -1.2,1,1,...,1)

24 Jacobian and 46 function evaluations with Nash-Marquardt	485.6 secs.
80 gradient and 240 function evaluations with mod. conjugate gradients	126 secs.
62 gradients and 116 function evaluations with variable metric	302 secs.
3907 functions with Nelder Mead polytope	3474 secs.

Medium sized problems:

Dynamic Linear Expenditure model, from starting values defined by ordinary least squares and "guesses" N=41 for general problem

Unconstrained problem took 2400 function evaluations of polytope	16 hrs.
Add barrier to stop parameter becoming positive(further 1500 evals.)	+ 9 hrs.
Remove offending parameter, use Marquardt 23 Jacobian and 42 fn evals	+ 3.5 hrs.

Above runs are sequential in taking last values of previous runs for starting point. Final Nash Marquardt calculation estimates standard errors.

Large problems: the function labelled S and T are all minimized using a modified conjugate gradients method. N is the order of the problem.

N	Function S time	gradients	functions	Function T time	gradients	functions
3				2.4s	2	5
5	6.6s	5	10			
10	8.3s	4	7	18.1s	7	19
20				50.9s	11	25
50				276.3s	29	64
100	67.7s	4	8	1837.6	100	211
500	246.8s	3	6			
1000	496.0s	3	6			
1250	620.8	3	6	44 hours	675	1359

(manually stopped; function value small)

STATISTICAL PROCEDURES FOR LOW DOSE EXTRAPOLATION OF QUANTAL RESPONSE TOXICITY DATA

John Kovar and Daniel Krewski
Health and Welfare Canada

Abstract

Maximum likelihood procedures for fitting the probit, logit, Weibull and gamma multi-hit dose response models with independent, additive or mixed independent/additive background to quantal assay toxicity data are reviewed. In addition to parameter estimation, the use of the above models for low dose extrapolation is indicated with both point estimates and lower confidence limits on the "safe" dose discussed. A computer program implementing these procedures is described and two sets of toxicity data are analyzed to illustrate its use.

Key words: virtually safe dose, risk estimation, dose response models, maximum likelihood estimation.

1. Introduction

As a result of the increasing awareness of the potential health hazards of environmental chemicals, considerable effort is currently being devoted to the identification and regulation of those compounds which are carcinogenic as well as those which have other toxic effects. Information on the toxic effects of such chemicals is necessarily derived primarily from bioassay studies conducted using animal models. Since the number of animals used must be compatible with the practical limitations imposed by space, time and cost, the dose levels employed must be sufficiently large so as to induce an observable rate of response. In order to estimate risks in the human population, it is thus necessary to first extrapolate from the results obtained at these high dose levels to lower levels consistent with the levels anticipated in the environment and then to make the conversion from animal to man.

Statistical procedures for low dose extrapolation involve a mathematical model under which the probability P(d) of an induced response at dose d is assumed to have a particular functional form, say P(d) = f(d). Here, we will consider only non-threshold models for which f(d) is strictly increasing in d for all $d \geq 0$. While absolute safety can only be guaranteed in this case when d = 0, a virtually safe level of exposure may be defined by $d^* = f^{-1}(p)$, where p is some suitably low level of risk.

In section 2, existing mathematical models are reviewed, with particular attention devoted to the probit, logit, Weibull and gamma multi-hit models. The accommodation of background response assuming either independence, additivity or a combination thereof is described. Maximum likelihood procedures which may be employed in fitting these models are then discussed along with procedures for assessing goodness-of-fit. Methods for obtaining upper confidence limits on the risk at a given dose and lower confidence limits on the dose corresponding to a specified risk are presented, along with a conservative linear procedure.

In section 3, the application of these techniques is illustrated using two sets of quantal response toxicity data taken from the literature. Features of the computer program used to obtain these results are summarized in the Appendix.

2. Low Dose Extrapolation

2.1 Mathematical Models

Statistical or tolerance distribution models are based on the notion that each animal in the population has its own tolerance to the test compound (Krewski and Van Ryzin, 1981). Three well known models of this type are the probit, logit and Weibull models defined respectively by

$$P(d) = (2\pi)^{-1/2}\int_{-\infty}^{\alpha+\beta\log d}\exp(-u^2/2)du, \quad (\beta>0) \qquad 2.1$$

$$P(d) = [1+\exp(-\alpha-\beta\log d)]^{-1}, (\beta>0) \text{ and} \qquad 2.2$$

$$P(d) = 1-\exp(-\lambda d^m), (\lambda, m>0) \qquad 2.3$$

$$= 1-\exp(-\exp(\alpha+\beta\log d)), (\beta=m>0, \alpha=\ln\lambda).$$

Stochastic or mechanistic models are based on the notion that for each animal, a positive response is the result of the random occurrence of one or more biological events. The gamma multi-hit model is based on the concept that a response will be induced after the target site has been hit k times by a single biologically effective unit of dose within a specified time interval. If the number of hits during this period follows a homogeneous Poisson process in d, then the probability that an individual will respond at dose d is

$$P(d) = \int_0^d [\Gamma(k)]^{-1}\lambda^k t^{k-1}\exp(-\lambda t)dt \qquad 2.4$$

(Rai & Van Ryzin, 1979, 1981), where $\Gamma(k)$ denotes the gamma function. P(d) in (2.4) may also be interpreted as a tolerance distribution model where the tolerance distribution is gamma with scale parameter $1/\lambda$ and shape parameter $k>0$.

Another stochastic model is based on the assumption that the induction of irreversible self-replicating toxic effects such as carcinogenesis is the result of the occurrence of a number of different random biological events. This and other assumptions (see Crump, Hoel, Langley & Peto (1976) or Crump (1979) for details) leads to a multi-stage model of the form

$$P(d) = 1-\exp[-\Sigma_{i=1}^{k}\beta_i d^i], (\beta_i \geq 0), \qquad 2.5$$

where k denotes the number of events or stages. While we will not consider this model here, we note that efficient statistical procedures for fitting this model have been developed by Guess and Crump (1976, 1978) and Hartley and Sielken (1977).

The shape of the dose-response curves for the above models in the low dose region will have considerable impact on estimates of risk associated with low levels of exposure (Figure 1). The logit, Weibull and multi-hit models are linear at low doses only when the shape parameters β, m and k in these models are equal to unity. When these parameters are greater than unity, the dose response curves for these models approach zero at a slower than linear or sublinear rate. The probit model is inherently sublinear at low doses and generally leads to relatively low estimates of risk at low dose levels.

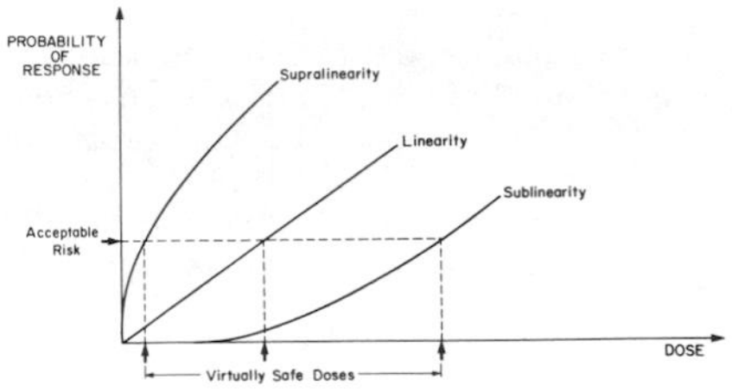

Figure 1. Linearity, sublinearity and supralinearity at low doses.

The dose response curves for the logit, Weibull and multi-hit models can approach zero at a faster than linear or supralinear rate when the slope parameters β, m and k are less than unity, although the biological plausibility of this behaviour seems questionable.

2.2 Incorporation of Background Response

All of the models discussed previously assume that the background response rate P(0) is zero. In many experiments, however, the response of interest also occurs spontaneously in control animals. This background may be assumed to be independent of the induced responses or additive in a mechanistic manner (Hoel, 1980). If the spontaneous and induced responses are assumed to be independent, then the probability of observing a response of either type at dose d is given by

$$P^*(d) = \gamma + (1-\gamma)P(d) \qquad 2.6$$

where $0 < \gamma < 1$ denotes the spontaneous response rate. Under the additivity assumption, the background response rate may be considered as arising from an effective background dose $\delta > 0$, with

$$P^*(d) = P(d + \delta) \qquad 2.7$$

A combination of both independent and additive background may be represented by the model

$$P^*(d) = \gamma + (1-\gamma)P(d + \delta). \qquad 2.8$$

In the presence of spontaneously occurring responses, a virtually safe level of exposure may be defined in terms of the acceptable increment Π in the added risk over background (Figure 2),

$$\Pi(d) = P^*(d) - P^*(0). \qquad 2.9$$

Assuming independence, the low dose behaviour of the added risk may be nonlinear as in the case of background. Assuming additivity, however, the added risk will quite generally be linear at low doses (Crump, Hoel, Langley & Peto, 1976). This conclusion is valid in those cases where the test compound increases the spontaneous rate of response through the acceleration of an already ongoing process.

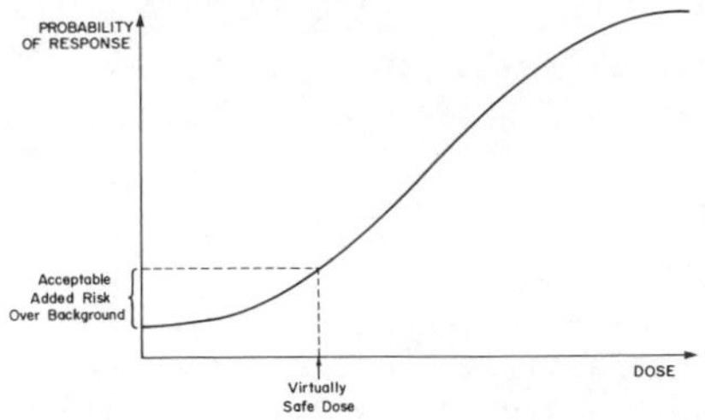

Figure 2. Determination of a virtually safe dose in the presence of background

2.3 Maximum Likelihood Estimation

Suppose that a total of n animals are used in an experiment involving dose levels $0 \le d_1 < \ldots < d_m$ and that x_i of the n_i animals at dose d_i respond (i=1,..., m). We now consider the estimation of the excess risk $\Pi(d)$ and the VSD d* on the basis of these data.

Let $P_i^* = P^*(d_i;\underset{\sim}{\theta})$ denote the probability of a positive response at dose d_i under any dose response model $P^*(d;\underset{\sim}{\theta})$ with (t dimensional) parameter $\underset{\sim}{\theta}$. For the probit, logit, Weibull and gamma multi-hit models, moreover, t=2, 3 or 4 according to whether no background, either additive or independent background, or mixed background is present. Assuming that each animal responds independently of all other animals in the experiment, the likelihood of the observed outcome is given by

$$L(\underset{\sim}{\theta}) = \Pi_{i=1}^{m} \binom{n_i}{x_i} (P_i^*)^{x_i}(Q_i^*)^{n_i-x_i}, \qquad 2.10$$

where $Q_i^* = 1 - P_i^*$. Since maximization of $L(\underset{\sim}{\theta})$ or $\ell(\underset{\sim}{\theta}) = \log L(\underset{\sim}{\theta})$ using direct analytical procedures is generally not possible, the maximum likelihood estimator $\hat{\underset{\sim}{\theta}}$ of $\underset{\sim}{\theta}$ must be obtained using iterative numerical procedures as discussed in the Appendix. The upper limit of $L(\underset{\sim}{\theta})$ may be obtained by replacing P_i^* by x_i/n_i and Q_i^* by $(n_i-x_i)/n_i$ in (2.10).

An estimator of $\Pi(d)$ is given by

$$\hat{\Pi}(d) = \hat{P}^*(d)-\hat{P}^*(0) \qquad 2.11$$

where $\hat{P}^*(d) = P^*(d_i;\hat{\underset{\sim}{\theta}})$, and that of d* by

$$d^* = \hat{\Pi}^{-1}(\pi). \qquad 2.12$$

Under reasonable regularity conditions, it can be shown (Krewski & Van Ryzin, 1981) that $\sqrt{n}(\hat{\Pi}(d) - \Pi(d))$, $\sqrt{n}(\hat{d}^* - d^*)$, $\sqrt{n}(\log\hat{d}^* - \log d^*)$ and $\sqrt{n}(1/\hat{d}^* - 1/d^*)$ are asymptotically normally distributed with zero means and variances

$$\sigma_1^2 = \sum_{r=1}^{t} \sum_{s=1}^{t} \frac{\delta\Pi}{\delta\theta_r} \frac{\delta\Pi}{\delta\theta_s} \sigma^{rs}, \quad \sigma_2^2 = (\delta\Pi/\delta d|_{d=d^*})^{-2}\sigma_1^2,$$

$$\sigma_3^2 = (d^*)^{-2}\,\sigma_2^2 \text{ and } \sigma_4^2 = (d^*)^{-4}\,\sigma_2^2 \text{ respect-}$$

ively, where $((\sigma^{rs}))$ is the information matrix for $\underset{\sim}{\theta}$. Thus approximate $100(1-\rho)\%$ upper confidence limit for $\Pi(d)$ at any fixed dose d>0 is given by

$$U(d) = \hat{\Pi}(d) + z_{1-\rho}\hat{\sigma}_1/\sqrt{n} \qquad 2.13$$

where $z_{1-\rho}$ is the $100(1-\rho)$ percentile of the cumulative normal distribution function. Similarly $100(1-\rho)\%$ lower confidence limits on d^* may be obtained by taking the anti-log of the lower confidence limit for $\log d^*$

$$L_1(d) = \exp\{\log \hat{d}^* - z_{1-\rho}\hat{\sigma}_3/\sqrt{n}\}. \qquad 2.14$$

or by inverting the upper confidence limit for $1/d^*$

$$L_2(d) = 1/\{1/\hat{d}^* + z_{1-\rho}\hat{\sigma}_4/\sqrt{n}\}. \qquad 2.15$$

Finally, we consider statistical goodness-of-fit tests which may be applied in order to determine whether a given model provides a reasonable fit to the experimental data in the observable range. It can be shown that the usual chi-square goodness-of-fit test

$$\chi^2 = \Sigma(x_i - n_i\hat{P}_i^*)^2/(n_i\hat{P}_i^*\hat{Q}_i^*) \qquad 2.16$$

has a chi-square distribution with m-t degrees of freedom provided that the assumed model $P^*(d;\theta)$ is correct (Krewski and Van Ryzin, 1981). Similarly, the G-square statistic

$$G^2 = 2(\sup \ell(\underset{\sim}{\theta}) - \ell(\hat{\underset{\sim}{\theta}})) \qquad 2.17$$

has the same asymptotic properties as χ^2 (Bishop, Fienberg and Holland, 1975). Then p, defined as the probability that a χ^2 random variable with m-t degrees of freedom will exceed the observed value of the test statistic, measures the degree to which the observed data are consistent with the assumed model, with small values of p suggesting lack of fit.

2.4 Linear Extrapolation

Because of the uncertainties involved in assessing risks at low levels of exposure, some regulatory authorities have advocated the use of conservative risk assessment procedures based on the assumption of low dose linearity (Interagency Regulatory Liaison Group, 1979; U.S. Environmental Protection Agency, 1980). A simple extrapolation procedure (Van Ryzin, 1980; Gaylor & Kodell, 1980) which provides for the possibility of low dose linearity involves fitting a suitable model to the experimental data and then extrapolating linearly from some point on the fitted curve where the excess risk is still within the observable range. This procedure not only accommodates low dose linearity, but will provide a conservative upper limit on risk at low levels of exposure whenever the true dose response curve is sublinear in the low dose region.

Letting $d^*_0 = \hat{\Pi}^{-1}(\pi_0)$, where π_0 denotes a level of risk such as 10^{-2} or 10^{-1} within the observable range, an estimate of the upper limit $\Pi_X(d)$ on the excess risk at any dose $d \le d^*_0$ is given by $\Pi_X(d) = d\pi_0/\hat{d}^*_0$. An estimate of the corresponding lower limit $d^*_X = \Pi_X^{-1}(\pi)$ on d^* is given by $\hat{d}^*_X = (\pi/\pi_0)\hat{d}^*_0$. Upper and lower confidence limits on $\Pi_X(d)$ and d^*_X respectively may be obtained through the use of a lower confidence limit on d^*_0 in place of the point estimate $\hat{d}^*_0$.

3. Applications and Examples

Application of the dose response models and statistical procedures discussed in section 2 will be illustrated using actual experimental data on the toxic responses induced by dieldrin and ethylenethiourea (ETU), as assembled by the Food Safety Council (1980). The data obtained for the two examples is presented in Table I.

Table I. Dose Response Data for Two Substances

Substance (Species)	Type of Response	Response Data: Dose	# responding / # on test
Dieldrin (Mouse)	Liver tumor (ppm)	0	(17 /156)
		1.25	(11 /60)
		2.50	(25 /58)
		5.00	(44 /60)
ETU (Rat)	Fetal Anomalies (mg/kg)	0	(0/167)
		5	(0/132)
		10	(1/138)
		20	(14/81)
		40	(142/178)
		80	(24/24)

The probit, logit, Weibull and gamma multi-hit models with independent, additive and mixed background were fitted to the two data sets using a program described in the Appendix. The results of the low dose extrapolation based on the models with independent background are presented pictorially in Figure 3 along with the fitted Weibull model.

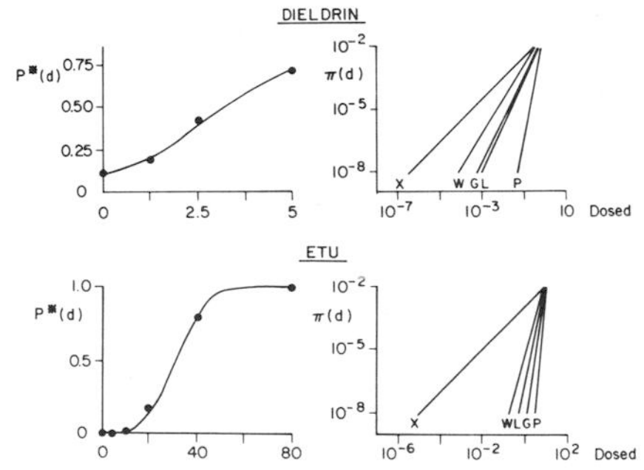

Figure 3. Fitted Weibull model and estimates of excess risk based on 5 extrapolation procedures

W - Weibull; L - Logit; X - Linear Extrapolation; G - Gamma multi-hit; P - Probit

The parameter estimates under all four models with independent or additive background are presented in Table II. In both cases the mixed background model selected the corresponding independent or additive background model as the best fit. Point estimates and lower confidence limits under five extrapolation procedures are given in Table III.

Appendix: RISK 81: A computer program for risk assessment

RISK 81 uses a modified Fletcher and Powell (1963) routine to globally maximize the log-likelihood function under one or more of the twelve models described in section 2.1. The program then computes the covariance matrix of the estimates and

Table II. Maximum likelihood estimates of the parameters for the probit, logit, Weibull and gamma multi-hit models with independent or additive background

Model (Background)*	Dieldrin α**	Dieldrin β**	Dieldrin γ or δ	ETU α	ETU β	ETU γ or δ
Probit (I)	-1.6	1.3	.11	- 8.3	2.5	0
Probit (A)	-39	11	27	-11.5	3.2	6.7
Logit (I)	-2.7	2.2	.11	-14.8	4.4	0
Logit (A)	-26	9.6	12.6	-14.8	4.4	0
Weibull (I)	-2.5	1.7	.11	-12.0	3.4	0
Weibull (A)	-6.3	3.0	3.9	-12.0	3.4	0
Gamma (I)	.57	2.3	.11	.23	7.2	0
Gamma (A)	1.4	16	7.8	.23	7.2	0

* I and A stand for independent and additive background respectively.

** For the gamma multi-hit models α and β stand for λ and k respectively.

Table III. Point estimates of the virtually safe doses (VSD's) at an added risk over background of 10^{-5} and 95% lower confidence limits (LCL's) based on 5 extrapolation procedures.

Substance	Model*	VSD	L_1	L_2
Dieldrin	Probit	.142	.0497	.0694
	Logit	.0199	.00329	.00712
	Weibull	.00462	.000441	.00138
	Gamma	.0184	.00236	.00602
	X**($\pi=10^{-2}$)	.000295	.000118	.000154
	X($\pi=10^{-1}$)	.000122	.0000780	.0000842
ETU	Probit	5.15	3.94	4.06
	Logit	2.13	1.40	1.50
	Weibull	1.16	.669	.749
	Gamma	3.28	2.15	2.30
	X($\pi=10^{-2}$)	.00893	.00710	.00727
	X($\pi=10^{-1}$)	.00179	.00158	.00159

* The models considered in this table are those with independent background only.

** X stands for linear extrapolation based on the Weibull model with independent background from a risk π.

their standard errors. After evaluating the goodness-of-fit, RISK 81 computes the VSD's for eight values (10^{-i}, i=1, 8) of added risk over background. Depending on the user instructions, the program will then compute lower confidence limits on the VSD's based on either L_1(2.14), L_2(2.15) or both. The confidence limits are evaluated at four critical levels. Finally, if any environmental doses were input, the program will evaluate the associated added risk over background for each such dose under all of the models that were fitted. As well, upper confidence limits U(2.13) on the risk will be computed.

A tape of the computer program along with the user instructions is available upon request at cost.

References

Bishop, Y.M.M., Fienberg, S.E. and Holland, P.W. (1975). Discrete Multivariate Analysis: Theory and Practice, MIT Press, Cambridge, Mass.

Crump, K.S., Hoel, D.G., Langley, C.H. & Peto, R. (1976). Fundamental carcinogenic processes and their implications for low dose risk assessment. Cancer Research 36, 2973-2979.

Crump, K.S. (1979). Dose response problems in carcinogenesis. Biometrics 35, 157-167.

Fletcher, R. and Powell, M.J.D. (1963). A rapidly convergent descent method for minimization. Computer Journal 6, 163-168.

Food Safety Council (1980). Proposed system for Food Safety Assessment. Food Safety Council, Washington, D.C.

Gaylor, D.W. & Kodell, R.L. (1980). Linear interpolation algorithm for low dose risk assessment of toxic substances. Journal of Toxicology and Environmental Health (to appear).

Guess, H.A. & Crump, K.S. (1976). Low dose extrapolation of data from animal carcinogenicity experiments - analysis of a new statistical technique. Mathematical Bioscience 32, 15-36.

Guess, H.A. & Crump, K.S. (1978). Maximum likelihood estimation of dose-response functions subject to absolutely monotonic constraints. Annals of Statistics 6, 101-111.

Hoel, D.G. (1980). Incorporation of background in dose-response models. Federation Proceedings 39, 73-75.

Hartley, H.O. & Sielken, R.L. (1977). Estimation of "safe doses" in carcinogenic experiments. Biometrics 33, 1-30.

Interagency Regulatory Liaison Group. (1979). Scientific bases for identification of potential carcinogens and estimation of risks. Journal of the National Cancer Inst. 64, 241-268.

Krewski, D., & J. Van Ryzin. (1981). Dose response models for quantal response toxicity data. In Current Topics in Probability and Statistics, Csorgo, M., D. Dawson, J.N.K. Rao, and E. Saleh (eds.), North-Holland, New York. (in press).

Rai, K. & Van Ryzin, J. (1979). Risk assessment of toxic environmental substances based on a generalized multi-hit model. In Energy and Health. N. Breslow & A. Whittemore (eds.), SIAM, Philadelphia, 99-117.

Rai, K. & Van Ryzin, J. (1981). A generalized multi-hit dose response model for low-dose extrapolation. Biometrics (to appear).

U.S. Environmental Protection Agency (1980). Water quality criteria documents: availability. Federal Register 45, 79318-79319.

Van Ryzin, J. (1980). Quantitative risk assessment. Journal of Occupational Medicine 22, 321-326.

AN ECONOMIC DESIGN OF $\bar{X}$-CHARTS WITH WARNING LIMITS TO CONTROL NON-NORMAL PROCESS MEANS

M.A. Rahim, University of Windsor
R.S. Lashkari, University of Windsor

ABSTRACT

In this paper, we develop an expected cost model for a production process under the surveillance of an $\bar{x}$-chart with warning limits for controlling the non-normal process mean. The economic design of control charts involves the optimal determination of the design parameters that minimize the expected total cost of monitoring the quality of the process output. The design parameters of a general control chart with warning limits are the sample size, the sampling interval, the action limit coefficient, the warning limit coefficient, and the critical run length. To develop the expected loss-cost function, expressions for the average run lengths, when the process is in control, and when the process is out of control are derived. A direct search technique is employed to obtain the optimal values of the design parameters. The effects of non-normality parameters on the loss-cost function and on the design parameters are discussed using a numerical example.

($\bar{X}$-CHARTS; NON-NORMAL PROCESSES; WARNING LIMITS; OPTIMIZATION)

INTRODUCTION

The control chart technique was first introduced by Shewhart [26] and it has since been used with considerable success in quality control situations. In Shewhart's $\bar{x}$-charts the control limits or action limits are set at $\pm$ k standard deviations of the sample mean from the target value. The process is subject to the occurrence of a single assignable cause of variation which takes the form of a shift in the process mean from $\mu \pm \delta\sigma$, where σ is the process standard deviation and δ is the shift parameter. Samples of fixed size are taken at regular intervals of time and the sample means are plotted on the $\bar{x}$-chart. If a sample mean falls outside the action limits, the process is considered to be out of control and some rectifying action is taken.

In order to increase the sensitivity of the Shewhart's $\bar{x}$-control charts, Page [22] introduced the $\bar{x}$-control charts with both action and warning limits. In an $\bar{x}$-chart with both warning and action limits, a search for the assignable cause is undertaken if the last sample mean falls outside the action limits, or if the last sample mean completes a run of specified length which lies in between the warning and action limits.

In the remainder of this paper, unless otherwise mentioned, we will call an $\bar{x}$-chart with action limits an "$\bar{x}$-chart", and an $\bar{x}$-chart with both action and warning limits an "$\bar{x}$-chart with warning limits".

To use an $\bar{x}$-chart with warning limits, the user specifies the sample size, the sampling interval, the coefficients of the action and warning limits, and the value of the critical run length. Selection of the parameters of a control chart such that an "income" measure is maximized is called the economic design of the chart. Economic design of $\bar{x}$-charts used to maintain the current control of a production process was first developed by Duncan [7], and was later extended by others [for example 1,5,8,11,12]. Economic design of $\bar{x}$-charts with warning limits was first investigated by Gordon and Weindling [13], and was later extended by Chiu and Cheung [4].

The assumption in the works mentioned above is that the process characteristic under investigation is a normally distributed random variable. However, in many industrial situations, process variables do not conform to the assumption of normality. In such cases, conventional control charts may wrongly indicate the state of the process. Several Studies [6,9,10,14,15,16,20,25,27] have dealt with the effects of non-normality on control charts, and the consequences of incorrectly assuming normality. But a quantitative assessment of the effects of non-normality on the design of control charts has been generally lacking.

Following Duncan's work for normal processes, Nagendra and Rai [21], Raouf et al. [24], and Lashkari and Rahim [18], investigated the economic design of $\bar{x}$-charts to control non-normal process means. Also, Lashkari and Rahim [19] developed an economic design of cusum charts under non-normality assumption.

In this paper we develop a per hour loss-cost function for an $\bar{x}$-chart with warning limits, in terms of the parameters of the chart (i.e., the sample size, sampling interval, the coefficients of action and warning limits, and the critical run length). A pattern search technique is employed to obtain the optimum values of the parameters by minimizing the loss-cost function. Finally, numerical examples are provided to demonstrate the effects of non-normality on the loss-cost function and design parameters of the chart.

FORMULATION OF LOSS-COST FUNCTION

A production process which has two states, in-control and out-of-control, is considered. The process is assumed to start in a state of in-control. The quality characteristic of the process variable is measurable on a continuous scale, and is assumed to have a non-normal distribution with probability density function $f(\mu_0, \sigma^2, \beta_1, \beta_2)$ with mean μ_0, variance σ^2, measure of skewness β_1, and measure of kurtosis β_2. The process is subject to the occurrence of an assignable cause, which shifts the process mean from μ_0 to $\mu_0 + \delta\sigma$, where δ is the shift parameter, and σ, β_1 and β_2 are assumed to remain stable throughout the working period. It is assumed that the assignable cause occurs according to a Poisson process with mean occurrence rate of λ.

The process is assumed to be shut-down during the search for the assignable cause. A sample of fixed size n is taken at regular intervals of time and the sample mean is plotted on a one-sided $\bar{x}$-chart with warning limits. The upper action limit is set at $\mu_0 + k_a\sigma/\sqrt{n}$, where k_a is the upper control limit coefficient. The upper warning limit is set at $\mu_0 + k_w\,\sigma/\sqrt{n}$ where $0 < k_w < k_a$. A search for the assignable cause is undertaken if the last sample mean falls outside the action limit, or if the last sample mean completes a critical run length R_c between the warning and action limits. An expected time of τ_s hours and an average search cost of K_s are required if the assignable cause does not exist in which case the production is resumed after the search. If the assignable cause actually exists, it can always be discovered and eliminated, but it takes a further average repair time of τ_r hours and a further repair cost of K_r to restore the process to the state of in-control. No assignable cause is assumed to occur while taking a sample. The cost of taking a sample and maintaining the control chart is assumed to be adequately represented by the linear relationship (b+cn). Finally, it is assumed that V_0 is the profit per hour earned by the process when operating in-control and V_1, the profit per hour earned by the process when operating out of control.

Following the general outlines of the works of Duncan [7] and, Chiu and Cheung [4], the loss-cost function of the process under the surveillance of an $\bar{x}$-chart with warning limits for controlling the non-normal process means can be formulated as follows.

Let T_a be the random time during which the process operates under the state of control. By assumption T_a has an exponential distribution with $E(T_a)=1/\lambda$. Let M be the number of samples taken before the process goes out of control, and G the number of samples taken after the Mth sample and up to the moment the chart signals lack of control. Let N be the number of false alarms occurring among the first M samples. Then it is straightforward to see that the expected length of the production cycle consists of four parts: a) the in-control period, b) the search times due to false alarms, c) the out of control period and d) the search and repair times due to true alarms.

Thus, the expected length of a production cycle is:

$$sE(M) + \tau_s E(N) + sE(G) + \tau_s + \tau_r \quad \ldots(1)$$

and the expected income from a production cycle is:

$$V_0/\lambda + V_1 E(Ms+Gs-T_a) - E(N)K_s - (b+cn)E(M+G) - K_s - K_r \quad (2)$$

Hence the average net income per hour is:

$$I = \text{equation (2)/equation(1)} \quad \ldots(3)$$

The assignable cause occurs t_1 time units after the Mth sample is taken, such that:

$$E(t_1)=E(T_a-Ms)=\{1-(1+\lambda s)\exp(-\lambda s)\}/\{\lambda-\lambda\exp(-\lambda s)\} \cong s/2 - \lambda s^2/12 \quad \ldots(4)$$

Thus from the equation (4):

$$E(M)=1/\lambda s-\tfrac{1}{2}+\lambda s/12 \quad \ldots(5)$$

To determine the expected number of false alarms during the first M samples, we have for fixed M [3]: $E(N|M) \cong M/R_0$ where R_0 is the average run length (ARL) of $\bar{x}$-chart with warning limits at the acceptable quality level μ_0. Thus from equation (5):

$$E(N) \cong E(M)/R_0 = \{1/\lambda s-\tfrac{1}{2}+\lambda s/12\}/R_0 \quad \ldots(6)$$

Taylor [28] has shown, by computer simulation, that the dependence of E(G) on M is negligible, and that it could be approximated by:

$$E(G) \cong R_1 \quad \ldots(7)$$

where R_1 is the ARL of the chart at the rejectable quality level, $\mu_1 = \mu_0 + \delta\sigma$. Thus, substituting equations (6) and (7) into equation (3) and defining $U = V_0 - V_1$; $V = K_s + V_0\tau_s$;

$W = K_r + K_s + V_0(\tau_r+\tau_s)$; $B_0 = (1/s-\lambda/2+\lambda^2 s/12)/R_0$;

$B_1 = (R_1-\tfrac{1}{2}+\lambda s/12)s$; $L = V_0 - 1$, we have,

$$L = \frac{\lambda U B_1 + V B_0 + \lambda W + (b+cn)(1+\lambda B_1)/s}{1+\lambda B_1 + \tau_s B_0 + \lambda(\tau_r+\tau_s)} \quad \ldots(8)$$

where L represents the average long run per-hour loss-cost of the process.

EFFECT OF NON-NORMALITY ON LOSS-COST FUNCTION

Before minimizing equation (8) to obtain the optimum design parameters of the $\bar{x}$-chart with warning limits, it is noted that the values of the average run lengths R_0 and R_1 are dependent on the probability density function of the process variable, which, by our assumption, is non-normal. To take into consideration the effects of non-normality on the control charts design it is assumed that the first four terms of the Edgeworth series expansion provide an adequate approximation to the statistical distribution of the quality characteristic [29]. Denoting the quality characteristic of the product by the random variable X, the probability density function (pdf) of the standardized sample average $y = (\bar{x}_n-\mu_0)/\,\sigma/\sqrt{n}$ is given by [10]:

$$g(y)=f_n(y)=\phi(y)-\frac{\gamma_1}{6\sqrt{n}}\phi^{(3)}(y)+\frac{\gamma_2}{24n}\phi^{(4)}(y)+\frac{\gamma_1^2}{72n}\phi^{(6)}(y) \quad (9)$$

where $\phi(x)$ is the pdf of the standardized normal variate x; $\phi^{(r)}(x) = (d/dx)^r\phi(x)$; $\gamma_1 = \sqrt{\beta_1}$; $\gamma_2 = \beta_2 - 3$.

It is assumed that the function f(x) is positive definite and unimodal, and the measures of skewness and kurtosis satisfy the condition: $\beta_2 \geq 1 + \beta_1$ as specified by Barton and Dennis [2].

Now the average run length, for a one-sided $\bar{x}$-chart with warning limits as given by Page [23], is:

$$ARL = (1-q^{R_c})/[1-q-p(1-q^{R_c})] \quad \ldots(10)$$

where R_c is the critical run lengths, p is the probability that a point falls below the warning limit, and q is the probability that a point falls between the warning and action limits. When the process is out of control, we have

$$p = \Phi(k_w-\delta\sqrt{n}) - \frac{\gamma_1}{6\sqrt{n}}\phi^{(2)}(k_w-\delta\sqrt{n})+\frac{\gamma_2}{24n}\phi^{(3)}(k_w-\delta\sqrt{n})+\frac{\gamma_1^2}{72n}\phi^{(5)}(k_w-\delta\sqrt{n}) \quad \ldots(11a)$$

and

$$q = \Phi(k_a-\delta\sqrt{n}) - \Phi(k_w-\delta\sqrt{n}) - \frac{\gamma_1}{6\sqrt{n}} [\phi^{(2)}(k_a-\delta\sqrt{n}) - \phi^{(2)}(k_w-\delta\sqrt{n})] + \frac{\gamma_2}{24n} [\phi^{(3)}(k_a-\delta\sqrt{n}) - \phi^{(3)}(k_w-\delta\sqrt{n})] + \frac{\gamma_1}{72n} [\phi^{(5)}(k_a-\delta\sqrt{n}) - \phi^{(5)}(k_w-\delta\sqrt{n})] \quad ..(11b)$$

where Φ denotes the distribution function of the unit normal variate. Thus, R_1 is obtained by substituting equations (11a,b) into equation (10). Similarly, when we let $\delta = 0$ in equations (11a,b) and substitute the resulting p and q into equation (10) we obtain R_0. These expressions will then be used in equation (8) for locating the minimum position of L.

DETERMINATION OF THE OPTIMAL DESIGN PARAMETERS

In order to obtain the optimum control plan, we minimize the objective function L, given by equation (8), with respect to the design variables, i.e., n and s, the coefficients of action and warning limits k_a and k_w, and the critical run length, R_c. The dependence of L on three parameters, k_a, k_w, and R_c, through equations (10) and (11a,b) precludes the use of any analytical optimization method. Rather, the direct search method of Hooke and Jeeves [17] is employed to minimize L with respect to the vector of variables (n, s, k_a, k_w, R_c), However, due to the characteristics of function L, some modifications to the method have to be made in order to account for the inherent constraints on some of the design variables. These modifications are as follows:

(i) n and R_c assume integer values.

(ii) k_a and k_w maintain the relationship such that $0 < k_w < k_a$.

(iii) the expressions for R_1 and R_0 are non-negative for given values of γ_1, γ_2, and δ.

From the past studies on the design of $\bar{x}$-charts with warning limits [4] under the normality assumption, it has been found that the value of the critical run length R_c takes either 1 or 2. Therefore, in the process of optimization, the range of values for R_c is from 1 to 4. Furthermore, in the conventional design of $\bar{x}$-charts with warning limits, normally $k_a = 3$ and $k_w = 2$ are selected [4]. Thus, to specify initial values for k_a and k_w, we use the relation $k_w=2/3\ k_a$.

Finally, after choosing the initial values of n, k_a, and k_w, an initial value for s is determined as [19]:

$$s \cong \{(\frac{V}{R_0} + b+cn)/[\lambda U(R_1 - \tfrac{1}{2})]\}^{\frac{1}{2}}. \quad ..(12)$$

A NUMERICAL EXAMPLE

To obtain the optimal design parameters, the search method assumes that the objective function is convex. Since it is not possible to analytically investigate the convexity of L, some analysis of its behaviour was conducted through numerical studies, which indicated that the surface of L is approximately convext in the region around the optimal value.

With the assumption of convexity of the objective function, and for a given set of cost and time parameters, optimal plans may be determined for a wide range of the non-normality parameters γ_1 and γ_2, and of the shift parameter δ. Table 1 presents a portion of the optimal results for a situation where γ_1 and γ_2 vary from -0.5 to 1.0 with increments of 0.5, parameter δ is fixed at 2.0, and the rate of the occurrence of the assignable cause $\lambda = 0.01$. Other parameters are assumed as follows: $V_0 = 150$, $V_1 = 50$, $k_r = 20$, $k_s = 10$, $\tau_r = 0.2$, $\tau_s = 0.1$, $b = 0.5$ and $c = 0.1$. The Table indicates that, for instance, when $\gamma_1=0.5$, $\gamma_2=1.0$, and $\delta=2$, the optimal plan is obtained at $R_c=2$, n=5, s=1.43 hours, $k_a=2.91$, $k_w=2.50$ and the objective function value is L=2.296.

Table 1 may also be used to study the effects of non-normality parameters on the loss-cost function and the design parameters. For instance, it is observed from the Table that the effect of skewness is more marked than that of kurtosis. For given γ_2, the values of s, k_a, k_w, and L increase as γ_1 increases. The same is true of L. However, the critical run length R_c remains unchanged, and the sample size n does not show marked variations.

Table 1. Optimal Values of Design Parameters and Loss-Cost Function for $\bar{X}$-Chart With Warning Limits. ($\lambda= 0.1$, $V_0=150$, $V_1=50$, $k_r=20$, $k_s=10$, $\tau_r=0.2$, $\tau_s=0$, $b=0.5$, $c=0.1$)

	δ 2.0				
γ_2	γ_1 -0.5	0.0	0.5	1.0	
-0.5	4	5	5	5	n
	1.28	1.40	1.42	1.46	s
	2.51	2.77	2.85	2.88	k_a
	2.17	2.36	2.48	2.51	k_w
	2	2	2	2	R_c
	2.203	2.253	2.276	2.294	L
0.0	4	5	5	5	n
	1.29	1.41	1.43	1.45	s
	2.53	2.79	2.87	2.90	k_a
	2.17	2.37	2.48	2.52	k_w
	2	2	2	2	R_c
	2.216	2.261	2.281	2.300	L
0.5	4	5	5	6	n
	1.30	1.41	1.43	1.52	s
	2.54	2.81	2.88	3.12	k_a
	2.17	2.37	2.48	2.73	k_w
	2	2	2	2	R_c
	2.229	2.269	2.289	2.304	L
1.0	4	5	5	6	n
	1.31	1.41	1.43	1.52	s
	2.55	2.83	2.91	3.13	k_a
	2.17	2.38	2.50	2.73	k_w
	2	2	2	2	R_c
	2.241	2.276	2.296	2.308	L

ACKNOWLEDGEMENT

This research was supported by the Natural Sciences and Engineering Research Council of Canada. Their financial assistance is gratefully acknowledged.

REFERENCES

1. Baker, K.R., "Two Process Models in the Design of $\bar{x}$-Charts", AIIE Transactions, Vol. 3, 1971, pp. 257-263.
2. Barton, D.E. and Dennis, K.E., "The Conditions Under Which Gram Charlier and Edgeworth Curves are Positive Definite and Unimodal" Biometrika, Vol. 39, 1952, pp. 425-427.
3. Chiu, W.K., "The Economic Design of Cusum Charts for Controlling Normal Means", Applied Statistics, Vol. 23, 1974, pp. 420-433.
4. Chiu, W.K. and Cheung, K.C., "An Economic Study of $\bar{x}$-Charts With Warning Limits", Journal of Quality Technology, Vol. 9, 1977, pp. 166-171.
5. Chiu, W.K. and Wetherill, G.B., "A Simplified Scheme for the Economic Design of $\bar{x}$-Charts", Journal of Quality Technology, Vol. 6, 1974, pp. 63-69.
6. Delaporte, P.J., "Control Statistique de Products Industrials dont la Quality n'a pas une Distribution de Laplace - Gauss", Proceedings of International Statistical Conference, India, 1951.
7. Duncan, A.J., "The Economic Design of $\bar{x}$-Charts Used to Maintain Current Control of a Process", Journal of American Statistical Association, Vol. 51, 1956, pp. 228-242.
8. Duncan, A.J., "The Economic Design of $\bar{x}$-Charts When There Is A Multiplicity of Assignable Causes", Journal of American Statistical Association, Vol. 66, 1971, pp. 363-380.
9. Ferrel,W.D. and Kemp, K.W., "Control Charts for Log-Normal Universes", Industrial Quality Control, Vol. 15, 1958, pp. 4-6.
10. Gayen, A.K., "On Setting Up Control Charts For Non-Normal Samples", Indian Society for Quality Control Bulletin, Vol. 53, 1953, pp. 43-47.
11. Gibra, I.N., "Economically Optimal Determination of the Parameters of an $\bar{x}$-Chart", Management Science, Vol. 17, 1971, pp. 635-646.
12. Goel, A.L., Jain, S.C., and Wu, S.M., "An Algorithm for the Determination of the Economic Design of $\bar{x}$-Charts Based on Duncan's Model", Journal of the American Statistical Association, Vol. 63, 1968, pp. 304-320.
13. Gordon, G.G. and Weindling, J.I., "A Cost Model for Economic Designs of Warning Limits Control Chart Schemes", AIIE Transactions, Vol. 7, 1975, pp. 319-329.
14. Gruska, G.R. , "Distributional Analysis of Non-Normal Multivariate Data", ASQC Technical Conference Transactions, Chicago, 1978, pp. 553-560.
15. Hahn, G.J., "How Abnormal is Normality?", Journal of Quality Technology, Vol. 3, 1971, pp. 18-22.
16. Heiks, "When Statistics Aren't Quite Normal", Industrial Engineering, Vol. 9, 1977, pp. 40-43.
17. Hooke, R., and Jevees, T.A., "Direct Search Solution of Numerical and Statistical Problems", Journal of the Association of Computing Machines, Vol. 8, 1961, pp. 212-229.
18. Lashkari, R.S. and Rahim, M.A., "An Economical Design of $\bar{x}$-Charts for Controlling Non-Normal Process Means Considering the Cost of Process Shut-Down", Proceedings, Computer Science and Statistics, 12th Symposium on the Interface, University of Waterloo, 1979, pp. 457-460.
19. Lashkari, R.S. and Rahim, M.A., "An Economic Design of Cumulative Sum Charts to Control Non-Normal Process Means", Presented at the 3rd National Conference on Computers & Industrial Engineering,Orlando, Oct. 22-24 1980.
20. Moore, P.G., "Non-Normality in Quality Control Chart", Applied Statistics, Vol. 6, 1957, pp. 171-179.
21. Nagendra, Y. and Rai, G., "Optimum Sample Size and Sampling Interval for Controlling the Mean of Non-Normal Variables", Journal of the American Statistical Association, Vol. 66, 1971, pp. 637-640.
22. Page, E.S., "Control Charts with Warning Lines" Biometrika, Vol. 42, 1955, pp. 243-257.
23. Page, E.S., "A Modified Control Chart with Warning Limits", Biometrika, Vol. 49, 1962, pp. 171-175.
24. Raouf, A., Lashkari, R.S. and Rahim, M., "Optimal Design of $\bar{x}$-Charts for Controlling Non-Normal Process Means Using Direct Search Technique", American Statistical Association Annual Meeting, Aug. 13-16, 1979, Washington, D.C.
25. Schilling, E.G. and Nelson, P.R., "The Effect of Non-Normality on the Control Limits of $\bar{x}$-Charts", Journal of Quality Technology, Vol. 8, 1976, pp. 183-188.
26. Shewhart, W.A., "Economic Control of Quality and Manufactured Product", Van Nostrand Reinhold Company, Inc., Princeton, N.J., 1931.
27. Singh, H.R., "Producer and Consumer Risks in Non-Normal Populations", Technometrics, Vol. 8, 1966, pp. 335-343.
28. Taylor, H.M., "The Economic Design of Cumulative Sum Control Charts", Technometrics, Vol. 10, 1968, pp. 479-488.
29. Wallace, D.L.,"Asymptotic Approximations to Distributions", Annals of Mathematical Statistics, Vol. 29, 1958, pp. 635-654.

PRIOR PROBABILITIES, MAXIMAL POSTERIOR, AND MINIMAL FIELD ERROR LOCALIZATION

G. E. Liepins, University of Georgia[1]

and

D. J. Pack, Union Carbide Corporation, Nuclear Division

Abstract

One of the significant difficulties with automatic edit and imputation is that any attempt to rigorously justify the methods used must confront the problem of the "error model": the observed record x is the true record y plus an error vector ε,

$$x = y + \varepsilon.$$

In practice, only partial information is available about either the distribution of the true y, or about the nature of ε, and without complete prior information, the model cannot be reliably decomposed into its component parts. As a result, to understand and estimate the performance of various editing techniques, it becomes imperative to characterize this performance in terms of various possible error models and data configurations. This paper reports on simulation studies to determine the effectiveness of various methods of error localization as the parameters, prior probability of error, dependence/independence of error, and data configuration are varied.

Keywords - error model, automatic data editing, error localization, "fields to impute," error detection, error localization.

Automatic edit and imputation can be considered as a three stage process:

1. Identify erroneous data records.
2. For each erroneous record, localize the error.
3. "Correct" the error by imputation.

That the error localization and imputation process can always be done mechanically that is, changes induced to render the modified record acceptable, is clear. Simply map each failing record onto some fixed acceptable record. What constitutes the major challenge to automatic edit and imputation is to establish a computationally tractable procedure for localization and imputation which reconstructs as well as possible the true data. That is, the procedure must be successful empirically, justifiable theoretically, and the consequence of its use must be understood statistically. Implicit in these requirements is that the data are multi-purpose, and hence that any "fix" should preserve as many statistical properties as possible. The purpose of this paper is to begin to explore the appropriateness of selected error localization procedures with particular emphasis on (weighted) minimum fields error localization.

It is sufficient to consider a data set subject to a system of m linear and n non-negativity constraints, (1) and (2).

$$\Sigma a_{ij}x_i \le b_j,\ j = 1,\ldots,m \tag{1}$$

$$x_i \ge 0,\ i = 1,\ldots,n \tag{2}$$

Any data record which fails one or more of the constraints is in error, and the process of checking the individual record against the constraints is the identification step of automatic data editing. Once a record x° has been identified to be in error, localization requires that an inference be made as to which components (fields) x_i° are wrong. Let $\bar{x}^\circ$ denote that the record x° is in error and $\bar{x}_i^\circ$ that the ith component (field) is wrong. The optimal inferential localization procedure would be maximal posterior probability error localization [Liepins and Pack (3)]: Find the index set J which maximizes the probability

$$P(\bigcap_{i \in J} \bar{x}_i^\circ | \bar{x}^\circ) \tag{3}$$

subject to the existence of an acceptable record y satisfying $y_i = x_i^\circ$ for $i \notin J$.

Unfortunately, the informational requirements and the computational difficulty associated with maximal posterior probability error localization almost inevitably preclude its use. Nonetheless, its performance (proportion of times error correctly localized) can be studied both theoretically and through simulation to provide insight into the limits of error localization. Simulation studies are still continuing, but there is evidence already that the generic performance curve is of the general form of fig. 1.

Conjectured Performance of Maximal Posterior Error Localization as a Function of the Prior Probability of Error in the Fields

Performance

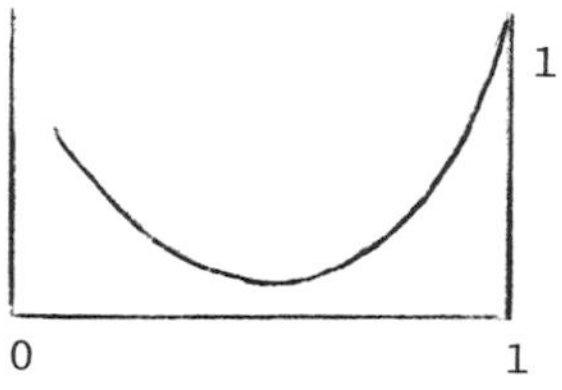

Prior Probability of Error

Figure 1

(The apparent perfect performance of maximal posterior probability error localization when the prior probability of error $p = p_i = 1$ for each of the fields is technically correct, yet of no practical significance. The error can be localized perfectly insofar as all fields are wrong, but there is little basis for imputation.) The information required for the implementation of maximal posterior probability error localization is complete distributional knowledge of the data process

$$x = y + \varepsilon \tag{4}$$

In this formulation, the observed record is x, the true record is y and the error vector is ε. Unfortunately, knowledge about the data process is generally limited, and the relationship between the many possible data processes and computationally tractable error localization procedures is not well established.

Under the rather restrictive assumptions (5) and (6) below, maximal prior probability error localization can be represented as a mathematical programming problem (7)-(11) [Liepins (2)], called the minimum weighted fields to impute (MWFI) problem.

$$P\{\varepsilon_i \neq 0 | \varepsilon_k = 0\} = P\{\varepsilon_i \neq 0\} = p_i \text{ for } i \neq k \tag{5}$$

ε_i has a uniform distribution for all i over the set of feasible value for ε_i (6)

The MWFI problem becomes: "Find the set J of indices which minimizes

$$\Sigma c_i \delta(\varepsilon_i) \tag{7}$$

suject to $x - \varepsilon$ is acceptable, (8)

$$\delta(\varepsilon_i) = \begin{cases} 0 \text{ if } \varepsilon_i = 0 \\ 1 \text{ otherwise,} \end{cases} \tag{9}$$

and $\varepsilon_i = 0$ for $i \notin J$ (10)

For the linear constraints (1) and (2), condition (8) can be rewritten [Sande (4)] as

$$\Sigma a_{ij}(x-\varepsilon) \leq b_j, \; j=1,\ldots,m \tag{11}$$

$$(x-\varepsilon)_i \geq 0, \; i=1,\ldots,n \tag{12}$$

Implicit in the above formulation (7)-(11) is the requirement that the coefficients c_i are inversely related to the p_i, the prior probabilities of error in the various fields.

In an attempt to improve the performance of MWFI localization and bring it more into line with maximal posterior probability error localization, other choices of coefficients have been suggested and have been studied through simulation [Liepins and Pack (3)]. For the examples studied and the weightings tested, it cannot be concluded that any weighting performs consistently better than equal weighting, that is, $c_i = 1$ for all i. In fact, the performance of MWFI error localization seems to be a complicated and poorly understood function of the data process, constraint configuration, and weighting. Selected results appear in tables 1 and 2 summarized from Liepins and Pack (3).

Performance, of MWFI Localization for two Examples as a Function of Prior Probability of Error and of Weighting.

Example 1

Prior Probability of Error: $p = p_i$	Weighting	Performance
.05	1	.399
	2	.398
	3	.430
.10	1	.332
	2	.331
	3	.357
.20	1	.220
	2	.219
	3	.239

Example 2

Prior Probability of Error: $p = p_i$	Weighting	Performance
.05	1	.366
	2	.366
	3	.286
.10	1	.321
	2	.320
	3	.251
.20	1	.240
	2	.239
	3	.188

Key: Weighting 1 - equal weighting. Weighting 2 - decrease those coefficients of fields active in failed constraints. Weighting 3 - Similar to weighting 2, but increase those coefficients of fields not active in failed constraints.

Table 1

(Table 2 appears on the following page)

Two of the issues that remain hidden in these tables are first, that the specific combination of fields which have high prior error rates and those which have low prior error rates affects the success of MWFI, and second that theoretical results support the contention that for all practical purposes weighting 1 and 2 will behave similarly. The one characterization of MWFI error localization which appears to be generally true is that in terms of any of the success measures studied, success appears to be inversely proportional to the prior error rate, and similarly, inversely proportional the frequency of fields in error.

Success Index 3 for Various Average Prior Error Rates and Differences in Rates

Example 1	all prior error rates p_i equal		medium differences in prior error rates		large differences in prior error rates	
average prior probability of error	proportion of correct fields	weighting 3, success index 3	proportion of correct fields	weighting 3, success index 3	proportion of correct fields	weighting 3, success index 3
.2	.705	.724	.710	.743	.722	.768
.3	.630	.659	.642	.678	.653	.697
.4	.552	.596	.562	.612	.578	.632
.5	.470	.535	.482	.549	.495	.564
.6	.383	.471	.396	.484	.412	.500
.7	.292	.403	.307	.419	.322	.433

Key: Success index 3 - the proportion of fields correctly categorized as either correct or in error.

Table 2

Performance of MWFI Error Localization as a Function of Prior Probability of Error in the Fields.

Performance of MWFI Error Localization

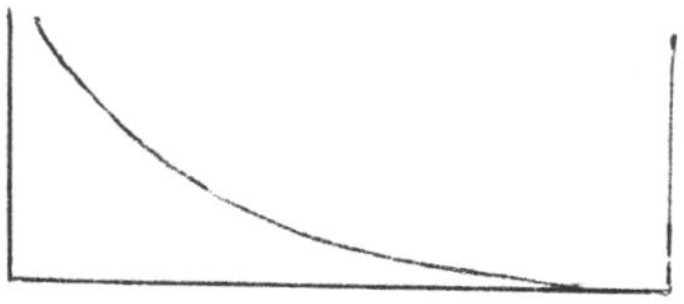

0 1

Prior Probability of Error p.

Figure 2

The tentative conclusion to be drawn from this is that only for those cases where the error rate is low, can MWFI be expected to perform with any success.

To close the circle, it is a matter of some interest to determine when maximal posterior probability error localization is of necessity a minimal error localization, that is, a feasible set S for MWFI such that no feasible set $S' \subset S$ exists for which $S' \neq S$.

For a discrete data satisfying conditions (5) and (6) and subject to normal form constraints [Fellegi and Holt (1)], necessary any sufficient conditions are given in (13) below. Moreover, as the prior probability of error p in the fields tends to zero, condition (13) is necessary and sufficient to insure that maximal posterior probability error localization for these data is actually a solution to MWFI with $c_i = 1$, that is, a minimum cardinality error localization.

The prerequisite definitions are given now. Definitions. Let $A \subset R^m$ be the acceptance region and let the orthogonal projection on the ith component of R^m be denoted Π_i. Then $\Pi_i(A) = A_i$ and for purposes of this paper A_i will be bounded and measurable with $|A_i|$ its measure (count of points). Further, for the distribution of true values, there is associated a probability measure P on A. Let $x^\circ(S) = \{x | x_i = x_i^\circ \; i \notin S, \; x_i \neq x_i^\circ \; i \in S\}$, the set of true values equal to x° in exactly the fields not indexed by S. Then $w[x^\circ(s)]$ is the probability of this set; for the case of uniform distribution of true values over the acceptance region $w[x^\circ(S)]$ would just be the relative number of points in the acceptance region with components equal to x_i° for $i \notin S$ and components not equal to x_i° for $i \in S$.

For any two index sets S and S', with S' derived from S by the augmentation of the index k, that is, $S' = S + k$, and whenever the sets $x^\circ(S) \neq \emptyset$ and $x^\circ(S') \neq \emptyset$, the index set S will be associated with a larger posterior probability than S' whenever

$$\frac{\prod_{i\in S}(|A_i|-1)^{-1} w[x^\circ(S)] \prod_{i\in S} p_i \prod_{i\notin S}(1-p_i)}{\prod_{i\in S'}(|A_i|-1)^{-1} w[x^\circ(S')] \prod_{i\in S'} p_i \prod_{i\notin S'}(1-p_i)} = \frac{(|A_k|-1)\, w[x^\circ(S)](1-p_k)}{w[x^\circ(S')]\, p_k} > 1 \qquad (13)$$

Unfortunately, at this time, condition (13) remains of little practical significance and will reamin so until an appropriate approximation to the function w[ ] can be determined.

[1] On leave of absence from Oak Ridge National Laboratory. Research supported by Oak Ridge National Laboratory, subcontract #000S7982.

References

1. I. P. Fellegi and D. Holt (1976), A Systematic Approach to Automatic Edit and Imputation," Journal of the American Statistical Association," 71, 17-35.

2. G. E. Liepins (1980), "A Rigorous Systematic Approach to Automatic Data Editing and its Statistical Basis," Oak Ridge National Laboratory Technical Manuscript, ORNL/TM-7126.

3. G. E. Liepins and D. J. Pack, unpublished paper.

4. G. Sande (1978), "An Algorithm for the Fields to Impute Problems of Numerical and Coded Data," Statistics Canada.

Lecture Notes in Statistics

Vol. 1: R. A. Fisher: An Appreciation. Edited by S. E. Fienberg and D. V. Hinkley. xi, 208 pages, 1980.

Vol. 2: Mathematical Statistics and Probability Theory. Proceedings 1978. Edited by W. Klonecki, A. Kozek, and J. Rosiński. xxiv, 373 pages, 1980.

Vol. 3: B. D. Spencer, Benefit-Cost Analysis of Data Used to Allocate Funds. viii, 296 pages, 1980.

Vol. 4: E. A. van Doorn: Stochastic Monotonicity and Queueing Applications of Birth-Death Processes. vi, 118 pages, 1981.

Vol. 5: T. Rolski, Stationary Random Processes Associated with Point Processes. vi, 139 pages, 1981.

Vol. 6: S. S. Gupta and D.-Y. Huang, Multiple Statistical Decision Theory: Recent Developments. viii, 104 pages, 1981.

Vol. 7: M. Akahira and K. Takeuchi, Asymptotic Efficiency of Statistical Estimators. viii, 242 pages, 1981.

Vol. 8: The First Pannonian Symposium on Mathematical Statistics. Edited by P. Révész, L. Schmetterer, and V. M. Zolotarev. vi, 308 pages, 1981.

Springer Series in Statistics

L. A. Goodman and W. H. Kruskal, Measures of Association for Cross Classifications. x, 146 pages, 1979.

J. O. Berger, Statistical Decision Theory: Foundations, Concepts, and Methods. xiv, 420 pages, 1980.

R. G. Miller, Jr., Simultaneous Statistical Inference, 2nd edition. 300 pages, 1981.

P. Brémaud, Point Processes and Queues: Martingale Dynamics. 352 pages, 1981.

ISBN 0-387-**90633**-9
ISBN 3-540-**90633**-9